THE THESAURUS OF CHEMICAL PRODUCTS

Volume I: Generic to Tradename

Compiled by

Michael and Irene Ash

Chemical Publishing Co.
New York, NY

Chemical Publishing Co.

ISBN 0-8206-0315-5

Printed in the United States of America

PREFACE

This two volume set accesses all of the tradenames by which a multitude of chemical products are sold. With Volume I, by knowing the generic name of a chemical product, the user can locate the other tradenames by which the product is known. With Volume II as the companion, it is possible for the user to find this information by just knowing one of the tradenames of the generic chemical product.

This reference set is the culmination of several years of research and compilation. No other reference includes this type of comprehensive coverage spanning the entire chemical industry. Surfactants, polymers, resins, plastics, pharmaceuticals, cosmetic additives, agricultural chemicals, and other industrial additives are all included in this unique compilation. This work will prove essential to all chemists, manufacturers, and sales people involved in the chemical industry.

In using these volumes, it should be noted that the numbers in brackets refer to the manufacturer of the tradename product and these manufacturers are listed at the end of each volume along with their addresses.

Abietic acid polyglycol ester.
Produkt RT 275 [948]

Acetamide MEA.
Lipamide MEAA [525]
Schercomid AME, AME-70 [771]

Acetaminophen.
Acetaco Tablets [518]
Acetaminophen Elixir, Tablets, Suppositories [751]
Acetaminophen Uniserts Suppositories [886]
Acetaminophen with Codeine Phosphate Tablets [751]
Algisin Capsules [718]
Amacodone Tablets [868]
Amaphen Capsules, with Codeine #3 [868]
Anacin-3, Children's Acetaminophen Chewable Tablets, Elixir, Drops, Maximum Strength Acetaminophen Tablets and Capsules, Regular Strength Acetaminophen Tablets [922]
Anacin-3 with Codeine Tablets [97]
Anatuss Tablets & Syrup, with Codeine Syrup [563]
Anoquan [541]
Apamide [63]
Apap 300 mg. with Codeine Tabs, with Codeine Elixir [769]
APAP w/Codeine Tablets, #3, #4 [344]
Asprin-Free Arthritis Pain Formula By the Makers of Anacin Analgesic Tablets [922]
Bancap Capsules, c Codeine Capsules, HC Capsules [661]
Blanex Capsules [281]
Capital with Codeine Suspension, with Codeine Tablets [153]
Children's Panadol Chewable Tablets, Liquid, Drops [355]
Chlorzone Forte Tablets [769]
Chlorzoxasone w/APAP Tablets [344]
Chlorzoxazone with APAP Tablets [240]
Codalan [510]
Co-Gesic Tablets [166]
Colrex Compound Capsules, Compound Elixir [750]
Compal Capsules [750]
Comtrex [134]
Congesprin Liquid Cold Medicine [134]
CoTylenol Cold Medication Tablets & Capsules, Liquid Cold Medication, Children's Liquid Cold Formula [569]
Darvocet-N 50, -N 100 [523]
Di-Gesic [166]
Double-A Tablets [281]
Dristan, Advanced Formula Decongestant/Antihistamine/Analgesic Capsules, Advanced Formula Decongestant/Antihistamine/Analgesic Tablets, Ultra Colds Formula Aspirin-Free Analgesic/Decongestant/Antihistamine/Cough Suppressant Nighttime Liquid, Ultra Colds Formula Asprin-Free Analgesic/Decongestant/Antihistamine/Cough Suppressant Tablets & Capsules [922]
Duradyne DHC Tablets [661]
Empracet with Codeine Phosphate [139]
Esgic Tablets & Capsules [352]
Espasmotex Tablets [78]
Excedrin Extra-Strength, P.M. [134]
Extra-Strength Datril Capsules & Tablets [134]
G-1 Capsules, G-2 Capsules, G-3 Capsules [394]
Gemnisyn Tablets [746]
Hycodaphen Tablets [86]
Hycomine Compound [269]
Hyco-Pap [512]
Hydrogesic Tablets [281]
Korigesic Tablets [868]
Maxigesic Capsules [560]
Maximum Strength Panadol Capsules & Tablets [355]
Medigesic Plus Capsules [888]
Midrin Capsules [153]
Migralam Capsules [505]
Norcet Tablets [402]
Norel Plus Capsules [888]
Oxycodone Hydrochloride & Acetaminophen Tablets [751]
Pacaps [512]
Parafon Forte Tablets [570]
Percocet [269]
Percogesic Analgesic Tablets [902]
Phenaphen w/Codeine Capsules, -650 with Codeine Tablets [737]
Phenate [541]
Phrenilin Tablets, Forte [153]
Propoxyphene & Apap Tablets 65/650 [769]
Protid [512]
Repan Tablets [304]
Saroflex [765]
Sedapap-10 Tablets [563]
Sine-Aid Sinus Headache Tablets [569]
Singlet [580]
Sinubid [674]
Sinulin Tablets [153]
SK-APAP with Codeine Tablets, -Oxycodone with Acetaminophen Tablets, -65 APAP Tablets [805]
Stopayne Capsules [814]
Supac [593]

Acetaminophen *(cont'd).*
T-Gesic Capsules [930]
Talacen [932]
Two-Dyne Capsules [436]
Tylenol acetaminophen Children's Chewable Tablets, Elixir, Drops, Extra-Strength, acetaminophen Liquid Pain Reliever, Extra-Strength, acetaminophen Tablets & Capsules [569]

Acetazolamide.
Acetazolamide Tablets [769]
Diamox Parenteral, Sequels, Tablets [516]

Acetic acid.
Aci-Jel Therapeutic Vaginal Jelly [666]
Fecal Straining Kit [468]
Otic Domeboro Solution, Tridesilon Solution 0.05% [585]
Otic-HC Ear Drops [394]
Otipyrin Otic Solution (sterile) [498]
VoSol HC Otic Solution, Otic Solution [911a]

Acetic acid esters.
Tegin E series [360]

Acetic acid fatty acid glyceride.
Axol E 61 [856]

Acetic acid monoalkanolamide.
Schercomid AME [771]

Aceto-glycerides.
Lamegin EE [716]

Acetohexamide.
Dymelor [523]

Acetohydroxamic acid.
Lithostat [593]

3-(A-Acetonyl-4-chlorobenzyl)-4-hydroxy-coumain.
Tomorin [190]

Acetophenone.
Dymex [777]

6-Acetoxy-2, 4-dimethyl-m-dioxane.
Giv-Gard DXN [353]

Acetylacetonate chelate.
Tyzor AA [269]

Acetyl cyclohexane sulfonyl peroxide.
Trigonox® ACS-M28 [16]

Acetyl cyclohexylsulfonyl peroxide.
Lupersol 228Z [536]

Acetylcysteine.
Mucomyst, with Isoproterenol [572]

Acetylenic alcohol alkoxylated.
Emcol D75-13, D75-33. [935]

N-Acetyl ethanolamide.
Carsamide AMEA [529]

6-Acetyl-1, 1, 2, 3, 3, 5-hexamethyl indan.
Phantolid - 263270 [406]

7-Acetyl, 1, 1, 3, 4, 4, 6-hexamethyltetralin.
Tonalid - 263400 [406]

Acetyl hexamethyl tetralin.
Tonalid [692]

Acetyl peroxide.
Acetyl Peroxide-IB25 [536]

0-Acetyl-2-sec-butyl-4, 6-dinitrophenol.
Aretit [416]

Acetyl sulfisoxazole.
Gantrisin Pediatric Suspension, Syrup, Lipo Gantrisin [738]
Pediazole [747]

Acetyl tributyl citrate.
ATBC [223]
Citroflex A-4 [691]

Acetyl triethyl citrate.
Citroflex A-2 [691]

Acetyl trioctyl citrate.
Citroflex A-8 [691]

Acrylic.
NeoCryl A-550, SR-276 [702]

Acyclovir.
Zovirax Ointment 5% [139]

Acyclovir sodium.
Zovirax Sterile Powder [139]

Acyl amido-amine.
Tequat RO [514]

Acylamino carboxylic acid.
Arkomon SO [416]

Acyl amino poly glycol ether sulfate sodium salt.
Genapol AMS [55]

Acyclaminopolyglycol ether sulfate triethanolamine salt.
Genapol AMS [55]

Acyl (coco fatty) amido glycine betaine.
Standapol CIM-40 [400]

Acyl lactylates.
Crolactil CSL, SISL, SSL [223]

Adenosine 5-monophosphate.
Adenosine [518]

Adenosine-5-phosphoric acid.
Lycedan [776]

Adenosine triphosphate.
Triphosaden [776]

Adipate ester.
Kemester 5651, 5653, 5654 [429]
Staflex DIBA, ODA [722]

Adrenocorticotropic hormone.
A.C.T.H. "40" Injectable, "80" Injectable [661]
Acthar, HP Acthar Gel [80]

Albumin (bovine).
Ultralog 02-3640-10 [172]

Albumin, normal serum.
Albuminar-5, Normal Serum Albumin (Human) U.S.P. 5%, -25 Normal Serum Albumin (Human) U.S.P. 25% [80]
Buminate 5%, Normal Serum Albumin (Human), U.S.P., 5% Solution, 25% Normal Serum Albumin (Human), U.S.P., 25% Solution [434]

Albuterol.
Proventil Inhaler, Tablets [772]
Ventolin Inhaler, Tablets [354]

Alcohol ether sulfate.
Witcolate 1247H, 1259, 1276, 1390, 7031, 7093, 7103 [934]

Alcohol ethoxy sulfate, ammonium salt.
Manro ALES 60 [545]

Alcohol ethoxy sulfates, sodium salt.
Manro BEC 28, BEC 70, BES 27, BES 60, BES 70, DES [545]

Alcohol polyethyleneoxy phosphate ester acid fatty.
Antara LB-400 [338]

Alcohol polyglycolether phosphate fatty.
Steinaphat EAK 8190 [274]

Alcohol sulfate.
Sunnol LDF-110, LL-103 [524]

Aldehyde amine.
Roy-AC 30 Rubber Accelerator, 40 Rubber Accelerator [752]

Algin.
Kelco-Gel [484]
Kelcosol [484]
Kelgin F, HV [484]
Kelzan [484]
Satialgine S20 [280]

Algin-carrageenan.
Cocoloid [484]

Alginic acid.
Kelacid [485]
Satialgine H8 [280]

Alkane sulfonate.
Hostapur SAS 60 [416]

Alkane sulfonate, sodium salt.
Hostapur SAS Brands [416]
Lutensit A-PS [106]

Alkanolamide.
Acetamide MEA [935]
Atlas EM-16, WA-104 [94]
Cedemide B, C [256]
Charlab Condensate K [159]
Clindrol 101CG, 200MS, 868 [197]
Condensate 640 [209]
Dergon SW [77]
Detergent Concentrate 840 [599]
EM-980, -983, -985 [483]
Emulser OM [77]
Ethylan KELD, LBC [509]
Isomul Extra [461]
Isorezoff [461]
Kerinol C 109 [253]
Monalube 29-78, 780 [599]
Monamid 7-153CS, 853 [599]
Monamine AC-100, ADS-100, ADY-100, ALX-100S, CF-100M, R8-26 [599]
Monamulse 748, -653-C [599]
Onyxol 201 [663]
Surco A-10 MM, B10 MA, SR 200 [663]
Witcamide 82, 272, 511, 512
Witcamide 1017, 5130, 5133, 5138, 5140, 5195 [934]
Witcamide 6445, AL69-58, CDA [935]
Witcamide M-3 [934]
Witcamide S-771
Witcamide S-780 [935]
Witco 934 [934]

Alkanolamide alkaline.
Isoslushoff, X-M-50, X-NP-25 [461]

Alkanolamide, back titrated.
Surco SR-100 [663]

Alkanolamide, C_{16}-C_{18}.
Geronol Aminox/3 [550]

Alkanolamides, ethoxylated.
Alkamidox C2, C5, L2, L5 [33]
Alkaminox O-2 [33]
Cyclomide EE [938]

Alkanolamide, ethoxylated fatty acid.
Eumulgin C4 [401]

Alkanolamide ethylene oxide condensate.
Ethylan LM2 [509]

Alkanolamide, fatty.
Consamine CA [207]
Ecconol B [300]
Progasol 230, 443, 457 [537]
Rycomid® 2120, 2120-B1, 2120-B2 [756]
Sterling Amide 374 [145]

Alkanolamide, fatty acid.
Alkamide 2110, 2204 [33]
Aminol HF-2C, VR-14 [320]
Becrosan 210A, 210D [150]

Alkanolamide, fatty acid *(cont'd).*
Kerinol 2012 F, AA 62 [253]
Marlamid KL [178]
Ninol 128 +, 128 Extra, 201, 1281, 1285, 1301, 2012 Extra, AA62, AA62 Extra [825a]
Surco WC Concentrate [663]

Alkanolamide (2:1), linoleic
Foamole L [895]

Alkanolamide, sodium sulfosuccinate ester.
Monamate CPA-40%, OPA-30 [599]

Alkanolamide, sulfosuccinate.
Zoharpon SM [946]

Alkanolamide sulfosuccinate, fatty.
Alkasurf SS-O-ME [33]
Cosmopon BL [514]
Foamer [390]

Alkanolamide sulfosuccinate, fatty acid.
Varsulf SBL 203 [796]

Alkanolamine fatty condensate.
Marlazin OK 1 [178]

Alkanol amine lauryl sulfate.
Emersal 6440 [290]
Zoharpon DT-80 [946]

Alkanol polyglycol ether.
Laventin CW [107]

n-Alkenyl dimethyl ethyl ammonium bromide.
Onyxide 75 [663]
SD-75 [319]

Alkenyl succinic acid, disodium salt.
Rewocor B 3010 [725]

Alkenyl succinic acid, TEA-salt.
Rewocor B 3032 [725]

Alkenyl succinic anhydride.
Produkt B 2045 [223]
Rewocor B 2045 [725]

Alkoxy acid, sodium salt.
Alkawet B [529]

Alkoxylate.
Agrilan AEC123 [509]
Spreading Agent ET0672 [223]

Alkoxypropyl-1,3-diaminopropane, C_{12-14}.
Tomah DA-62 [866]

Alkyamine alkylaryl sulfonate.
Stepantex DA-52 [825]

Alkylacetal.
Degressal® SD 10 [106]

Alkyl alkoxylate, fluorinated.
Fluorad FC-171 [862]

Alkylamide betaine.
Merpalyt L-15-1 [285]

Alkylamide ester compound.
Polysofter N-606 [630]

Alkyl amide imidazoline compounds, acetate
Sofnon HG-180 [864]

Alkyl amide propyl dimethyl amine.
Empigen AT [25]

Alkyl amido amine.
Arzoline 215 [76]
Empigen AS, AT [23]
Steramine FPA 197, PNA 75 [253]
Witcamine 209, 210, 211 [934]

Alkyl amido amine oxide.
Empigen OS [23]

Alkylamidoamine, oxyethylenated.
Base 3059 E [253]

Alkyl amido betaine.
Amphosol AA [825a]
Empigen BS, BT [25]
Lorapon AM-B 13 [274]
Rewopon AM-B 13 [274]
Steinapon AM-B 13 [274]

Alkyamidoethyl alkyl imidazolinium methyl methosulfate.
Carsosoft S-75, S-90, S-90M [155]

Alkylamidoethylimidazoline.
Servamine KOO 330 [788]

Alkylamido glyceryl phosphobetaine.
Monateric P-CDL [599]

Alkylamido phosphobetaine.
Monateric P-1013, P-1023 Modified [599]

Alkyl amido propyl dimethyl amine.
Empigen AS, AS/H, AT [23]

Alkyl amido propyl dimethyl amine betaine.
Empigen BS, BS/H, BT [23]

Alkyl amido propyl dimethyl amine oxide.
Empigen OS [23]

Alkylamido-propyldimethylamine oxide.
Rewominoxid B 204 [274]
Steinaminoxid B 204 [274]

N-alkyl (C_{16}-C_{22}), amidropropyl, N-N dimethyl, N-benzyl ammonium chloride.
Schercoquat ROAB [771]

N-alkyl (C_{16}-C_{22})-(3-amidopropyl)-N-N dimethyl-N-ethyl ammonium ethyl sulfate.
Schercoquat ROAS [771]

Alkyl-amine.
Avitex NA, R [269]

Alkylamine.
Amine B11 [487]
Lilamin B11 [487]

Alkyl amine.
Lutostat MSW 30, MSW 88 [253]

n-Alkylamine acetate.
Acetamin 24, 86 [478]

N-alkyl amine acetate.
Catisol AO/100 [825a]

n-alkyl amines acetic acid salts.
Armac C, CD, HT, T, TDD, HT Flake [16]
Duomac T [16]

Alkylamine dodecylbenzene sulfonate.
Ninate 411 [825]

Alkyl amine ethoxylated.
Berol 28, 303 [16]

Alkylamine guanidine polyoxyethanol.
Aerosol C 61 [233]

Alkylamine oxides.
Empigen OB, OH, OY [25]

Alkylamine poly-alkylene oxide ether.
Merpoxen S-5-96/10, S-5-96/20, S-5-96/40, S-5-96/70 [285]

Alkylamine polyglycol ether.
Marlowet 5400 [178]

Alkyl-amine polyoxyethylated.
Katapol VP-532 [338]

Alkyl amine, primary.
Synprolam 35 [441]

Alkyl amine, primary (cetyl).
Crodamine 1.16D [223]

Alkyl amine primary (coconut).
Crodamine 1.C [223]

Alkyl amine, primary (lauryl).
Crodamine 1.12D [223]

Alkyl amine, primary (oleyl).
Crodamine 1.0 [223]

Alkyl amine, primary (stearyl).
Crodamine 1.18 [223]

Alkyl amine, primary (tallow).
Crodamine 1.T [223]

Alkyl amine, primary (tallow hydrogenated)
Crodamine 1.HT [223]

Alkylamino acid, alkylolamine salt.
Amphotensid D1 [948]

Alkylaryl compound.
Eccotex P. Conc. [277]

Alkylaryl ether sulfate.
Syntopon S 493, S 630, S 1030 [934]

Alkyl aryl ether.
Triton CF10, CF-87 [743]

Alkyl aryl, ethoxylated.
Texo LP 528A, 147 [855]

Alkylaryl ethoxylate, phosphate ester.
Emphos CS-121, CS-136, CS-147, CS-330, CS-733, CS-1361 [935]

Alkylaryl phosphate esters.
Hodag PE-005, PE-104, PE-106, PE-109, PE-206, PE-209 [415]
Rexophos 25/67, 25/97 [390]

Alkylaryl phosphite.
Wytox® 320 [659]

Alkylaryl polyalkylene oxide ether.
Merpoxen T-8-3 [285]

Alkylaryl polyether.
Triton® CF-10, CF-21 [742]

Alkyl aryl polyether alcohol.
Agrimul 70-A
Triton® X-120, X-155, X-155-90%, X-207 [742]
Witconol NP-40, NP-60, NP-300 [935]

Alkylaryl polyether ethanol.
Triton® X-363M [742]

Alkylaryl polyether sulfate, sodium salt.
Triton® 770, W-30, X-301 [742]

Alkylaryl polyether sulfonate, sodium salt.
Triton® X-200, X-202 [742]

Alkylarylpolyethoxycarboxylate.
Merpemul 8122, 8610 [285]

Alkylaryl polyethoxyethanol.
Witcosperse 201 [935]

Alkyl aryl polyglycol ether.
Basopon LN [109]
Dehydrophen 65, 100 [401]
Emulsogen N-060, N-090 [55]
Eumulgin 286 [401]
Hostapal 3634, CV [416]
Hostapur CX [416]
Marlowet ISM, TM [178]
Produkt RT 237 [948]
Standapon 95, 100 [400]

Alkyl aryl polyglycol ether sulfate, sodium salt.
Hostapal BV [55]

Alkylaryl polyglycolic ethers.
Geropon 40/D [550]

Alkylaryl polyoxyalkylene ethers.
Teric 160, 161, 163 [439]

Alkyl aryl polyoxyethylene ether.
Chemcol Agwet 110 [171]

Alkylaryl polyoxyethylene glycol.
Triton® X-114SB [742]
Trycol TP-6 [289]

Alkyl aryl sulfate.
Hexaryl D 60 L [937a]
Hexaryl L 30 [938]
Witcolate D51-51 [935]

Alkylaryl sulfonate.
Ahco A-117 [445]
Atlas G711, G-3300 [438]
Atlas G-3300B [92]
Atlox 4861B [92]
Cenegen 7 [225]
Compound 8-S [269]
Consoscour 47 [207]
Cycloryl ABSA [234]
Detergent Slurry [179]
Dymsol LP [252]
Ecconol 606 [300]
Emkane HAD, HAL [292]
Emkatex AA, AA-80 [292]
Endet [837]
Hipochem FNL [410]
Iberwet B.I.G. [386]
Idet-5L, -5L spl. NF, -5LP, -10, -20P [837]
Lanitol A [77]
Nacconol 35SL, 40F, 90F [825]
Osimol [176]
Pantex [855]
Pentex 40 [200]
Progasol 40 [537]
Sandozin AM [762]
Serdet DDK 31 [778]
Siponate 330 [29]
Solar 40, 80, 40 BD Flakes & Granules, 80 Flakes & Granules [840]
Sulfamin [116]
Texo 227, 1060 [855]
Udet® 950 [688]
Witco 918 [935]
Witconate 605T [935]
Witconate 1250 [937a]

Alkyl aryl sulfonate, amine neutralized.
Arylan PWS [253]
Emkane HAX [292]
Manro HCS [545]

Alkylaryl sulfonate amine salt.
Cresterge-AAS Spec [222]
Eccoterge ASB [277]
Rueterg 97-S, 97-T, IPA-HP [320]

Alkyl aryl sulfonate, salt calcium.
Phenyl Sulphonate CA & CAL [416]

Alkylaryl sulfonate lethoxylated.
Triton X-200, X-202, X-305, X-405, X-705 [743]

Alkyl aryl sulfonate/formaldehyde condensates.
Hi-Fluid [849]

Alkyl aryl sulfonate, salt.
Elecut S-507 [849]

Alkylaryl sulfonate, sodium salt.
Acto 450, 500, 630, 632, 636, 639 [306]
Alconox [30]
Alcotabs [30]
DET-Washmatic [837]
Iberpon W [386]
Idet 10, 20-P [386]
Kleen-Paste [937]
Sulframin 40, 40DA, 40RA, 45, 45LX, 85, 90, 1240, 1245, 1250 [935]
Terg-a-zyme [30]

Alkyl aryl sulfonate, triethanolamine salt.
Manro TDBS 60 [545]
Sulframin 40T, 60T [935]

Alkyl aryl sulfonic acid.
Dobanic Acid 83 [792]
Rueterg Sulfonic Acid [320]
Witco 97H Acid, 1298 Hard Acid, 1298 Soft Acid [935]
Witco Acide B, Acide TPB [937a]
Witco CHB Acid, D 51-24 [935]

Alkylaryl sulfonic acid, amine salt.
Emcol P-1045, P-1049, P-1059, P-1059B, P-1073 [937]
Sulfotex IPL-A [399]

Alkylaryl sulfonic acid concentrate.
Emkane Acid [292]

Alkylaryl sulfonic acid, linear.
Bio-Soft S-100 [825]
Cycloryl ABSA [234]
Sulframin 1288, 1298, 1388 [935]

Alkylaryl sulfonite, TEA salt.
Mazon 60T [564]

Alkylate sulfonate, linear.
Nacconol 35SL. [825]
Ultrawet 60L [72]

Alkylate sulfonic acid, linear.
Bio Soft S-100 [825]
Siponate SA [29]
Ultrawet 99LS [72]

Alkyl benzene.
Alkylate A215, A225, A230 [602]

Alkyl benzene-cumene sulfonic acid, SO_3 co-sulfonate.
Nansa 1906 [24]

Alkyl benzenes, linear.
Alkylate 215, 225, 230 [602]
Dobane 45, 80 [791]
Dobane 83, 102, 103, 113 [792]
Dobane 123, 945, HP, JN, 055 [791]
Nalkylene 500, 515, 550, 600 [206]
Wibarco [177]

Alkyl benzene sulfonate.
Lipon P-106 [524]
Marlon A 350, A 360, A 365, A 375, A 390, A 396, ARL, AFM 40, AFM 50 N, AS_3 [178]
Nissan Newrex, Ohsen A [636]
Protosan, EP [710]
Soft Detergent 60 [524]

Alkylbenzene sulfonate, alkanolamine salt.
Lumo 1683 [948]
Lutensit A-LBA [106]

Alkylbenzene sulfonate, amine salt.
Merpemul 1088, 1088 L, 1165, 1178 [285]

Alkyl benzene sulfonate, linear.
Nissan Soft Osen 550A [636]
Zoharlab [946]

Alkyl benzene sulfonate, sodium branched.
Polystep A-16 [825a]

Alkyl benzene sulfonic acid.
A.B.S. 87% [858]
Dobanic Acid 83 [792]
Dobanic Acid 103 [795]
Nansa 1042, 1909, SSA, SSA/L, SSA/P [24]
Neopelex FS [478]
Polyfac ABS-95 [918]
Sulphonic Acid LS [390]
Witcor® Acid 94H, 97H [934]

Alkyl benzene sulfonic acid, amine salt.
Cedepon AM [256]

Alkyl benzene sulfonic acid, branched.
DDBS 100 [946]
Nansa SBA [23]

Alkyl benzene sulfonic acid, isopropylamine salt.
Alkasurf IPAM [34]

Alkyl benzene sulfonic acid, linear.
Calsoft LAS-99 [697]
Carsosulf UL-100 Acid [155]
Cedepon Acid 100 [256]
DI-SA-97 [248]
Dobanic Acid 102, JN [791]
LABS-100 [946]
Polystep A-13 [825]
Sulfonic Acid LS [390]
Sulfosoft [116]
Ufacid K [874]

Alkyl benzene sulfonic acid (propylene tetramer).
Witco® 1298 Hard Acid [935]

Alkyl benzene sulfonic acid, sulfuric acid sulfonated.
Nansa HSA/L, SSAL [26]

Alkyl benzene sulfonic acid, triethanolamine salt.
Carsofoam T-60-L [155]

Alkyl benzyl ammonium salt.
Texnol R 5 [635]

Alkyl benzyl dimethyl ammonium chloride.
Empigen BCB50, BCF 80 [25]
Gardiquat 12H [24]
Sanisol C [478]

Alkyl benzyl dimethyl quaternary compound.
Alkaquat DMB-451, 80% [33]

Alkyl benzyl imidazolinium chloride.
Uniquat® CB-50 [529]

Alkyl benzyl phthalate.
Santicizer 261, 278 [602]

N-alkylbetaine.
Amphitol 24B, 86B [478]
Amphosol DM [825]
Chimin CB [514]
Empigen BB [25]
Empigen BB-AU [26]
Nissan Anon BF [636]
Obazoline LB-40 [864]

Alkyl dialkoxy amine oxides.
Varox 185E [796]

Alkyl dialkoxy ether amine oxides.
Varox 188E, 191E [796]

Alkyl diamine.
Nissan Asphasol 10, Asphasol 20 [636]

Alkyl di(aminoethyl) glycine.
Nissan Anon LG [636]

Alkyl diarylamine.
Irganox LO-6 [190]

Alkyl diaryl sulfonate.
Alkanol ND [269]
Hipochem No. 641 [410]

Alkyldihydroxy ethyl amine oxide.
Nissan Unisafe A-LE [636]

Alkyl (C_8-C_{18}) dimethyl amine.
Armeen DMCD, DMMCD [16]
Barlene® 12, 14 [529]
Empigen AB, AB/T, AD, AF, AG, AH, AM, 5086 [23]
Farmin DMC, DM20, DM24, DM40, DM5-24, DM60, DM68, DM86 [478]

Alkyl (C_8-C_{18}) dimethyl amine *(cont'd)*.
Onamine 65, 835, 1214, 1416 [663]

Alkyl dimethyl amine betaine.
Empigen BB [23]

Alkyl (C_8-C_{18}) dimethylamine oxide.
Aromox DMCD [16]
Conco XA-S, XA-T [209]
Empigen OB, OH, 5083 [23]

Alkyl dimethyl amine oxide.
Nissan Unisafe A-LM [636]
Schercamox DMA [771]

Alkyl dimethyl ammonium chloride.
Carsoquat 621 (80%) [155]

Alkyl dimethyl benzyl ammonium bromide.
Catigene BR/80B, T/80 + T/50 [825a]

Alkyl dimethyl benzyl ammonium chloride.
Alacsan 7LUF [29]
Alkaquat DMB 451, 50%, DMB-451, 80% [33]
Arquad B-50% USP, B-90 USP, DMMCB-50, DMMCB-75 [16]
Barquat 1552, MB-50, MB-80, MX-50, MX-80, OJ-50 [529]
Bio-Quat 50-24, 50-25, 50-28, 50-30, 50-40, 50-42, 50-60, 50-65, 50-MAB, 80-24, 80-28, 80-35, 80-40, 80-42 [120]
BTC 50, 50 USP, 65, 65 USP, 100, 824, 835, 2565, 8248, 8249, E-8358 [663]
Carsoquat 621 [155]
Dimanin A [112]
Empigen BAC, BAC90, BCB50, BCF80, BCM75, BCM75/A [23]
Gardiquat 1450, 1480, SV 480 [24]
Germ-I-TOL [408]
Hyamine 3500 [742]
Maquat LC-125 50% and 80%, MC-1412 50% and 80%, MC-1416 50% and 80% [558]
Neo-Germ I-TOL [408]
Nissan Cation F_2-50, Cation M_2-100 [636]
Onyxide 3300 [662]
Quatrene CB, CB-80, MB-80, -50 [400]
Querton 246 [487]
Retarder CA [390]
Roccal 50% Technical [413]
Sanisol CPR, CR, CR-80%, HTPR, OPR, TPR [803]
Variquat 60LC, LC80, 50MC, 80MC [796]
Vikol RQ [903]

Alkyl dimethylbenzyl ammonium saccharinate.
Loraquat QA 100 [274]
Onyxide 3300 [663]

Alkyl dimethyl betaine.
Alkateric BC [33]
Empigen BB [26]

Alkyl dimethyl dichloro benzyl ammonium chloride.
Bio-Quat 50-MAC [120]
Maquat DLC-1214 50 % and 80% [58]
Tetrosan 3, 4D [663]

Alkyl dimethyl ethyl ammonium bromide.
Bio-Quat 50-MAB [120]

Alkyl dimethyl ammonium chloride.
BTC 65 [663]

Alkyl dimethyl ethyl benzyl ammonium chloride.
BTC 471 [663]

Alkyl dimethyl ethyl benzyl ammonium cyclohexylsulfamate.
Onyxide 172 [663]

N-alkyl dimethyl 1-naphthylmethyl ammonium chlorides.
BTC 1100 [663]

Alkyl diphenyl ether sodium disulfonic acid.
Sandet ALH [764]

Alkyl disodium sulfosuccinamate.
Rewopol SMS 35 [725]
Secolat [825a]

Alkylene oxide addition products.
Marlox FK 14, FK 64, L6, LM 25/30, LM 75/30, LP 90/20, NP 109, OD 69, OD 105, S 58 [178]

Alkylene oxide adducts.
Chimipal PE 500, PE 510, PE 520, PE 530, PE 540, PE 550, PE 560 [514]

Alkyl ether phosphates.
Crodafos [223]

Alkyl ether sulfate.
Empicol MD [25]
Empimin LAM 50 [23]
Genapol PGM, TSM [416]

Alkyl ether sulfate, ammonium salt.
Steol CA 460, KA 460 [825]
Steposol CA-207 [825]

Alkyl ether sulfate, sodium salt.
Steol KS 460 [825]

Alkylether sulfosuccinate.
Kohacool L-400 [864]

Alkyl ethoxy benzyl dimethyl ammonium chloride.
Empigen BCY 40 [25]

Alkyl ethoxy dimethyl amine.
Empigen AY, OY [25]

Alkyl ethoxylate.
Liptol 40C [524]
Polystep F-12, F-13 [825a]

Alkyl ethoxy sulfosuccinate, sodium salt.
Cosmopon BT [514]

Alkyl glucoside.
Lutensol GD 70 [106]

Alkyl imidazoline.
Empigen 5078 [23]
Monateric LF-100 [599]

Alkyl imidazoline chloride.
Nissan Cation AR-4 [636]

Alkyl imidazoline laurate.
Texnol IL [635]

Alkyl imidazoline methosulphate.
Empigen FRB 75S, FRC 75S, RCS 75L [23]

Alkyl imidazolinium chloride.
Quaternary O [190]

Alkyl imidazolinium methosulfate.
Carsosoft CFI-75 [529]

Alkyl isoquinolinium bromide.
Catinal LQ-75 [864]
Isothan Q-75 [663]

n-Alkylmercaptan.
Thio Kalchol 08, 20, 24, 40 [478]

Alkyl methy amine, secondary.
Synprolam 35M [441]

Alkyl (C_{12}-C_{14}-C_{16}) methyl benzyl ammonium chloride.
Alkaquat DMB-451 [33]

Alkyl methy isoquinolinium chloride.
Ammonyx 781 [663]

Alkyl methyl tauride.
Lipolan TE [524]
Lipolan TE (P) [525]

Alkyl naphthalene, linear.
Petro ULF [688]

Alkyl-naphthalene sulfonate.
Cordon N-400 [320]
Nekal NF [388]
Sellogen [252]
Sorbit P [190]

Alkyl naphthalene sulfonate, linear.
Petro BA [688]

Alkyl naphthalene sodium sulfonate, salt.
Petro P, S, WP Modified [688]

Alkyl naphthalene sulfonic acid.
Sellogen WL Acid [252]

Alkyl naphthalene sulfonic acids, sodium salt.
Daxad® 11, 11G, 11KLS, 13, 15, 16, 17, 19 [363]
Harol RG-71L [365]
Nekal BX Conc. Paste, BX Dry [106]

Alkylolamide.
Actramide 189 [811]
Hyonic LA [252]
Ninol P-621 [825]
Rewocor AC 28 [725]
Rewomid SD [725]
Solar F342 [840]
Stepan P-621 [825a]
Swanic 52L [837]
Witcamide CDS, LDTS, LM, LMT, N, NA, NA1 [937a]

Alkylolamide ether sulfate, fatty.
Nissan Sunamide C-3, CF-3, CF-10 [636]
Perlankrol TM [509]

Alkylolamides, ethoxylated.
Amidox C_2, C5, L2, L5 [825a]

Alkylolamide ethoxylates, fatty acid.
Empilan LP2, MAA [23]

Alkylolamide, ethoxylated fatty acid, sulfosuccinic acid, semi-ester.
Elfanol 850 [16]

Alkylolamide ethylene oxide condensate.
Ethylan CH, CL, CRS [509]

Alkylolamide, fatty acid.
Emulan FJ [106]
Marlamid A 18, D 1218, D 1885, KL, KLA, M 18, M 1218, OS 18 [178]
Ninol HDA-7 [825]

Alkylolamide polyglycol ether, fatty acid.
Genagen CA-050 [416]

Alkylolamide sulfosuccinate, fatty acid.
Rewopol SBL 203 [725]

Alkyloxyalkyl sodium sulfate.
Sellogen 641 [252]

Alkyloxypropylamine.
Tomah PA-62, PA-1820 [866]

Alkylphenols
J-Sperse 35 [842]

Alkyl phenol adduct with EO 9-10 moles.
Nekanil 910 [106]

Alkyl phenol disulfide.
Vultac 2, 3, 7 [682]

Alkyl phenol with EO 6.
Lutensol AP 6 [106]

Alkyl phenol with EO 7.
Lutensol AP 7 [106]

Alkyl phenol with EO 8.
Lutensol AP 8 [106]

Alkyl phenol with EO 9.
Lutensol AP 9 [106]

Alkyl phenol with EO 10.
Lutensol AP 10 [106]

Alkyl phenol with EO 14.
Lutensol AP 14 [106]

Alkyl phenol with EO 20.
Lutensol AP 20 [106]

Alkyl phenol with EO 30.
Lutensol AP 30 [106]

Alkylphenol ether, ethoxylated.
Liponox NC2Y [524]

Alkyl phenol ether sulfate.
Cyclosol AP 90 [938]

Alkylphenol ether sulfate, ammonium salt.
Chimipon NA [514]
Perlankrol FD-63, FF, PA Conc. [509]

Alkyl phenol ether sulfate, sodium salt.
Lutensit A-ES [106]
Perlankrol FN-65, RN-75, SN [509]

Alkyl phenol ethoxylate.
Cedepal CA-210, CA-520, CA-630, CA-720, CA-890, CA-897 Cirrasol AEN-XZ [93]
Empilan NP's [23]
Ethylan BZA, ENTX, GMF, PQ [509]
Geronol TZ/B [550]
Iconol OP-1.5, OP-3, OP-5, OP-7, OP-10, OP-13, OP-16-70, OP-40, OP-40-70, OP-70 [109]
Igepal CA-887, CA-890, CA-897, CO-210, CO-430, CO-630, CO-850, CO-880, CO-890, CO-897, CO-970, CO-977, CO-980, CO-987, CO-990, CO-997, DM-730, DM-880 [338]
Levelan P307 [509]
Monolan PM 7 [509]
Nekanil 907, LN [106]
Polystep F-1, F-4, F-6, F-9, F-10, F-95-B [825]
Rexol 25/4, 25/6, 25/7, 25/8, 25/11, 25/14, 25/15, 25/407, 25 CA, 25 J, 25JM1, 25JWC [390]
Sandoxylate PN-9, PN-10.5, PO-5 [761]
Sterox DF, DJ, ND, NE, NG, NJ, NK, NM [602]
Swanic CD-48, CD-72 [837]
Tex-Wet 1155 [855]

Alkyl phenol, ethoxylated sulfate.
Eccoscour D-7 [277]

Alkyl phenol ethoxylate, sodium salt.
Emulphor OPS 25 [106]

Alkylphenol ethylene oxide adduct.
Berol 02, 09, 26, 277, 278, 281, 291, 295, 716, WASC [116]
Witco 936, 960, 980 [938]

Alkyl phenol ethylene oxide adduct, fatty acid ester.
Witco 910, 924 [938]

Alkylphenol ethylene oxide ether.
Iberscour W Conc. [386]

Alkyl phenolformaldehyde alkoxylates.
Landemul DCR Series [509]

Alkylphenol, oxalkylation product.
Wettol EM 2 [106]

Alkylphenol polyethoxide ether.
Hipochem AR-100 [410]

Alkylphenol polyglycol ethers.
Akyporox NP 15, NP 40, NP 70, NP 105, NP 150, NP 200, NP 300, NP 300V, NP 1200V, OP 40, OP 100, OP 115, OP 250, OP 400V [182]
Arbylen Conc. [176]
Hostapal CVS, N040, N060, N080, N090, N100, N110, N130, N150, N230, N300, N400 [55]
Marlophen 83, 84, 85, 86, 86 S, 87, 88, 84, 810, 812, 814, 820, 825, 830, 850, X [178]
Mulsifan RT 37 [948]
Serdox NNPQ 7/11 [788]
Sterling NPX [145]

Alkylphenol polyglycol ether carboxylic acids.
Akypo OP 80, OP 115, OP 190 [182]

Alkylphenol polyglycol ether ethylene oxide, 9.5 moles.
Neutronyx 600 [662]

Alkylphenol polyglycol ether ethylene oxide, 11 moles.
Neutronyx 656 [662]

Alkylphenol polyglycolether phosphate.
Loraphat E 1027 [274]
Rewophat E 1027 [274]
Steinaphat E 1027 [274]

Alkylphenol polyglycol ether, sulfated, ammonium salt.
Neutronyx S-60 [662]

Alkylphenol polymethylene sulfonate.
Iberpenetrant-114

Alkylphenol, polyoxylthylated.
Agepal Surfactants [338]

Alkyl phenol polyoxyethylene ether.
Swanic 6L, 7L, CM, D-16, D-24, D-52, D-60, D-72, D-80, D-120, J, X-100 [837]

Alkylphenol surfactant.
Sunmorl NP Conc. [630]

Alkyl phenoxy ethanol, sulfated.
Carsonol ANS [529]

Alkylphenoxypoly (ethylenoxy) ethanol.
Iconol NP-915 [109]
Igepal CTA-639 [338]

Alkyl phenoxy poly (ethyleneoxy) ethanol, sulfated, ammonium salt.
Concopal A, AS [209]

Alkyl phenoxy polyoxyethylene ethanol.
Hyonic PE-40, PE-60, PE-90, PE-100, PE-120, PE-240, PE-250, PE-260 [252]
Makon 4, 6, 8, 10, 12, 14, 20, 30 [825]

Alkyl phenoxy polyoxyethylene sulfate, sodium salt.
Gardisperse AC [24]

Alkylphenylpoly-ethoxyether.
Nonarox 575, 730, 1030, 1230 [711]
Simulsol NP 575 [711]

Alkyl phenylpolyglycol ester.
Solegal W Conc. [416]

Alkyl phosphate.
Atcopen SD [228]
Clink PR-46
Deatron N [630]
Delion 964, 6067 [849]
Marpol C-180
Neorate NA-30 [711]
Paracol OP, SV [635]

Alkylphosphate, acid.
Geronol FAT/4 [550]

Alkyl phosphate ester.
Monafax 939 [599]

Alkyl phosphate, ethoxylated.
Permax PM 503 [944]

Alkyl phosphate polyoxyethylene alkyl ether.
Delion 342 Conc. [849]

Alkyl phosphate, potassium salt.
Emalox C 102, OEP-1 [944]

Alkyl phosphate, salt.
Electrostripper F [478]

Alkyl phosphate surfactant.
Delion 964 [849]

Alkyl phosphonate, sodium salt.
Fostex SN [400]

Alkyl phosphonic acid.
Fostex-E, S, U [400]
Hostaphat OPS [416]

Alkyl polyamide surfactant.
Sunsoflon CK, K-2, MT-100 [630]

Alkyl polyamine ethyl glycine hydrochloride.
Salabon 50 [849]

Alkyl polyether.
Trycol LF-1 [289]
Witconol NP-100, NP-300 [934]

Alkylpolyether alcohol.
Triton® DN-14, X-180, X-185, X-190 [742]

Alkyl polyether, carboxylated.
Emcol® CN-6 [934]

Alkylpolyethoxyester.
Simulsol A, A 686, O, P4 [711]

Alkylpolyethoxy ethanol.
Triton® X-67 [742]

Alkylpolyethoxyether.
Alkylox P1904 [711]
Desadipol AX Extra [711]
Montegal 150 RG, AP 80, DZ [711]
Simulsol A, A 686, O, P4 [711]

Alkylpolyethoxyethersulfate.
Montelane CL 2588, L 2088, L 4088, LCT Supra [711]

Alkylpolyethylenimine.
Corfax™ 712 [215]

Alkyl polyglycolamine.
Produkt RT 63 [948]

Alkylpolyglycolether.
Emulgane O [253]
Genapol T-500, T-800 [416]
Hostacerin T-3 [55]
Hostapur CX Highly Conc. [416]
Marlowet 4800, 4857, 4862, 5001, BL, FOX, GFN, GFW, ISM, PA, PW, WOE [178]
Serdox NBSQ 4/3, NBSQ 5/5, NTST 9/12 [788]
Zusolat 1005/85, 1008/85, 1009/85, 1010/85, 1012/85 [948]

Alkyl polyglycol ether ammonium methyl sulfate.
Berol 563 [116]

Alkyl alcohol polyglycol ether, 2.2 mo (C_9-C_{11}) EO; nonionic.
Merpoxen UN 22 [285]

Alkyl polyglycol ether oil.
Emulsogen LP [55]

Alkyl polyglycol ether sulfate, sodium salt.
Genapol LRO and ZRO Brands [416]

Alkyl polyglycolic ether.
Geropon TBS [550]

Alkyl polyglycolic ether.
Soprofor RHS [550]

Alkyl polyoxyalkylene ether.
Mazawet 77, DF [564]
Surfonic J-4 [470]
Surfonic JL-80X, LF-17 [854]

Alkyl polyoxyalkyline ether.
Mazon 21 [564]

Alkyl polyoxyethylene ethanol, fluorinated.
Fluorad FC-170-C [862]

Alkylpolyoxyethylene ether.
Adsee® 799 [935]
Cerfak 1400 [425]
Consamine 15 [207]

Alkyl polyoxyethylene glycol.
Trycol 5951 [289]

Alkyl polyoxyethylene glycol ether.
Chemcol Agwet 115-M [171]
Witconol 1206, 1207 [934]

Alkylpolyoxyethylene phosphate, amine salt.
Geronol TZ/C [550]

Alkyl polyphosphate.
Atcowet C [228]
Chimin P10 [514]

N-s-alkyl-1,3-propanediamine.
Duomeen L-11, L-15 [79]
Synprolam 35N3 [441]

Alkyl-1,3-propylene diamine.
Diamine B11 [487]

Alkyl-sulfamido carboxylic acid.
Emulsogen H [55]

Alkylsulfamido carboxylic acid, sodium salt.
Bohrmittel Hoechst [55]
Emulsogen STH [55]

Alkyl sulfate.
Cortapol V [711]
Migregal 2N, NC-2 [637]
Sipex ADS, FA30 Modified [29]
Solanos AD Supra [711]
Stantex Pen 111 [400]
Texapon CS-Paste, P [401]

Alkylsulfate, sodium salt.
Sulfetal 4069, C 38 [948]

Alkyl sulfate, triethanolamine ammonium salt.
Genapol CRT 40 [55]

Alkyl sulfonate.
Base 11-T [156]
J-Lev FH-7 [842]

Alkylsulphonate, sodium salt.
Lutensit A-PS [106]

Alkyl sulfonate, triethanolamine salt.
Arodet 60 T Soft [82]

Alkyl sulfonic acid alkyl phenyl ester.
Vulkanol SF [112]

N-(alkylsulfonyl)glycine; 88% active sodium salt.
Emulsifer STH [338]

Alkyl sulfosuccinamide.
Rewopol TMS/F [725]

Alkylsulfosuccinate.
Alphenate TH 454 [253]

Alkyl sulfosuccinate, sodium salt.
Cosmopon BLM [514]

Alkyl trimethyl ammonium bromide.
Catinal HTB, HTB-70 [864]
Empigen CHB [23]
Vantoc AL [442]

Alkyl trimethyl ammonium chloride.
Arquad S [79]
Empigen 5080 [23]
Genamin KDM [55]
Genamin KDM-F [416]
Quartamin 24P, 86P [478]
Quartamin CPR, HTPR, TPR [803]
Synprolam 35TMQC [441]
Variquat A-200 [796]
Varisoft E228 [796]

Alkyl trimethylammonium halide.
Dupont Retarder LAN [269]

Alkyl trimethyl ammonium methosulphate.
Empigen CM [25]

Allantoin.
Alphosyl Lotion, Cream [719]
Herpecin-L Cold Sore Lip Balm [144]
Vagilia Cream, Suppositories [519]
Vagimide Cream [519]
Vaginal Sulfa Suppositories [769]

Allantoin acetyl methionine.
Almeth [775]

Allantoin ascorbate.
Allantoin Vit. C [775]

Allantoin galacturonic acid.
Allantoin Galacturonic Acid Complex [775]

Allantoin glycyrrhetinic acid.
Allantoin Glycyrrhetinic Acid Complex [775]

Allantoin PABA.
ALPABA [775]

Allantoin polygalacturonic acid.
ALPOLYGAL [775]

Allopurinol.
Allopurinol Tablets [769]
Lopurin [127]
Zyloprim [139]

Allylacrylate.
SR-200 [766]

2,6-allyl-dichlorosilaneresorcinol.
Nolane [680]

Allyl diglycol carbonate.
CR-39 [705]

N-allyl-n-n-dimethylamine.
Sequalog 02-7520-00 [172]

Allyl glycidyl ether.
MON-ARC® AGE [69]

2-allyl-4-hydroxy-3-methyl-2-cyclopenten-1-one ester of 2,2-dimethyl-3-(2-methylpropenyl)-cyclopropanecarboxylic acid.
MGK® Allethrin [568]

Allyl methacrylate.
MON-ARC® AM [69]
SR-201 [766]

d1-3 allyl-2-methyl-4-oxo-2-cyclopenten-1-yl-2,2-dimethyl 3-(2 methyl-1-propenyl)cyclopropanecarboxylate.
D-TRANS® Allethrin [568]

Allyl trimethyl ammonium chloride.
Variquat A200 [796]

Almond meal.
Lipo AM [525]

Aloe vera gel.
Terra-Dry [853]

Alphaprodine hydrochloride.
Nisentil injectable [738]

Alprazolam.
Xanax Tablets [884]

Alprostadil.
Prostin VR Pediatric Sterile Solution [884]

Alseroxylon.
Rauwiloid Tablets [734]

Alumina.
Al-0104 T 3/16″, -0109 P, -1401 P(MS), -1404 T 3/16″, -1609 P, -3438 T 1/8″, -3916 P, -3945 E 1/16″, -3970 P, -3980 T 5/32″, -3996 R 3.5×1.5 mm, -4028 T 3/16″, -4126 E 1/16″ [389]
Alon [141]
Aluminum Oxide C [244]
Catapal SB [206]
Cosmetic Alumina Hydrate [496]
KA-101 [475]
Lucalox [342]
Magnesia & Alumina Oral Suspension [751]

Alumina hydrate.
Alumina Hydrate [846]

Alumina hydrated.
Micral 932 [392]

Alumina-silica.
Kaocast [98]
Kaowool [98]

Alumina silicated.
A1-1602 T 1/8″, -1605 P [389]

Alumina trihydrate.
CM-66 [808]
FRE [808]
G-431 [808]
H-36 [808]
LV-336 [808]
SB-30, -31, -31C, -136, -331, -332, -335, -336, -431, -432, -632, -932 [808]
Solem ATH [808]

Alumina zirconia-silica refractory.
Zirmul [169]

Aluminosilicate gel.
Decalso [685]

Aluminosilicate, hydrated.
Pyrax RG [894]

Aluminum acetate.
Acid Mantle Creme & Lotion [257]
Osti-Derm Lotion [679]
Otic Domeboro Solution [585]

Aluminum butoxide (secondary).
Aluminum Chelak BEA-1 [389]

Aluminum carbonate gel.
Basaljel Capsules & Swallow Tablets, Suspension & Suspension, Extra Strength [942]

Aluminum chloride.
Drysol [686]
Xerac AC [686]

Aluminum chlorohydrate.
Aluminum Hydroxychloride 23 Hoechst, 47 Hoechst [55]
Breezee Mist Foot Powder [679]
Chlorhydrol, Granular, Impalpable, 50% Solution [721]
Locron Extra, Flakes, L, P, P Extra, Powder, S, Solution [55]
Micro Dry [721]
Wickenol 303, [926]
Wickenol® 308 [926]
Wickenol 321, 323, 326 [926]
Wickenol CPS® 325, 331 [926]

Aluminum chlorohydrex.
Rehydrol (Reheis) [721]

Aluminum chlorhydroxide.
Pedi-Dri Foot Powder [679]

Aluminum citrate.
Alutrat [901]

Aluminum dichlorohydrate.
Wickenol 307 [926]

Aluminum distearate.
Aluminum Stearate EA Food Grade [935]

Aluminum glycinate.
Bufferin with Codeine No 3 [134]

Aluminum hydroxide.
C-Weiss 1 [351]

Aluminum hydroxide.
Aluminum & Magnesium Hydroxides with Simethicone I & II [751]
Camalox [746]
Gaviscon Liquid Antacid [551]
Gelusil, -M, —II [674]
Maalox, Plus, TC Suspension [746]
Mygel Suspension [344]
Trickup [543]

Aluminum hydroxide gel.
ALternaGEL Liquid [831]
Aludrox Oral Suspension [942]
Aluminum Hydroxide Gel, Gel-Concentrated [751]
Amphojel Suspension [942]
Arthritis Pain Formula By the Makers of Anacin Analgesic Tablets [922]
Kudrox Suspension (Double Strength) [499]
Mylanta Liquid, -II Liquid [831]
Simeco [942]

Aluminim hydroxide gel, dried.
Alu-Cap Capsules [734]
Aludrox Tablets [924]
Alu-Tab Tablets [924]
Amphojel Tablets [734]
Cama Arthritis Pain Reliever [257]
Gaviscon Antacid Tablets, -2 Antacid Tablets [551]
Mylanta Tablets, -II Tablets [831]

Aluminum isopropylate.
Aluminum Chelate PEA-1, Chelate PEA-2 [389]

Aluminum oxide.
Aloxite [383]
Oxide C [244]

Aluminum oxide, hydrated.
Hydral 700 Series [39]

Aluminum phosphide.
Degesck Phostoxin [243]
Fumitoxin [687]

Aluminum potassium silicate.
C-1000 Micro-Mica, C-3000 Micro-Mica [379]
Mica Waterground [392]

Aluminum powder.
C-Pigment 1 [351]
Silver #11 [91]

Aluminum sesquichlorohydrate.
Wickenol 306, 308 [926]

Aluminum silicate.
Allen R [236]
Kaopolite SF [479]
Mercer R Clay [236]
Pyrax [894]
Smithko No. 1 Clay, No. 2 Clay [804]
Soft Clay/Hard Clay [164]
Type 50 Clay [804]
Veecote [894]
Windsor [236]

Aluminum silicate, amorphous.
Polestar 200R Calcined Clay, 501 Calcined Clay [381]

Aluminum silicates, anhydrous.
Burgess Optivhite, Optivhite P, Icecap K, Iceberg, No. 30-P [138]

Aluminum silicate, calcined.
Altowhite LL [347]
Glomax JDF, LL [347]
Harvick Clay 5-C & 7-C [392]
Polyfil 40, 70, 80 [428]
Satintone No. 1, No. 2, No. 5, Special [295]
Tisyn [138]

Aluminum silicate, dehydroxylated.
Al-Sil-Ates [336]
An-Hy-Drol [336]
Nuopaque [336]

Aluminum silicate, delaminated.
Kaopaque [387]

Aluminum silicate, delaminated hydrated.
Polyfil DL [428]

Aluminum silicate, hydrated.
Barden [428]
Champion Hard Clay [392]
GK Soft Clay [392]
Harvick 2, 6, Clay 15, 42, 50R, 59 & 59F Clay [392]
Hi-White R [428]
Hydrite [387]
Hydrite 121-S, Flat D, PX, PXS, R, RS, UF [347]
LGB Clay [428]
Natka 1200 Hard Clay [342]
Nopak [428]
Nuflo [428]

Aluminum silicate, hydrated *(cont'd)*.
Polyfil F, FB, HG-90, X, XB [428]
Royal Queen, Royal Duchess, Royal Countess [752]
Struex-80 [347]
Suprex Clay [428]
Type Clay [810]

Aluminum silicate hydrous.
ASP 072, 100, 170, 172, 200, 352, 400, 400P, 600, 800, 900 [295]
Burgess Poly Clay, No. 10, No. 20 [138]
Kaolex [428]

Aluminum silicate hydrous with stearate, 0.5.
ASP 101 [295]

Aluminum powdered mica.
Mineralite Mica [752]

Aluminum starch octenylsuccinate.
Dry Flo [619]

Aluminium stearate.
Radiastar 1208 [658]
Witco Aluminum Monostearate N.F. [935]

Aluminum sulfate.
Domeboro Powder Packets & Tablets [585]
Pedi-Boro Soak Paks [679]

Aluminum tris-o-ethyl phosophonate.
Aliette [730]

Aluminum tristearate.
Witco Aluminum Stearate 132 [935]

Aluminum zirconium tetrachlorohydrate.
Wickenol 373 [926]

Aluminum zirconium tetrachlorohydrex GLY.
Wickenol 368, 369, 374 [926]

Amantadine hydrochloride.
Symmetrel [269]

Amblygonite (q.v.).
Alumalith [333]

Amcinonide.
Cyclocort Cream, Ointment [516]

Amide amine.
Nissan Softer-706, Softer 1000 [636]

Amide amine, oxethylated.
Lamephan 303 [176]

Amide amino condensate.
Emkagen Concentrate [292]

Amide ether sulfate.
Genapol AMS [416]

Amide ethoxylate.
Eccoful FC Conc. [277]
Eccoterge SCH [277]
Ethomid HT/15, HT/60, O/15 [16]

Amide ethoxylated fatty acid.
Lutensol FSA 10 [106]

Amide fatty.
Oramide DL 200, ML 200 [711]

Amide polyglycol ethers, fatty acid.
Dionil OC, SD, SH 100, W 100 [178]

Amide sulfate.
Celopon [77]
Emkapon K [292]
Emkatex PX29 [292]

Amide sulfonate.
Alkamide 2112, 2122 [33]
Emkapon L, SS, TS [292]
Ninex 24 [825]

Amide sulfonate, fatty.
Emkapon Jel BS [292]

Amides, sulfonated, fatty acid.
Humectol C [426]
Humectol C Highly Conc. [416]

Amido alkylamine acetate.
Sulfostat KNT [948]

Amido alkyl betaine, fatty acid.
Amphotensid B4 [948]
Tego-Betain L7, L10, T [856]

Amido alkyldimethylamine oxide, fatty acid.
Aminoxid WS 35 [856]

Amidoamine.
Trymeen 6657 [289]
Uni-Rez 2880, 2828-D, 2810, 2850 [876]

Amidoamine, cationic, fatty.
Textamine 1839 [400]

Amidoamine, salt, fatty.
Rycofax 618 [249]

Amido-ether sulfate.
Monamine 779 [559]

Amido sulfosuccinate salt.
Witco 938 [938]

Aminacrine hydrochloride.
Vagilia Cream, Suppositories [519]
Vagimide Cream [518]

Aminacrine preparations.
AVC Cream, Suppositories [580]
Vaginal Sulfa Suppositories [769]

Amine.
Tinuvin 144, 622, 770 [190]
Wingstay Antioxidants/Antiozonants [363]

Amine, alkoxylated, carboxylic salt.
Marlowet 5529 [178]

Amiloride hydrochloride.
Midamor Tablets [578]
Moduretic Tablets [578]

Amine alkylaryl sulfonate.
Ninate 411 [825a]
Witco 918 [934]
Witconate 795, 93S, 702, 1075X, P10-59, TAB, YLA [935]

Amine alkyl sulfate.
Richonol 2310, 3715 [732]

Amine catalyst,
Dabco R-595, T, TMR, TMR-2, WT, X-DM [13]

Amine condensate.
ESI-Terge AH 20 [892]
Gafsoft 325 [338]

Amine-dialtyl-dithiophosphate.
Rhenocure AT [728]

Amine dodecylbenzene sulfonate.
Conco AAS Special 3 [209]
Hetsulf 60T, IPA [407]
Marlopon AMS 60 [178]

Amine, ethoxylated.
Armostat 310, 350, 375, 410, 450, 475, 550, 575 [653]
Triton CF32 [743]

Amines, ethoxylated fatty.
Ethomeen C/12, C/15, C/25, HT/12, HT/15 [16]
Ethoxamine C5, SF11 [938]

Amine ethoxylate, fatty.
Genamin C,O,S, and T Brands [416]
Leomin TR [416]
Lutensol FA 12 [106]
Neolisal HCN [711]
Zusomin O 102, O 105, O 112, S 110, S 125 [948]

Amine ethylene oxide condensate, fatty.
Ethylan TC, TF, TH-2, TN-10, TT-15 [509]

Amine lauryl sulfate, anhydrous.
Texapon Conc. 7021 [399]

Amine, phenolic.
Antioxidant 703 [302]

Amine oxide.
Alkamox LO, CAPO [33]
Aromox C/12-W, DMCD, DM14D-W, DMMCD-W, T/12 [16]
Berol 305 [116]
Emcol® L, M, [934]
Hartox DMCD [391]
Noxamine 02/30, CA 30 [162]
Sochamine OX 30 [937a]
Tomah AO-14-2, AO-17-2 [886]

Amine phosphate.
Witcor 3192 [935]

Amine, polyalkoxylated.
Mazeen DBA-1 [564]

Amine polyethoxy carboxylate, fatty, potassium salt.
Ampho T-35 [81]

Amine polyglycol condensate.
Triton® CF-32 [742]

Amine polyglycol ethers, fatty.
Marlazin L 10, OL 20, S 10, S 15, S 20, S 25, S 40, T 10 [178]

Amines, polyoxyethylated.
Witcamine RAD-0500, RAD-0515, RAD-1100, RAD-1110 [935]

Amine sulfonate.
Alkamuls AG-IPAM [33]
Witconate 79S, 93S, [935]
Witconate P10 59 [937a]

Aminoacetic acid.
Glytinic [132]

Amino acid preparations.
Added Protection III Multi-Vitamin & Multi-Mineral Supplement [589]
Amino-Cerv [586]
Aminoplex Capsules & Powder [872]
Aminostasis Capsules & Powder [872]
Aminotate Capsules & Powder [872]
Carnitine - Amino Acid Preparation [872]
Endorphenyl [872]
Geravite Elixir [394]
Marlyn Formula 50 [553]
Vivonex High Nitrogen Diet, Standard Diet [652]

Aminoalkyl phosphonic acid.
Fostey S [400]

P-aminobenzoate, potassium.
Potaba [356]

Aminobenzoic acid.
Potaba [356]
PreSun 15 Sunscreen Lotion [920]

P-aminobenzoic acid.
Hill-Shade Lotion [412]
Mega-B [72]
RVPaba Lip Stick [284]

Aminobenzoic preparations.
Cetacaine Topical Anesthetic [168]
Pabalate Tablets, SF Tablets [737]
Potaba [356]

2-aminobutane.
Tutane [282]

Aminocaproic acid.
Amicar [516]
Aminocaproic Acid Injection [287]

Amino carboxylic acid.
Levelan NKS [556]

5-amino-4-chlorl-2-phenyl-3-(2H)-pyridazenone.
Pyramin [106]

6-amino-4-chloro-m-toluenesulfonic acid 3-hydroxy-2-naphthoic acid.
Atlas B3070 Monotherm Red Toner [496]

2-amino-5-chloro-p-tol-uenesulfonic acid 2-naphtol.
Atlas A2511 Red Lake C Toner [496]

3-amino-2,5-dichlorobenzoic acid.
Amiben [877]

Aminoethane.
Sodium N Methyl Taurine [181]

2-(2-amino-ethoxy)-ethanol.
Diglycolamine [470]

1-(2-amino-ethyl)-2-n-alkyl-2-imidazoline.
Witcamine 209 [925]

Amino-ethyl-imidazoline.
Servamine KOO 330 [788]

Aminoethylpiperazine.
D.E.H. 39 [261]

2-amino-2-ethyl-1,3-propanediol.
AEPD [449a]

Aminoethyl sulfonate, sodium salt.
Iberpol Gel [386]

4-Amino-3-methyl-6-phenyl-1,2,4-triazin-5(4H)-one.
Goltix [112]

2-Amino-2-methyl-1-propanol.
AMP, 95 [449a]

Amino (methylene) phosphonic acid.
Fostex AMP [400]

Aminoglutethamide.
Cytadren [189]

4-Amino-2-hydroxytoluene.
Rodol PAOC [531]

2-Amino-6-mercaptopurine.
Tabloid Brand Thioguanine [139]

Aminomethyl propanediol.
AMPD, 95, Regular [456]

2-Amino-5-nitrophenol.
Rodol YBA [531]

m-Aminophenol,
Rodol EG, 2G, P Base [531]

m-Aminophenol, hydrochloric acid.
Rodol EGC [531]

m-Aminophenol sulfate.
Rodol EGS, PS [531]

Aminophylline.
Aminophyllin Injection [134]
Aminophyllin Injection [783]
Aminophyllin Tablets [782]
Aminophylline Injection [287]
Aminophylline Oral Liquid, Suppositories & Tabs [769]
Aminophylline Tablets [344]
Aminophylline Tablets & Oral Solution [751]
Mudrane GG Tablets, GG-2, Tablets [703]
Phyllocontin Tablets [714]
Somophyllin & Somophyllin-DF Oral Liquids, Rectal Solution [324]

Amino polycarboxylic acid.
Varkon HL [77]

N-Amino propionate, fatty.
Sipoteric CAPA [29]

Amino-propyl amines, fatty.
Adogen 560, 570-S, 572 [796]

Aminopropyltriethoxy silane.
Union Carbide Organofunctional Silane A-1100 [878]

Aminosalicylic acid.
Pamisyl [674]

6-Amino-m-toluenesulfonic acid-3-hydroxy-2-naphthoic acid.
Atlas A6225 Lithol Rubine Toner, B3319 [496]

3-Amino-1, 2, 4-triazole.
Cytrol Amitrol-T [50]

Aminotri (methylene phosphonic acid).
Fostex AMP [400]
Unihib 305 [529]

Amino tris (methylene phosphonic acid).
Dequest 2000® [602]

Amitraz.
BAAM® [884]
MITABAN® [884]

Amitriptyline.
Amitriptyline HCI Tablets [344]
SK-Amitriptyline Tablets [805]

Amitriptyline hydrochloride.
Amitriptyline HCI Tablets [119]
Amitriptyline Hydrochloride Tablets [751]

Ammonia dynamite.
UNIMITE™ [406]

Ammonia gelatin.
Hercogel A [406]

Ammonia neutralized, condensed naphthalene sulfonic acid.
Lomar PWA [252]

Ammonium acetate.
Ultralog 99.99% 05-9560-00 [172]

Ammonium acetate, polypropoxylated quaternary.
Emcol CC 55 [937a]

Ammonium alcohol ether sulfate.
Calfoam NEL-60 [697]
Witcolate AE-3 [935]
Witcolate S1300C [934]

Ammonium alginate.
Algum B2A [280]
Marex [484]
Superloid [485]

Ammonium alkylaryl ether sulfate.
Cedepal CO-436 [256]
Progalan X-13 [537]

Ammonium alkyl benzene sulphonate.
Nansa AS40 [23]

Ammonium alkyl ether sulfate.
Cedepal FA-406 [256]
Polystep, B-20, B-22 [825a]
Sterling ES-600 [145]

Ammonium alkyl ethoxy sulphate.
Empimin LAM30, LAM 50 [23]

Ammonium alkyl phenol ethoxylate sulfate.
Polystep B-1 [825a]

Ammonium alkyl polyether ethoxylate sulfate.
Polystep B-13 [825]

Ammonium alkyl sulfate.
Orvus K Liquid [709]

Ammonium alkyl sulfosuccinamate.
Empimin MSS [23]

Ammonium alkyl tri-ethoxysulfate.
Empimin AQ60 [25]

Ammonium bacteriostat, quaternary.
Diasan [496]

Ammonium benzoate.
Ultralog 99.99+% 05-9720-00 [172]

Ammonium bisulfate.
Ultralog 99.99% 05-9840-00 [172]

Ammonium bromide, quaternary.
Fog Herx-R [190]

Ammonium chloride.
Neutrolox [682]
Ultralog 99.99% 06-0220-00 [172]

Ammonium chloride, quaternary.
Adogen 432 [796]
Bina QAT-43 [190]
Emulsifier Three, Four, Five [866]
Jet Quat [471]
Katapone VV-328 [338]
Tomah Q-14-2, Q-17-2. Q-D-T [866]

Ammonium chloride quaternary polyethoxylated.
Ethoquad 18/12, C/12, O/12 [79]

Ammonium chloride, polypropoxylated quaternary.
Emcol CC 9, CC 36 [937a]

Ammonium chromate.
Ultralog 99.99% 06-0520-00 [172]

Ammonium citrate dibasic.
Ultralog 99.9% 06-0560-00 [172]

Ammonium compound, quaternary.
Acryloft, Acryloft Conc. [338]
Berol Fintex 577 [116]
Hipochem M-51 [410]

Ammonium cumene sulfonate.
Ammonium Cumene-sulfonate 60 [178]
Eltesol ACS 60 [25]
Naxonate 6AC [754]
Reworyl ACS 60 [725]
Ultra NCS Liquid [935]
Witconate NCS [935]

Ammonium decyl ether sulfate.
Carsonol ADES [529]

Ammonium dichloride, diquaternary.
Duoquad T-50 [79]

Ammonium dichromate.
Ultralog 99.9% 06-0880-00 [172]

Ammonium dodecyl benzene sulfonate.
Conco AAS-50, AAS-50E, AAS-50S [209]
Hetsulf 50A [407]
Nansa AS 40 [25]
Newcol 210 [635]

Ammonium ethoxy ether sulfate.
Texapon 230E [399]

Ammonium 2-ethylhexyl sulfate.
Serdet DSN 50 [788]

Ammonium ferric ethylenediaminetetraacetate.
Sequestrene NH4Fe [190]

Ammonium flouride.
Ultralog 99.99% 06-1360-00 [172]

Ammonium formate.
Ultralog 99.9% 06-1460-00 [172]

Ammonium hydroxide.
Acculog 06-2100-00 [172]

Ammonium iodide.
Ultralog 99.99% 06-2220-00 [172]

Ammonium laureth sulfate.
Alkasurf EA-60 [33]
Carsonol ALES-4, ALES-5, SES-A [529]
Conco Sulfate 216, WEM, WM [209]
Cycloryl M3D, M3D2, M3D60 [939]
Cycloryl MA 330, MA 360 [234]
Cycloryl MD [939]
Empicol EAB [548]
Maprofix LES60A, MB, MBO [663]
Rewopol AL1, AL2, AL3 [725]
Richonol S-1300, S-1300C [732]
Sipon 201-20, EA, EA-2, EAY [29]
Standapol 230E Conc., EA 1, EA 2, EA 3 [400]
Steol CA-460, CA-760 [825]
Sterling AM, ES 600 [145]
Texapon EA-1, EA-2, EA-3, EA-40 [399]
Witcolate S1300C [935]

Ammonium lauroyl sarcosinate.
Hamposyl AL-30 [363]

Ammonium lauryl ether sulfate.
Alkasurf EA-60 [33]
Calfoam NEL-60 [697]
Carsonol SES-A [155]
Cedepal SA-406 [256]
Conco Sulfate WM [209]
Cycloryl MA 330, MA 360 [234]
Drewpon EKZ [265]
Empicol EAA70, EAC 60 [25]
Equex AEM [709]
Manro ALES 27 [545]
Nutrapon AL 60, KF 3846 [198]
Polyfac AES-60A [918]
Polystep B-11 [825a]
Rewopol AL3, 60% [725]
Sactipon 2 OA [521]
Sactol 2 OA [521]
Sipon EA, EAY [29]
Texapon EA-1, EA 2 [399]
Texapon NA [398]
Tylorol A [858]
Ungerol AM3-60 [874]

Ammonium lauryl ethoxy sulphate.
Empicol EAA, EAA70, EAB, EAB70, EAC60, EAC70 [23]
Empimin AQ-60 [24]

Ammonium lauryl polyether sulfate.
Lakeway 201-20 [126]

Ammonium lauryl sulfate.
Akyposal ALS 33 [182]
Alkasurf ALS [33]
Avirol 200 [400]
Carsonol ALS, ALS Special [155]
Cedepon LA-30 [256]
Conco Sulfate A [209]
Cycloryl MA [234]
Drewpon NH, NHAC [265]
Emal AD-25 [478]
Emersal 6430 [290]
Empicol 0931 [23]
Empicol AL30 [24]
Empicol AL30/T [23]
Empicol AL70 [25]
Lakeway 101-20 [126]
Lonzol LA-300 [529]
Lorol Liquid NH Sulphonated [745]
Manro ALS 30 [545]
Maprofix ER, LES-60A, NH, NH 54, NHL, NHL-22 [663]
Mars AMLS [554]
Melanol LP 30 [785]
Montopol LA 20 [711]
Norfox ALS [650]
Nutrapon HA 3841 [198]
Polystep B-7 [825]
Rewopol ALS [725]
Richonol AM [732]
Sactipon 2 A [521]
Sactol 2 A, 2 OA [521]
Sermul EA 129 [788]
Sipon L22, L-22HV [29]
Standapol A, AMS-100 [400]
Stepanol AM, AM-V [825]
Sterling AM, AM-HV [145]
Sulfotex 7122 [399]
Sulfotex WAA [400]
Texapon A [399]
Texapon A 400, Special [401]
Witcolate AM [935]
Zoharpon LAA [946]

Ammonium lauryl sulfate, ethoxylated.
Empicol EAB [25]

Ammonium lauryl sulfurs.
Jordanol AL 300 [474]

Ammonium lauryl triethoxysulphate.
Empimin AQ60 [23]

Ammonium lignin sulfonate.
Greenz [195]
Tanz [195]

Ammonium metavanadate.
KM® [490]

Ammonium monooleamido PEG-2 sulfosuccinate.
Standapol SH-200 [400]

Ammonium myreth sulfate.
Conco Sulfate 444 [209]
Cycloryl ME 361 [234]
Rewopol AM1 [725]

Ammonium myreth sulfate *(cont'd).*
Richonol S-1300M [732]
Standapol EA-40 [400]

Ammonium myristyl ether sulfate.
Texapon EA-40 [399]

Ammonium nitrate.
Herco-Prills® [406]
Hydrolin [659]
Niox [270]

Ammonium nitrate and coating agents 98%:2%
Ammo-nite [298]

Ammonium nitrate and water
Feran [35]

Ammonium nonoxynol-4 sulfate.
Alipal CO-436 [338]
Neutronyx S-60 [663]

Ammonium nonyl phenol ether sulfate.
Merpoxal NOA 4080 [285]

Ammonium oxalate.
Ultralog 99.9% 06-3100-00 [172]

Ammonium pareth-25 sulfate.
Neodol 25-3A [791]
Standapol AP-60 [400]
Sterling ES 600 [145]
Witcolate AE-3 [935]

Ammonium perchlorate.
KM® [490]
Ultralog 99.95% 06-3220-00 [172]

Ammonium perfluoroalkyl carboxylate.
Fluorad FC-143 [862]

Ammonium perfluoroalkyl sulfonate.
Fluorad FC-93 [862]

Ammonium phosphate.
Fyrex [821]
Ultralog 99.9% 06-3480-00 [172]

Ammonium phosphate, polypropoxylated quaternary.
Emcol CC 57 [937a]

Ammonium polyacrylate.
Acrysol G-110 [742]
Hycryl [818]

Ammonium polyoxyethylene nonylphenyl ether sulfate.
Newcol 560SF [635]

Ammonium, quaternary.
Vantoc CL [439]

Ammonium salt, polyquaternary.
Romie 802 [865]

Ammonium salt, quaternary.
Eleton KT-8 [865]
Ospin TAN [865]

Ammonium sulfamate.
Ammate [270]
Ultralog 06-4180-00 [172]

Ammonium sulfate.
Ammonium Sulfate [100a]
Ultralog 99.99% 06-4220-00 [172]

Ammonium sulfite.
Ultralog 06-4300-00 [172]

Ammonium stearate.
Ammonium Stearate 33% Liquid [390]
Denlube 33-B [365]

Ammonium tartrate 99.9%.
Ultralog 06-4360-00 [172]

Ammonium thiocyanate.
Acculog 06-4700-00 [172]
Ultralog 06-4600-00 [172]

Ammonium thiosulfate.
Jiffix [542]

Ammonium xylene sulfonate.
Conoco AXS [206]
Eltesol AX 40 [23]
Naxonate 4AX [754]
Naxonate AX [621]
Reworyl AXS [725]
Richonate AXS [732]
Stepanate AM [825]
Ultra NXS Liquid [935]
Witconate NXS [935]

Amniotic fluid.
Liquide Amniotique Special-LAP2 [501]

Amobarbital.
Amytal [523]

Amobarbital sodium.
Amytal Sodium Pulvules, Sodium Vials [523]
Tuinal [523]

Amoxapine.
Asendin [516]

Amoxicillin.
Amoxicillin Capsules [119]
Amoxicillin Capsules, for Oral Suspension [674]
Amoxicillin Suspension [119]
Amoxicillin Trihydrate Capsules & Powder for Oral Suspension [769]
Amoxil® Capsules, 250 mg, 500 mg, Chewable Tablets, 125 mg, 250 mg, for Oral Suspension, 125 mg/5 ml, 250 mg/5 ml, Pediatric Drops for Oral Suspension 50 mg per ml [114]
Polymox Capsules, for Oral Suspension, Pediatric Drops [134]
Trimox Capsules & for Oral Suspension [817]

Amoxicillin *(cont'd).*
Wymox Capsules & Oral Suspension [942]

Amoxicillin/potassium clavulanate.
Augmentin® 125, 250 for Oral Suspension, 250 Tablets, 500 Tablets [114]

Amphetamine aspartate.
Obetrol-10, -20 [726]

Amphetamine sulfate.
Benzedrine [804a]
Obetrol-10, -20 [726]

Amphotericin B.
Fungizone Cream/Lotion/Ointment, Intravenous [817]
Mysteclin-F Capsules, Syrup [817]

Ampicillin.
Amcill Capsules, Oral Suspension [674]
Ampicillin Capsules, Suspension, -Probenecid Suspension [119]
Omnipen Capsules, Oral Suspension, Pediatric Drops [942]
Polycillin, -N for Injection, -PRB [134]
SK-Ampicillin Capsules & for Oral Suspension [805]
Totacillin® Capsules, 250 mg, 500 mg, for Oral Suspension, 125 mg/5 ml [114]

Ampicillin sodium.
Omnipen-N [942]
Polycillin-N for Injection [134]
SK-Ampicillin-N for Injection [805]
Totacillin® -N, 1 Gm, 2 Gm, 10 Gm, 250 mg, 500 mg [114]

Ampicillin trihydrate.
Ampicillin Trihydrate Capsules & Powder for Oral Suspension [769]
Principen Capsules, for Oral Suspension, with Probenecid Capsules [817]

Amyl acetate.
Pentacetate [682]

Amylase.
Amerzyme Alpha Amylase 3.2 [54]
BAN 240 L, 120 L, 1000 S, 360 S [654]
Fermalpha™ [313]
Spezyme® AA, BBA [313]

Amylase (bacterial).
Amizyme [707]
HT-44 [584]
Premierzymes I and II [584]
Rapidase [671]
Spezyme AA [313]

Amylase (barley).
Spezyme BBA [313]

Amylase enzyme.
Zymetec™ AA-170 [301]

Amylase (fungal).
Amylase-AO [54]
Clarase [584]
Dextrinase A [584]
Fermalph [313]
Fungamyl [654]
Mylase [913]

n-Amylcinnamic alcohol.
Buxinol [353]

Amylcinnamic aldehyde.
Buxine [353]

Amyloglucosidase.
AMG 200 L, 150 L [654]
Diazyme [584]

Amylolytic enzyme.
Arco-Lase, Plus [72]
Cotazym, -S [664]
Festal II [418]
Festalan [418]
Gustase [350]
Kutrase Capsules [499]
Ku-Zyme Capsules [499]

Amyl para-dimethyl-aminobenzoate.
Escalol 506 [895]

t-Amyl peroxyneode-canoate.
Lupersol TA-46, TA-46M75 [536]

t-Amyl peroxy pivalate.
Lupersol TA-54M75 [536]

Anethole.
Anethole USP [12]

Anisaldehyde, sodium bisulfite.
Anoplex [800]

Anisotropine methylbromide.
Valpin 50, 50-PB [269]

Antazoline phosphate.
Antistine Phosphate [191]

Anthracene 99.9%.
Ultralog 07-3800-00 [172]

Anthracite.
Carb-O-fil [789]
163 Filler [752]

Anthranilic acid.
AA [797]

Antimony dialkyldi-thiocarbamate.
Vanlube 73, 622, 648 [894]

Antimony mercaptide.
Stanclere A-121, A-121 C, A-221 [459]
Synpron 1027, 1034 [845]

Antimony oxide
Mastermix Antimony Oxide-200-PD [392]
Thermoguard S [596]
Ultralog 07-5785-00 [172]

Antipyrine.
Auralgan Otic Solution [97]
Collyrium Eye Lotion, w/ Ephedrine Eye Drops [942]
Otipyrin Otic Solution (sterile) [498]
Tympagesic Otic Solution [7]

Apricot extract.
Lipofruit Apricot [525]

Apricot kernel oil.
Lipovol P [525]

Aprobarbital.
Alurate Elixir [738]

N-arachidyl-be-henyl 1,3-propylenediamine, 90%.
Kemamine D-190

Arachidyl propionate.
Waxenol 801 [926]

Arylphosphate, acid.
Geronol FAT/7 [550]

Aryl polyethoxy condensate.
Cenegen NWA [225]

Aryl sodium sulfonate, alkylated.
Titanole RMA [863]

Aryl-sulfamido carboxylic acid.
Hostacor H Liquid [55]

Ascorbic acid.
Ascorbic Acid Ampul Type, No. 604065700, Fine Powder, No. 604565200, Fine Granular, No. 6045655, Granular, No. 604065400, Super-Fine Granular, No. 604065900, Type S, No. 604566000, Ultra-Fine Powder, No. 604565300 [738]
C-Span [281]

Ascorbyl dipalmitate.
Nikkol Dipalmitoyl Ascorbic Acid [631]

Ascorbyl palmitate.
Ascorbyl Palmitate No. 604120000 [738]

Asparaginase.
Elspar [578]

Aspirin.
A.P.C. with Codeine, Tabloid brand [139]
Anacin Analgesic Capsules, Analgesic Tablets, Maximum Strength Analgesic Capsules, Maximum Strength Analgesic Tablets [922]
Arthritis Bayer Timed-Release Aspirin [355]
Arthritis Pain Formula By the Makers of Anacin Analgesic Tablets [922]
Arthritis Strength Bufferin [134]
Ascriptin, A/D, with Codeine [746]
Aspirin 325 mg with Codeine Tabs [769]
Aspirin Suppositories [751]
Aspirin w/ Codeine Tablets [344]
Axotal [7]
Bayer Aspirin and Bayer Children's Chewable Aspirin, Children's Cold Tablets [355]
Buff-A Comp Tablets, No. 3 Tablets (with Codeine) [563]
Bufferin, with Codeine No. 3 [134]
Cama Arthritis Pain Reliever [257]
Congesprin [134]
Cosprin, Cosprin 650 [355]
Darvon with A.S.A., -N with A.S.A. [523]
Di-Gesic [166]
Dihydrocodeine Compound Tablets [769]
Double-A Tablets [281]
Easprin [674]
Ecotrin Duentric Coated Aspirin, Maximum Strength Safety-Coated Aspirin Capsules, Maximum Strength Tablets, Regular Strength Safety-Coated Capsules [805]
Empirin with Codeine [139]
Equagesic [942]
Excedrin Extra-Strength [134]
Extra-Strength Bufferin Capsules & Tablets [134]
Fiorinal, w/ Codeine [258]
4-Way Cold Tablets [134]
Gemnisyn Tablets [746]
Hyco-Pap [512]
Maximum Bayer Aspirin [355]
Mepro Compound Tablets [769]
Methocarbamol with Aspirin Tablets [769]
Midol [355]
Norgesic & Norgesic Forte [734]
Oxycodone Hydrochloride, Oxycodone Terephthalate & Aspirin Tablets (Half & Full Strengths) [751]
Percodan & Percodan-Demi Tablets [269]
Propoxyphene Compound 65 [769]
Robaxisal Tablets [737]
SK-65 Compound Capsules [805]
SK-Oxycodone with Aspirin Tablets [805]
Supac [593]
Synalgos-DC Capsules [463]

Aspirin *(cont'd).*
Talwin Compound [932]
Vanquish [355]
Verin [900]
Zorprin [127]

Atenolol.
Tenormin Tablets [831]

Atropine nitrate, methyl.
Festalan [418]

Atropine sulfate.
Antrocol Tablets, Capsules & Elixir [703]
Arco-Lase Plus [72]
Atropine Sulfate Injection [134]
Atropine Sulfate Injection [287]
Butabell HMB Tablets [765]
Comhist Tablets [625]
Di-Atro Tablets [518]
Diphenoxylate Hydrochloride & Atropine Sulfate Tablets & Oral Solution [751]
Probocon [518]
Ru-Tuss Tablets [127]
SK-Diphenoxylate Tablets [805]
Trac Tabs, Tabs 2X [436]
Urised Tablets [915]
Urogesic Tablets [281]

Attapulgite.
Attaclay [295]
Attacote [295]
Attaflow [295]
Attagel 150, 350 [295]
Dilvex [330]
Dilvexa [330]
Florcox [330]
Florex [330]
Micro-Cote [330]
Micro-Sorb [330]
Min-U-Gel [330]
Pharmasorb [295]
Refinex [330]

Aurothioglucose.
Solganal Suspension [772]

Avocado oil.
Lipovol A [525]

Azatadine maleate.
Optimine Tablets [772]
Trinalin Repetabs Tablets [772]

Azathioprine.
Imuran Tablets [139]

2-Azido-4-isopropylamino-6-methylthio-3-triazine.
Mesoranil [190]

Aziocillin sodium.
Azlin [585]

2,2′-Azobis(2,4-dimethylvaleronitrile).
Vazo 52 [269]

1,1′-Azobisformamide.
Azocel [307]

2,2′-Azobisisobutyronitrile.
Poly-Zole® AZDN [659]
Vazo 64 [269]

Azodicarbonamide.
Azoblow [708]
Celogen 754 [881]
Ficel AC, AC-SP4 [118]
Kempore® 125, FF, MC [659]
Porofor ADC/E, ADC/F, ADC/K, ADC/M, ADC/R, ADC/S [112]

Azodicarbonamide 1,1′ azobisformamide.
Azocel 504, 506, 508, 525 [307]

Azodicarbonamide and barium-zinc complex stabilizer, 93.5%:65%
Onifine-CE [668]

Azodiisobutyronitrile.
Poly-Zole AZDN [618]

1-Azo-2-hydroxy-3-(2,4-dimethylcarboxanilido)naphthalene-1′-(2-hydroxybenzene).
Magon [506]

Azoisobutyric dinitrile.
Porofor N [112]

Azosulfisoxazole.
Azo-Sulfisoxazole Tablets [344]
Azo-Sulfisoxazole Tablets [769]

Azo yellow pigment.
Dalamar [270]

Bacampicillin hydrochloride.
Spectrobid Tablets & Oral Suspension [739]

Bacillus thuringienses.
SOK® [884]

Bacitracin.
Baciguent® [884]

Bacitracin methylene disalicylate.
DPS [297]
Fortracin [680]

Bacitracin-polymyxin-neomycin.
Mycitracin® [884]

Bacitracin zinc.
Cortisporin Ointment, Ophthalmic Ointment [139]
Neo-Polycin [580]
Neosporin Aerosol, Ointment, Ophthalmic Ointment Sterile, Powder [139]
Polysporin Ointment, Ophthalmic Ointment [139]

Baclofen.
Lioresal Tablets [340]

Barbiturate preparations.
Alurate Elixir [738]
Buff-A Comp Tablets, No. 3 Tablets (with Codeine) [563]
Donnatal Capsules, Elixir, Extentabs, Tablets [737]
G-1 Capsules, G-2 Capsules, G-3 Capsules [394]
Kinesed Tablets [831]
Levsin/Phenobarbital Tablets, Elixir & Drops [499]
Levsinex/Phenobarbital Timecaps [499]
Lotusate Caplets [932]
Mebaral [133]
Nembutal Sodium Capsules, Sodium Solution, Sodium Suppositories [3]
Repan Tablets [304]

Barium-cadmium laurate, coprecipitated.
Mark XI [74]

Barium-cadmium phenate.
Mark LL [74]

Barium carbonate.
Ceramix [705]
Kenmix [488]
Ultralog 08-5700-00 [172]

Barium chloride.
Ultralog 08-5985-00 [172]

Barium dinonylnaphthalene sulfonate.
Vanplast 201 (corrosion inhibitor) [894]

Barium dinonylnaphthalene sulfonate in microcrystalline wax.
Na-Sul BSN/W765, BSN/W780 [894]

Barium dinonylnaphthalene sulfonate in light min. oil.
Na-Sul BSB, BSN [894]

Barium dinonylnaphthalene sulfonate in light min. oil, carbonated.
Na-Sul 611 [894]

Barium lead stabilizer.
Mark 232B, 550, 556 [74]

Barium nitrate 99.99%.
Ultralog 08-7040-00 [172]

Barium petroleum sulfonate.
Petronate Basic [936]
Petrosul Neutral Barium [684]

Barium sulfate.
Barimite, Barimite-XF [861]
Barotrast® [80]
Barytes [387]
No. 22 Barytes [861]
Blanc Fixe [102], [109], [804]
Esophotrast® [80]
Oratrast® [80]

Batyl alcohol.
Nikkol Batyl Alcohol [631]

Batyl isostearate.
Nikkol GM-18IS [631]

Batyl monoisostearate.
Nikkol GM-18IS [631]

Batyl monostearate.
Nikkol GM-18S [631]

Batyl stearate.
Nikkol GM-18S [631]

Bauxite.
Porocel [590]

Bayberry wax.
Bayberry Wax-STRALPITZ [830]

Beclomethasone dipropionate.
Beclovent Inhaler [354]
Beconase Nasal Inhaler [354]
Vancenase Nasal Inhaler [772]
Vanceril Inhaler [772]

Beeswax.
Apifil [339]
BB [855a]
Beeswax-STRALPITZ, White Bleached-STRALPITZ, White-STRALPITZ, Yellow Refined-STRALPITZ, Yellow-STRALPITZ [830]
Mismo Beeswax [458]

Beeswax, hydrophilic.
Apifil [339]

Beeswax, synthetic.
Beeswax, Synthetic-STRALPITZ [830]

Beeswax, synthetic *(cont'd)*.
Lipo BEE 102 [525]
Lipowax 6138G [525]

Behenalkonium chloride.
Kemamine BQ-2802C [429]

Behenamide.
Kemamide B [429]

Behenamidopropyl dimethylamine.
Lexamine B-13 [453]
Schercodine B [771]

Beheneth-5.
Nikkol BB-5 [631]

Beheneth-10.
Nikkol BB-10 [631]

Beheneth-20.
Nikkol BB-20 [631]

Beheneth-30.
Nikkol BB-30 [631]

Behenic acid.
Crodacid PG 3440 [223]
Hystrene 9022 [429]

Behenic imidazoline.
Schercozoline B [771]

Behentrimonium chloride.
Genamin KDM [55]

Behenyl alcohol.
Nikkol Behenyl Alcohol 65, Behenyl Alcohol 80 [631]

Behenylamine.
Nissan Amine VB [636]

Behenyl dimethyl benzyl ammonium chloride.
Incroquat BDBC 25P [223]

Behenyl erucate.
Schercemol BE [771]

Behenyl trimethyl ammonium chloride.
Incroquat BTC 25P [223]

Behenyl trimethyl ammonium methosulphate.
Incroquat BTQ 25C, BTQ 25P [223]

Benactyzine hydrochloride.
Deprol [911a]

Bendroflumethiazide.
Corzide [817]
Naturetin Tablets [817]
Rauzide Tablets [817]

Benefin.
Balan [282]

Bentonite.
Aquagel [616]
Bentolite H4430 [924]
Mineral Colloid BP 2430 [924]

Benzalkonium chloride.
Barquat MB-50, MB-80 [529]
Cation G-40 [764]
Empigen BAC, BCM 75 [25]
Kemamine BAC, BQ-6502C [429]
Lebon GM [764]
Rewoquat B 50 [725]
Roccal II 50% [413]
Roccal II 70-3076 [414]
Swanol CA-101 [631]
Zephiran Chloride [932]

Benzene-1,3-disulfonyl hydrazide and chlorinated paraffin, 50 pbw: 50 pbw.
Porofor B 13/CP 50 [112]

Benzene hexachloride.
Gamtox [261]
Hexadow [261]
Kwell Cream, Lotion, Shampoo [719]
Scabene Lotion, Shampoo [827]

Benzene sulfohydrazide.
Porofor BSH Paste [597]

Benzene sulfonic acid.
Reworyl B 70 [725]

Benzene sulfonyl hydrazide.
Porofor BSH Powder [112]

Benzene sulfonyl hydrazide and paraffin oil, 75 pbw: 25 pbw.
Porofor BSH Paste [597]
Porofor BSH Paste M [112]

Benzethonium chloride.
Dalidyne [239]

Benzoate, C_{12}-C_{15}.
Finsolv TN [320]

Benzocaine.
Anbesol Gel Antiseptic Anesthetic, Liquid Antiseptic Anesthetic [922]
Auralgan Otic Solution [97]
Cetacaine Topical Anesthetic [168]
Children's Chloraseptic Lozenges [709]
Dalidyne [239]
Derma Medicone-HC Ointment [575]
Dermoplast Aerosol Spray [97]
Dieutrim Capsules [518]
Ger-O-Foam [356]
Hedal H-C Suppositories [78]
Hurricaine Liquid 1/4cc Unit Dose, Oral, Topical Anesthetic Gel, Liquid, Spray [117]
Otipyrin Otic Solution (sterile) [498]
Pazo Hemorrhoid Ointment/Suppositories [134]
Rectal Medicone-HC Suppositories [575]
Tympagesic Otic Solution [7]

Benzocaine *(cont'd).*
Wyanoids Hemorrhoidal Suppositories [942]

Benzodihydropyrone.
Dolocotone [337]

Benzoic acid.
Trac Tabs, Tabs 2X [436]

Benzoic acid, sodium salt.
Sodium Benzoate NF, FCC [691]

Benzonatate.
Tessalon [269]

Benzophenone.
Mark 202A [75]
Mark 1413 UV Absorber, 1535 UV Absorber [74]

Benzophenone-1.
Uvinul 400 [109]

Benzophenone-2.
Uvinul D-50 [109]

Benzophenone-3.
Spectra-Sorb UV-9 [50]
Uvinul M-40 [109]

Benzophenone-4.
Spectra-Sorb UV-284 [50]
Uvinul MS-40 [109]

Benzophenone-6
Uvinul D-49 [109]

Benzophenone-8.
Spectra-Sorb UV-24 [50]

Benzophenone-9.
Uvinul DS-49 [109]

Benzophenone-11.
Uvinul 490 [109]

1,4 Benzoquinone dioxime.
BQDD [708]

n-(2 Benzothiazolyl)-n′-methylurea.
Gatnon [112]

Benzothiazyl-2-t butyl sulfenamide.
Vulkacit NZ [597]

Benzothiazyl-2-cyclohexyl sulfenamide.
Vulkacit CZ/MGC, NZ/EGC [597]

Benzothiazyl-2-dicyclohexyl sulfenamide.
Vulkacit DZ/C [597]

2-Benzothiaz N,N-diethylthiocarbamylsulfide.
Ethylac [682]

2,2′Benzothiazyl disulfide.
Akrochem MBTS [15]
Altax [894]
MBTS [15]
Pennac MBTS-0 [682]
Royal MBTS [752]

2-Benzothiazyl-N-morpholine disultide.
Akrochem Accelerator MF [15]

Benzothiazyl-2-sulfene morpholide.
Vulkacit MOZ/LG [597]

Benzatriazole
Cobratec® 99 [797]

Benzphetamine hydrochloride.
Didrex™ [884]

Benzthiazide.
Aquatag [723]
Exna Tablets [737]
Hydrex Tablets [868]

Benztropine mesylate.
Cogentin 1249 [578]

Benzoyl peroxide.
5 Benzagel (5% benzoyl peroxide) & 10 Benzagel (10% benzoyl peroxide), Acne Gels, Microgel Formula [246]
BZQ-25, BZQ-40, BZQ-50, BZQ-55, BZW-70, BZW-80 [890]
Cadet™ BPO-70, BPO-78 [16]
Cadox 40E, BCP, BEP-50, BFF-50, BFF-60W, BP-55, BS [16]
Desquam-X 5 Gel, 10 Gel, 4 Wash, 10 Wash [920]
Fostex 5% Benzoyl Peroxide Gel, 10% Benzoyl Peroxide Cleansing Bar, 10% Benzoyl Peroxide Gel, 10% Benzoyl Peroxide Wash [920]
Lucidol-70, -78, -98 [536]
Luperco AFR-250, AFR-500, 501 [536]
Persa-Gel 5% & 10%, W 5% & 10% [666]
Superox 744 [722]
Vanoxide-HC Acne Lotion [246]
Xerac BP5 & Xerac BP10 [686]

Benzyl acetate.
Cyclochem BA [234]

Benzyl acrylate.
SR-432 [766]

6-Benzyladenine.
BAP [699a]

Benzylalkylammonium chloride.
Arquad B-100 [79]

Benzyl benzoate.
Rewolub BBE [725]

Benzylbutyl phthalate.
Unimoll BB [112]

o-Benzyl-p-chlorophenol.
Santophen® 1 Germicide [602]

(o-Benzyl-p-chlorophenol), isopropanol 75:25.
Santophen® 1 Germicide Solution 75% [602]

Benzyl dimethyl coconut ammonium chloride.
Kemamine BQ 6502 C [429]

Benzyl dimethyl dodecyl ammonium chloride.
Kemamine BQ 6902 C [429]

Benzyl dimethyl hexadecyl ammonium chloride.
Kemamine BQ 8802 C [429]

Benzyl dimethyl hydrogenated tallow ammonium.
Kemamine BQ 9702 C [429]

Benzyl dimethyl tallow ammonium chloride.
Kemamine BQ 9742 C [429]

s-Benzyl n,n-di-sec-butylthiolcarbamate or n,n-disecbutyl-s-benzylthiolcarbamate.
Drepamor [310]

Benzyloctyl adipate.
Adimoll BO [112]

n-(Benzyloxycarbonyloxy)succinimide.
Sequalog 10-4190-00 [172]

Benzylparaben.
Nipabenzyl [632]

Benzylpenicilloyl-polylysine.
Pre-Pen [499]

Benzyl phthalate.
Santicizer 278 [602]

Benzyl trimethyl ammonium chloride.
Hipochem Migrator J [410]
Variquat B200 [796]

Beryllium oxide.
Ultralog 10-7320-00 [172]

Betaiminodipropionate, partial sodium salt.
Deriphat 160-C [400]

Betaine.
Armoteric LB [16]
Betafin AP (monohydrate, pharm. grade), BC (anhydrous, tech. grade), BP (anhydrouss, pharm. grade) [313]
Emcol® DG, NA-30 [935]
Hartaine CB-40 [391]
TEGO® Betaines [856]

Betaine, amphoteric
Ampholan B-171 [509]
Cycloteric BET C-30, BET I-30, BET O-30 [234]
Estat EFB 171 [253]

Betaine hydrochloride.
Zypan Tablets [819]

Betaines, phosphated.
Jorphotaine [474]

Betamethasone.
Celestone Syrup & Tablets [772]

Betamethasone acetate.
Celestone Soluspan Suspension [772]

Betamethasone benzoate.
Benisone Gel/Cream/Lotion/Ointment [214]
Uticort Cream, Gel, Lotion & Ointment [674]

Betamethasone dipropionate.
Diprolene Ointment 0.05% [772]
Diprosone Cream 0.05%, Lotion 0.05% w/w, Ointment 0.05%, Topical Aerosol 0.1% w/w [772]

Betamethasone sodium phosphate.
B-S-P [518]
Celestone Phosphate Injection, Soluspan Suspension [772]

Betamethasone valerate.
Beta-Val Cream 0.1% [519]
Valisone Cream 0.1%, Lotion 0.1%, Ointment 0.1%, Reduced Strength Cream [772]

Bethanechol chloride.
Bethanechol Chloride Tablets [240]
Duvoid [652]
Myotonachol [356]
Urecholine [578]

BHA.
Tenox BHA [278]

BHA; BHT; ethyl alcohol 25:25:50%.
Sustane P [883]

BHA; BHT; propyl gallate; citric acid; propylene glycol; veg. oil; glycerol monooleate 10:10:6:6:8:28:32%.
Sustane W [883]

BHA; BHT; veg. oil 20:20:60%.
Sustane HW-4 [883]

BHA; citric acid; propylene glycol 20%:20%:60%.
Sustane 8 [883]

BHA; citric acid; propylene glycol 40:8:52%.
Sustane Q [883]

BHA; ethyl alcohol 50:50.
Sustane PA [883]

BHA; TBHQ; citric acid; propylene glycol 20:6:4:70%.
Sustane 31 [883]

BHA; veg. oil: 30:70%.
Sustane 4A [883]

BHT.
Sustane BHT [883]

BHT *(cont'd)*.
Tenox BHT [278]

BHT; BHA; veg. oil: 22%:18%:60%.
Sustane 6 [883]

Bioflavonoids.
Lemon Bioflavonoid Complex [834]

Biotin.
Mega-B [72]
Megadose [72]

Biperiden.
Akineton [495]

Biphenyl.
Chemcryl BLC [537]

Biphenyl, alkylated.
Katanol 387 [338]

Biphenyl flake.
ChemKar WC [537]

Biphenyl, liquid.
Amacarrier LBP [4]

B(1,1-Biphenyl)-4 yloxy)-n(1,1 dimethylethyl)-1 H-1,2,4 triazole-1-ethanol.
Baycor [112]

3-(3-1,1′-Biphenyl-4yl-1,2,3,4-tetrahydro-1-naphthylenyl)-4-hydroxy-2H-1-benzopyran-2-one.
Ratak [444]

Bisabolol.
Dragosantol 2/012681 [263]

Bisacodyl.
Bisacodyl Patient Pack, Suppositories, Tablets [751]
Bisacodyl Suppositories [344]
Dulcolax Suppositories, Tablets [125]
Evac-Q-Kwik [7]
Fleet Bisacodyl Enema, Prep Kits [325]
Theralax® Suppositories, 10 mg [114]

Bisacodyl tannex.
Clysodrast® [80]

Bis alkylamido phosphobetaine.
Monateric P-BSA [549]

Bis-amide, fatty.
Drewax 110 [265]

(Bis-2-aminoethyl) cocoamine oxide.
Schercamox CMA [771]

1,3-Bis-(2-benzothiazolyl-mercaptomethyl)urea.
El-Sixty [602a]

1,4-Bis(bromoacetoxy)-2-butene.
Slimacide V-10 [904]

Bis(4-t-butyl-cyclohexyl) peroxydicarbonate.
Percadox® 16, 16N, 16-W40 [16]

2,3,4,5 Bis(2-butylene) tetrahydrofurfural.
MGK® Repellent 11 [568]

α, α′-Bis (t-butylperoxy) diisopropylbenzene
Percadox® 14, 14/40 [16]
Peroximon F [604]
Vul Cup R [406]

1,1-Bis (t-butylperoxy)-3,3,5-trimethyl cyclohexane.
Luperco 231-XL [392]
Lupersol 231 [392], [682]

Bis (o-chloro-benzoyl) peroxide.
Cadox OS, PS [16]

5-[Bis(2-chloro-ethyl)amino]uracil.
Uracil Mustard [884]

1,1-Bis (chlorophenyl)-2,2-dichloro-ethane.
Rothane [742]

1,1-Bis (4-chlorophenyl) ethanol.
Zikron [638]

1,2-Bis (3,5-di-t-butyl-4-hydroxy hydrocinnamoyl).
Irganox MD-1024 [190]

Bis (2,4-di-t-butylphenyl) pentaerythritol diphosphite containing 1.0% triisopropanolamine.
Ultranox 626 [129]

Bis (2,4-dichloro-benzoyl) peroxide.
Cadox® TS-50 [16]

Bis diisobutyl maleate (diisobutyl carbitol maleate).
Staflex DIBCM [722]

N,N′-Bis (1,4-dimethyl pentyl)-p-phenylenediamine.
Flexzone 4L [881]
Santoflex 77 Antiozonant [602]
UOP 788 [883]
Vulkanox 4030 [597]

Bis (dimethylthiocarbamyl) disulfide.
Thiurad Vulcanization Accelerator [602]

Bis(2-ethylhexyl)peroxydicarbonate.
Trigonox EHP [653]
Trigonox EHP-AT70, EHP-C75 [16]

N,N′-Bis (1-ethyl, 3-methyl pentyl)-p-phenylene diamine.
Flexzone 8L [881]
UOP 88 [883]

1,4-Bis(2-hydroxpropyl) 2-methylpiperazine.
DHP-MP [109]

Bishydroxycoumarin (q.v.).
Dicumarol [933]

N,N-Bis (2-hydroxy-ethyl alkylamine).
Alacstat C-2 [29]
Chemstat 122, 182 [170]
Varstat K22, T22 [796]

Bis (2-hydroxyethyl) cocoamine.
Ethomeen C/12 [79]

Bis (2-hydroxyethyl) coco amine oxide.
Alkamox C2-O [33]
Aromox C/12, C/12-W, CD/12 [79]
Jorphox KCAO [474]

N,N-Bis(2-hydroxy ethyl)-N-(3′-dodecyloxy-2-hydroxy propyl) methyl-ammonium chloride.
Cyastat® 609 [52]

N-N-Bis-(2-hydroxyethyl) glycine.
(Chemalog Grade) 11-6780-00 [172]

Bis (2-hydroxyethyl) octadecylamine.
Ethomeen 18/12 [79]

Bis (2-hydroxyethyl) oleylamine.
Ethomeen O/12 [79]

Bis (2-hydroxyethyl) soyaamine.
Ethomeen S/12 [79]

Bis (2-hydroxyethyl) tallowamine.
Ethomeen T/12 [79]

Bis(2-hydroxy ethyl) tallowamine oxide.
Aromox T/12 [79]
Jorphox KTAO [474]

Bis(2-hydroxyethyl)-tallowammonium ethanoate.
Mirataine TM [592]

2,4- Bis(isopropylamine)-6-methoxy-s-triazine.
Pramitol [190]

2,4-Bis (isopropylamino)-6-(methylthio)-s-triazine.
Caparol [190]

Bis melaminium pentate tripenta-erythritol, 90:10.
XP-1668 [129]

2,3:4,6-Bis-O (1-methyl ethylidene)-O (-L-xylo-2-hexulofuranosonic acid), sodium salt.
Atrinal [537a]

N,N′-Bis-(1-methylheptyl)-p-phenylenediamine.
UOP 288 [883]

Bismuth dimethyl dithiocarbamate.
Akrochem Bismet [15]
BDMC [402]
Bismate [894]

Bismuth oxide.
Ultralog 11-9420-00 [172]

Bismuth oxychloride.
Biron Extra, Fines, HB, NLF-D, NRN, SP, SPLS [744]
C-Weiss 10 [351]
Depthin B-5-Dry, B-50-Dry [744]
Mearlite GBU, GSU, LBU, LSU, MBU [573]
Pearl-Glo, 1085; F, 1096; M, 1098; UVR, 1086 [542]

Bismuth oxyiodide.
Wyanoids HC Rectal Suppositories, Hemorrhoidal Suppositories [942]

Bismuth subcarbonate.
Wyanoids HC Rectal Suppositories, Hemorrhoidal Suppositories [942]

Bismuth subgallate.
Anusol Suppositories, -HC [674]
Hedal H-C Suppositories [78]

Bismuth subsalicylate.
Pepto-Bismol Liquid & Tablets [709]

2,4-Bis (n-octylthio)-6-14-hydroxy-3,5-di-t-butyl-aanilino)1,3,5-triazine.
Irganox 565 [190]

Bis(pentachloro-2,4-cyclopentadien-l-yl).
Pentac [421]

Bisperoxide; 40KE grade is supported on Burgess KE clay.
Vul-Cup R, 40KE [406]

Bisphenol A.
Ucar® Bisphenol A [878]

Bisphenol A diacrylate, ethoxylated.
SR-349 [766]

Bisphenol A dimethacrylate, ethoxylated.
SR-348 [766]

Bisphenol A, hydrogenated.
HBPA [602a]

Bisphenol A, para-para grade.
Parabis [261]

Bisphenol, hindered.
Naugawhite, Naugawhite Powder [881]

Bis-phenolic.
Naugard K [881]

Bis-phenolic phosphite, alkylated-arylated.
Agerite Geltrol [894]
Vanox 1005 [894]

Bis-phenol, sterically hindered.
Vulkanox NKF [597]

n,n-Bis (phosphoro-methyl) glycine.
Polaris [601]

(trans)-1,2-Bis(n-propyl-sulfonyl) ethene.
Vancide PA [894]

Bispyrithione.
Omadine Disulfide [659]

Bisstearamide.
Alkamide STEDA [33]

Bis (tribromophenoxy) ethane.
FF 680 [369]

Bis (tri-n-butyltin) oxide.
Vikol #AF-25, LO-25, PX-15 [903]

Bis (tri-chloromethyl) sulfone.
Amerstat 294 [264]

Bis-tridecyl sodium sulfosuccinate.
Geropon BIS/SODICO [550]

Bleomycin sulfate
Blenoxane [134]

Bonechar.
Fluokarb [685]

Borax, anhydrous.
Trona [490]

Borax, dehydrated.
Three Elephant® Pyrobor® [490]

Borax pentahydrate.
Tronabor [59]

Borax-propylene-glycol-butynediol condensate.
Bor-guard [821]

Boric acid.
Clear Eyes Eye Drops [3]
Collyrium Eye Lotion, w/ Ephedrine Eye Drops [942]
Murine Plus Eye Drops [3]
Three Elephant® [490]
Wyanoids HC Rectal Suppositories, Hemorrhoidal Suppositories [942]

Boric acid ointment.
Ocu-Boracin [655a]

Boric diethanolamide.
Rewocor RA-B 90 [725]

Bornelone.
Prosolal S-9 2/066133 [263]

Boron carbide.
B4C [8]

Boron, elemental.
Trona® [490]

Boron nitride.
Boralloy [411]
Borazon [342]

Boron tribromide.
Extrema Grade IC Grace [774]
Trona® [490]

Boron trichloride.
Trona® [490]

Borosilicate glass.
Pyrex Glass Brand No. 7740 [216]

Bretylium tosylate.
Bretylol [49]

Bromelains.
Ananase [746]

Bromoacetic acid.
Bromoacetic Acid [904]

B-(p-Bromobenzhydryloxy) ethyl dimethylamine hydrochloride.
Ambodryl [674]

2-Bromobutyric acid.
BBA [369]
Hydrogen Bromide, Anhydrous [369]

Bromochlorodimethyl-hydantoin.
Di-Halo [135]
°GSD 550 [358]

o-(4-Bromo-2-chlorophenyl)-o-ethyl s-propyl phosphorothioate.
Curacros [190]

Bromocriptine mesylate.
Parlodel Capsules & Tablets [762]

Bromodi ethylaetylurea.
Adalin [932]

Bromodiphenhydramine hydrochloride.
Ambenyl Expectorant [551]
Bromanyl Expectorant [769]

5-(Bromomethyl)-1,2,3,4,7,7-hexachloro-2-norbornene.
Bromodan [416]

2 Bromo-2-nitropropane-1,3-diol.
Bronopol [453]
Lexgard Bronopol [453]
Onyxide 500 [663]

3-(4-Bromophenyl)-1-methoxy-1-methylurea.
Patoran [106]

Brompheniramine maleate.
Brocon C.R. Tablets [334]
Bromfed Capsules (Timed Release), -PD Capsules (Timed Release), Tablets [612]
Bromphen Compound Elixir - Sugar Free, Compound Tablets, DC Expectorant, Expectorant [769]
Dimetapp Elixir, Extentabs [737]
Dura Tap-PD [271]
E.N.T. Syrup [814]
Poly-Histine-DX Capsules, Syrup [124]
Tamine S.R. Tablets [344]

Brompheniramine maleate elixir.
Veltap Elixir [510]

Brompheniramine maleate expectorant, and tablets.
Veltane Tablets and Expectorant [510]

Brompheniramine maleate, phenylephrine HCL, phenylpropanolamine HCL, 4 mg: 5 mg: 5 mg per ml.
E-Tapp Elixir [281]

Brompheniramine maleate tablets, T.R.
Veltap Lanatabs [510]

Bronze powder.
Light Lemon #200 Bronze [91]
Mirrorgold #125 [91]
Richpale gold 760 [91]

Buclizine hydrochloride.
Bucladin-S Softab Tablets [831]

Bumetanide.
Bumex Injection, Tablets [738]

Bupivacaine hydrochloride.
Marcaine Hydrochloride, Hydrochloride with Epi [133]
Sensorcaine Solutions [90]

Busulfan.
Myleran [139]

Butabarbital.
Butabell HMB Tablets [765]
Pyridium Plus [674]
Quibron Plus [572]
Tedral-25 Tablets [674]

Butabarbital sodium tablets.
Butisol Tablets [510]

Butadiene acrylonitrile latices.
Perbunan N Latex 1590, N Latex 2818, N Latex 3090, N Latex 3415 M, N Latex T, & N Latex VT [112]

Butalan.
Butabarbital Sodium Elixir [510]

Butalbital.
Amaphen Capsules, with Codeine #3 [868]
Anoquan [541]
Axotal [7]
Bancap Capsules, w/ Codeine Capsules [661]
Buff-A Comp Tablets, No. 3 Tablets (with Codeine) [563]
Butalbital Compound [769]
Esgic Tablets & Capsules [352]
Fiorinal, w/ Codeine [258]
G-1 Capsules, G-2 Capsules, G-3 Capsules [394]
Medigesic Plus Capsules [888]
Pacaps [512]
Phrenilin Tablets, Forte [153]
Repan Tablets [304]
Sedapap-10 Tablets [563]
T-Gesic Capsules [930]
Two-Dyne Capsules [436]

Butamben.
Cetacaine Topical Anesthetic [168]

Butamben picrate.
Butesin Picrate Ointment [3]

Butane.
Hydrocarbon Propellant A-17 [695]

Butanedioic acid mono(2,2-dimethyl hydrazide).
Alar [881]

1,4-Butanediol diacrylate/hydroquinone inhibitor, 100-150 ppm.
SR-213 [766]

Butane-1,4-diol succinate.
HI-EFF 4B [67]

Butorphanol tartrate.
Stadol [134]

Butoxydiglycol.
Butyl Carbitol [878]
Dowanol DB [261]

Butoxyethanol.
Butyl Cellosolve [878]
Dowanol EB [261]

a-[2-(Butoxyethoxy-ethoxy]-4,5-methylene-2-propyltoluene.
Pyrocide® 175 [568]

Butoxyethyl oleate.
Plasthall 325 P [379]

N-Butoxy polyoxyethylene polyoxypropylene glycol.
Macol 660, 3520, 5100 [564]

B-Butoxy-B′-thio-cyanodiethyl ether.
Lethane 384 [742]

Butyl acetyl ricinoleate.
Flexricin P-6 [640a]

Butyl acrylate.
Norsucryl [651]

n-Butyl alcohol.
Alfol 4 [206]

4-t Butylamino-2-chloro-6-ethylamino-s-triazene.
Gardoprim [190]

2-t-Butylamino-4-ethylamino-6-methoxy-s-triazine.
Caragard [190]

t-Butylaminoethyl methacrylate.
Mon-Arc® MA-3 [69]

2-t-Butylazo 2-cyano-butane.
Luazo 82 [536]

2-t-Butylazo-2-cyano-4-methoxy-4-methylpentane.
Luazo-55 [536]

2-t-Butylazo 2-cyano-propane.
Luazo 79 [536]

Butyl benzoate.
Anthrapole AZ, BM [77]
Chemcryl C-101-N [537]
Dai Cari XBN [903]
Servon XRL 174 [788]

Butyl benzoate, emulsified.
Hipochem B-3-M [410]

N-t-Butyl-2 benzothiazole sulfenamide.
Akrochem BBTS [15]
Delac-NS [881]
Pennac TBBS-0 [682]
Santocure NS [392], [602]
Vanax NS [894]

Butyl benzyl phthalate.
Hatcol BBP [393]
Santicizer 160 [602]

n-Butyl-4,4-bis(t-butylperoxy)valerate.
Lupersol 230 [682]
Trigonox 17/40 [16]

t-Butylcarbazate.
(Sequalog Reagent) 15-2620-10 [172]

6-Butyl-o-cresol.
Prodox® 441, 641 [315]

2-Butyl-p-cresol.
Prodox® 640 [315]

t-Butyl cumyl peroxide.
Lupersol 801 [536]
Trigonox T [16]

5-n-Butyl-2-dimethylamino-4-hydroxy-6-methylpyrimidine.
Milcurb [444]

2-(sec)-Butyl-4,6-dinitrophenyl-3-methyl-2-butenoate.
Morocide [416]

Butylene glycol polyester.
Staflex 550, 802, 804 [722]

Butyl esters.
Butyl Esters [223]

Butyl ether, polyalkoxylated.
Witconol NS500K [935]

Butyl ether, polyproboxylated.
Witconol APEB [937a], [938]

5-Butyl-2-ethylamino-6-methylpyrmidin-4-yl dimethylsulfamate.
Nimrod [444]

OO-t-Butyl 0-(2-ethyl-hexyl) monoperoxycarbonate.
Lupersol TBEC [536]

n-Butyl-n-ethyl-a,a,a-trifluro-2,b-dinitro-p-toluidine.
Balan [282]

i-Butyl fatty acid ester.
Rilanit IBTiOK, IBTi [401]

t-Butyl hydroperoxide.
Trigonox A-W70 [16]

t-Butyl hydroquinone.
Tenox TBHQ [278]

4,4′-Butylidenebis (6-t-butyl-m-cresol).
Santowhite® Powder Antioxidant [602]

00-t-Butyl 0-isopropyl monoperoxycarbonate.
Lupersol TBIC-M75 [536]

Butyl isostearate.
Nikkol GM-18IS [631]

n-Butyl mesityl oxide oxalate.
Indalone [332]

Butyl methoxydibenzoylmethane.
Parsol 1789 [353]

Butyl 2-methyl-4-chlorophlnoxy acetate.
Yamaclian M [639]

p=t-Butyl- α -methylhydrocinnamaldehyde.
Lillial [353]

Butyl myristate.
Bumyr [44]
Radia 7070 [658]
Wickenol 141 [926]

Butyl naphthalene sodium sulfonate.
Emkal BNS, BNX Powder [292]

Butyl naphthalene sulfonic acid.
Emkal BNS Acid [292]

Butyl octyl phthalate.
Hatcol BOP [393]
Kodaflex HS-3 [278]
Staflex BOP [722]

Butyl oleate.
Butyl Oleate [223]
Emerest 2328 [289]
Graden Butyl Oleate [365]
Grocor 4000 [374]
Iberwet—BO [386]
JA-FA BO [465]
Plasthall 503 [379]
Radia 7040 [658]
Rilanit IBO [401]
Uniflex BYO [876]

Butyl oleate, sulfated.
Chemax SBO [170]
Hipochem Dispersal SB [410]

Butyl oleate, sulfated *(cont'd).*
Marvanol SBO 60% [552]

Butyl PABA.
Butesin [3]

Butylparaben.
Lexgard B [453]
Nipabutyl [632]
Preservaben B [867]
Protaben B [713]

t-Butyl perbenzoate.
Trigonox C [16]

T-Butyl peroctoate.
Trigonox 21, 21-C50, 21-OP50 [16]

t-Butyl peroxyacetate.
Lupersol 70, 75M [536]

t-Butyl peroxy-benzoate.
Esperox-10 [890]
Norox [644]

t-Butyl peroxy 2-ethyl hexanoate.
Esperox-28 [890]
Lupersol PMS, PDO [536]

t-Butyl peroxyisobutyrate.
Lupersol 80 [536]

t-Butyl peroxy isopropyl carbonate.
BPIC [705]

t-Butyl peroxy-maleic acid.
Esperox-41-25 [890]
Luperco PMA-40 [536]
Lupersol PMA [536]

t-Butyl peroxyneo-decanoate.
Lupersol 10-M75, 10 [536]

t-Butyl peroxy neohexanoate.
Lupersol 601 [536]

t-Butyl peroxypivalate.
Lupersol 11 [536]

t-Butyl perpivalate.
Trigonox 25-C75 [16]

2-t-Butylphenol.
Prodox® 142 [315]

p-t-Butylphenol.
Prodox® 144, 144A [315]

2-(p-t-Butylphenoxy) isopropyl 2-chloroethyl sulfite.
Aramite [881]

t-Butylphenyl diphenyl phosphate.
Santicizer 154 [602]

Butyl phosphate, acid form.
Servoxyl VPIZ 100 [788]

Butyl stearate.
Butyl Stearate [223]
Crodamol BS [223]
Emerest 2325 [291]
Emerest 2326 [284]
Estol 1456 [875]
Grocor 5410, 5510 [374]
JA-FA BS [465]
Kessco Butyl Stearate [79]
Nikkol GM-18S [631]
Radia 7051 [658]
Rilanit BS, IBS [401]
Uniflex BYS [876]
Unimate BYS [876]
Wickenol 122 [926]

Butyltin carboxylate.
Advastab T-52N Concentrate, T-290 [192]
Stanclere T-85, TL, T-876, T-877, T-878, T-55 [459]

Butyltin mercaptide.
Advastab TM-180 [192]
Stanclere T-94 C, T-126, T-801 [459]

Butyl 2-ξ4-(5-trifluoromethyl-2-pyridinyl oxy) phenoxyξ propanate.
Fusilade [438]

6-t-Butyl-2,4-xylenol.
Prodox® 340 [315]

2-Butyne-1,4-diol.
Butynediol [338]
Korantin® BH Liquid, BH Solid [106]

Butyraldenyde oxime.
Troykyd Anti IN BTO [869]

Butyraldoxime.
Skino #1 [605]
Troykyd Anti-Skin BTO [605]

T-Butyrolactone.
BLO [338]

Butyrophenone.
Haldol Tablets, Concentrate, Injection [570]
Inapsine Injection [467]
Innovar Injection [467]

Cadmium chloride.
Ultralog 16-6900-00 [172]

Cadmium diamyldithio-carbamate.
Amyl Cadmate [894]

Cadmium diethyldithio-carbamate.
Cadmate [894]
Ethyl Cadmate [894]

Cadmium oxide.
Ultralog 16-7125-00 [172]

Cadmium sebecate, potassium chromate, thiram, 5%:5%:16%.
Kromad [542]

Cadmium sulfide.
Cadmium Yellow Lithopones, Yellow Toners [496]

Cadmium sulfoselenide.
Cadmium Orange Lithopones, Orange Toners, Red Toners [496]

Caffeine.
A.P.C. with Codeine, Tabloid Brand [139]
Amaphen Capsules, with Codeine #3 [868]
Anacin Analgesic Capsules, Analgesic Tablets, Maximum Strength Analgesic Capsules, Maximum Strength Analgesic Tablets [922]
Anoquan [541]
Buff-A Comp Tablets, No. 3 Tablets (with Codeine) [563]
Cafamine T.D. 2X Capsules [518]
Cafergot, Cafergot P-B [258]
Cafetrate-PB Suppositories [769]
Compal Capsules [723]
Dexatrim 18 Hour [860]
Di-Gesic [166]
Dihydrocodeine Compound Tablets [769]
Efed II Capsules [38]
Esgic Tablets & Capsules [352]
Excedrin Extra-Strength [134]
Fiorinal, Fiorinal w/ Codeine [258]
G-1 Capsules [394]
Korigesic Tablets [868]
Medigesic Plus Capsules [888]
Migralam Capsules [505]
No Doz [134]
Pacaps [512]
Propoxyphene Compound 65 [769]
Repan Tablets [304]
SK-65 Compound Capsules [805]
Soma Compound, Compound w/ Codeine [911a]
Synalgos-DC Capsules [463]
T-Gesic Capsules [430]
Two-Dyne Capsules [436]
Vanquish [355]

Calamine.
Caladryl [674]
Dome Paste Bandage [585]

Calcifediol.
Calderol Capsules [884]

Calciferol.
Calciferol Drops (Oral Solution); in Oil, Injection; Tablets [499]

Calcitonin, synthetic.
Calcimar Solution [891]

Calcitriol.
Rocaltrol Capsules [738]

Calcium acetate.
Domeboro Powder Packets & Tablets [585]
Pedi-Boro Soak Paks [679]

Calcium acid methanearsonate.
Super Dal-E-Rad Calar [904]

Calcium alginate.
Satialgine C [280]

Calcium alkylaryl sulfonate.
Alkasurf AG-CA, AG-CAHF, CA [33]
Berol 822 [116]
Casul® 70 HF [859]
Flo Mo 50H, 60 H [247]
Ninate 401 [825a]
Wettol EM 1 [106]
Witconate 605A [935]

Calcium alkyl benzene sulphonate.
Nansa EVM70P [23]
Ninate 401-HF [825]

Calcium alkylbenzene sulfonate, (branched).
Merpemul 4065, 4070 [285]
Nansa EVM62H [23]

Calcium alkylbenzene sulfonate, (linear)
Merpemul 1065, 1070 [285]

Calcium arsenate.
Kilmag [35]

Calcium bis (0-ethyl (3-5-di-t-butyl-4-hydroxybenzyl) phosphonate).
Irganox 1425 [190]

Calcium carbide, precipitated.
Albaglos [691]

Calcium carbonate.
Allied Whiting 2.5, A-1 [15]
Calcene [705]
Calcet [593]
Calcium Carbonate (Surface Treated) [843]
Calcium Carbonate (Whiting) [381], [392], [428], [660], [752], [804]

Calcium carbonate *(cont'd).*
Calofort S [402]
Cal-Sup [734]
Calwhite [348], [804]
Calwhite II [348]
Camel-Cal [328]
Camel-CARB, -FIL [345]
Camel-TEX [345], [752]
Camel-Wite [328]
Carbokup [543]
Carnalox [746]
CC-100, 105 [843]
DAY/CAL Percipitated Calcium Carbonate [388]
Dowcarb [261]
40-200 [348]
Fosfree [593]
GamaCo, GamaCo II, Gama-Sperse [348]
G-White [428]
H-White [428]
Iromin-G [593]
L-220, L303, L306, L3002 [203]
Micro-white [387]
Micro White 20, 25, 25/SAM, 50, 95 [843]
Mission Prenatal, Prenatal F.A., Prenatal H.P. [593]
Natacomp-FA Tablets [868]
Natalins Rx, Tablets [572]
Nu-Iron-V Tablets [563]
OMYA BLH, Hydrocarb [660]
Omyalene [660]
Omyalite 95T [660]
Os-Cal 250 Tablets, 500 Tablets [551]
Pramet FA [747]
Pramilet FA [747]
Precipitated Calcium Carbonate Extra Heavy USP, Precipitated Calcium Carbonate Extra Light USP, Precipitated Calcium Carbonate Heavy USP, Precipitated Calcium Carbonate Light USP, Precipitated Calcium Carbonate Medium USP [924]
Prenate 90 Tablets [124]
Purecal [109]
Q-White (70% Slurry), (Spray Dried]) [428]
Smithko Kalkarb Whiting [804]
Stan-White, 325, 350, 400, 450, 500, 600 Coated Grades [392]
Supercoat [861]
Supermite [861]
Troykyd Anti-Float Powder [869]
Ultralog 16-9100-00 [172]
Vitaflex C [872]
York White [154]
Zenate Tablets [723]

Calcium carbonate, precipitated.
Day Cal [392]
Hakuenka CC [752]
Millical Plus [691]
Multiflex MM, SC [691]
Royal White Light [752]
Ultra-Pflex [691]

Calcium carbonate & diatomaceous earth, 80%:20%.
Lorite [617]

Calcium chloride.
Calcium Chloride Injection [134]
Calcium Chloride Injection [287]

Calcium cyanamide.
Cyanamid [712]

Calcium dinonylnaphthalene sulfonate in light min. oil.
Na-Sul 729 [894]

Calcium disodium edetate.
Calcium Disodium Versenate Injection [734]

Calcium disodium EDTA.
Versene CA [261]

Calcium dodecylbenzene sulfonate.
Emulsifier C-6 [170]
Nansa EVM 62H, EVM 70P [25]
Polyfac ABS-60C [918]
Richonate CS-6021H [732]
Sermul EA 88 [788]

Calcium dodecyl benzene sulfonic acid.
Alkasurf CA [33]

Calcium glubionate.
Neo-Calglucon Syrup [258]

Calcium gluconate.
Calcet [593]
Calcium Gluconate Injection [287]
Calcium Gluconate Tablets [751]
Fosfree [593]
Iromin-G [593]
Mission Prenatal, Prenatal F.A., Prenatal H.P. [593]

Calcium glycerophosphate.
Calphosan, Calphosan B-12 [152]

Calcium hydroxide.
Calcium Hydroxide, USP 802 [924]
Rhenofit CF [728]

Calcium hypochlorite.
HTH [659]
Pittclor [705]

Calcium iodide.
Calcidrine Syrup [3]
Norisodrine w/ Calcium Iodide Syrup [3]

Calcium lactate.
Cacet [593]
Calphosan, Calphosan B-12 [152]
Fosfree [593]
Iromin-G [593]
Mevanin-C Capsules [117]
Mission Prenatal, Prenatal F.A., Prenatal H.P. [593]

Calcium lignosulfonate.
Amberlig [47]
Glutrin [47]
Goulac [47]
Lignosite [752]
Lignosol B, BD [720]
Lignox [616]
Marabond [47]
Marasperse C-21 [47]
Norlig 11 [47]

Calcium lignosulfate, desugared.
Toranil [504]

Calcium limestone.
Pel-Lime [58a]

Calcium metasilicate wollastonite.
VANSIL W-10, 20, 30 [894]

Calcium nitrate 99+%.
Ultralog 17-1360-00 [172]

Calcium oxide.
Calcium Oxide, FCC 801 [924]
Kenmix [488]
Rhenosorb F [728]

Calcium (oyster shell).
Os-Cal 250 Tablets, 500 Tablets, Forte Tablets, Plus Tablets, -Gesic Tablets [551]

Calcium pantothenate.
Al-Vite [267]
B-C-Bid Capsules [350]
Besta Capsules [394]
Calcium Pantothenate, No. 63918 [738]
Pantholin [286]

Calcium petroleum sulfonates.
Calcium Petronate [936]
Petronate 25C, 25H, HMW [936]
Petrosul Neutral Calcium [684]

Calcium polycarbophil.
Mitrolan [737]

Calcium polysilicate.
Flo-Gard [705]

Calcium propionate.
Guard Calcium Propionate [65]

Calcium silicate.
Micro-Cel A, B, C, E, T-38, T-70 [472]

Calcium silicate-wollastonite.
Vansil W-9, W-10, W-20, W-30, W-50 [894]

Calcium stearate.
Calcium Stearate [392]
Flexichem [299]
Interstab CA-18-1
Petrac® CP-11, CP-11 LS, CP-11 LSG [688]
Radiacid 1060 [658]
Radiastar 1060 [658]
Witco Calcium Stearate EA, Calcium Stearate F.P. [935]

Calcium stearoyl lactylate.
Admul CSL 2007, CSL 2008 [704]
Crodactil CSL [223]
Grindtek FAL 2 [373]
Lamegin CSL [176]
Lisat C [856]
Pationic CSL [735]
Radiamuls CSL [658]
Stearolac C [672]
Verv [675]

Calcium sulfate.
Calcium Sulfate, Anhydrous NF 164 [924]
Cal-Sul [582]
C-Weiss 6 [351]
'Extra super' English Terra Alba [188]

Calcium sulfonate.
Alkamuls AG-CA [33]
Emcol D 24-25 [934]
Emcol P-1020 BU [938]
Hybase C-300 [936]
Witconate™ 605A [935]

Calendula extract.
Pot Marigold HS, Marigold LS [20]

Cantharidin.
Cantharone, Cantharone Plus [786]
Verr-Canth [199]
Verrusol [199]

Capramide DEA.
Alrosol C [190]
Cyclomide DV170 [234]
Monamid 150-CW [599]
Product LT 10-45-1, 10-45-2 [197]
Standamid CD [400]

Capramide DEA, 2:1 kritchesvsky.
Comperlan CD [399]

Capreomycyn sulfate.
Capastat Sulfate [523]

Capric diethanolamide.
Emid 6544 [289]
Monamid 150-CW [599]

Capric fatty acid, 90%.
Hystrene 9010 [430]

Capric imidazoline dicarboxylate.
Amphoterge J-2, KJ-2 [529]

Caproamphocarboxyglycinate.
Miranol S2M Conc. [592]

Caproamphocarboxypropionate.
Miranol S2M-SF Conc. [592]

Caproamphoglycinate.
Miranol SM Conc. [592]

Caproamphopropionate.
Miranol SM-SF Conc. [592]

Caproamphopropylsulfonate.
Miranol SS Conc. [592]

Caprolactone monomer.
CAPA Monomer [809]

Capryl alcohol.
Victawet 58B [821]

Caprylic alcohol.
Alfol 8 [206]
Laurex 8 [548]

Capryl diethanolamide, 2:1.
Standamid CD [400]

2-Caprylic-1 (ethyl β-oxipropanoic acid) imidazoline, sodium salt.
Monateric Cy-Na 50% [594]

Caprylic fatty acid, 90%.
Hystrene 9008 [430]

Caprylic fatty acids, substituted imidazoline.
Miramine® CPC [592]

Capryloamphocarboxyglycinate.
Miranol J2M Conc. [592]

Capryloamphocarboxypropionate.
Miranol J2M-SF Conc. [592]

Capryloamphoglycinate.
Emery 5418 [291]
Rewoteric AM-V [725]

Capryloamphopropionate.
Monateric Cy Na-50 [599]

Capryloamphopropylsulfonate.
Miranol JS Conc. [592]

Caprylocapric glycerides, ethoxylated.
Labrafac Hydro [339]
Labrasol [339]

Caprylyl dimethyl amine.
Lilamin 308 D [522]

Capsicum extract.
Pod Pepper HS [20]

Captopril.
Capoten [817]

Caramiphen edisylate.
Rescaps-D T.D. Capsules [344]
Tuss-Ade Timed Capsules [769]
Tuss-Ornade Liquid, Spansule Capsules [805]

Carbamate.
LZ 5022 [532]

Carbamazepine.
Tegretol Chewable Tablets, Tablets [340]

Carbamide, fatty.
Ahcovel Base 500, Base OB Solid [438]
Intrasoft® 49 [225]

Carbamide peroxide.
Ear Drops by Murine, Murine Ear Wax Removal System/Murine Ear Drops [3]
Proxigel [714]

Carbaryl insecticide (1-naphthyl N-methyl carbamate).
Sevin [878]

(Carbazochrome salicylate).
Adrenosem® Salicylate, 2.5 mg, Syrup & Tablets [114]

(Carbazochrome salicylate) + benzyl alcohol, 2%.
Adrenosem® Salicylate, 5 mg, 10 mg [114]

Carbenicillin disodium.
Pyopen®, 1 Gm, 2 Gm, 5 Gm, 10 Gm, 20 Gm [114]

Carbenicillin indanyl sodium.
Geocillin Tablets [739]

Carbetapentane citrate.
Tussar SF, Tussar-2 [891]

Carbetapentane tannate.
Rynatuss Tablets & Pediatric Suspension [911a]

Carbidopa.
Sinemet Tablets [578]

Carbinoxamine maleate.
Brexin Capsules & Liquid [767]
Cardec DM Drops & Syrup [769]
Rondec Drops, Syrup, Tablet, -DM Drops, -TR Tablet [747]

Carbon.
Continex N339, N347, N550, N650, N660, N683, N762 [208]
Darco [438]
Hydro Darco [438]
N220 (ISAF), N299, N347 (HAF-HS), N330 (HAF), N339, N375, N660 (GPF) [801]

Carbon *(cont'd).*
Shawinigan Acetylene Black 99.5% [790]
Stainless Thermax Floform (N907), Powder (N908) [449]
Thermax Floform (N990), (N991) Powder [449]
United N110 (SAF) 99.5%, N234 99.5%, N326 (HAF-LS) 99.5%, N330 (HAF) 99.5%, N339 99.5%, N347 (HAF-HS) 99.5%, N351 99.5%, N358 (SPF) 99.5%, N375 99.5%, N539 (FEF-LS) 99.5%, N550 (FEF) 99.5%, N650 (GPF-HS) 99.5%, N660 (GPF) 99.5%, N754 (SRF-LS) 99.5%, N762 (SRF-LM) 99.5%, N774 99.5%, N787 99.5% [87]

Carbon black.
Floform [204]
Furnex [202]
Galvan [141]
Huber ARO-30 [428]
Kenmix [488]
Ketjenblack E.C. [653]
Permablak [600]
Royal Spectra [202]
Stainless Thermax Powder N-908 [894]
Sterling VL-N642 [141]
Thermax [894]
Thermax Powder N-990, N-991 [146]
Thermax Stainless Floform N-907, N-908 [146]

Carbon, decolorizing.
CAL [142]

Carbon dioxide.
Evac-Q-Kit [7]

Carbon tetrachloride.
Extrema Grade [774]

Carboprost tromethamine.
Prostin/15M® [884]

2-(2-Carboxyethoxy) ethyl 2-[2-(2-hydroxy)] ethoxyethyl methyldodecylammonium methyl sulfate, potassium salt.
Sanac C [81]

[2-(2-Carboxyethoxy) ethyl] [2-[2-(hydroxy) ethoxyethyl] methyloctadecenylammonium methyl sulfate, potassium salt.
Sanac S [81]

2-Carboxy-2′-hydroxy-5′-sulfoformazylbenzene.
Zincon [506]

Carboxylic acid.
Tamol SAC Liquid [106]

Carboxylic acid alkanolamide.
Marlowet 5439, 5480 [178]

Carboxylic acid alkylolamine ester.
Marlowet 5436 [178]

Carboxylic acid, amine salt.
Marlowet 5600, 5606, 5609, 5622, T [178]

Carboxylic acid, ammonium salt.
Gradol 250-A [365]

Carboxylic acid anhydride, chlorinated.
Cloran [882a]

Carboxylic acid, hydroxylated.
Chupol EX [849]

Carboxylic acid, nitrogenous.
Emulan SH, SH 85 [106]

Carboxylic acid polyglycol ester.
Marlowet 4603, 4702, 4703, OTS, SLS [178]

Carboxylic acid, salt.
Poiz 530 [478]

Carboxylic acid, sodium salt.
Gradol 250, 300 [365]
Nissan Polystar OM, OMP [636]
Tamol® MD [106]
Tamol PA Powder, Liquid [106]
Tamol S Liquid [106]

Carboxymethylcellulose, sodium salt.
Cellulose Gum [406]
Dieutrim Capsules [518]

Carboxymethyl hydroxyethylcellulose.
CMHEC [406]

Carboxymethyl hydroxyethyl cellulose, sodium salt.
CMHEC-37L [406]

Carboxymethylmercaptosuccinic acid.
Evanacid 3CS [303]

Carisoprodol.
Carisoprodol Compound Tablets [240]
Carisoprodol Compound Tablets [344]
Carisoprodol Tablets [240]
Carisoprodol Tablets [344]
Soma, Compound, Compound w/ Codeine [911a]
Soprodol Tablets (Carisoprodol) [769]

Carmine.
Karmin [838]

Carnauba.
No. 1 Yellow Carnauba Wax [271]

Carnitine hydrochloride.
Carnitine [148]

β -Carotene.
Added Protection III Multi-Vitamin & Multi-Mineral Supplement [589]
Karotin [838]
Solatene Capsules [738]

Carrageenan.
Genu® [406]
Seakem 3, LCM [549]
Seaspen PF [549]
Viscarin 402, TP-4, XLV [549]

Casanthranol.
Dialose Plus Capsules [831]
Docusate Sodium with Casanthranol Capsules [751]
Peri-Colace [572]

Cascara sagrada.
Aromatic Cascara Fluidextract, Milk of Magnesia-Cascara Suspension Concentrated [751]
Peri-Colace [572]

Castor oil.
Castor Oil, Castor Oil Flavored [751]
DB Oil [100]
Fleet Flavored Castor Oil Emulsion [325]
Fleet Prep Kits [325]
Granulex [409]
Hydrisinol Creme & Lotion [679]
Neoloid [516]
Stimuzyme Plus [615]
Surfactol-13 [641a]

Castor oil amidopropyl dimethylamine.
Lexamine R-13 [453]

Castor oil, dehydrated.
Isoline [941]
Dehydrol [797]

Castor oil + E.O. 36 moles.
Teric C12 Series [439]

Castor oil, ethoxylated.
Alkasurf CO-5, CO-10, CO-15, CO-20, CO-25, CO-30, CO-40M, CO-200 [33]
Arnox TGE-05, 16, 36, 40 and 80 [76]
Chemax CO-5, CO-15, CO-30, CO-40, CO-80, CO-200/50 [170]
Emulpon EL 18, EL 20 [938]
Emulpon EL 33 [937a]
Emulpon EL 40, HT [938]
Etocas 30, 100 [223]
Eumulgin RO 40 [401]
Polyfac CO-36 [918]
Rexol C14, CCN [390]
Sandoxylate C-10, C-15, C-32 [761]
Surfactol® 318, 365, 590 [158]
Trylox CO-5, CO-16, CO-30, CO-40, CO-80, CO-200 [290]

Castor oil, ethoxylated hydrogenated.
Chemax HCO-5, HCO-16, HCO-25, HCO-200/50 [170]

Castor oil ethylene oxide, 40 moles.
T-Det C-40 [859]

Castor oil, hydrated.
Cutina® HR [401]
Radia 3200 [658]
Trylox HCO-5, HCO-16, HCO-25 [290]
Wochem #850 [941]

Castor oil oxyethylate.
Emulan EL [106]

Castor oil polyglycol esters, EO, (17 mol).
Servirox OEG 45 [788]

Castor oil polyglycol esters, EO, (26 mol).
Servirox OEG 55 [788]

Castor oil polyglycol esters, EO, (32 mol).
Servirox OEG 60 [788]

Castor oil polyglycol esters, EO, (180 mol).
Servirox OEG 90/50 [788]

Castor oil polyglycolether.
Merpemul 5200, 5400, 5500 [285]
Merpoxen RO 200, RO 300, RO 400, RO 800 [285]

Castor oil, polyoxyethylated (200).
Emulphor EL-980 [338]
Flo Mo 30C, 36C, 40C, 54C [247]
T-DET C-40 [859]

Castor oil, sulfated.
Actrasol C50, C75, C85, PSR [811]
Ahco AJ-110 [445]
Castor Oil Sulfated 50%, Sulfated 68% [198]
Chemax SCO [170]
Consos Castor Oil [207]
Cordon NU 890/75 [320]
Eureka 102 [94]
Haroil SCO-50, SCO-65, SCO-75, SCO-7525 [365]
Hartenol V-63 [391]
Hipochem Dispersol SCO [410]
Marvanol SCO 75% [552]
Monopole Oil [252]
Monosulf [252]
Nopcocastor, L [252]
Nopcosulf CA-60, CA-70 [252]
Standapol SCO [400]
Turkey Red Oil 100% [948]

Castor oil, sulfonated.
Cordon NU 890/75 [320]
Hartex V63, V64 [390]
Monosulph [252]
Sulfonated Castor Oil 50%, LC, NA [77]
Supratol VF [390]

Castor oil, sulfonated *(cont'd)*.
Titam Castor No. 75 [863]

Castor oil, sulfonated ester.
Montaline RH [711]

Catalase (fungal).
Fermcolase [313]

Cefaclor.
Ceclor [523]

Cefadroxil monohydrate.
Duricef [572]
Ultracef Capsules, Tablets & Oral Suspension [134]

Cefamandole nafate.
Mandol [523]

Cefazolin sodium.
Ancef [805]
Kefzol [523]

Cefoperazone sodium.
Cefobid [739]

Cefotaxime sodium.
Claforan [418]

Cefoxitin sodium.
Mefoxin [578]

Ceftizoxime sodium.
Cefizox [805]

Cefuroxime sodium.
Zinacef [354]

Cellulase.
Celluclast 200 L Type N, 2.0 L Type X [654]
Celluferm™ [313]
Celluzyme Chewable Tablets [239]
Enzobile Improved Formula [541]
Kanulase [257]

Cellulolytic enzyme.
Arco-Lase, Plus [72]
Celluzyme Chewable Tablets [239]
Festal II [418]
Festalan [418]
Gustase [350]
Kutrase Capsules [499]
Ku-Zyme Capsules [499]

Cellulose.
Avicel PH-101, PH-102, PH-105 [332]
Elcema G-250, P 100 [244]
Solka-Floc BW-20, BW-40, BW-100, BW-100 Special, BW-200, BW-2030 [136]

Cellulose acetate.
Cast Cellulose Acetate Sheet 867, Film 904 [943]
Plastacele [270]

Cellulose acetate butyrate ester.
CAB-171-15S, CAB-381-0.1, CAB 381-0.5, CAB 381-2, CAB 381-20, CAB-500-1, CAB 500-5, CAB-531-1, CAB-551-0.2, CAB-551-0.01, CAB-553-0.4 [278]
Cellidor B 500-05 (formerly BH), B 500-10 (formerly BM), B 500-15 (formerly BW), B 500-20 (formerly BWW), B 700-04 (formerly BHH) [112]

Cellulose acetate ester.
CA-394-60 [278]
Cellidor A, S 100-25 (formerly AM), S 100-30 (formerly AW), S 100-33 (formerly AWW), S 200-17 (formerly SH-17), S 200-22 (formerly SM), S 200-27 (formerly SW), S 200-32 (formerly SWW) [112]

Cellulose acetate propionate ester.
CAP-482-0.5, CAP-482-20, CAP-504-0.2 [278]

Cellulose, cyanoethylated.
Cyanocel [50]

Cellulose diacetate.
Cast Cellulose Di Acetate Film 912 [349]

Cellulose ether.
Natrosol 250 [406]

Cellulose, ethyl ether.
Ethyl Cellulose NF Grade [406]

Cellulose propionate ester.
Cellidor CP 300-08 (formerly CPHH), CP 400-05 (formerly CPH), CP 400-10 (formerly CPM), CP 400-15 (formerly CPW), CP 400-20 (formerly CPWW), CP 790-19 (formerly Cp-Ro) [112]

Cellulose sodium phosphate.
Calcibind [593]

Cellulose triacetate.
Cast Cellulose Triacetate Sheet HP-550, Cast Cellulose Triacetate HP-556 [943]

Cephalexin.
Keflex Oral Preparations [254]

Cephalothin sodium.
Keflin, Neutral, Vials & Faspak [523]
Seffin® [354]

Cephapirin sodium.
Cefadyl [134]

Cephradine.
Anspor [805]
Velosef Capsules, for Infusion (Sodium-Free), for Injection, for Oral Suspension, Tablets [817]

Ceresin.
Mobil Wax [598]

Cerfuroxime sodium.
Zinacef® [354]

Ceric chloride.
Tego®-Chlorid [856]

Cerium oxide.
Rareox [364a]

Ceruletide diethylamine.
Tymtran Injection [7]

Cesium acetate 99.9%.
Ultralog 18-6550-00 [172]

Cesium bromide 99.9%.
Ultralog 18-6870-00 [172]

Cesium chloride 99.9%.
Ultralog 18-7060-00 [172]

Cesium fluoride 99.5%.
Ultralog 18-7360-00 [172]

Cesium formate 99.9%.
Ultralog 18-7400-00 [172]

Cesium hydroxide 99+%.
Ultralog 18-7540-00 [172]

Cesium iodide 99.9%.
Ultralog 18-7580-00 [172]

Cesium nitrate 99+%.
Ultralog 18-7740-00 [172]

Cesium sulfate 99+%.
Ultralog 18-8100-00 [172]

Cetalkonium chloride.
Cetol [319]

Cetamine oxide.
Jordamox CDA [474]

Ceteareth-2.
Lowenol C-279 [531]
Macol CSA-2 [564]

Ceteareth-3.
Hostacerin T-3 [55]
Procol CS-3 [713]

Ceteareth-4.
Lipal 4CSA [715]
Lipocol SC-4 [525]
Procol CS-4 [713]

Ceteareth-5.
Macol CSA-5 [564]
Procol CS-5 [713]

Ceteareth-6.
Lipocol SC-6 [525]
Procol CS-6 [713]
Siponic E-3 [29]

Ceteareth-8.
Lipocol SC-8 [525]
Procol CS-8 [713]

Ceteareth-10.
Lipocol SC-10 [525]
Macol CSA-10 [564]
Procol CS-10 [713]
Rewomul CSF 11 [725]
Siponic E-5 [29]

Ceteareth-12.
Eumulgin B1 [400]
Lipocol SC-12 [525]
Procol CS-12 [713]
Standamul B-1 [400]

Ceteareth-15.
Lipal 15CSA [715]
Lipocol SC-15 [525]
Macol CSA-15 [564]
Procol CS-15 [713]

Ceteareth-17.
Procol CS-17 [713]

Ceteareth-20.
Cyclogol 1000 [939]
Empilan KM20 [548]
Eumulgin B-2 [400]
Hetoxol CS-20 [407]
Lipocol SC-20 [525]
Macol CSA-20 [564]
Procol CS-20 [713]
Rewomul CSF 20 [725]
Standamul B-2 [400]

Ceteareth-27.
Plurafac A-38 [109]
Procol CS-27 [713]

Ceteareth-30.
Eumulgin B-3 [400]
Hetoxol CS-30 [407]
Lipocol SC-30 [525]
Procol CS-30 [713]
Siponic E-15 [29]
Standamul B-3 [400]

Ceteareth-40.
Macol CSA-40 [564]

Ceteareth-55.
Plurafac A-39 [109]

Cetearyl alcohol.
CO-1670 [709]
Crodacol CS-50 [223]
Cyclal Cetostearyl Alcohol [939]
Cyclogol Cetylstearyl Alcohol [234]
Dehydag Wax O [400]
Lanette O [400]
Lanol CS [785]
Laurex CS [548]
TA-1618 [709]

Cetearyl alcohol, polyethylene glycol.
Standamul B-1 [400]

Cetearyl alcohol (EO 20), polyethylene glycol ether.
Standamul B-2 [400]

Cetearyl octanoate.
Crodamol CAP [223]
Purcellin Oil 2/066210 [263]
Schercemol 1688 [771]

Cetearyl palmitate.
Crodamol CSP [223]

Ceteth-1.
Procol CA-1 [713]

Ceteth-2.
Brij 52 [438]
Hetoxol CA-2 [407]
Industrol CA-2 [923]
Lipocol C-2 [525]
Macol CA-2 [564]
Procol CA-2 [713]
Simulsol 52 [785]

Ceteth-4.
Lipocol C-4 [525]
Macol CA-4 [564]
Procol CA-4 [713]

Ceteth-6.
Nikkol BC-5.5 [631]
Procol CA-6 [713]

Ceteth-10.
Brij 56 [438]
Lipocol C-10 [525]
Macol CA-10 [564]
Procol CA-10 [713]
Simulsol 56 [785]

Ceteth-16.
Lamacit CA [375]
Procol CA-16 [713]

Ceteth-20.
Brij 58 [438]
Hetoxol CA-20 [407]
Lipocol C-20 [525]
Macol CA-20 [564]
Nikkol BC-20 [631]
Procol CA-20 [713]
Simulsol 58 [785]

Ceteth-25.
Emalex 125 [628]
Procol CA-25 [713]

Ceteth-30.
Lipocol C-30 [525]
Procol CA-30 [713]

Ceteth-45.
Emulgator K 752 [375]

Cetethyldimonium bromide.
Ammonyx DME [663]
Bretol [408]

Cetethyl morpholinium ethosulfate.
Barquat CME-35 [529]

Cetostearyl alcohol, ethoxylated.
Cetomacrogol 1000 BP [223]
Cyclogol NI [938]

Ceto stearyl alcohol, polyethoxylated.
Emulsifier CS [223]

Ceto stearyl lactate.
Crodamol CSL [223]

Ceto-stearyl methacrylate.
Empicryl TM [23]

Cetostearyl palmitate.
Crodamol CSP [223]

Ceto stearyl stearate.
Crodamol CSS [223]

Cetrimonium bromide.
Acetoquat CTAB [5]
Bromat [408]
Cetrimide [408]
Cycloton V [939]

Cetrimonium chloride.
Ammonyx CETAC [663]
Arquad 16-25W, 16-29W [79]
Barquat CT-29, CT-429 [529]
Dehyquart A [400]
Genamin CTAC [55]
M-Quat 1630 [564]
Variquat E228 [796]

Cetrimonium tosylate.
Cetats [408]

Cetyl acetate.
Ritacyl [735]

Cetyl alcohol.
Adol 52, 54, 520 [796]
Alfol 16 [206]
Cachalot C-50, C-51, C-52 [581]
Cetal [44]
CO-1695 [709]
Crodacol C, C-70, C-95 [223]
Cyclal Cetyl Alcohol [939]
Cyclogol Cetyl Alcohol [234]
Dehydag Wax 16 [400]
Kalcohl 60 [478]
Lanol C [711]
Laurex 16 [548]
Lipocol C [525]
Lorol 24 [270]

Cetyl alcohol, alkoxylated.
Procetyl AWS [223]

Cetyl alcohol, alkoxylate, phosphated ester.
Crodafos SG [223]

Cetyl alcohol ethoxylate.
Siponic C-20 [29]

Cetyl alcohol, polyethoxylated (2).
Simulsol 52 [711]

Cetyl alcohol, polyethoxylated (10).
Simulsol 56 [711]

Cetyl alcohol, polyethoxylated (20).
Simulsol 58 [711]

Cetyl betaine.
Lonzaine 16S [529]
Product BCO [269]

Cetyl chloride.
Barchlor 16S [529]

Cetyl dimethyl amine.
Baircat B-16 [529]
Barlene 16S [529]

Cetyl dimethyl amine oxide.
Ammonyx CO [663]
Barlox 16S [529]

Cetyldimethylammonium chloride.
Servamine KAZ 422 [788]

Cetyl dimethyl benzyl ammonium chloride.
Ammonyx DME, T [663]
Cetol [408]
Dehyquart CDB [401]

Cetyl dimethyl ethyl ammonium bromide.
Ammonyx DME [663]
Bretol [408]
Cetylcide Solution [168]

Cetyl ether, alkoxylated.
Procetyl AWS [223]

Cetyl ether, phosphated.
Crodafos CAP, SG [223]

N-Cetyl N-ethyl morpholinium ethosulfate (35%).
Atlas G-263 [92]
Atlas G-263 Anhydrous [445]
Barquat CME Anhydrous [529]
Barquat CME 35 [529]

Cetyl isooctanoate.
Nikkol CIO [631]

Cetyl lactate.
Cegesoft C19 [375]
Ceraphyl 28 [895]
Crodamol CL [223]
Cyclochem CL [234]
Liponate CL [525]
Nikkol Cetyl Lactate [631]
Schercemol CL [771]
Wickenol 507 [926]

Cetyl morpholinium ethosulfate.
Barquat CME-35, CME-A [529]

Cetyl myristate.
Kessco 654 [79]
Schercemol CM [771]

Cetyl octanoate.
Nikko CI0, CIO-P [631]

Cetyl-oleyl alcohol + EO 8 moles.
Teric 17AS [439]

Cetyl-oleyl alcohol + EO 10 moles.
Teric 17A10 [439]

Cetyl-oleyl alcohol + EO 13 moles.
Teric 17A13 [439]

Cetyl-oleyl alcohol + EO 25 moles.
Teric 17A25 [439]

Cetyl oleyl alcohol ethylene oxide condensate.
Ethylan 172, ME, OE, R [509]

Cetyl oleyl ether phosphate, acid form, (EO 7).
Servoxyl VPGZ 7/100 [788]

Cetyl palmitate.
CEP-33 [397]
Cutina CP [400]
Cutina CP-A [401], [934]
Cyclochem CP [234]
Kessco 653 [79]
Kessco-Cetylpalmitat [16]
Nikkol N-SP [631]
Radia 7500 [658]
Schercemol CP [771]
Standamul 1616 [400]
Waxenol 816 [926]

Cetyl pyridinium bromide.
Acetoquat CPB [5]

Cetyl pyridinium chloride.
Acetoquat CPC [5]
CPC [408]
Fungoid Creme & Solution, Tincture [679]

Cetyl stearate.
Radia 7501 [658]
Schercemol CS [771]

Cetyl stearyl alcohol.
Dehydag Wax O [401]
Lanette O [399]
Lanol CS [711]

Cetyl stearyl alcohol with EO 12 mol.
Eumulgin B1 [401]

Cetyl-stearyl alcohol + EO 16 moles.
Teric 16A16 [439]

Cetyl stearyl alcohol with EO 20 mol.
Eumulgin B2 [401]

Cetyl-stearyl alcohol + EO 22 moles.
Teric 16A22 [439]

Cetyl-stearyl alcohol + EO 29 moles.
Teric 16A29 [439]

Cetyl stearyl alcohol with EO 30 moles.
Eumulgin B3 [401]

Cetyl stearyl alcohol and sodium lauryl sulfate, 90:10.
Dehydag Wax SX [401]

Cetyl trimethyl ammonium bromide.
Acetoquat CTAB [5]
Bromat [408]

Cetyl trimethyl ammonium chloride.
Ammonyx CETAC, CETAC-30 [663]
Barquat CT-29, CT-429 [529]
Carsoquat CTM-29, CTM-429 [155]
Chemquat 16-50 [170]
Dehyquart A [401]
Genamin CTAC [416]
Querton 16CL 29, 24CL 35 [487]
Servamine KAZ412 [788]
Tequat BC [514]
Variquat E228 [796]

Cetyl trimethyl ammonium p-toluene sulfonate.
Cetats [408]

Cetyltrimethylammonium tosylate.
Cetats [319]

Charcoal, activated.
Charcoal (Activated) USP in Liquid Base [131]
Karbokoff Tablets [78]

Chenodiol.
Chenix Tablets [750]

Cherry pit oil.
Lipovol CP [525]

Chlophedianol hydrochloride.
ULO Syrup [734]

Chloral hydrate.
Chloral Hydrate Capsules [344]
Chloral Hydrate Capsules, Syrup [751]
Chloral Hydrate Capsules [769]
Noctec Capsules & Syrup [817]
SK-Chloral Hydrate Capsules [805]

Chlorambucil.
Leukeran [139]

Chloramphenicol.
Chloramphenicol Ophthalmic Solution 5% [769]
Chloromycetin Cream, 1%, Kapseals, Opthalmic Ointment, Otic [674]
Chloromyxin [674]
Ocu-Chlor [655a]
Opthochlor, 5% [674]
Opthocort [674]

Chloramphenicol palmitate.
Chloromycetin Palmitate [674]

Chloramphenicol sodium succinate.
Chloramphenicol Sodium Succinate Injection, Sterile [287]
Chloromycetin Sodium Succinate [674]

Chlorcyclizine hydrochloride.
Mantadil Cream [139]

Chlordiazepoxide.
Libritabs Tablets [738]
Limbitrol Tablets [738]
Menrium Tablets [738]

Chlordizepoxide hydrochloride.
A-Poxide [3]
Chlordiazepoxide Capsules [344]
Chlordiazepoxide HCl Capsules [769]
Chlordiazepoxide Hydrochloride Capsules [751]
Clipoxide Capsules [769]
Librax Capsules [738]
Librium Capsules [738]
Librium Hydrochloride [419]
Librium Injectable [738]
SK-Lygen Capsules [805]

Chlorendic anhydride.
CA-57 [369]

Chlorhexidine gluconate.
Chlorhexidine Gluconate 20% Solution [529]
Hibiclens Anntimicrobial Skin Cleanser [831]
Hibistat Germicidal Hand Rinse [831]
Hibitane Tincture (Tinted & Non-tinted) [831]

3-Chlor-2 hydroxy propyl trimethyl ammonium chloride.
Servo XRK 53 [788]

Chlorine, min 95%.
Daxan [251]

Chlorine dioxide.
Anthium Dioxide [455]

Chlormethoxypropyl mercuric acetate.
Troysan CMP Acetate [869]

Chlormezanone.
Trancopal [133]

n-Chloroacetyl-n (2,6-diethylphenyl) glycine ethyl ester.
Antor [118]

1-(3-Chloroallyl)-3,5,7-triaza-1-azoniaadamantane chloride.
Dowicil 75, 200 [261]

p-Chlorobenzyl p-chlorophenyl sulfide.
Mitox [183]

Chlorobutanol.
Chloretone [674]

5-Chloro-N-(2-chloro-4-nitrophenyl)-2-hydroxybenzamide 2 aminoethanol, 1:1.
Bayluscid [112]

p-Chloro-m-cresol.
Ottafect [669]
PCMC [426]

Chlorodeceth-14.
Dextrol XL-15 [250]

2-Chloro-N,N-dially acetamide.
Randox® CDAA [602]

2-Chloro-A-ξ diethoxyphosphinothioxyloxy)imino ξ-phenylacetonitrile.
Baythion C [112]

2-Chloro-2′,6′-diethyl-N-(methoxymethyl) acetanilide.
Lasso® alachlor [602]

2-Chloro-2′,6′-diethyl-N-(methoxymethyl) acetanilide/glyphosate, isoprylamine salt :2.6 lbs:1.4 lbs/gallon.
Bronco* [602]

2-Chloro-n-(2,6-dimethylpheny)-n-(1H-pyra-zol-1-methyl)-acetamide.
Butisans [107]

2-Chloro-N-ξ2,6-dinitro-4-((trifluoro-methyl)-phenyl)-N-ethyl-6-florobenzenemethanamine.
Prime + [190]

2-Chloro-4-ethylamine-6-isopropylamino-5-triazine.
Aatrex [190]

2-ξξ4-Chloro-6-(ethylamino)-s-triazin-2-ylξaminoξ-2-methylpropionitrile.
Bladex [791]

2-chloroethylphosphonic acid.
Ethrel [42]
Florel [877]

n-(2-Chloroethyl)-A,A,A,-trifluoro-2,6-dinitro-n-propyl-p-toluidine.
Basalin [108]

(2-Chloroethyltrimethyl ammonium chloride).
Cycocel [50]

2-Chloroethyl-tris (2-methoxy)silane.
Alsol [190]

2-Chloro-N-isopropylacetanilide.
Ramrod® Propachlor [602]

0-(5-Chloro-1-ξmethylethylξ1H1,2,4-triazol-3-yl)0,0-Diethyl phosphorothioate.
Miral [190]

5-Chloro-2-methyl-4-isothiazolin 3-one.
Kathon® 86 MW, CG, LX, WT [742]

5-Chloro-3-methyl-4-nitro-1H pyrazole.
Release [329]

4-Chloro-2-nitroaniline ⇒ o-chloroacetoacetanilide.
Atlas A8089 Monofast Yellow 10G Toner [496]

2-Chloro-4-nitroaniline ⇒ 2-naphthol.
Atlas A6460 Blazing Red Toner [496]

3-Chloro-n-(2-oxoperhydro-3-funyl)-cyclopropane-carbonanilide.
Vinicur [772]

B-(4-Chlorophenoxy)-A-(1,1-dimethylethyl)1H-1,2,4-triazole-1-ethanol.
Baytan [112]

1-(4-Chlorophenoxy)-3,3-dimethyl-1-(1H-1,2,4-trizol-1-yl)-2-butanone.
Bayleton [112]

3-ξp-(p-Chlorophenoxy) phenylξ-1,1-Di-methylurea.
Tenoran [190]

2(m-Chlorophenoxy) propionate.
Fruitone CPA [877]

p-Chlorophenyl p-chlorobenzene sulfo-nate; miticide.
Estonmite [59]

n(2-Chlorophenyl)-A-(4-chlorophemyl-5-(pyrimidinemethanol).
Rubigan [282]

1-(4-Chlorophenyl) 3-(2,6 difluouroben-zoyl) urea.
Dimilin [268]

3-(p-Chlorophenyl) 1-1-Dimethylurea trichloroacetate.
Urox [423]

n-(4-Chlorophenyl)-2,2-dimethylvaleramide.
Potablan [772]

2-Chloro-p-phenylenediamine sulfate.
Rodol Brown SO [531]

n-(4 Chloro-phenyl)-n′methoxy-n methylurea.
Aresin [416]

2-Chlorophenyl-n-methylcarbamate.
Etrofol [112]

Chlorophenylmethyl silicone fluid.
SF 1029 [341]

Chlorophenyl methyl siloxane.
SF 1250 [341]

s-((p-Chlorophenyl-thio) methyl) 0,0-diethyl phosphorodithioate.
Trithion [821]

4-Chlorophenyl2, 4,5-trichlorophenyl sulfide; =2,4,5,4'tetrachlorodiphenyl sulfide.
Animert V-101 [268]

Chlorophyllin-copper complex.
Klorofyll [838]

Chlorophyll preparations.
Chloresium Ointment, Solution [759]
Chlorophyll Complex Peries [819]
Derifil Tablets & Powder [759]
Panafil Ointment [759]

Chloroprene rubber.
Baypren 100, 112, 115, 124, 130, 210, 211, 213, 214, 215, 220, 230, 233, 235, 236, 243, 320, 320 GR, 321, 321 GR, 330, 331 [112]

Chloroprene rubber, sulfer-modified grades.
Baypren 610, 710 [112]

Chloroprocaine hydrochloride.
Nesacaine Solutions, -CE Solutions [90]

3-Chloro-1,2-propanediol.
Epibioc Rodenticide [687]

Chloroquine hydrochloride.
Aralen Hydrochloride [932]

Chloroquine phosphate.
Aralen Phosphate, Phosphate w/ Primaquine Phosphate [932]
Chloroquine Phosphate Tablets [119]
Chloroquine Phosphate Tablets [240]

2-Chloro-1-(2,4-dichlorophenyl)-vinyl diethyl phosphate.
Compound 4072 [791]

4-Chlororesorcinol.
Rodol CRS [531]

Chlorothalonil (tetrachloroisophthalonitrile).
Bravo W-75 [252]
Daconil 2787 [252]

Chlorothiazide.
Aldoclor Tablets [578]
Chloroserpine 250 & 500 Tablets [769]
Chlorothiazide Tablets [240]
Chlorothiazide Tablets [344]
Chlorothiazide Tablets [769]
Chlorothiazide w/ Reserpine Tablets [344]
Diupres [578]
Diuril Tablets & Oral Suspension [578]
SK-Chlorothiazide Tablets [805]

Chlorothiazide sodium.
Diuril Intravenous Sodium [578]

Chlorotrianisene.
TACE 12 mg Capsules, 25 mg Capsules, 72 mg Capsules [580]

2-Chloro-1-(2,4,5-trichlorophenyl) vinyl dimethyl phosphate, z isomer.
Gardona [794]
Rabon [791]

Chlorotrifluoroethylene.
Kel-F 81 Plastic [862]

Chlorotrifluoromethane.
Isotron 13 [682]

Chloroxine.
Capitrol Cream Shampoo [920]

Chloroxylenol.
Anti-Sept [781]
Fungoid Creme & Solution, Tincture [679]
Micro-Guard [839]
Ottasept Extra [669]
Sween Prep, Soft Touch [839]

p-Chloro-m-xylenol.
Ottasept [315]

Chlorphenesin carbamate.
Maolate® [884]

Chlorpheniramine.
Dallergy Syrup, Tablets and Capsules [513]
Decongestant Elixir [769]
Donatussin Drops [513]
Probahist Capsules [518]
Quadrahist Pediatric Syrup, Syrup & Timed Release Tablets [769]

Chlorpheniramine maleate.
AL-R 6 TD Capsules, AL-R 12 TD Capsules [765]
Anafed Capsules & Syrup [304]
Anamine Syrup, T.D. Caps [563]
Anatuss Tablets & Syrup, Anatuss with Codeine Syrup [563]
Brexin L.A. Capsules [767]
Chlorafed H.S. Timecelles, Liquid, Timecelles [394]
Chlorpheniramine Maleate T.D. Capsules [344]
Chlorpheniramine Maleate Tablets [751]
Citra Forte Capsules, Syrup [132]
Codimal-L.A. Capsules [166]
Colrex Compound Capsules, Elixir [750]
Comhist LA Capsules, Liquid, Tablets [652]
Comtrex [134]
Co-Pyronil 2 [254]
Coryban-D Capsules [690]
CoTylenol Cold Medication Tablets & Capsules, Liquid Cold Medication, Children's Liquid Cold Formula [569]
Deconamine Tablets, Elixir, SR Capsules, Syrup [115]

Chlorpheniramine maleate *(cont'd).*
Decongestant-AT (Antitussive) Liquid [769]
Dehist [661]
Dristan, Advanced Formula Decongestant/Antihistamine/Analgesic Capsules, Tablets, Ultra Colds Formula Aspirin-Free Analgesic/Decongestant/Antihistamine/Cough Suppressant Nighttime Liquid, Tablets and Capsules [922]
Drize Capsules [86]
Dura-Vent/A, -Vent/DA [271]
E.N.T. Tablets [814]
Extendryl Chewable Tablets, Sr. & Jr. T.D. Capsules, Syrup [326]
Fedahist Expectorant, Gyrocaps, Syrup & Tablets [746]
4-Way Cold Tablets [134]
Histalet DM Syrup, Forte Tablets, Syrup, X Syrup & Tablets [723a]
Histaspan-D Capsules, -Plus Capsules [891]
Histor-D Timecelles [394]
Hycomine Compound [269]
Iophen-C Liquid [769]
Isoclor Timesule Capsules [49]
Korigesic Tablets [868]
Kronofed-A Jr. Kronocaps, Kronocaps [314]
Kronohist Kronocaps [314]
Naldecon [134]
Neotep Granucaps [723]
Nolamine Tablets [153]
Novafed A Capsules, Liquid [580]
Novahistine DH [580]
Ornade Spansule Capsules [805]
P-V-Tussin Syrup [723a]
Pediacof [133]
Phenate [541]
Probocon [518]
Protid [512]
Pseudo-Hist Capsules, Expectorant, Liquid [420]
Quelidrine Syrup [3]
Resaid T.D. Capsules [344]
Rescaps-D T.D. Capsules [344]
Rhinafed Capsules, -EX Capsules [566]
Rhinolar Capsules, -EX Capsules, -EX 12 Capsules [566]
Ru-Tuss Tablets, II Capsules [127]
Ryna Liquid, -C Liquid [911a]
Scot-Tussin Sugar-Free, Syrup [779]
Singlet [580]
Sinovan Timed [267]
Sinulin Tablets [153]
Symptrol TD Capsules [765]
T-Dry Capsules, Jr. Capsules [930]
Triaminic Cold Syrup, Cold Tablets, -12 Tablets [257]
Triaminicol Multi-Symptom Cold Syrup, Multi-Symptom Cold Tablets [257]
Trinex Tablets [560]
Tussar DM, SF, -2 [891]
Tussi-Organidin, DM [911a]

Chlorpheniramine maleate products.
Phenetron Products [510]

Chlorpheniramine preparations.
Deconamine Tablets, Elixir, SR Capsules, Syrup [115]
Fedahist Expectorant, Gyrocaps, Syrup & Tablets [746]
Norel Plus Capsules [888]

Chlorpheniramine tannate.
Rynatan Tablets & Pediatric Suspension [911a]
Rynatuss Tablets & Pediatric Suspension [911a]

Chlorpromazine.
BayClor [111]
Chlorpromazine Tablets & Concentrate Syrup [344]
Thorazine [805]

Chlorpromazine hydrochloride.
Chlorpromazine HCl Injection [287]
Chlorpromazine HCl Tablets [769]
Chlorpromazine HCl in Tubex [942]
Thorazine Tablets [510]

Chlorpropamide.
Diabinese [691]

Chlorprothixene.
Taractan Tablets [738]

Chlorprothixene hydrochloride.
Taractan Concentrate, Injectable [738]

Chlorprothixene lactate.
Taractan Concentrate [738]

Chlortetracycline hydrochloride.
Aureomycin Ointment 3% [516]

Chlorthalidone.
Chlorthalidone Tablets [3]
Chlorthalidone Tablets [240]
Chlorthalidone Tablets [769]
Chlorthalidone Tablets, USP [674]
Combipres Tablets [125]
Demi-Regroton Tablets [891]
Hygroton Tablets [891]
Regroton Tablets [891]
Thalitone Tablets [125]

Chlorzoxazone.
Algisin Capsules [718]
Blanex Capsules [281]

Clorazepate dipotassium.
Tranxene Capsules & Tablets, -SD, -SD Half Strength [3]

Clotrimazole.
Gyne-Lotrimin Vaginal Cream 1%, Vaginal Tablets [772]
Lotrimin Cream 1%, Solution 1% [772]
Mycelex 1% Cream, 1% Solution, Troches, -G, -G 1% Vaginal Cream [585]

Cloxacillin.
Cloxacillin Capsules [119]
Cloxacillin Sodium Capsules [769]
Cloxacillin Solution [119]

Cloxacillin sodium.
Cloxapen® Capsules, 250 mg, Capsules, 500 mg [114]

Cloxacillin sodium monohydrate.
Tegopen [134]

Coal.
Kofil 500 OT, 500 R, 1000 [392]

Coal-tar.
BRV [35]
Denorex Medicated Shampoo and Conditioner, Medicated Shampoo, Regular & Mountain Fresh Herbal Scent [922]
Fototar Cream 1.6%, Stik 5% [284]
Pentrax Tar Shampoo [214]
Zetar Emulsion, Shampoo [246]

Cobalt(II) chloride 6 H20.
Ultralog 23-1400-00 [172]

Cobalt molybdate.
HT-400 E 1/8″ [389]

Cobalt(II-III) oxide.
Ultralog 23-1580-00 [172]

Cocaine hydrochloride.
Cocaine Hydrochloride Topical Solution [751]

Cocamide.
Armid C [79]
Nitrene 230 OA, 250 BS [399]

Cocamide DEA.
Active #2 [123]
Alkamide CDE, CDO [33]
Alrosol B [190]
Amide CD 2:1 [437]
Aminol COR-2, HCA [320]
Calamide C [697]
Carsamide CA, C-3, C-7644, C-7649, LE, SAC [529]
Clindrol Superamide 100C, Superamide 100CG, 101CG, 200CG, 200CGN, 200RC, 200HC, 202CGN, 203CG, 204CG, 206CGN, 207CGN, 209CGN [197]
Comperlan KD [400]
Comperlan KDO [399], [400]
Comperlan PD, SDO [399]
Condensate CO [209]
Condensate HCA [174]
Condensate NP, PA, PN, PO, PS [209]
Condensate X-45R, X-45W [174]
Cyclomide CD [939]
Cyclomide DC 212, DC 212/S, DC 212/SE, KD [234]
Emid 6514, 6515, 6530, 6531, 6533, 6534, 6575 [291]
Empilan CDE, CDE/FF, CDX [548]
Gafamide CDD-518 [338]
Hetamide MC [407]
Hyamide 1:1 [437]
Lanamid 50, 120 [508]
Mackamide 100-A, C [567]
Mazamide 70, 80, CA-20, CS-148 [564]
Monamid 150-ADD, 705, 759, 815, ADD-633-LE [599]
Monamine AA-100, AA-100PG, ADD-100, ADD-100-LE, ADS-100 [599]
Ninol 128 Extra, 2012 Extra, 4821 [825]
Oramide DL 200 [785]
Orapol DL 210 [785]
Protamide CKD, DCA, DCAW, HCA [713]
Rewo-Amid DC 212/LS, DC 212/S, DC 212/SE, DC 220/SE, DL 240 [725]
Richamide 6404, 6445, 6625, CD, CDA, LIQ, M-3, S-771, S-780 [732]
Schercomid 1-102, 1-107, CCD, CDA, CDA-H, CDO-Extra, SCE, SCO-Extra [771]
Serdolamine PPF 67 [175]
Sipomide 1500 [29]
Standamid KD, KDO, PD, SD, SDO [399]
Sterling Amide 374, Sterling DEA [145]
Super Amide B-5, GR [663]
Surco 128T, SR-200-Para Coco Special, W.C. Concentrate [663]
Synotol 119 N, CN 60, CN 80, CN 90, Detergent E [715]
Unamide CDX, JJ-35, JJ-35 Special, LDL, N-72-3, RDX [529]
Varamide A-1 [796]
Witcamide 82 [934]
Witcamide 1017 [935]
Witcamide 5130, 5133 [934]
Witcamide M-3 [935]

Cocamide DEA, coconut oil base.
Ninol 128 Extra [825]

Cocamide DEA, methyl cocoate base.
Ninol 2012 Extra [825]

Cocamide DEA lauryl sulfate.
Emery® 6731 [290]

Cocamide MEA.
Alkamide CME, CMO [33]
Amide No. 27 [709]
Aminol CM, CM Flakes, CM-D Flakes [320]
Carsamide CMEA [529]
Clindrol 100MC, 100MCG [197]
Comperlan 100, KM [400]
Comperlan SM [399]
Cyclomide C212 [234]
Cyclomide CM [939]
Emid 6500 [291]
Empilan CME [548]
Foamole M [895]
Hetamide CME [407]
Mackamide CMA [567]
Monamid CMA [599]
Nitrene 185 S [399]
Onyxol 12 [663]
Oramide ML 100, ML 115 [785]
Rewo-Amid C212 [725]
Richamide MX [732]
Schercomid CME [771]
Standamid CM, CMG [400]
Standamid KM, SM [399]
Sterling Granulated Wax [145]
Witcamide MEAC [935]

Cocamide MIPA.
Cyclomide CP [939]
Empilan CIS [548]
Monamid CIPA [599]
Rewo-Amid IPP 240 [725]
Schercomid CMI [771]
Serdolamine PPG 72 [175]

Cocamido amine oxide.
Mazox CA [564]

Cocamidopropylamine oxide.
Alkamox CAPO [33]
Ammonyx CDO [663]
Barlox C, CO [529]
Carsamine Oxide CB [529]
Finamine CO [320]
Mackamine CAO [567]
Mazox CA [564]
Monalux CAO [599]
Ninox CA [825]
Rewominoxid B 204 [725]
Schercamox C-AA [771]
Standamox CAW [399]
Varox 1770 [796]

Cocamidopropyl betaine.
Aerosol 30 [50]
Amonyl 380 BA [785]
Amphosol CA, CG [825]
Carsonam 3, 32, 33S, 3147 [529]
Chemadene NA-30 [732]
Cycloteric BET-C30, BET-CW [234]
Dehyton K [400]
Deriphat BAW [399]
Emcol DG, NA-30 [935]
Emery 5430, 6748 [291]
Lexaine C, CG30, CG50, J [453]
Lonzaine C, CO [529]
Mackam 35 [567]
Mafo CAB [564]
Maprolyte C [663]
Mirataine® CB, CBC, CBM, CBR [592]
Monateric ADA, CAB, COAB, MCB [599]
Rewoteric AM-B13 [725]
Schercotaine CAB [771]
Standapol BAW, BC-35 [400]
Sterling CAB [145]
Surco Coco Betaine [663]
Tego-Betaine C, L-7, L10, S, T [360]
Varion CADG [796]
Velvetex BA-35 [400]
Velvetex BC-35, BK-35 [399]

Cocamidopropyl dimethylamine.
Carsamine CB [529]
Cyclomide CODI [234]
Lexamine C13 [453]
Mackine 101 [567]
Schercodine C [771]

Cocamidopropyl dimethylamine dihydroxymethylpropionate.
Richamate 2955 [732]

Cocamidopropyl dimethylamine lactate.
Mackalene 116 [567]

Cocamido propyl dimethyl amine oxide.
Mackamine CAO [567]

Cocamidopropyl dimethylamine propionate.
Emcol 1655 [935]
Richamate 1655 [732]

Cocamidopropyl hydroxysultaine.
Lonzaine CS, JS [529]
Mirataine® CBS [592]
Schercotaine SCAB [771]

Cocamidopropyl lauryl ether.
Marlamid KL [178]

Cocamidopropyl PG-dimonium chloride phosphate.
Monaquat P-TC [599]

Cocamine.
Armeen C, CD [79]
Kemamine P-650 [429]

Cocamine oxide.
Aromox DMC-W [79]
Barlox 12 [529]
Hoe S 2319 [55]
Mackamine CO [567]
Mazox CO [564]
Monalux CO [599]
Ninox C [825]

Cocaminobutyric acid.
Armeen Z [79]

Cocaminopropionic acid.
Deriphat 151C [399]

Coceth-7 carboxylic acid.
Nu-Mole CM-7A [41]

Cochin diethanolamide, 1:1.
Surco 128T [663]

Cochin monoethanolamide.
Surco CMEA flaked [663]

Cochin oil soap.
ESI-Terge 40% coconut oil soap [294]

Coco alkanolamide.
Unamide D-10 [529]

Coco alkanolamide, ethoxylated.
Unamide C-2, C-5 [529]

Coco-alkyl betaine.
Amphoram CB A30 [162]

Coco-alkyl dimethylamine.
Nissan Tertiary Amine FB [636]

Coco-alkyl dimethyl benzyl ammonium chloride.
Nissan Cation F_2-10R, F_2-20R, F_2-40E, F_2-50 [636]

Coco-alkyl taurine.
Amphoram CT 30 [162]

Coco-alkyl trimethyl ammonium chloride.
Nissan Cation FB, FB-500 [636]

Cocoamide.
Armid C [79]

Coco amide betaine.
Mafo CAB [564]

Cocoamide DEA.
Jordamide CCO Cond., JT128, WC Conc. [474]
Nitrene 200 SD [399]
Schercomid CDO-Extra [771]
Super Amide GR [663]
Surco WC Conc., 128T [663]

Cocoamide EO + 15 moles.
Teric 12M15 [439]

Coco amide, ethoxylated.
Unamide C-5 [529]

Cocoamide MEA.
Jordamide CMEA [474]
Surco CMEA [663]

Coco amidoalkyl betaine.
Emery 5430 [290]
Lexaine CG-30 [453]
Maprolyte C [663]
Norfox Coco Betaine [650]
Rewoteric AM-B 13 [725]

Coco amido alkyl dimethyl amine.
Schercodine C [771]

Coco amido amine.
Mackine 101 [567]

Coco amido amine oxide.
Barlox C [529]
Standamox C 30 [398]

Coco amido betaine.
Alkateric CAB [33]
Amonyl 380 BA [711]
Amphosol CA, CG [825]
Carsonam 32 [155]
Lonzaine C, CO [529]
Varion CADG, CADG-HS, CADG-LS [796]

Cocamidopropylamine oxide.
Jordamox CAPA [474]
Varox 1770 [796]

Cocoamidopropyl betaine.
Aerosol 30 [50a]
Aremsol A [745]
Cycloteric BET C-30 [234]
Jortaine CAB-35, CAB-40 [474]
Lebon 2000 [764]
Lexaine C [453]
Maprolyte C [663]
Monateric ADA, ADFA, CAB, MCB [599]
Sipoteric CB, COB [29]
Velvetex BA [400]
Velvetex BA-35 [399]

Coco amido propyl dimethyl amine oxide.
Alkamox CAPO [33]
Ammonyx CDO [663]
Barlox® C [529]
Ninox CA, FCA [825]
Schercamox C-AA [771]
Varox 1770 [796]

Cocamido propyl dimethyl benzyl ammonium chloride.
Aremsan C40 [745]
Quatrene CA [400]

Coco amido propyl dimethyl betaine.
Lonzaine® C, Lonzaine CO (Carsonam 3), CS [529]

Coco amido sulfobetaine.
Jortaine CSB [474]
Lonzaine-CS, JS [529]
Sandobet SC [761]

Cocoamine.
Amine KKD [487]
Crodamine 1.CD [223]
Lilamin KKD [487]
Noram C [162]

n-Cocoamine acetate.
Acetamin C [803]
Armac C [79]
Crodamac 1.CA [223]
Noramac C [162]

Cocoamine + EO 2 moles.
Teric 12M2 [439]

Cocoamine + EO 5 moles.
Teric 12M5 [439]

Coco-amine + EO 15 moles.
Teric 12M Series [439]

Coco amine, ethoxylated.
Accomeen C2, C5, C10, C15 [81]
Chemeen C-2, C-5, C-10, C-15 [170]
Noramox C2 to 15 [162]
Varonic K202, K205, K210, K215 [796]

Coco amine, ethoxylated, amine oxide.
Alkamox C20 [33]

Coco amine oxide.
Mackamine CO [567]

Coco amine polyglycol ether, EO 11.
Serdox NCA 11 [788]

Cocoamine (primary).
Armeen C, CD [79]

Coco amines, primary ethoxylates.
Alkaminox C-2, C-5, C-10 [33]

N-Coco amino acid, potassium salt.
Mafo C-12 [564]

Cocoamino betaine.
Lonzaine 12C [529]

N-Coco- β -aminobutyric acid.
Armeen Z [79]
Armeen Z-9 [16]

N-Coco β -amino butyric acid, sodium salt.
Armeen SZ [16]

N-Coco β -amino propionate, sodium salt.
Deriphat 151 [399], [400]

N-Coco amino propionic acid.
Amphoram CP1 [162]
Deriphat 151C [400]

Cocoamphocarboxyglycinate.
Alkateric 2CIB [33]
Amphoterge W-2 [529]
Lexoteric 2 [453]
Miranol C2M Conc. O.P. [592]
Monateric CDX-38, CSH-32 [599]
Rewoteric AM-2C [725]
Sipoteric 1398 [29]

Cocoamphocarboxypropionate.
Amphoterge K-2 [529]
Cycloteric DC-SF [234]
Miranol C2M-SF Conc., C2M-SF 70%, C2M-SF 75%, C2M-SFE Conc., C2M-SFP [592]
Monateric 805-P2 [599]
Rewoteric AM-2CSF [725]

Cocoamphocarboxypropionic acid.
Miranol C2M Anhydrous Acid [592]

Cocoamphocarboxypropylsulfonate.
Amphoterge KS [529]

Cocoamphoglycinate.
Alkateric CIB [33]
Amphoterge W [529]
Dehyton® G [401]
Emery 5412 [291]
Miranol CM Conc. N.P. [592]
Monateric CM-36S [599]
Rewoteric AM-2C/NM [725]
Rexoteric XCO-30, XCO-40, XOO-30 [727]
Standapol CIM-40 [400]
Velvetex GC-88 [400]

Cocoamphopropionate.
Amphoterge K [529]
Cycloteric MC-SF, MV-SF [234]
Miranol CM-SF Conc., CM-SFX Conc. [592]
Monateric CA-35, CAM-40 [599]
Rewoteric AM-KSF [725]

Cocoamphopropylsulfonate.
Amphoterge SB [529]
Miranol CS 60%, CS Conc. [592]
Rewoteric AM-CS [725]

Coco benzyl imidazolinium chloride.
Quatrene C-5-6 [400]

Coco betaine.
Accobetaine CL [81]
Alkateric BC [33]
Amonyl 265 BA [711]
Ampho B11-34 [829]
Amphosol C [825]
Dehyton AB-30 [399]
Deriphat BCW [399]
Emcol CC37-18 [935]

Coco betaine *(cont'd)*.
Hartaine CB-40 [391]
Jortaine CB-40 [474]
Lonzaine 12C [529]
Mackam CB [567]
Marsetaine C-30 [554]
Mirataine CBS, CDMB [592]
Monateric CB [599]
Protachem CB 45 [713]
Sipoteric CB, COB [29]
Standapol AB-45 [400]
Varion CDG [796]
Velvetex AB-45, BC [400]

Coco chloride.
Barchlor 12C [529]

Coco DEA super-amide.
Carsamide SAC [529]

Coco-1,3 di-amini propane.
Duomeen CD [16]

N-Coco-1,3 diaminopropane.
Duomeen C, C Special [79]

Coco-1,3-diaminopropane diacetate.
Duomac C [16]

Cocodiethanolamide.
Carsamide C-3, CA, LE [529]
Chimipal DCL [514]
Comperlan ID, PD, SD [399]
Emid 6531, 6533, 6534, 6538 [289]
Iconol 28, COA [109]
Lakeway 100-CA [126]
Mackamide C [567]
Nitrene N [399]
Super-Amide GR [663]
Unamide C-72-3 [529]
Unamide C-7645, C-7649, C-7944 [72], [529]
Unamide CDX, D-10, GG-75, N-72-3 [529]
Varamide A10, A12, A83 [796]
Witcamide 82 [935]

Coco diethanolamide, super.
Carsamide SAC [155]
Emid 6514 [289]
Unamide CDX, JJ-35, LDL [529]

Coco dimethyl amine.
Barlene 12C [529]
Noram DMC [162]

Coco dimethyl amine oxide.
Genaminox CS, KC [416]
Hartox DMCD [391]

Coco dimethyl ammonium carboxylic acid betaine.
Deriphat BC, BCW [400]

Cocodimethylbenzyl-ammonium chloride.
Merpiquat K-8-2 [285]
Noramium C85 [162]
Querton KKBCL50 [487]
Servamine KAC 422 [788]

Coco dimethyl betaine.
Lonzaine 12C [529]

N-Coco dipropylene triamine.
Trinoram C [162]

Coco, ethoxylated.
Varonic K202 [796]

Coco-ethyl dimonium ethosulfate.
Dextrol AS-150 [250]

Coco fatty acid alkanolamide, 2:1.
Mazamide CA-20, LM-20 [564]

Coco fatty acid condensate, collagen polypeptide, potassium salt.
Maypon 4C [825a]

Coco fatty acid condensate, collagen polypeptide, triethanolamine salt.
Maypon 4CT [825a]
Monteine LCT [711]

Coco fatty acid, diethanol amid.
Comperlan COD, KD [401]
Lauridit KD, KDG [17]
Loropan KD [858]
Stamid HT [198]

Coco fatty acid ethanolamide.
Comperlan KM [401]

Coco fatty acid, ethoxylated.
Chemax E200ML, E400ML, E600ML [170]

Coco fatty acid, hydrogenated.
Hystrene 5012 [429]
Industrene 333 [429]

Coco fatty acids, imidazoline.
Miramine® CC [592]

Coco fatty acid isopropyl ester.
Cetiol IP 448 [401]

Coco fatty acid monoethanolamide.
Comperlan 100 [401]
Lauridit KM [17]
Loropan CME, KM [858]

Coco fatty acid polydiethanolamide.
Comperlan PD [401]

Coco-fatty acid, sodium sulfosuccinate, amido-ester.
Schercopol CMIS-Na [771]

Coco fatty alcohol sulfate, sodium salt.
Sulfetal C90 [948]

Coco fatty diethanolamide, super.
Carsamide LE [155]

Coco hydroxyethyl imidazoline.
Alkazine C [33]

Coco hydroxyethyl imidazoline *(cont'd).*
Unamine C [529]
Varine C [796]

Cocohydroxyethyl PG-imidazolinium chloride phosphate.
Monaquat P-TZ [599]

Cocoimidazoline.
Amphoterge K, W [529]

Coco imidazoline benzyl chloride, quaternary.
Uniquat CB-50 [529]

Coco imidazoline betaine.
Carsonam C, C-SF, C-SPCL [155]

Coco imidazoline betaine dicarboxylate.
Carsonam DC, DC-SF, DC-70%-SF [155]

Coco imidazoline carboxylate.
Mackam 2C-SF, 1C-SF [567]

Coco imidazoline dicarboxylate.
Amphoterge K-2 (Carsonam DC-SF), W-2 (Carsonam DC) [529]
Mackam-2C, 2C-75 [567]

Coco imidazoline monocarboxylate.
Amphoterge K (Carsonam C-SF), W (Carsonam C) [529]
Mackam-1C [567]

Coco imidazoline sulfonate.
Amphoterge SB [529]

Coco methanolamide.
Carsamide CHEA [529]

Coco monoethanolamide.
Carsamide CMEA [529]
Clindrol 100MCG [197]
Comperlan 100 [399]
Emid® 6500 [290]
Intermediate 325 [429]
Unamide CMX [529]

Coco monoglyceride, ethoxylated.
Varonic LI-63, LI-67 [796]

N-Coco morpholine.
Armeen N-CMD [16]
Baircat® NCM [529]

Coco-morpholine oxide.
Aromox N-CM-W [79]
Barlox NCM [529]

Coco nitrile.
Arneel C [79]

Coconut acid.
Neo-Fat 255, 265 [79]
PRIFAC 5901, 7901 [875]
Wochem #200 [941]

Coconut Alcohol.
CO-618 [709]
Laurex CH [548]

Coconut alcohol sulfosuccinate, sodium salt.
Setacin F Special Paste [948]

Coconut alkanolamide 2:1.
Alkamide 2106, 2110, 2204A, CL-63 [33]
Aramide CDS, CDX [76]
Clindrol 201CGN, 209CGN, Superamide 100C [197]
Dergon OM [77]
Hartamide AD [391]
Monamid 7-100, 150-AD, 150-ADD, 150-DR, 664 [599]
Monamine AA-100, AD-100, ADD-100, ALX-80SS [599]
Tex-Wet 1131 [460]

Coconut alkyl dimethyl ammonium betaine.
Dehyton AB-30 [399]

Coconut alkylolamide.
Kerapol 791 [253]

Coconut amide polyglycolether.
Lorapal C 6 [274]
Steinapal C 6 [274]

Coconut amido alkyl betaine.
Amido Betaine C, C Conc. [946]

Coconut amido betaine.
Schercotaine CAB [771]

Coconut amine acetate.
Kemamine A 650 [429a]

Coconut amine condensate.
Cindet C-12 [193]
Titanterge 100, CAC [863]

Coconut amine ethoxylate.
Alkaminox C-5, C-10 [33]

Coconut amine, primary.
Kemamine P-650, P-650 D [429a]

Coconut ampholytic (Syndet).
Schercoteric MS [771]

Coconut amphoteric.
Monateric CA-35%, CEM-38%, CDX-38, C-M36S, C Na-40% [599]
Sipon-1127 [29]

Coconut DEA.
Alkamide CDE, CDO [33]

Coconut dialkanolamide.
Calamide CW-100 [697]

Coconut diamine.
Dicrodamine 1.C [223]
Lilamin 560 [522]

Coconut diethanolamide.
Alkamide 1002, 2104, 2106, 2204, CDE, CDO [33]
Ardet CB, DC, DCAE, DCX, DMA, DMC, DMF, WO [73]
Calamide C, CW-100 [697]
Calsuds CD-6 [697]
Carsamide 7644 [155]
Clindrol Superamide 100C, 100CG [197]
Clindrol Superamide 100CG [197]
Conco Emulsifier K [209]
Crillon CDY [223]
Cyclomide CD, CDR, CDU [938]
Cyclomide DC212, DC212M, DC212/S, DC212/SE, KD [234]
Emid® 6575 [290]
Empilan 2502 [25]
Empilan 2544 [23]
Empilan CDE [26]
Empilan CDE/FF, CDE/X, CDEY, CDX, FD, FD20, FE [23]
ESI-Terge 10 [294]
ESI-Terge B-15, C-5, S-10 [892]
Ethulan A15 [253]
Ethylan Al5, LD, LDG, LDS [509]
Felsamid-KPD [823]
Hartamide OD [391]
Hymolon CWC, K-90 [390]
Lauramide 11, ME [946]
Manro CD, CDS [545]
Marlamid D 1218 [178]
Marsamid 10, 40, 50 [554]
Monamid 7-100, 7-153CS, 150-AD, 150-ADD, 150-DR, 150-GLT, 770 [599]
Monamine AA-100, AD-100, ADD-100, ALX-80 SS, ALX-100 S, I-76 [599]
Ninol LD [825a]
Nissan Stafoam DF [636]
Purton CFD [948]
Rewomid CD, DC 212/S, DC 212/SE [725]
Serdolamide PPF 51, PPF 67, PPFG 68 [788]
Sterling DEA [145]
Surco Coco Condensate [663]
Tohol N-220, N-220X [864]
Ufanon K-80, KD-S [874]
Witcamide 5130, M-3 [935]

Coconut diethanolamide, super.
Felsamid-KDG [823]
Gafamide® CDD-518 [338]

Coconut-diethanolamine condensation product.
Condensate PC [710]

Coconut diethanol superamide.
Mazamide 80, 159A [564]

Coconut dimethyl t amine.
Lilamin 367 D [522]

Coconut ether sulfate, ammonium salt.
Steol CA 460 [825a]

Coconut ether sulfate, sodium salt.
Steol CS 460 [825a]

Coconut fatty acid.
Emery 621 [289]
Hetamide MC [407]
Industrene 325, 328 [429]
Merpoxen KM 45 [285]

Coconut fatty acid alkanolamide.
Clindrol Superamide 100C, 200RC [197]
Condensate PA. PO [209]
Hartamide CE-90 [391]
Infrasan C [724]
Surco 128T, CMEA, WC Conc. [663]

Coconut fatty acid alkanolamide, hydrogenated.
Clindrol 200 HC [197]

Coconut fatty acid alkanolamine condensate.
Standamid KDM [400]

Coconut fatty acid alkylamide.
Nopcogen 14-LT [253]

Coconut fatty acid alkylamido propyldimethylamine oxide.
Loramine B 204 [274]

Coconut fatty acid alkylolamide, ethoxylated.
Ethylan CH, CRS, LM2 [509]

Coconut fatty acid amidoalkyldimethylamine oxide.
Rewominox B 204 [725]

Coconut fatty acid amine condensate.
Solar-Air, Solar Regular [840]

Coconut fatty acid condensate, protein, potassium salt.
Merpinol CK 30 [285]

Coconut fatty acid diethanolamide.
Comperlan LS [401]
Lauridit KD, KD G [16]
Loramine DC 212/S, DC 220/SE [274]
Manro CD, CDS [545]
Marlamid D 1218, DF 1218 [178]
Merpinamid KD 11, LSD/E [285]
Onyxol 345 [663]
Rewomid DC 212/S
Schercomid CDA [771]
Standamid KD [400]
Steinamid DC 212/S, DC 220/SE [274]
Synotol CN-60, CN-80, CN-90 [265]
Teric CDE [439]

Coconut fatty acid diethanolamine.
Loramine DC 212/SE [274]

Coconut fatty acid ester.
Ethylan CF71 [509]

Coconut fatty acid ethanolamide ethoxylate.
Elfapur KA 45 [16], [764]

Coconut fatty acids, ethoxylated.
Alkasurf L-2, L-5, L-9, L-14 [33]

Coconut fatty acids, imidazoline.
Monazoline C [599]

Coconut fatty acid isopropanol amide.
Serdolamide PPG 72 [788]

Coconut fatty acid MEA.
Surco CMEA Flake [663]

Coconut fatty acid, monoethanolamide.
Clindrol 100MC [197]
Lauridit KM [16]
Loramine C 212 [274]
Manro CMEA [545]
Marlamid M 1218 [178]
Merpinamid KM, LSM [285]
P&G Amide No. 27 [709]
Rewomid C 212 [274]
Schercomid CME [771]
Serdolamide PPE 87 [788]
Standamid 100 [398]
Steinamid C 212 [274]

Coconut fatty acid monoethanolamide + E.O. 3 moles.
Teric CME Series [439]

Coconut fatty acid monoethanolamide + E.O. 7 moles.
Teric CME Series [439]

Coconut fatty acid monoethanolamide-ethoxylate.
Merpoxen KM 60 [285]

Coconut fatty acid monoethanolamide, polyglycoether 4.5.
Merpoxen KM 45 [285]

Coconut fatty acid monoethanolamide, polyglycolether 6.
Merpoxen KM 60 [285]

Coconut fatty acid monoethanolamide-polyglycolether 9.
Merpoxen KM 90 [285]

Coconut fatty acid monoisopropanolamide.
Loramine IPP 240 [274]
Schercomid CMI [771]
Steinamid IPP 240 [274]

Coconut fatty acid polydiethanolamide.
Loramine DL 240 [274]
Merpinamid KD 12, LSD [285]
Rewomid DL 240 [274]
Steinamid DL 240 [274]

Coconut fatty acids, polytheylene glycol 400 ester.
Emulsan K [724]

Coconut fatty alcohol polyglycol ether.
Genapol C-100, C-200, C-250 [55]

Coconut fatty alcohol, sulfated, sodium salt.
Elfan 280 [16]
Elfan 280 Powder Concentrated [16], [764]

Coconut fatty amine ethoxylate.
Ethylan TC, TF, TN-10 [509]

Coconut fatty amine oxethylate.
Genamine C-020, C-050, C-080, C-100, C-150, C-200 [55]

Coconut hydroxyethyl imidazoline.
Alkazine C [33]

Coconut imidazoline betaine.
Empigen CDR10 [23]
Empigen CDR 30 [25]

Coconut isopropanolamide.
Empilan CIS [25]
Felsamid-KI [823]
Serdolamide PPG 72 [788]

Coconut (50% lauric)-amidopropyldimethyl-amine oxide.
Felsoxid-KP-50 [823]

Coconut (70% lauric)-amidopropyldimethyl amine oxide.
Felsoxid-KP-70 [823]

Coconut lauric DEA.
Alkamide CL 63 [33]

Coconut-lauric diethanolamine.
Alkamide CL63 [33]

Coconut MEA.
Alkamide CME [33]

Coconut MEA ethoxylate.
Alkamidox C-2, C-5 [33]

Coconut monoalkanolamide.
Ethylan LM [509]

Coconut monoethanolamide.
Alkamide CME, CMO [33]
Ardet CEA [73]
Chimipal MC [514]
Cyclomide C212 [234]
Emid 6500 [289]
Empilan CM [24]
Empilan CME [26]
Felsamid-KM [823]
Manro CMEA [545]

Coconut monoethanolamide *(cont'd).*
Monamid CMA [599]
Monamide [946]
Ninol CNR [825a]
Nissan Stafoam MF [636]
Rewomid C212 [725]
Surco CMEA [663]
Swanic 51 [837]

Coconut monoethanolamide + EO 3 moles.
Teric CME3 [439]

Coconut monoethanolamide + EO 7 moles.
Teric CME7 [439]

Coconut monoethanolamide, ethoxylated.
Alkamidox C-2, C-5 [33]
Empilan LP2, LP10, MAA [23]

Coconut monoethanolamide ethoxylate + ethylene oxide 2 moles.
Alkamidox CME-2 [33]

Coconut monoethanolamide ethoxylate/ ethylene oxide 5 moles.
Alkamidox CME-5 [33]

Coconut monoisopropanolamide.
Empilan CIS [23]
Rewomid IPP 240 [725]

Coconut oil.
Hetamide RC [407]
Pureco 76 [147]

Coconut oil alkanolamide.
Aminol COR-2, COR-2C [320]
Clindrol 200 CG, 202CGN, 204CGN [197]
Condensate PS [209]
Mazamide 70, 80, CS-148 [564]

Coconut oil alkylolamide.
Gradonic FA-20 [365]

Coconut oil, amides.
Amides of Coconut Oil [612a]
Ecconol 628 [300]

Coconut oil amido compound.
Surco Coco Betaine [663]

Coconut oil amine.
Radiamine 6160, 6161 [658]

Coconut oil amine acetate.
Radiamac 6169 [658]

Coconut oil DEA condensate.
Norfox DC [650]

Coconut oil DEA superamide.
Aminol COR-4 [320]
Unamide LDL [529]

Coconut oil diamine.
Radiamine 6560 [658]

Coconut oil diethanolamide.
Accomid C [81]
Cedemide D [256]
Clindrol Superamide 100CG, 200 CGN [197]
Sactilan LDC, LDE [521]
Schercomid CDO Extra [771]
Varamide A1, A-2, A-10, A-12, A-83, MA-1 [796]

Coconut oil diethanolamine condensate.
Gafamide CDD 518 [338]

Coconut oil, ester.
Aldo DC, TC, TCL [358]

Coconut oil fatty acid amine condensate.
Solar 25, CO [840]

Coconut oil fatty acid diethanolamide 1:1.
Profan 24 Extra, 128 Extra, 2012E [764]
Stafoam F [636]

Coconut oil fatty acid monoethanolamide.
Profan AB20 [764]

Coconut oil, hydrogenated.
Cobee 92, 110 [715]

Coconut oil soap.
Hipochem LH—Soap [410]

Coconut oil superamide.
Aminol HCA [320]

Coconut pentaethoxy methylammonium methylsulfate.
Rewoquat CPEM [725]

Coconut polydiethanolamide.
Cyclomide Pinemulse [234]
Cycloryl DCA [234]
Rewomid DL 240 [725]

N-Coconut 1,3-propylenediamine.
Kemamine D-650 [430]

Coconut sulfonic acid alkanolamide 2:1.
Mazamide 25 [564]

Coconut sulfosuccinate.
Mackanate CM [567]

Coco oil diethanolamide 1:1.
Unamide JJ-35, LDL [529]

Coco oil diethanolamide, super.
Emid 6515 [289]

Coco-penta-ethoxy methyl-ammonium metho-sulfate.
Loraquat CPEM [274]

Coco-5-polyglycol-ethylammonium ethosulphate.
Merpiquat K-26 [285]

N-Coco-1,3-propanediamine.
Duomeen C, C Special [79]

Coco-1,3-propylene diamine.
Diamine KKP [487]
Dinoram C [162]

Coco-1,3-propylene diamine *(cont'd).*
Kemamine D 650, D 650 D [429a]
Lilamin KKP [487]

N-Coco propylene diamine acetate.
Dinoramac C [162]

N-Coco 1,3 propylene diamine diacetate.
Kemamine AD 650 [429a]

Coco quaternary, carboxylated.
Sanac C [829]

Coco sulfo-betaine.
Varion CAS [796]

Cocosulfoimidazoline.
Amphoterge KS, SB [529]

Coco-sultaine.
Lonzaine 12CS [529]

Coco superamide.
Accomid C [829]
Varamide MA1 [796]

Cocotriglyceride.
Coconut Butter [529]

Cocotrimethyl ammonium chloride.
Arquad C-33-W [16]
Noramium MC 50 [162]
Servamine KAC 412 [788]

Cocotrimonium chloride.
Arquad C-33W [79]
Kemamine Q-6503B [429a]

Cocoyl amidoalkylamine oxide.
Textamine Oxide CA, CAW [399]

Cocoyl amido betaine.
Mackam 35, 35 HP [567]

Cocoyl diethanolamide.
Nitrene 11230, 13026, A-309, C, C Extra, S-114 [399]

Cocoyl imidazoline.
Miramine CC [592]

Cocoyl isethionate.
Tauranol I-78 [320]

Cocoyl sarcosinamide DEA.
Cyclomide DS [939]
Rewo-Amid SD [725]

Cocoyl sarcosine.
Carsonate C [529]
Hamposyl C [363]
Maprosyl C [663]
Sarkosyl LC, NC [190]

Cocyl diethanolamide.
Nitrene C, N [400]

Codeine.
APAP w/ Codeine Tablets, #3, #4 [344]
Apap 300 mg. with Codeine Tabs, with Codeine Elixir [769]
Aspirin 325 mg. with Codeine Tabs [769]
Calcidrine Syrup [3]
Empracet with Codeine Phosphate Nos. 3 & 4 [139]
Terpin Hydrate & Codeine Elixir [751]

Codeine phosphate.
A.P.C. with Codeine, Tabloid brand [139]
Acetaco Tablets [518]
Acetaminophen with Codeine Phosphate Tablets [751]
Actifed-C Expectorant [139]
Amaphen with Codeine #3 [868]
Anacin-3 with Codeine Tablets [97]
Anatuss with Codeine Syrup [563]
Ascriptin with Codeine [746]
Bancap c Codeine Capsules [661]
Bromanyl Expectorant [769]
Bromphen DC Expectorant [769]
Buff-A Comp No. 3 Tablets (with Codeine) [563]
Bufferin with Codeine No. 3 [134]
Capital with Codeine Suspension, with Codeine Tablets [153]
Codalan [510]
Codeine Phosphate Injection [287]
Codeine Phosphate Oral Solution [751]
Codeine Phosphate in Tubex [942]
Codimal PH [166]
Colrex Compound Capsules, Compound Elixir [750]
Conex with Codeine [661]
Decongestant Expectorant, -AT (Antitussive) Liquid [769]
Deproist Expectorant w/ Codeine [344]
Empirin with Codeine [139]
Fiorinal w/ Codeine [258]
G-2 Capsules, G-3 Capsules [394]
Guiatuss A-C Syrup [769]
Iophen--C Liquid [769]
Maxigesic Capsules [560]
Naldecon-CX Suspension [134]
Novahistine DH, Expectorant [580]
Nucofed Capsules, Expectorant, Pediatric Expectorant, Syrup [114]
Pediacof [133]
Phenaphen w/ Codeine Capsules, -650 with Codeine Tablets [737]
Phenergan Expectorant w/ Codeine, VC Expectorant w/ Codeine [942]
Poly-Histine Expectorant with Codeine [124]
Robitussin A-C, -DAC [737]
Ru-Tuss Expectorant [127]
Ryna-C Liquid, -CX Liquid [911a]
SK—APAP with Codeine Tablets [805]
Soma Compound w/ Codeine [911a]

Codeine phosphate *(cont'd).*
Stopayne Capsules, Syrup [814]
Triafed-C Expectorant [769]
Triaminic Expectorant w/ Codeine [257]
Tussar SF, -2 [891]
Tussi-Organidin [911a]
Tylenol w/ Codeine Elixir, w/ Codeine Tablets, Capsules [570]

Codeine phosphate + aspirin, 30 mg:325 mg.
Anexsia® with Codeine [114]

Codeine phosphate + pseudophedrine HCI, 20 mg:60 mg.
Nucofed® Capsules, Syrup [114]

Codeine sulfate.
Ambenyl Expectorant [551]
Aspirin w/ Codeine Tablets [344]
Codeine Sulfate Tablets [751]

Colchicine.
ColBENEMID [578]
Colchicine Ampoules [523]
Colchicine Tablets [240]
Colchicine Tablets [523]
Col-Probenecid Tablets [240]
Probenecid w/ Colchicine Tablets [344]
Probenecid with Colchicine Tablets [769]

Colestipol hydrochloride.
Colestid Granules [884]

Colistimethate sodium.
Coly-Mycin M Parenteral [674]

Colistin sulfate.
Coly-Mycin S Oral Suspension, s Otic w/ Neomycin & Hydrocortisone [674]

Collagen.
Lencoll [520]

Collagenase.
Biozyme®-C [80]
Santyl Ointment [495]

Collagen hydrolysates.
Crotein A, C, O [223]

Collagen protein hydrolysate mol weight 1300.
Remanol 1300, 1300/35 [285]

Collagen, soluble
Collasol [223]
Pancogene S [339]
Secol BA1 [453]
Sollagen [424]

Collectotrichum gloeosporioides.
Collego™ [884]

Copper.
Added Protection III Multi-Vitamin & Multi-Mineral Supplement [589]

Copper arsenic, chromated.
Oxcel [591]

Copper(II) chloride.
Ultralog 23-5500-00 [172]

Copper chromite.
Cu-0203 T 1/8″, -0211 P, -0223 P, -0233 T 1/8″, -1413 P, -1422 T 1/8″, -1800 P, -1803 P, -1808 T 1/8″, -1809 T 1/8″, -3829 P [389]

Copper chromite catalyst.
Marchon C21 [23]

Copper dihydrazinium sulfate.
Omazene [659]

Copper dimethyl dithiocarbamate.
Akrochem Accelerator CUDD [15]
CDMC [402]
Cumate [894]

Copper and manganese, 87:13%, resistance alloy.
Manganin [927]

Copper naphthenate.
Troysan Copper 8 [869]

Copper(I) oxide.
Fungi-Rhap CU-75 [220]

Copper(II) oxide.
Ultralog 23-5800-00 [172]

Copper, oxygen-free.
OFHC [58]

Copper phthalocyanine.
Atlas A5712 Monofast Blue Toner, C1828 Monofast Blue Toner [496]
Monastral Blue C BT-443-D [269]

Copper powder.
Natural Copper #200 [91]

Copper 8 quinolinolate.
Isotrol CQ-8, CQ-WR [315]

Copper sulfate.
Hemo-Vite [267]
Ultralog 23-7060-00 [172]
Vio-Bec Forte [750]

Corn starch.
Melogel [619]
Staysize 109 [818]
Stayzyme [818]

Corn syrup, hydrogenated.
Polyol 3070, 7000 [529]

Cortex rhamni frangulae.
Movicol [649]

Corticotropin.
Acthar® [80]

Corticotropin, repository.
H.P. Acthar® Gel [80]

Cottonseed glyceride.
Myverol 18-85K Distilled Monoglycerides [278]

Cottonseed glyceride, hydrogenated.
Myverol 18-07 Distilled Monoglycerides, 18-07K Distilled Monoglycerides [278]

Cottonseed oil, hydrogenated.
Duratex [272]
Durkee 07 [272]

Cottonseed oil monoglyceride, unsaturated.
Dimodan CP [373]

Crataegus extract.
Midland Hawthorn HS [20]

Cresol, butylated, styrenated.
Wingstay V [362]

Cromolyn sodium.
Intal, Intal Nebulizer Solution [324]
Nasalcrom Nasal Solution [324]

Cryoflex plasticizer.
SR-660 [766]

Cryptenamine preparations.
Diutensen Tablets [911a]

(-)-A-Cubebene, (-)-4-methyl-3-heptanol, (-)-A-multistriatin.
N-Trap Elm Bark Beetle [21]

Cumene sulphonic acid.
Eltesol CA65, CA96 [25]
Sulframin CSA [935]

Cumylperoxy neo-decanoate.
Lupersol 188M75 [536]

Cumylperoxy pivalate.
Lupersol 47M75 [536]

Cupric chloride.
Coppertrace [80]

Cuprous oxide.
Copper-Sandoz [762]

1-Cyano-1-(t-butylazo) cyclohexane.
Luazo 96 [536]

Cyanocobalamin.
Al-Vite [267]
BayBee-12 [111]
B-C-Bid Capsules [350]
Berubigen® [884]
Chromagen Capsules [767]
Cyanocobalamin (Vit. B_{12}) Injection [287]
Cyanocobalamin No. 69932 [738]
Cyanocobalamin in Tubex [942]
Eldertonic [563]
Hemocyte Injection [888]
Hemo-Vite [267]
I.L.X. B_{12} Elixir Crystalline, Tablets [489]
Neuro B-12 Forte Injectable, Injectable [505]
Niferex-150 Forte Capsules [166]
Nu-Iron-V Tablets [563]
Tia-Doce Injectable Monovial [105]
Trinsicon/Trinsicon M Capsules [354]
Vicon Forte Capsules [354]

Cyanoethyl acrylate.
Reactomer RC®-100 [608]

Cyano (4-fluro-3-phenoxyphenyl) methyl-3-(2,2-dichloroethenyl)-2,2-dimethyl-cyclo-propane carboxylate.
Baythroid [112]

A-(Cyanomethoximiar)-benzacetonitrile.
Concep [190]

(s)-A-Cyano-m-phenoxybenzl (1R, 3R)-3(2,2-dibromovinyl)-2,2 dimethylcyclopropane-carboxylate.
Decis [749]

(±)-Cyano (3-phlnoxyphenyl) methyl (+)-4-(difluoromethoxy) -A-(1-methylethyl) benzeneacetate.
Pay-Off [50]

O-P-Cyanophenyl 0-ethyl phenyl-phosphorothioate.
Surecide [832]

Cyclacillin.
Cyclapen-W [942]

Cyclandelate.
Cyclandelate Capsules [344]
Cyclandelate Capsules [769]
Cyclospasmol [463]
Cyclospasmol Capsules [510]

Cyclizine preparations.
Marezine [139]

Cyclobenzaprine hydrochloride.
Flexeril Tablets [578]

Cyclohexanediamine tetraacetic acid.
Chel CD [190]

Cyclohexanedimethanol diglycidyl ether.
Heloxy MK-107 [931]

Cyclohexanol with methanol 2.25%.
Hexalin [270]

Cycloheximide.
Acti-Dione™ [884]

Cyclohexyl acrylate.
SR-220 [766]

Cyclohexylamine (99% purity).
Cyclohexylamine (C.H.A.) [906]

Cyclohexyl amine acetate.
Coagulant CHA [112]

2-(Cyclohexylamino)ethanesulfonic acid.
Chemalog 24-9450-00 [172]

n-Cyclohexyl-2-benzothiazole sulfenamide.
Akrochem C.B.T.S. [15]
CBTS [15]
Delac S [881]
Durax [362], [894]
Pennac CBS-O [682]
Royal CBTS [752]
Santocure [392], [602]
Sulfene I [708]
Vulcafor CBS [G tinted) [212], [910]
Vulkacit CZ/MG/C [597]

Cyclohexyl ethyl amine.
Vulkacit HX [112]

Cyclohexyl methacrylate.
SR-208 [766]

N-(Cyclohexylthio) phthalimide.
Santogard PVI Prevulcanization Inhibitor [602]

Cyclomethicone.
Dow Corning 344 Fluid, 345 Fluid [262]
Masil SF-V [564]
SF-1202 Silicone Fluid [341]
Silicone 344 Fluid, 345 Fluid [262]
Silicone SF-1173 [341]
Siloxane F-222, F-223, F-250, F-251, SWS-03314, SWS-03400 [814]
Volatile Silicone 7158, 7207, 7349 [878]

Cyclopentamine hydrochloride.
Clopane Hydrochloride[100] [286]

1,2,3,4-Cyclopentanecarboxylic acid.
CPTA [882a]

1,2,3,4-Cyclopentanetetracarboxylic acid, dianhydride.
CPDA [882a]

Cyclopentolate HCI.
Ocu-Pentolate [655a]

Cyclophosphamide.
Cytoxan [572]
Neosar for Injection [7]

Cyclopropanecarboxylate.
Li-Ban Spray [690]

A-Cyclopropyl-A-(p-methoxypheny)-5-pyrimidine methanol.
A-Rest [282]

Cycloserine.
Seromycin [523]

Cyclothiazide.
Anydron [523]
Fluidil [7]

o-Cymen-5-Ol.
Biosol [667]

Cyproheptadine hydrochloride.
Cyproheptadine HCI Syrup & Tablets [769]
Cyproheptadine HCI Tablets [240]
Cyproheptadine HCI Tablets [344]
Periactin Syrup, Tablets [578]

Cytarabine.
Cytosar-U® [884]

Cytosine arabinoside.
Cytosar [884]

1-(1'-Cytosinyl)-4-ξL-3' amino-5'-(1"-n-methyl-guanidino)-valerylaminoξ-1,2,3,4-tetradeoxy-B-D-erythro-hex-2-ene-uronic acid.
BLA-S [476]

Danthron.
Modane Tablets [7]

DCNA.
Botran® [884]

DEA-cocoamphocarboxypropionate.
Cycloteric CD Conc. [234]
Miranol C2M-DEA 60% [592]

DEA-laureth sulfate.
Rewopol DL3 [725]

DEA-lauryl sulfate.
Cycloryl DA [234]
Maprofix DLS-35 [663]
Maprolyte 101 [663]
Melanol LP 20 D [785]
Rewopol DLS [725]
Richonol DLS [732]
Sipon LD [29]
Standapol DEA [400]
Stepanol DEA [825]
Sterling WADE [145]
Texapon DEA [399]

DEA-linoleate.
Linamine HC-2 [494]

DEA-methoxycinnamate.
Parsol Hydro [353]

Decabromodiphenyl oxide.
DE-83R [369]
FR-300-BA [261]
Saytex 102, 102E [768]

Decanoyl peroxide.
Decanox, -F [536]

Decenylsuccinic acid.
Leafseal [431]

Deceth-4 phosphate.
Gafac RA-600 [338]

Deceyl diphenyl ether disulfonate, sodium salt.
Conco Sulfate 3B2 [209]

n-Decyl alcohol.
Kalcohl 10H [478]

Decyl alcohol alkoxylate.
Iconol PD-8-90% [109]

Decyl alcohol (EO 4), ethoxylated.
Iconol DA-4 [109]

Decyl alcohol (EO 5.5), ethoxylated.
Iconol DA-5.5 [109]

Decyl alcohol (EO 6), ethoxylated.
Iconol DA-6, DA-6-90% [109]

Decyl alcohol (EO 9), ethoxylated.
Iconol DA-9 [109]

Decyl alcohol, ethoxylated, sodium sulfosuccinate ester.
Alkasurf SS-DA-3 [34]

n-Decylamine.
Amine 10D [487]
Lilamin 10D [487]

Decyl betaine.
Lonzaine 10S [529]

Decyl chloride.
Barchlor 10S [529]

n-Decyl dephenyl oxide disulfonic acid.
Conco Sulfate 3B2 [209]

Decyl dimethyl amine.
Barlene 10S [529]

Decyl dimethyl amine oxide.
Barlox 10S [529]
Conco XA-C [209]

Decyl diphenyl oxide disulfonic acid.
Conco Sulfate 3B2 Acid [209]

n-Decyl ether, ethoxylated.
Nikkol BD-1SY thru BD-8SY [631]

Decyl isostearate.
Schercemol DEIS [771]

Decyl oleate.
Cetiol V [399], [400]
Crodamol DO [223]
Schercemol DO [771]
Standamul CTV [400]

Decyloxypoly-(ethyleneoxy)-ethanol.
Emulphogene DA-530, DA-630, DA-639 [338]

Decyloxypropylamine.
Tomah PA-14 [866]

N-Decyloxypropyl 1,3-diaminopropane.
Tomah DA-14 [866]

Decyloxypropyl dihydroxyethyl methyl ammonium chloride.
Tomah Q-14-2 [866]

Decyltetradecanol.
Michel XO-147 [581]
NJCOL-240A [625]

Decyltetradeceth-30.
Nikkol BEG-2430 [631]

Decaglycerin decaisostearate.
Nikkol Decaglyn 10-IS [631]

Decaglycerin decaoleate.
Nikkol Decaglyn 10-O [631]

Decaglycerin decastearate.
Nikkol Decaglyn 10-S [631]

Decaglycerin heptaisostearate.
Nikkol Decaglyn 7-IS [631]

Decaglycerin heptaoleate.
Nikkol Decaglyn 7-O [631]

Decaglycerin heptastearate.
Nikkol Decaglyn 7-S [631]

Decaglycerin monoisostearate.
Nikkol Decaglyn 1-IS [631]

Decaglycerin monolaurate.
Nikkol Decaglyn 1-L [631]

Decaglycerin monomyristate.
Nikkol Decaglyn 1—M [631]

Decaglycerin monooleate.
Nikkol Decaglyn 1-O [631]

Decaglycerin monostearate.
Nikkol Decaglyn 1-S [631]

Decaglycerin pentaisostearate.
Nikkol Decaglyn 5-IS [631]

Decaglycerin pentaoleate.
Nikkol Decaglyn 5-O [631]

Decaglycerin pentastearate.
Nikkol Decaglyn 5-S [631]

Decaglycerol decaoleate.
Aldo DGDO [358]
Caprol 10G100 [829]
Drewmulse 10-10-O [715]
Polyaldo DGDO [358]

Decaglycerol decastearate.
Drewmulse 10-10-S [715]
Drewpol 10-10-S [715]

Decaglycerol decostearate.
Caprol JB [829]

Decaglycerol dioleate.
Caprol 10G20 [829]

Decaglycerol mono-dioleate.
Caprol PGE860 [829]

Decaglycerol octaoleate.
Drewmulse 10-8-O [715]

Decaglycerol tetraoleate.
Caprol 10G40 [829]
Drewmulse 10-4-O [715]
Hodag SVO-1047 [415]
Mazol PGO-104 [564]

Decahydronaphthalene.
Decalin [270]

Decamethyl cyclopentasiloxane.
Abil B8839 [856]

DEG dioleate.
Alkamuls DEG-DO [33]

DEG ditallate.
Alkamuls DEG-DT [33], [34]

DEG monolaurate.
Alkamuls DEG-ML [33], [34]

DEG monooleate.
Alkamuls DEG-MO [33]

DEG monostearate.
Alkamuls DEG-MS [33]

Dehydroabietic acid.
Galex G-75 [392], [620]
Galex NXD [392]
Galex W-100 [620]
Galex W-100-D [392]

Dehydroabietyl amine 8-hydroxyquinolinium 2-ethylhexoate.
Cunimene [777]

Dehydroacetic acid.
DHA [261]

Dehydrocholic acid.
Decholin [63]

Demethylchlortetracycline hydrochloride.
Declomycin [51]

Demeton.
Systox [184]

Desmopressin acetate.
Stimate™ [80]

Dexamethason sodium phosphate.
Ocu-Dex [655a]

Dextran.
Sephadex G-25 [693]

Dextranase, fungal.
Dextranase Novo 25 L [654]

Dextrin.
Nadex 360 [619]

Dextroamphetamine sulfate.
Dexedrine [804a]

Dextrose, refined.
Cerelose [220a]

Diacetyl tartaric acid monoglyceride.
Datamuls 43, 47, 4720, 4741 [856]

Diacetyl tartaric monoglycerides.
Radiamuls DATA, DATA 2003, DATA 2004, DATA 2008 [658]

Dialkyl adipate.
Santicizer 97 [602]

Di alkylamine C_{13}.
AZamine S13B [554]

Di-n-alkyl amines.
Farmin D-86 [478a], [803]

Dialkyl aminoethyl stearate.
Cerasynt 303 [895]

Dialkyl diamido amine acetate.
Empigen FKC75K, FKC75KY, FKC100K [23]

Dialkyl diamido amine lactate.
Empigen FKH75L, FKH75LY, FKH100L [23]

Dialkyl diether glutarate.
Plasthall 7050 [379]

Dialkyl dimethyl ammonium chloride.
Adogen 432, 432CG [796]
Bardac 2050, 2080 [529]
Genamin DSAC [416]
Quartamin D 86P [478]
Radiaquat 6443, 6470 [658]

Dialkyl dimethyl benzyl, ammonium chloride.
BTC 812 [663]

Dialkyl dimethyl methosulfate quaternary compound.
Alkaquat DHTS [33]

α,ω, Dialkyl-dimethylpolysiloxane.
Abil ZP 2434 [856]

Dialkyl imidazolinium methosulfate.
Ammonyx 4080 [663]

Dialkyimidazolinium methosulfate, hydrogenated, quaternary.
Rewoquat W 7500 H [725]

Dialkyl imidazolinium quaternary.
Alkaquat S [33]
Alkaquat ST [33], [34]

Di-n-alkyl-methyl amines.
Farmin M2C, M2R86 [803]

Dialkyl naphthalene sulfonates.
Leonil DB, DB Powder [416]

Di-alkyl phenol ethoxylate.
Alkasurf DNP-10 [33]

Dialkylphenoxypoly-(ethyleneoxy) ethanol.
Igepal DM-430, DM-530, DM-710, DM-730, DM-880, DM-970 [338]

Dialkyl phthalate.
Palatinol FF21, FF31, FF41 [99]
Santicizer 711 [602]

Dialkyl sulfimide.
Leophen RBD [107]

Dialkyl sulfosuccinate.
Airrol CT-1 [864]
Supermontaline SLT [711]

Dialkyl sulfosuccinate polyoxyethylene alkylphosphate, sodium salt.
Pelex OT-P, RP [478]

Dialkyl sulfosuccinate, sodium salt.
Pelex OT-P [478]

n,n-Diallyl-2-chloroacetamide.
Rondox [601]

Diallyldimethylammonium chloride.
Cat-Floc [142]

Diallyl fumarate.
SR-204 [766]

Diallyl phthalate.
Diall [35]
Diallyl Phthalate (DAP) [379]

Diamide, fatty.
Carnapol 77X [724]

Diamines.
Duomeen C, CS, O, T, TDO, TX [253]
Jet Amine D-C, D-O, D-T [471]

Diamine dioleate salts.
Diamin DO, DT, HT, O, S, T [803]

Diamine, ethoxylated fatty.
Ethoduomeen T/13, T/20, T/25 [16]

Diamine fatty acid condensate.
Michelene 10, 15 [581]

Diaminochrysazin.
Boran [506]

(trans)-1,2-Diaminocyclohexane tetraacetic acid monohydrate.
ChelCD [190]

p,p-Diaminodiphenyl-methane.
Tonox [881]

2,6-Diaminopyridine.
Rodol 26PYR [531]

Diammonium N-cocoyl sulfosuccinamate.
Empimin MSS [25]

Diammonium EDTA.
Sequestrene Diammonium [190]

Diammonium ethylene bis dithiocarbamate.
Amobam [318]

Diammonium N-lauryl sulfosuccinamate.
Empimin MSS [23]

Diammonium phosphate.
Di-MoN [852]

2,5-Di(t amyl) hydroquinane.
Santovar A Antioxidant [602]

1,1-Di-(t-amylperoxy) cyclohexane.
Lupersol 431-80B [536]

2,4-Di-t-amylphenol.
Prodox® 156 [315]

Di-90% arachidyl-behenyl secondary amine.
Kemamine S-190 [429]

Dibasic lead phthalate.
Vanstay 90 [894]

Dibehenyldimonium chloride.
Kemamine Q-2802C [429a]

Dibenzo-p-quinone dioxime.
Dibenzo PQD [845]

2.2′-Dibenzothiozyl disulphide.
Accelerator DM [5]
MBTS [881]
Vulcafor MBTS [212], [910]
Vulkacit DM, DM/MGC [597]

Dibenzoyl-p-quinonedioxime.
Dibenzo G-M-F [881]

Dibenzyl azelate.
Plasthall DBZZ [379]

Dibenzyl ether.
Vulkanol BA [112]

Dibenzyl toluene.
Lipinol T [178]

Dibetanaphthyl-phenylenediamine.
Agerite White [894]

Dibromodimethylhydantoin.
Brom 55 [135]
Bromine™, Bromine Chloride™ [369]
Bromoacetic Acid™ [369]
Bromobenzene™ [369]
Dibromantin [68]

Dibromoethylbenzene.
Alkazene 42 [261]

Dibromoethyldibromocyclohexane.
Saytex BCL-462 [768]

3,5-Dibromo-4-hydrorybenzaldehyde-0-(2′,4′-dinitrophenyl) oxime.
Faneron [190]

Dibromoneopentyl glycol.
Emery 9336 [289]
FR-1138 [261]

2,2-Dibromo-3-nitrilo-propionamide.
Dow Antimicrobial 7287, 8536 [261]

Dibromophenol.
Emery 9331 [289]

2,3 Dibromopropanol.
DBP [369]

Di(butoxy-ethoxy-ethyl)adipate.
Natro-Flex BCA [392], [620]
PlastHall DBEEA [379]
SR-650 [766]
TP-95 Plasticizer [857]

Dibutoxy ethoxy ethyl forinate.
SR-660 [766]

Di(butoxy-ethoxy-ethyl)formal.
TP-90B Plasticizer [857]

Dibutoxyethoxyethyl glutarate.
PlastHall DBEEG [379]

Dibutoxyethyl adipate.
PlastHall DBEA [379]
Staflex DBEA [722]

Dibutoxyethyl azelate.
PlastHall DBEZ P [379]

Dibutoxyethyl glutarate.
PlastHall DBEG [379]

Dibutoxyethyl phthalate.
PlastHall DBEP [379]

Dibutoxyethyl sebacate.
PlastHall DBES [379]

Di-n-butyl adipate.
Cetiol B [401]
Rilanit DBA [401]

Dibutylamine pyrophosphate.
D.P. Solution [602a]

Dibutylammonium oleate.
Barak [270]

2,6-Di-t-butyl-p-cresol.
BHT (Food Grade) [12], [497]
CAO-1, -3 [796]
Dalpac [406]
DBPC [497]
Deenax [296]
Naugard BHT [881]
Vanlube PC [894]
Vulkanox KB [597]

2,6-Di-t-butyl- α -dimethylamino-p-cresol.
Ethanox 703 [302]

Di-t-butyl diperoxyazelate.
Lupersol 99 [536]

Di-t-butyl diperoxyphthalate.
Lupersol KDB [536]

Dibutyldithiocarbamic acid, N,N-dimethylcycolexylamine salt.
RZ-100 [392], [602]

Dibutyl fumarate.
Staflex DBF [722]

2,5-Di-t-butylhydroquinone.
2,5-Di-t-butylhydroquinone [278]

2-(3′,5′-Di-t-butyl-2′-hydroxyphenyl)-5-chlorobenzotriazole.
Tinuvin 327 [190]

Dibutyl maleate.
Staflex DBM [722]

Dibutyl methylene bis-thioglycolate.
Vulkanol 88 [112]

2,6-Di-t-butyl-4-methylphenol.
Ionol [791]

Di-t-butyl peroxide.
Trigonox B [16], [653]

2,2-Di (t-butylperoxy) butane.
Lupersol 220-D50 [536]

1,1-Di (t-butylperoxy) cyclohexane.
Lupersol 331-80B [536]

1,1-Di (t-butylperoxy) 3,3,5-trimethyl cyclohexane.
Lupersol 231, 231-XL [536]
Trigonox® 29/40, 29-B75 [16], [653]

2,6-Di-t-butylphenol.
Ethanox 701 [302]

2,4-Di-t-butylphenyl 3,5-di-t-butyl-4-hydroxybenzoate.
UV-Chek AM-340 [315]

Di (2,4-di-t-butylphenyl) pentaerythritol diphosphite.
Mark 5070 [934]

Dibutyl phthalate.
Hatcol DBP [393]
Kodaflex DBP [278]
Palatinol DBP [99]
(Polycizer) DBP [392]
Staflex DBP [722]
Unimate DBP [876]
Unimoll DB [112]
Vestinol C [178]

Dibutyl sebacate.
Hallco DBS [379]
Hatcol DBS [393]
(Polycizer) DBS [392]
Rilanit DBS [401]
Uniflex DBS [876]
Unimate DBS [876]

2,6-Di-t-butyl-4-sec-butylphenol.
Isonox 132 [770]
Vanox 1320 [894]

1.3 Dibutyl thiourea.
DBTU [708], [833]
Thiate U [894]

Dibutyl tin bis iso-octyl thioglycolate.
Therm-Chek 840 [315]

Dibutyltin diacetate.
Fomrez SUL-3 [935]

Dibutyl tin dilaurate.
Cata-Chek 820 [315]
Fomrez SUL-4 [935]

Dibutyltin maleate.
Advastab T-340 [192]

Dibutyl tin maleate ester.
Therm-Chek 837 [315]

2,6-Di-t-butyl-p-tolyl methylcarbamate.
Azak [406]

Di-t-butyl dicarbonate.
Sequalog 99% 29-0870-00 [172]

Dibutyl xanthogen disulfide.
C-P-B [881]

Dicalcium phosphate.
Phos-Feed [46]

Dicapryl adipate.
Uniflex DCA [876]
Unimate 600 [876]

Dicapryloyl cystine.
Lipacide C8CY [730]

Dicapryl phthalate.
Uniflex DCP [876]

N-(1,2-Dicarboxyethyl) N-alkyl (C_{18}) sulfosuccinamate.
Alconate 2CSA [29]

Dicarboxylate, amphoteric.
Cycloteric DC-SF [234]

Dicarboxyl coco imidazoline compound.
Antaron MC-44 [338]

Dicarboxylic coco imidazoline, sodium salt.
Antaron MC-44 [338]

Dicarboxylic acid, C21.
Diacid 1550 [918]

Dicarboxylic caprylic.
Sochamine A 8955 [937a], [938]

Dicarboxylic organic acid (C_{21}) 6-carboxy-4-hexyl-2-cyclohexenel-octanoic acid.
Westvaco Di Acid [917a]

Dicatechol borate, di-o-tolylguanidine salt.
Permalux [270]

Dicetyl peroxydicarbonate.
Liladox [486]

s-(2,3-Dichloroallyl) diisopropylthiocarbamate.
Avadex [601]

o-Dichlorobenzene.
Dizene [705]
Ozene [35]

3,3′-Dichlorobenzidine⇒o-acetoacetanisidide.
Atlas B3503 Diarylide Yellow Toner [496]

3,3′-Dichlorobenzidine]o-acetoacetotoluidide.
Atlas A9744 Diarylide Yellow Toner [496]

3,3′-Dichlorobenzidine⇒3-methyl-1-phenyl-5-pyrazolone.
Atlas A8868 Diarylide Orange Toner [496]

3,3′-Dichlorobenzindine⇒acetoacetanilide.
Atlas A9145 Diarylide Yellow Toner [496]

2,4-Dichlorobenzoyl peroxide.
Cadox TDP [16], [653]
Luperco CST [536]

1,1-Dichloro-2,2-bis(p-ethylphenyl)-ethane.
Perthane [742]

4,6-Dichloro-n-(2-chlorophenyl) 1,3,5-triazin-2-amine.
Dyrene [597]

Dichlorodifluromethane.
Isotron 12 [682]

Dichlorodimethylhydantoin.
Halane [109]

3,5-Dichloro (n-1,1-dimethyl-2-propynl) benzamide.
Kerb [742]

2,2′-Dichloroethyl ether.
Chlorex [878]

n′-Dichlorofluoromethylthio-n,n-dimethyl-n′-phenylsulfamide.
Euparen [112]

n′-Dichlorofluoromethylthio-n,n-dimethyl-n′-(4-totyl) sulfamide.
Euparen M [112]

2,3-Dichloro-n-(4-fluorophenyl) maleimide.
Spartcide [500]

2,3-Dichloro-1,4-napthoquinone.
Quintar 540F [423]

Dichlorophene.
G-4 Pure, G-4 Technical [353]

2,4-Dichlorophenoxybutyric acid.
Butoxone [186]

2-(2,4-Dichlorophenoxy)propionic acid,2-ethylhexyl ester.
Propi-Rhap [881]

O-[2,4-Dichlorophenyl]-O-methyl isopropylphosphoramidothioate.
Zytron [261]

2,4-Dichlorophenyl-4-nitrophenyl ether product.
Tok [742]

1-ξ2-(2,4-Dichlorophenyl) 4-propyl-1,3-dioxolan-2-ylmethyl ξ 1H-1,2,4-triazole.
Tilt [190]

3,6-Dichloro-2-pyridinecarboxylic acid.
Dowco 290 [261]

Dichlorotetrafluoroethane.
Isotron 114 [682]

Dichloro-m-xylenol.
Ottacide 4 [669]

N,N′-Dicinnanylidene-1,6-hexanediamine.
Diak No. 3 [269]

Dicloxacillin sodium.
Dycill® Capsules, 250 mg, 500 mg [114]

Dicocoamine.
Armeen 2C [79]
Noram 2C [162]

Dicoco dimethyl ammonium chloride.
Accoquat 2C-75, 2C-75—H [81]
Adogen 462 [796]
Arquad 2C-75 [79]
Jet Quat 2C-75 [471]
M-Quat 2475 [564]
Noramium M2C [162]
Quartamin DCP [803]

Dicoco imidazolinium quaternary.
Alkaquat C [33], [34]

N-Dicoco methylamine.
Noram M2C [162]

Dicoconut amine, secondary.
Kemamine S-650 [429a]

Dicoconut oil dimethyl ammonium chloride.
Radiaquat 6462 [658]

Dicoconut oil dimethyl quaternary ammonium chloride.
Radiaquat 6462 [658]

Dicumyl peroxide.
Di-Cup®, 40 KE, R, T [406]
Luperco 500-40C, 500-40KE [536]
Luperox 500R, 500T [536]
Percadox BC [653]
Peroximon DC [604]
Varox DCP-R, DCP-T [894]

Dicumyl peroxide on calcium carbonate.
Di-Cup 40C [406]
Varox DCP-40C [894]

Dicumyl peroxide on KE clay.
Varox DCP-40KE [894]

Dicyandiamide, accelerated.
D.E.H. 40 [261]

Dicyclohexylamine.
Sequalog Grade 30-8180-00 [172]

N,N, Dicyclohexyl-2-benzothiazol sulfenamide.
Sulfene 3 [708]
Vulcafor DCBS [212], [910]
Vulkacit DZ/C [597]

N-N′-Dicyclohexylcarbodiimide.
Sequalog Grade 30-8380-10 [172]

Dicyclohexyl phthalate.
Unimoll 66 M [597]

Dicyclohexyl sodium sulfosuccinate.
Alconate 2CH [29]

Dicyclomine hydrochloride.
Bentyl Tablets and Capsules [510]

Didecyl dimethyl ammonium chloride.
Arquad 2-10/50 [16]
Bardac 2250, 2280 [529]
Bio-Dac 50-22 [120]
BTC 1010, BTCO 1010 [663]
Querton 210CL50 [487]

Didecyl hydrogen phosphite.
Mark DDHP [75]

Didecyl glutarate.
PlastHall DDG P [379]

Didecylmethylamine.
Dama™ 10 [302]

Didecyl mono phenyl phosphite.
Mark DDMPP [75]

Di-eicosanyl-docosanyl secondary amine.
Kemamine S-190 [429a]

Diester sodium dioctyl sulfosuccinate.
Emcol 4500 [935]

Diethanolamide.
Serdolamide PPFG 68 [788]

Diethanolamide, fatty.
Orapol HC [711]
Seriken 90 [390]

Diethanolamide, fatty acid.
Dianol, Dianol 300 [237]
Ethylan GD [252], [509]
Rewomid F [725]
Schercomul H, K [771]

Diethanolamidooleamide DEA.
Witcamide 5168 [935]

Diethanolamine coco superamide.
Emid® 6514, 6515 [290]

Diethanolamine dodecylbenzene sulfonate.
Marlopon ADS 50 [178]

Diethanolamine lauric superamide.
Emid® 6511 [290]

Diethanolamine lauryl ether sulfate.
Cedepal SD-409 [256]

Diethanolamine lauryl sulfate.
Alkasurf DLS, WADX [33]
Carsonol DLS [155], [529]
Conco Sulfate EP [209]
Condanol DLS 35 [274]
Cycloryl DA [234]
Duponol EP [269]
Empicol 0031/T [23]
Empicol DA [23], [24]
Empicol DLS [25], [26]
Maprofix DLS 35 [663]
Nutrapon DE 3796 [198]
Polystep B-8 [825]
Sipon LD [29]
Standapol DEA [399], [400]
Stepanol DEA [825]
Sterling WADE [145]
Sulfotex WAD [399], [400]
Texapon DEA [399]

N,N-Diethanol lauramide.
Norfox DLSA [650]

2-(Diethoxyphosphinylimino)-1,3-dithiolane.
Cyolane [50]

2-(Diethoxyphosyphinylimino)-4-methyl-1,3-dithiolane.
Cytrolane [50]

p-Diethyl-aminoazobenzene.
Oil Yellow NB [270]

Diethylaminoethyl acrylate.
Mon-Arc® A-1 [69]

Diethylaminoethyl methacrylate.
Mon-Arc® MA-2 [69]

2-Diethylamino-6-methyl-4-pyrimidinyl diethyl phosphorothioate.
Primicid [444]

O-(2-Diethylamino-6-methylpyrimidin-4-yl) 0,0-dimethyl phosphorothioate.
Actellic [444]

N,N-Diethyl-m-aminophenol.
Rodol DEMAP [531]

N,N-Diethyl-m-aminophenol sulfate.
Rodol DEMAPS [531]

O,O-Diethyl 2,3-dihydro-3-oxo-2-phenyl-6-pyridazinyl phosphorothioate.
Ofunack [595]

Diethyldithiocarbamate.
Nicon [506]

Diethylene glycol adipate.
Hi-Eff [67]

Diethylene glycol diacrylate.
SR-230 [766]

Diethylene glycol dibenzoate.
Benzoflex 2-45 [898]

Diethylene glycol dibenzoate/dipropylene glycol dibenzoate (50/50 wt. blend).
Benzoflex 50 [898]

Diethylene glycol dibutyl ether.
Butyl Diglyme [367]
Dibutyl Carbitol [878]

Diethylene glycol diethyl ether.
Diethyl Carbitol [878]
Ethyl Diglyme [367]

Diethylene glycol dilaurate.
Cithrol DGDL N/E, DGDL S/E [223]

Diethylene glycol dilaurate *(cont'd).*
CPH-70-N [379]

Diethylene glycol dimethacrylate.
SR-231 [766]

Diethylene glycol dimethyl ether.
Diglyme [367]

Diethylene glycol dioleate.
Cithrol DGDO N/E, DGDO S/E [223]
CPH-84-N, CPH-344-N [379]

Diethylene glycol distearate.
Cithrol DGDS N/E, DGDS S/E [223]
CPH-200-N [379]

Diethylene glycol dithiocyanate.
Lethane A-70 [742]

Diethylene glycol dodecyl ether.
Nikkol BL-2SY [631]

Diethylene glycol laurate.
Pegosperse® 100L [358]

Diethylene glycol monobutyl ether.
Butyl Carbitol [878]
Butyl Dioxitol [791]
Ektasolve DB Solvent [278]

Diethylene glycol monobutyl ether acetate.
Butyl Carbitol Acetate [878]
Ektasolve DB Acetate [278]

Diethylene glycol mono-distearate.
Diglykolstearat [856]
Tegin A 422 [856]

Diethylene glycol monoethyl ether.
Dioxitol, Dioxitol-High Gravity, -Low Gravity [791]
Ektasolve DE [278]

Diethylene glycol monohexyl ether.
n-Hexyl Carbitol [878]

Diethylene glycol monolaurate.
Cithrol DGML N/E, DGML S/E [223]
CPH-4-SE, CPH-28-N, CPH-364-N [379]
Hodag DGL [415]
Mapeg DGLD [564]
Nopalcol I-L [252]
Pegosperse® 100 L, 100 ML [358]
Radiasurf 7420, 7421 [658]

Diethylene glycol monomethyl ether.
Ektasolve DM [278]
Methyl Carbitol [878]
Methyl Dioxitol [791]

Diethylene glycol monooleate.
Cithrol DGMO N/E, DGMO S/E [223]
CPH-1-SPEC.-SE, CPH-62-N, CPH-224-N [379]
DMO-33 [397]
Hodag DGO [415]
Radiasurf 7400 [658]

Diethylene glycol monopropyl ether.
Ektasolve DP [278]

Diethylene glycol monoricinoleate.
Pegosperse 100 MR [358]

Diethylene glycol monostearate.
Cithrol DGMS N/E, DGMS S/E [223]
Clindrol SDG [197]
CPH-104-DC-SE, 104-DG-SE, 130-N [379]
DMS-33 [397]
Drewmulse DGMS [266]
Hodag DGS [415]
Kessco-Diethylen-glykol-monostearat [16]
Kessco Glycol Esters [825]
Nopalcol 1-S [252]
Radiasurf 7410, 7411 [658]
Schercemol DEGMS [771]

Diethylene glycol monotallowate.
Nopalcol 1-TW [252]

Diethylene glycol oleate.
Pegosperse® 100 O [358]
Witconol DOS [934], [935]

Diethylene glycol stearate.
Hydrine [339]
Pegosperse® 100 S [358]
Witconol™ CAD [935], [938]

Diethylenetriamine.
D.E.H. 20 [261]

Diethylenetriamine dinonylnaphthalene sulfonate in light min. oil.
Na-Sul DTA [894]

Diethylenetriamine penta-acetate, pentasodium salt.
Chelon 80 [821]

Diethylenetriamine-pentaacetic acid.
Hamp-ex® Acid [364b]

Diethylene-triamine-pentaacetic acid, pentasodium salt.
Hamp-ex® 80 [364b]
Trilon C Liquid [106]
Versenex 80 [261]

Diethylene triamine penta (methylene phosphonic acid).
Dequest 2060 [602]

Di-2-ethylhexyl adipate.
Diocytl Adipate (DOA) [379]
Hatcol DOA [393]
Monoplex DOA [379]
Palatinol DOA [99]
PlastHall DOA [379]
(Polycizer) DOA [392]
Vestinol OA [178]

Di-2-ethylhexyl adipate *(cont'd).*
Wickenol 158 [926]

Di-2-ethyl-hexyl azelate.
Staflex DO2 [722]

Di-2-ethyl hexyl fumarate.
Staflex DOF [722]

Di-2-ethyl hexyl maleate.
Staflex DOM [722]

Di (2-ethylhexyl) peroxydicarbonate.
EHP [705]
Lupersol 223, 223-M75, 223-M [536]

Di 2-ethylhexyl phthalate.
Hatcol DOP [393]
Kodaflex DOP [278]
Palatinol DOP [99]
(Polycizer) DOP [392]
Staflex DOP [722]
Vestinol AH Spezial [178]

Di-2-ethylhexyl sebacate.
Hatcol DOS [393]
Rilanit DEHS [401]

Di (2-ethylhexyl) sodium sulfosuccinate.
AZdry 40, 70 [554]
Serwet WH 170 [788]

Di (2-ethylhexyl) succinate.
Wickenol 159 [926]

Di-(2-ethylhexyl)terephthalate.
Kodaflex DOTP [278]
Polycizer DOTP [392]

Diethylhydroxylamine.
Pennstop 1866 [682]

o,o-Diethyl o-(2-isopropyl-4-methyl-6-pyrimidinyl) phosphorothiote.
Diazinon [190]

Diethyl mercaptosuccinate, n-trichloromethylthio-4-cyclohexene-1, 2-dicarboximide/o, o-dimethyl dithiophosphate.
EVERShield CM Seed Proctectant [149]

o,o-Diethyl-o-(5-methyl-6-ethoxy-carbonylpyrazolo-ξ1.5.-A)-pyrimid-2-yl)-thionophosphate.
Afugan [416]

o,o-Diethyl o-ξ4-(methylsulfiny) phenyl ξ phosphorothioate.
Dasanid [112]

Diethyl palmitoyl aspartate.
PAP [481]

o,o-Diethyl-o-(5-phenyl-3-isoxazolyl) phosphorothioate.
Karphos [763]

Diethyl phthalate.
Deps [482]
Kodaflex DEP [278]

N,N'Diethylthiourea.
Akrochem® DETU Accelerator [15]
DETU [15], [833]
Pennzone E [682]
Thiate H [894]

N,N-Diethyl-m-toluamide.
Metadelphene [406]
MGK® Diethyltoluamide [568]

Diglycerine monostearate.
Nikkol DGMS [631]

Diglycerine sesquioleate.
Hostacerin DGO [55]

Di(glycerol)borate polyoxyethylene.
Emulbon T-80 [864]

Di(glycerol)borate polyoxyethylene monolaurate.
Emulbon T-20 [864]

Di(glycerol)borate polyoxyethylene monooleate.
Emulbon T-80 [864]

Di(glycerol)borate sesquioleate.
Emulbon S-83 [864]

Di(glycerol)borate sesquistearate.
Emulbon S-260 [864]

Diglycerol tetrastearate.
Caprol 2G4S [829]

Diglyceryl dioleate.
Nikkol DGDO [631]

Diglyceryl monooleate.
Nikkol DGMO-C [631]

Diglyceryl monostearate.
Nikkol DGMS [631]

Diglyceryl stearate malate.
Sunsoft 601 [847]

Diglycol coconate.
Witconol™ RDC-D [934], [935]

Diglycol laurate.
Lipo Diglycol Laurate [525]
Sole-Onic CDS [415], [840]

Diglycol stearate.
Grocor 5221SE [374]
Kessco Glycol Esters [825]
Mapeg DGS [564]
Witconol CAD [934]

Digoxin tablets.
Lanoxin Tablets [510]

Dihexyl sodium sulfosuccinate.
Monawet MM-80 [599]
Rewopol SBDH 65 [725]

Dihexyl sulfosuccinate, sodium salt.
Astrowet H-80 [28]

Dihydrocholeth-15.
Nikkol DHC-15 [631]

Dihydrocholeth-30.
Nikkol DHC-30 [631]

Dihydrocodeinone.
Dicodid [495]

2,3-Dihydro-2,2-dimethyl-7-benzofurany ξdibutylamino)thioξ methyl carbamate.
Advantage [331a]

Dihydrogenated-tallow dimethyl ammonium chloride.
Adogen 442 [796]
Arquad 2HT-75 [79]

Dihydrogenated tallow dimethyl ammonium methosulfate.
Carsosoft V-90 [155]

Dihydrogenated-tallow dimethyl ammonium methyl sulfate.
Varisoft 137 [796]

Dihydrogenated tallow dimethyl ammonium sulfate.
Carsosoft V-100 [155]

Di-hydrogenated tallow secondary amine.
Kemamine S-970 [429a]

5,6-Dihydro-2-methyl-n-phenyl-1,4-oxathiin-3-carboxamide.
Vitavax [881]

5,6-Dihydro-2-methyl-N-phenyl-1,4-oxathiin-3-carboxamide 4,4-dioxide.
Plantvax [881]

2,5-Dihydroperoxy-2,5-dimethylhexane.
Luperox 2,5-2,5 [536]

1,2-Dihydro-2,2,4 trimethylquinoline, polymerized.
AgeRite MA, Resin D [894]
Flectol Flakes [392]
Flectol® H, H Antioxidant, Pastilles Antioxidant [602]
Naugard Q [881]
Vanlube RD [894]

2,4-Dihydroxybenzophenone.
Syntase 100 [624]
Uvinul 400, D [338]

Dihydroxy-dichloro-diphenyl-methane (fungicide)
Fungicide M [77]

2,2′-Dihydroxy-4,4′-dimethoxybenzophenone.
Uvinol D-49 [338]

Dihydroxyethyl alkoxypropylamine oxide, C12-15.
Varox 185E [796]

Dihydroxyethyl cocamine oxide.
Schercamox CMA [771]

Dihydroxyethyl soyamine dioleate.
Lowenol S-216 [531]

Dihydroxyethyl tallowamine hydrochloride.
Miramine TA 26 [592]

Dihydroxyethyl tallowamine oxide.
Schercamox T-12 [771]

Dihydroxyethyl tallow glycinate.
Mirataine TM [592]

Dihydroxy diphenyl sulfone.
Eltesol TPA [25]

Dihydroxyethyl tallow amine oleate.
Necon 655 [41]

2,2′-Dihydroxy-4-methoxy-benzophenone.
Cyasorb UV 24 Light Absorber [50]

1,8-di(hydroxyphenyl)pentadecane.
Cardanol Bis-Phenol [862]

Diiodomethyl p-tolyl sulfone.
Amical 48 Powder and Dispersion, 50 Powder and Dispersion [3]

Di-isobutyl adipate.
DIBA [223]
Hatcol DIBA [393]
PlastHall DIBA [379]
Radia 7197 [658]

Diisobutyl azelate.
PlastHall DIBZ [379]

Diisobutyl cresoxy ethoxy ethyl dimethyl benzyl ammonium chloride.
Hyamine® 10-x [725]

Diisobutyl phenoxy ethoxy ethyl dimethylbenzyl ammonium chloride.
Benzethonium Chloride NF [725]
Hyamine® 1622 [725]

Diisobutyl sodium sulfosuccinate.
Monawet MB-45 [599]

Diisobutyl sulfosuccinate, sodium salt.
Astrowet B-45 [28]

Diisocetyl adipate.
Schercemol DICA [771]

Diisocyanate.
PDI Product 3997 [696]

Diisocyanate, aromatic.
Desmodur TT [112]

Diisodecyl adipate.
Hatcol DIDA [393]
Monoplex DDA [379]

Diisodecyl adipate *(cont'd).*
PlastHall DIBA, DIDA [379]
Staflex DIDA [722]

Diisodecyl glutarate.
PlastHall DIDG [379]

Diisodecyl pentaerythritol diphosphite.
Weston 600 [129]

Diisodecyl phthalate.
Hatcol DIDP [393]
Palatinol DIDP [99]
Polycizer DIDP [392]
Staflex DIDP [722]
Vestinol DZ [178]

Diisononanoyl peroxide.
Lupersol 219-M75 [536]

Diisononyl adipate.
Adimoll DN [112]
Staflex DINA [722]

Diisononyl maleate.
Staflex DINM [722]

Diisononyl phthalate.
Vestinol N [178]

Diisooctyl adipate.
PlastHall DIOA [379]
Staflex DIOA [722]

Diisooctyl dodecanedioate.
PlastHall DIODD [379]

Diisooctyl maleate.
Staflex DIOM [722]

Diisooctyl octylphenyl phosphite.
Weston 494 [129]

Diisooctyl phosphite.
Weston DOPI [129]

Diisooctyl phthalate.
Staflex DIOP [722]

Diisooctyl sulfosuccinate.
Triumphnetzer ZSN [948]

Di-isopropyl adipate.
Crodamol DA [223]
Lexol DIA [453]
Prodipate [94]
Radia 7194 [658]
Schercemol DIA [771]
Wickenol 116 [926]

Diisopropyl benzothiazole-2-sulfenamide.
Dipac [682]
Santocure IPS, IPS Vulcanization Accelerator [602]
Sulfene 7 [708]

Diisopropyl dimerate.
Schercemol DID [771]

Di-isopropyl 1,3-dithiolane-2-ylidenemalorate.
Fuji-One [629]

N- β-O, O-Diisopropyl dithiophosphoryl ethyl benzene sufonamide.
Betasan [821]

N-N-Diisopropylethylamine.
Sequalog Grade 33-6110-00 [172]

Diisopropyl fluorophosphate.
Floropryl [577]

Diisopropyl peroxydicarbonate.
IPP [705]
Percadox IPP-AT50 [16], [653]

Diisopropyl sebacate.
Prosebate [44]
Schercemol DIS [771]
Unimate DIPS [876]

Diisotridecyl phthalate.
Vestinol TD [178]

Dilaureth-10 phosphate.
Nikkol DLP-10 [631]

Dilauryldimonium chloride.
Kemamine Q-6902C [429a]

Dilauryl phosphite.
Weston DLP [129]

Dilauryl thiodiopropionate.
Argus DLTDP [74]
Carstab DLTDP [192], [857]
Cyanox® LTDP [52]
Evanstab 12 [303]

Dimenhydrinate.
Dramamine Tablets [510]
Nico-Vert [281]

Di-1-p-menthene.
Pinolene [587]

Dimer acid bis (N-amidopropyl, N-N dimethyl, N-benzyl ammonium chloride.
Schercoquat DAB [771]

Dimer acid bis (amido-propyl-N-dimethyl-N-ethyl ammonium ethyl sulfate).
Schercoquat DAS [771]

Dimethicone.
Foamkill 810F, 830F [230]
Masil SF (Series) [564]
Medical Fluid 360 [262]
Protachem Antifoam BM [713]
Rhodorsil Oils 70047 [730]
SF18(350) Silicone Fluid [341]
Silicone 200 Fluid, 225 Fluid, Medical 360 Fluid [262]
Silicone Fluid SF-96 [341]
Silicone Fluid SWS-101 [841]
Silicone L-45 [878]

Dimethicone *(cont'd)*.
Siloxane F-221 [841]
Viscasil [341]

Dimethicone copolyol.
Rhodorsil Oils 70646 [730]
Silicone 190 Surfactant, 193 Surfactant [262]
Silicone Copolymer F-754 [841]
Silicone L-720, L-722, L-7002, L-7500, L-7600 [878]

Dimethiconol.
Masil SF-R [564]
Silicone Fluid F-212 [841]

Dimethisoquin hydrochloride.
Quotane [804a]

Di-p-methoxydiphenylamine, diphenyl-p-phenylenediamine, phenyl-β-naphthylamine, 25%:25%:50%.
Thermoflex A [270]

Dimethoxyethyl phthalate [bis(2-methoxyethyl)phthalate].
Kodaflex DMEP [278]

N-N-dimethylacetamide 99+%.
Ultralog 34-1060-00 [172]

O,S-Dimethyl acetylphosphoramidothinate.
Orthene [183]

N,N,-Dimethyl N-(3-alkylamidopropylamine).
Product P7 [856]

Dimethyl alkyl amines.
Armeen DM12D, DM14D, DM16D, DM18D, DMCD, DMHTD, DMOD, DMSD, DMTD [79]

Dimethyl alkylbenzyl ammonium chloride.
Arquad B-100 [79]
Lutensit K-OC [106]
Variquat 50 MC, 60 LC, 80 MC, LC 80 [796]

N,N-Dimethyl amide.
Hallcomid M8-10 [379]

Dimethylamine sulfonate.
Calimulse DMS [697]

Dimethyl amine, synthetic (C_{13}-C_{15}).
Synprolam 35DM [441]

p-Dimethylaminobenzenediazo sodium sulfonate.
Dexon [184]

2-Dimethylamino-5,-6-dimethylpyrimidin-4-yl dimethylcarbamate.
Pirimor [444]

Dimethylaminoethyl acrylate.
Mon-Arc® EA-1 [69]

Dimethylaminoethyl methacrylate.
Mon-Arc® MA-1 [69]

(3-Dimethylamino-(methyleneimino phenyl)-n-mulylcarbamate hydrochloride.
Carzol [645]

4-(Dimethylamino)-3-methylphenol-methylcarbamate.
Matacil [112]

Dimethylaminomethyl phenols, substituted.
DMP [742]

2-Dimethylamino-2-methyl-1-propanol.
DMAMP, DMAMP-80 [449]

Dimethylaminopropylamine.
AZamine DMAPA-D, DMAPA-T [554]

Dimethylaminopropyl ricinolamide benzyl chloride.
ES-1239 [158]

S,S′-ξ2-(Dimethylamino)trimethylene ξ bis (benzenethiosulfonate.
Bancol [848]

S,S′-ξ2-Dimethylamino) trimethylene ξ bis (thiocarbamate) hydrochloride.
Padan [848]

4-Dimethylamino-3,5-xylyl N-methylcarbamate.
Zectran [261]

Dimethylammonium 2,4-dichlorophenoxyacetate.
Defy [477]

Dimethyl ammonium dimethyldithiocarbamate.
Vulnopol MM [27]

Dimethylammonium hydrogen isophthalate.
Vanax CPA [894]

Dimethyl amyl maleate.
Staflex DMAM [722]

Dimethyl arachidyl-behenyl amine, distilled, 90%.
Kemamine T-1902D [429], [430]

Dimethyl behenamine.
Kemamine T-2802D [429a]

1,3-Dimethyl-3-(2-benzothiazolyl)urea.
Tribunil [112]

2,5-Dimethyl-2,5-bis-(benzoylperoxy)hexane.
Luperox 118 [536]

2,5-Dimethyl-2,5-bis(t-butylperoxy) hexane.
Luperco 101-XL [392]

2,5-Dimethyl-2,5-bis(t-butylperoxy) hexane *(cont'd).*
Lupersol 101 [392], [682]

2,5-Dimethyl-2,5-bis(2-ethylhexanoylperoxy)hexane.
Lupersol 256 [536]

2,4-Dimethyl-6-t-butyl-phenol liquid.
MBX [497]

N-1,3-Dimethylbutyl-N′-phenol-p-phenylenediamine.
Permanox 6PPD (F) [212]

N-(1,3-Dimethylbutyl)-N′-phenyl-p-phenylene diamine.
Akrochem Antioxidant PD-2 [15]
Antozite 67 [894]
Santoflex 13 [392], [602]
Vulkanox 4020 (flaked or liquid) [597]
Wingstay 300 [362]

N,N-Dimethyl caproamide.
Hallcomid M-6 [379]

N,N, Dimethyl caprylate caprate.
Hallcomid® M-8-10 [379]

O,O-Dimethyl-O-(2,5-chloro-4-iodophenyl) phosphorthioate.
Nuvanol N [190]

Dimethyl cocoalkyl betaine.
Nissan Anon BF [636]

Dimethylcocoamine.
Amine 2MKK [16], [487]
Kemamine T-6502 [429a]
Lilamin 2MKK [487]

Dimethyl cocoamine, distilled.
Crodamine 3.ACD [223]
Kemamine T-6502D [429]

Dimethylcoco-amine oxide.
Aromox DMC, DMC-W, DMMC-W [79]

Dimethyl coconut t-amine, distilled.
Kemamine T 6502 D [429a], [430]

3,6-Dimethyl-3-cyclohexene-1-carbaldehyde.
Cyclovertal [399], [400]

N,N Dimethyl cyclohexyl ammonium dibutyl dithiocarbamate.
Akrochem Accelerator CZ-1 [15]

Dimethyl dialkyl ammonium chloride.
Querton 442L [487]

Dimethyl di-90% arachidyl-behenyl quaternary ammonium chloride.
Kemamine Q-1902C [429], [430]

2,5-Dimethyl-2,5-di (t-butylperoxy) hexane.
Luperco 101-XL [536]
Lupersol 101 [392], [682]
Varox [894]

2,5-Dimethyl-2,5-di(t-butyl peroxy) hexyne-3.
Lupersol 130 [392], [682]
Lupersol 130-XL [536]

Dimethyl dicoco ammonium chloride.
Adogen 462 [796]
Arquad 2C-75 [79]

Dimethyl dicoconut ammonium chloride.
Kemamine Q 6502 B [429a]

Dimethyl dicoco quaternary ammonium chloride.
Accoquat 2C-75, 2C-75-H [829]
Kemamine Q-6502C [429], [430]
Variquat K300 [796]

Dimethyl dieicosanyl docosanyl ammonium chloride.
Kemamine Q 1902 C [429a]

Dimethyl dihydrogenated tallow ammonium chloride.
Accosoft 707 [81]
Adogen 442 [796]
Kemamine Q-9702C [429], [429a]
Kemamine QSML/2 [429a]
Querton 442 [487]

Dimethyl di (hydrogenated tallow) ammonium methyl sulfate.
Accosoft 748 [81], [829]
Alkaquat DHTS [33]
Varisoft 137 [796]

Dimethyl di-hyd. tallow quaternary ammonium chloride.
Kemamine Q-9702C [429], [430]

N,N′-Dimethyl-N,N′dinitrosoterephthalamide/white mineral oil, 70:30%.
Nitrosan [270]

1,2-Dimethyl-3,5-diphenyl-1H-pyrazolium methyl sulfate.
Avenge [50]

Dimethyldisoya ammonium chloride.
Arquad 2S-75 [79]

Dimethyl distearyl ammonium chloride.
Arosurf TA 100 [79]

Dimethyl distearyl ammonium methyl sulfate.
Varisoft 190-100P [796]

Dimethyl ditallow ammonium chloride.
Querton 470 [487]

Dimethyldodecylamine.
Amine 2M12 [487]
Lilamin 2M12 [487]

Dimethyl dodecyl amine, distilled.
Kemamine T 6902 D [429a]

Dimethyl dodecyl betaine.
Nissan Anon BL [636]

Dimethyl eicosanyl-docosanyl amine, distilled.
Kemamine T 1902 D [429a]

Dimethyl ester 2,3,5,6-tetra-chloroterephthalic acid.
Dacthal [251]

2-ξ4-(1,1-Dimethyl-ethyl) phenoxy ξ cyclohexyl-2-propynl sulfite.
Omite [881]

O,O-Dimethyl-S-2-(ethylsulfinyl)ethyl phosphorothioate.
Meta-Systox R [184]

O,O-Dimethyl S-ξ(2,ethylsulfinyl) iso-propyl ξ phosphorothioate.
Metasystax-S [112]

O,O-Dimethyl S-ξ2(ethylthio)ethyl ξ phosphorothioate.
Metasystax (i) [112]

S-ξξ(1.1-Dimethyl-ethyl)thio ξ methyl ξ O,O-di-ethyl phosphorodithioate.
Counter [50]

Dimethyl fatty alkyl benzyl ammonium chloride, C_{12} C_{14}.
Lutensit K-LC, K-LC 80 [106]

Dimethyl fluid, silicone emulsion.
SM 2155, 2162 [341]

2,6-Dimethyl-5-hepten-1-al.
Melonal [353]

N,N′-Di-3(5-methylheptyl)p-phenylenediamine.
Antozite 2 [894]

Dimethylhexadecylamine.
Amine 2M16 [16], [487]
Lilamin 2M16 [487]

Dimethyl hexadecyl-amine, distilled.
Kemamine T 8802 D [429a]

Dimethylhexadecylamine oxide.
Aromox DM16 [79]

2,5-2,5. 2,5-Dimethylhexane-2,5-dihydroperoxide.
Luperox [536]

2,5-Dimethyl hexane-2,5-diperoxybenzoate.
Luperox 118 [536]

Dimethyl hexynol.
Surfynol 61 [13]

5,5-Dimethylhydantoin.
Ethyl Bromide (Bromethane™) [369]
Ethylene Dibromide (1,2-Dibromoethane™) [369]

Dimethylhydrogenated tallow amine.
Amine 2MHBG [16], [487]
Kemamine T-9702 [429a]
Lilamin 2MHBG [487]

Dimethyl hyd. tallow amine, distilled.
Kemamine T-9702D [429], [429a], [430]

Dimethyl (hydrogenated tallow) amine oxide.
Aromox DMHT [79]

Dimethyl hydrogenated tallow benzyl quaternary ammonium chloride.
Kemamine BQ-9702C [430]

Dimethyl 3-hydroxy-glutaconate dimethyl phosphate.
Bomyl [35], [423]

N,N-Dimethyl lauramide.
Hallcomid M-12 [379]

Dimethyl lauramine.
Kemamine T-6902 [429a]

Dimethyl lauramine oleate.
Necon LO [41]

Dimethyl lauryl amine.
Armeen DMMCD [16]

Dimethyl laurylamine, distilled.
Crodamine 3.A12D [223]

Dimethyl lauryl amine oxide.
Schercamox DML [771]
Varox 365 [796]

Dimethyl lauryl amino-betaine.
Lorapon AM-DML [274]

Dimethyl maleate.
Mon-Arc® MM-2 [69]

O,O-Dimethyl O-ξ3-methyl-4-(methylthio) phenyl ξ phosphorothioate.
Baytex [112]

N,N-Dimethyl-9[3-4-methyl-1-piperazinyl) propylidene]-thioxanthene-2-sulfonamide.
Navane [691]

3,3-Dimethyl-1-(methylthio)-2-butanone O-ξmethylamino ξ-carbomyl oxime.
Decamox [780]

3,5-Dimethyl-4-(methylthio)phenol methylcarbamate.
Mesurol [112]

O,O-Dimethyl (O-(4-(methylthio)-m-tolyl)phosphorothioate.
Baytex [184]

Dimethyl myristamine.
Kemamine T-7902 [429a]

Dimethyl myristyl amine oxide.
Schercamox DMM [771]

Dimethyloctadecylamine.
Amine 2M18 [487]
Lilamin 2M18 [487]

Dimethyl octadecyl t-amine, distilled.
Kemamine T 9902 D [429a]

Dimethyl octynediol.
Surfynol 82 [13]

N,N-Dimethyl oleamide.
Hallcomid M-18-OL [379]

Dimethyl oleyl-t-amine, distilled.
Kemamine T 9892 D [429a]

O,O-Dimethyl S-4-oxo-1,2,3-benzotriazin-3(4H)-ylmethyl phosphorodithioate.
Guthion [112], [184]

3-ξ2(3,5-Dimethyl-2-oxocyclohexyl)2-hydroxyethyl-ʒ glutarimide.
Acti-dione [870]

Dimethyl palmitamine.
Kemamine T-8902 [429a]

Dimethyl palmitylamine, distilled.
Crodamine 3.A16D [223]

Dimethyl 4,4-o-phenylenebis (3-thioallophanate).
Fungo 50 [542]

N,N-Dimethyl-p-phenylenediamine sulfate.
Rodol Gray DMS [531]

1,1-Dimethyl-3-phenylurea trichloroacetate.
Dozer [423]

Dimethyl phthalate.
Kodaflex DMP [278]
Staflex DMP [722]

Dimethyl phthalate, emulsified.
Marvanol Carrier RCK [552]

1,1-Dimethyl-piperidinium chloride.
Pix [106]

Dimethyl polysiloxane.
Akrochem Silicone Emulsion 350 Conc. [15]
Defoamer [477]
Dow Corning F1-1630 [262]

Dimethyl sebacate.
Dimethyl Sebacate (DMS) [393]

Dimethyl soyamine.
Kemamine T-9972 [429a]

Dimethyl soya t-amine, distilled.
Kemamine T 9972 D [429a], [430]

N,N-Dimethyl stearamide.
Hallcomid M-18, M 18-OL [379]

Dimethyl stearamine.
Kemamine T-9902 [429a]

Dimethyl stearylamine, distilled.
Crodamine 3.A18D [223]
Kemamine T-9902D [429], [430]

Dimethyl stearyl benzyl ammonium chloride.
Varisoft SDC [796]

2,4-Di- α -methyl-styrylphenol.
Prodox® 122 [315]

Dimethyl sulfone.
MSM [273]

Dimethyl tallowamine.
Kemamine T-9742 [429a]

Dimethyl tallow-t-amine, distilled.
Kemamine T-9742D [429], [429a], [430]

Dimethyl tallow benzyl quaternary ammonium chloride.
Kemamine BQ-9742C [429], [430]

[1R[1(S*)3(RS*)]]-2,2-Dimethyl-3-(1,2,2,2-tetrabromolthyl) cyclopropanecarboxylic acid ac-cyano-3(3-phenoxyphenyl)methyl ester.
Scout [55]

Dimethyl tetrachloroterephthalate.
Dacthal [780]

Dimethyltetradecylamine.
Amine 2M14 [487]
Lilamin 2M14 [487]

Dimethyl tetradecyl t-amine, distilled.
Kemamine T 7902 D [429a]

3,5-Dimethyltetrahydro 1,3,5,2H-thiadiazine-2-thione.
Troysan 142 [869]

Dimethyl tin mercaptide.
Mark 1900 [74]

n-ξ2,4-Dimethyl-5-ʒξ trifluoromethyl)sulfonylʒaminoʒphenylʒ acetamide.
Mefluidide [862]

Dimyristyl thiodipropionate.
Carstab DMTDP [192]
Cyanox MTDP [52]

2,6-Dinitro-N,n-dipropyl cumidine.
Paarlan [282]

3,5-Dinitro-n_4-n_4-dipropylsulfanilamide.
Surflan [282]

Dinitro(1-methylheptyl)phenyl crotonate.
Karathane [742]

Dinitroctyl-phenyl crotonate.
Karathane Liquid Conc., Karathane® WD [742]

Dinitrosopenta methylene tetramine.
Opex® [659]
Vulcacel BN 94 [212], [910]

3,5-Dinitro-o-toluamide.
Zoamix [261]

Dinonylphenol.
Berol Dinonylphenyl [116]

Dinonyl phenol ethoxylate.
Berol 269, 272 [116]
Iconol DNP-4, DNP-7, DNP-8, DNP-9, DNP-24, DNP-50, DNP-150 [109]

Dinonylphenol-70 mole ethylene oxide adduct.
T-DET D-70 [859]

Dinonylphenol-150 mole ethylene oxide adduct.
T-DET® D-150 [859]

Dinonylphenol polyglycol ethers.
Marlophen DNP 8, DNP 16, DNP 18, DNP 30 [178]
Serdox NDI 100 [788]

Dinonylphenol polyglycol ether phosphate, acid form (EO 9).
Servoxyl VPQZ 9/100 [788]

Dinonyldiphenylamine.
Vanlube DND [894]

Dinoprostone.
Prostin E2® [884]

Dinoprost tromethamine.
Lutalyse® [884]
Prostin F2 Alpha® [884]

O,O-di-n-octadecyl-3,5-di-t-butyl-4-hydroxybenzyl phosphonate.
Irganox 1093 [190]

Dioctyl adipate.
Adimoll DO [112]
Kodaflex DOA [278]
PlastHall DOA P [379]
Rilanit DOA [401]
Staflex DOA [722]
Uniflex DOA [876]
Wickenol 158 [926]

Dioctyl adipate [bis (2-ethylhexyl) adipate].
Kodaflex DOA [278]

Dioctyl azelate.
PlastHall DOZ [379]
Staflex DOZ [722]

Dioctyldecyl dimethyl ammonium chloride.
BTC 818 [663]

Dioctyl dimethyl ammonium chloride.
2050 BAR DAC [529]
Bardac LF, LF-80 [529]

p,p′-Dioctyldiphenylamine.
Vanlube 81 [894]
Vanox 1081, 12 [894]

Dioctyl dodecanedioate.
PlastHall DODD [379]

N,N-Di(2-octyl)p-phenylenediamine.
Antozite 1 [894]

Di-n-octyl-phthalate.
Rilanit DNOP [401]
Stave DOP [822]

Dioctyl phthalate [bis (2-ethylhexyl) phthalate].
Kodaflex DOP [278]

Dioctyl phthalate, dibutyl phthalate, 60%:5%.
Kodaflex HS-4 [278]

Dioctyl sebacate.
Monoplex® DOS [379]
PlastHall DOS [379]
Polycizer DOS [392]
Unimate DOS [876]

Dioctyl sodium sulfosuccinate.
Alrowet D-65 [190]
Atcowet W [130], [228]
Chemax DOSS-70, DOSS-75 [170]
Complemix-100 [233]
Gemtex PA-70, PA-75, PA-85P, PAX-60, SC-40, SC-70, SC-75 [320]
Kara Wet 70 [537]
Mackanate DOS [567]
Manoxol OT [544]
Monawet MO65-150, MO-70, MO-70-150, MO-70E, MO-70R, MO-70RP, MO-70S, MO-75-E. MO-84 R2W, MO-85P [599]
Nekal WT-27 [338]
Rewopol NEHS, SBDO 70 [725]
Solusol 75%, 100% [50]
Triton® GR-5M, GR-7M [742]

Dioctyl sodium sulfosuccinate, liquid form.
Pentex 99 [200]

Dioctyl succinate.
Wickenol 159 [926]

Dioctyl sulfosuccinate.
Arylene M40, M60 [390]
Atlas WA-100 [94]
Cyclopol SBD0 [234]
Rewopol SB-DO 70 [725]
Varsulf SB-DO-70 [796]

Dioctyl sulfosuccinate, sodium salt.
Astrowet 0-70-PG, 0-75 [28]
Merpasol DIO 60, DIO 75 [285]

Dioctyl terephthalate [bis (2-ethylhexyl) terephthalate].
Kodaflex DOTP [278]

Dioleate 400, polyethylene glycol ester.
Lipal 400 DW [715]

Dioleic imidazolinium quaternary.
Alkaquat O [33]

Dioleth-8 phosphate.
Nikkol DOP-8N [631]

Dioleyl imidazolinium methosulfate.
Rewoquat W 3690 [725]

Dioxathion.
Delnav [406]

n-(1,3-Dioxolan-2-yl-methyoxy)-iminobenzeneacetonitrile.
Concep II [190]

2-(1,3-Dioxolan-2-yl)phenyl-n-methyl-carbamate.
Elocron [190]

Dipalmitoyl hydroxyproline.
Lipacide DPHP [730]

Dipalmitoyl hydroxyprolinic acid.
Lipacide DPHP [730]

Dipentaerythritol.
Dipentek [850]

Dipentaerythritol ester.
GC-44-14 [715]

Dipentaerythritol monohydroxypenta acrylate.
SR-399 [766]

Dipentamethylene thiuram hexasulfide.
Akrochem DPTT [15]
Mastermix DPTH 4120 MB, DPTH 4186 MB [392]
Sulfads [894]
Tetrone A [269]

Dipentamethylene thiuram tetrasulphide.
DPTT [402], [708]
Tetrone A [270]

Dipentene.
Dipentene 200 [850]
Nesol [876]
Unitene D [110, [876]

Dipeptide coconut fatty acid condensate, potassium salt.
Lamepon S 2 [176]

Dipeptide coconut fatty acid condensate, triethanolamine salt.
Lamepon S 2 TR [176]

Diperester 2,5-dimethyl-2,5-bis (benzoyl peroxy) hexane.
USP-711 [890]

Diperester 2,5-di-methyl-2,5-bis (2-ethyl hexanoyl peroxy) hexane.
USP-245 [890]

Diphenamid.
Enide® [884]

Diphenhydramine hydrochloride, USP.
Benadryl Products [510]

Di (2-phenoxyethyl) peroxydicarbonate.
Luperox 204 [536]

Diphenylamine, alkylated.
Permanax HD [212], [910]
Wytox® ADP-F [659]

Diphenylamine antioxidant, alkylated.
OA 502 [678]

Diphenylamine, nonylated.
Polylite [881]

Diphenylamine, octylated.
Akrochem Antioxidant S [15]
Flectol ODP [392], [602]
Permanax OD [212], [910]
Vulkanoz OCD [597]

Diphenylamine, substituted.
Naugard 445 [881]

Diphenylcresyl phosphate.
Disflamoll DPK [597]

Diphenyl didecyl (2,2,4-trimethyl-1,3-pentanediol) diphosphite.
Weston 491 [129]

Diphenylene-oxide-4,4′-disulfohydrazide.
Porofor S 44 [112]

Diphenylene sulfide.
Dibenzothiophene [303]

Diphenylguanidine.
Akrochem DPG [15]
DPG [50], [392], [602]
DPG Vulcanization Accelerator [602]
Vanax DPG [894]
Vulcafor DPG [212], [910]
Vulkacit D/C [597]

Diphenylguanidine phthalate.
Guantal [602a]

Diphenyl hydrogen phosphite.
Mark DPHP [75]

Diphenyl isodecyl phosphite.
Weston DPDP [129]

Diphenyl isooctyl phosphite.
Weston ODPP [129]

Di (phenyl mercury) dodecenyl succinate.
Troysan PMDS-10 [869]

Diphenyloctyl phosphate.
Disflamoll DPO [597]

Diphenyloxazole dimethyl-POPOP per liter 5g:.1g.
Liquid Scintillator HF [68]

Diphenyl oxide, emulsified.
Marvanol Carrier ODP [552]

Diphenylpentaerythritol diphosphite.
Dipentite [421]

Diphenyl-p-phenylene-diamine.
AgeRite DPPD [894]
DPPD [708]
LZF [881]
Permanax DPPD (F) [212], [910]

Diphenylphosphinyl chloride.
Sequalog 38-1670-00 [172]

Diphenyl phosphite.
Weston DPP [129]

Diphenyl-sulfon-3,3′-disulfohydrazide.
Porofor D33 [112]

Diphenyl-thiourea.
A-1 Thiocarbanillide [392], [602]
DPTU [708]
Rhenocure CA [728]

1,2 Diphenyl-2-thiourea thiocarbonilide.
Akrochem Accelerator Thio No. 1 [15]

Di-POE (2) alkyl ether phosphate.
Nikkol DDP-2 [631]

Di-POE (4) alkyl ether phosphate.
Nikkol DDP-4 [631]

Di-POE (6) alkyl ether phosphate.
Nikkol DDP-6 [631]

Di-POE (8) alkyl ether phosphate.
Nikkol DDP-8 [631]

Di-POE (10) alkyl ether phosphate.
Nikkol DDP-10 [631]

Dipropylene glycol dibenzoate.
Benzoflex 2-45, 9-88, 9-88 SG [898]

Dipropylene glycol di laurate.
CPH-163-N [379]

Dipropylene glycol monolaurate.
Cithrol DPGML N/E, DPGML S/E [223]
CPH-162-N [379]

Dipropylene glycol monooleate.
Cithrol DPGMO N/E, DPGMO S/E [223]
CPH-312-N [379]

Dipropylene glycol monostearate.
Cithrol DPGMS N/E, DPGMS S/E [223]

Dipropylene glycol salicylate.
Dipsal [771]

Dipropylene triamine.
AZamine DPTA-T [554]

Di-n-propyl isocinchomeronate.
MGK® Repellent 326 [568]

Di-n-propyl 6,7-methylene-dioxy-3-methyl-1,2,3,4-tetrahydronaphthalene-1,2-dicarboxylate.
Isome [680]

Di (n-propyl) peroxy-dicarbonate.
Lupersol 221, 221-M85 [536]
NPP [705]

Dipyridamole.
Persantine Tablets [510]

Disalicylalpropylenediamine/aromatic solvent, 50:50%.
Copper Inhibitor 50 [270]

N,N′-Disalicylidene-1,2-diaminopropane/tolune solvent, 80:20%.
Ethyl Metal Deactivator [302]

Di (sec-butyl) peroxy-dicarbonate.
Lupersol 225, 225-M, 225-M60 [536]
SBP [705]
Trigonox® SBP, SBP-AX30, SBP-C75 [16], [653]

Di-sodium alkanolamide ethoxylate sulphosuccinate.
Empicol SGG [23]

Di-sodium alkanolamide sulphosuccinate.
Empicol SBB, SCC, SEE [23]

Disodium alkenyl succinate.
Produkt B 3010 [223]

Disodium alkylamido polyethoxy sulfosuccinate.
Avirol® SS-5100 [399], [400]

Disodium alkyl ethoxy sulfosuccinate.
Empicol SDD, SFF [25]

Disodium alkylolamide ethoxy sulfosuccinate.
Empicol SGG [25]

Disodium alkylolamide sulfosuccinate.
Empicol SBB, SCC [25]

Disodium N-alkylsulfosuccinamate.
Aerosol 19 [50]
Alcamate FA 82 [29]

Disodium alkyl sulfosuccinamide.
Condanol TMS [274]
Rewomat TMS [274]
Steinamat TMS [274]

Disodium alkyl sulfosuccinate.
Aerosol 501 [50]
Empicol SLL, STT [25]

Di-sodium-castor oil-fattyacid ethanolamide sulphosuccinate.
Merpasol L 53 [285]

Disodium cetearyl sulfosuccinate.
Rewopol SBF 1618 [725]

Disodium N-cetyl stearyl sulfosuccinamate.
Empimin MKK, MKK98, MKK/L [25]

Disodium cocamide MIPA sulfosuccinate.
Alconate CPA [29]

Disodium cocamido MEA-sulfosuccinate.
Mackanate CM [567]
Rewopol 1026, SBC 212 [725]
Schercopol CMS-Na [771]

Disodium cocamido MIPA-sulfosuccinate.
Mackanate CP [567]
Monamate CPA-40, CPA-100 [599]
Rewopol SB-IP 240 [725]
Schercopol CMIS-Na [771]

Disodium coco fatty acid ethanolamide sulfosuccinate.
Merpasol L 52 [285]

Disodium N-cocoyl sulfosuccinamate.
Empimin MHH [25]

Disodium copper EDTA.
Sequestrene NA2Cu [190]

Disodium deceth-6 sulfosuccinate.
Alconate D-6 [29]

Disodium-1,2-dihydroxy-benzene-3,5-disulfonate.
Tiron [506]

Disodium dihydroxyethyl ethylenediamine diacetate.
Kalex OH [390]

Disodium 2,5-dimercaptothiadiazole.
Vanchem NATD [894]

Disodium EDTA.
Perma Kleer Di Crystals [663]
Sequestrene Na 2 [190]
Versene NA [261]

Disodium EDTA-copper.
Sequestrene Na2Cu [190]

Disodium ethanol alkyl (C_{18}) amide sulfosuccinate.
Aerosol 413 [50]

Disodium (ethoxylated alkylamide) sulfosuccinate.
Aerosol 200 [50]

Disodium ethylene-1,2-bisdithiocarbamate.
Nabam [730]

Disodium ethylenediaminetetraacetate, dihydrate.
Hamp-Ene Na_2 [364b]

Disodium 2-ethyl hexyl-sulfosuccinate.
Servoxyl VLA 2170 [788]

Disodium fatty alcohol polyglycol ether-sulfosuccinate.
Merpasol L 29 [285]

Disodium N-hydroxy-ethylimino-diacetic acid.
Hamsphire EDG [364b]

Disodium isodecyl sulfosuccinate.
Aerosol A-268 [50]

Disodium laneth-5 sulfosuccinate.
Rewolan SB LAN/5, SB LAN/E [725]

Disodium lauramide MEA sulfosuccinate.
Alconate LEA [29]

Disodium lauramido MEA-sulfosuccinate.
Mackanate LM-40 [567]

Disodium laureth sulfosuccinate.
Alconate L-3 [29]
Mackanate EL [567]
Rewopol SBFA 1, SBFA 3, SBFA 30 [725]
Schercopol LPS [771]
Serdet VLB 1123 [175]

Disodium lauric acid ethanolamide sulfosuccinate.
Merpasol L 47 [285]
Rewopol SBL 203 [274]
Steinapol SBL 203 [274]

Disodium lauriminodipropionate.
Deriphat 160 [399]
Mirataine H2C [592]

Disodium lauriminopropionate.
Mirataine® H2C [592]

Disodium lauryl alcohol polyglycol-ether sulfosuccinate.
Condanol SBFA 30, 40% [274]
Steinapol SBFA 30, 40% [274]

Disodium lauryl alcohol sulfosuccinate.
Steinapol SBF 12, SBR 12-Powder [274]

Di-sodium lauryl ethoxy sulphosuccinate.
Empicol SDD [23]

Disodium N-lauryl β-iminodipropionate.
Deriphat 160 [400]

Disodium N-lauryl sulfosuccinamate.
Empimin MHH [23]

Disodium lauryl sulfosuccinate.
Emcol 4400-1 [934], [935]
Merpasol L 44 [285]
Miranate LSS [592]

Disodium methanearsonate.
Weed-E-Rad 360 [904]

Disodium monococoamide, sulfosuccinate.
Cyclopol SBG 212 [234]

Disodium monococoamido MEA sulfosuccinate.
Schercopol CMS-Na [771]

Disodium monococoamido sulfosuccinate.
Schercopol CMS-Na [771]

Disodium monoglyceride sulfosuccinate.
Emcol 4072 [934], [935]

Disodium monolauramido MIPA-sulfosuccinate.
Schercopol CMIS-Na [771]

Disodium monolauryl amidoethyl sulfosuccinate.
Servoxyl VLC 1200 [788]

Disodium mono oleamide PEG 2-sulfosuccinate.
Texapon SH 100, SH-135 Special [399]

Disodium monooleamide, sulfosuccinate.
Cyclopol SBG 280 [234]

Disodium monooleamido diglycol sulfosuccinate.
Anionyx 12S [663]

Disodium monooleamido MEA sulfosuccinate.
Schercopol OMS-Na [771]

Disodium monooleamido MIPA sulfosuccinate.
Schercopol OMIS-Na [771]

Disodium mono-oleamido PEG-2 sulfosuccinate.
Standapol SH-100, SH-135 Special [399], [400]

Disodium mono-oleamido polyethylene glycol sulfosuccinate.
Sulfotex DOS [399]

Disodium mono-oleamido sulfosuccinate.
Jordawet DMDS [474]

Disodium myristamido MEA-sulfosuccinate.
Emcol 4100M [934]

Disodium octaborate tetrahydrate.
FR28 [887]
Polybor [887]
Solubor [887]
Tim-Bor [887]

Disodium N-octadecyl sulfosuccinamate.
Aerosol 18 [50], [233]
Aerosol 19 [50a]
Alcopol FA [37]
Alkasurf SS-TA [33]
Lankropol ODS/LS [253], [509]
Lankropol ODS/PT [509]

Disodium N-octadecyl sulfosuccinate.
Cyanasol 18-Paste Surface [50]
Manro MA 35 [545]

Disodium oleamido MEA-sulfosuccinate.
Mackanate OM [567]
Schercopol OMS-Na [771]

Disodium oleamido MIPA-sulfosuccinate.
Emcol 4161L [934]
Mackanate OP [567]
Rewopol SB-IPE-280 [725]
Schercopol OMIS-Na [771]
Soleterge 8 [415]

Disodium oleamido PEG-2 sulfosuccinate.
Mackanate OD [567]
Monamate OPA-30, OPA-100 [599]
Standapol SH-100, SH-135 Special [400]
Surfactant S-100 [453]

Disodium N-oleyl sulfosuccinamate.
Alkasurf SS-OA [33]
Empimin MTT [23]
Empimin MTT/A [25]

Disodium pareth-3 sulfosuccinate.
Emcol 4300 [934]

Disodium ricinoleamido MEA-sulfosuccinate.
Mackanate RM [567]
Rewoderm S1333 [725]

Disodium N-stearyl sulfosuccinamate.
Empimin MKK [23]
Rewopol SBF 18, TMS [725]

Disodium sulfosuccinate.
Emcol 4100M [934], [935]
Emcol 4161-L [934], [937], [938]
Emcol 4300 [934], [935], [937], [938]
Emcol® K-8300 [935]

Disodium tallowiminodipropionate.
Deriphat 154 [399], [400]
Mirataine T2C [592]

Disodium undecylenamido MEA-sulfosuccinate.
Mackanate UM [567]
Rewocid SBU 185 [725]

Disodium undecylenic monoethanolamide sulfosuccinate.
Empicol SEE [23]

Disodium wheat germamido MEA-sulfosuccinate.
Mackanate WG [567]

Disodium wheat germamido PEG-2 sulfosuccinate.
Mackanate WGD [567]

Distarch phosphate.
National 78-1898 [619]

Distearyl dimethyl ammonium chloride.
Accosoft 707 [829]
Arosurf TA-100 [796]
Cation DS [764]
Quartamin D.86.P [803]
Varisoft DHT [796]

Distearyl dimethyl ammonium methyl sulfate.
Varisoft 190 [796]

Distearyldimonium chloride.
Genamin DSAC [55]
Kemamine Q-9902C [429a]

Distearyl ketone.
Stearone [74]

Distearyl pentaerythritol diphosphite.
Mark 5050, 5060 [75]
Weston 618, 619 [129]

Distearyl pentaerythritol diphosphite calcium stearate, 5.0% wt.
Weston 732 [129]

Distearyl phosphite.
Weston DSP [129]

Distearylphthalate.
Radiasurf 7505 [658]

Distearyl thiodipropionate.
Argus DSTDP [74]
Carstab DSTDP [192], [857]
Cyanox® STDP [52]
Disterdap [151]
Evanstab 18 [303]

Disulfoton.
Di-Syston [184]

N,N-Ditallowamidoethyl-N-hydroxyethyl-N-methyl ammonium methosulfate.
Carsosoft T-90 [529]

Ditallow amine, hydrogenated.
Kemamine S-970 [429a]
Noram 2 SH [162]

Ditallow diamido methosulfate.
Carsosoft T-75, T-90 [529]

Ditallow diamido, quaternary.
Carsosoft T-90 [155]

Ditallow dimethyl ammonium chloride.
Armosoft L [79]

Ditallow dimethyl ammonium methosulfate.
Carsosoft V-90 [529]

Di-tallow fatty alcohol phthalate.
Rilanit DTP [401]

Ditallow imidazolinium quaternary.
Alkaquat T [33], [34]
Sterling Snow White 750 [145]

Ditallow quaternary.
Accosoft 550, 620, A-155 [829]
Alkaquat DAET, DAPT [33]

Di-TEA-palmitoyl aspartate.
PAT-30 [481]

2,2′-Dithiobis (benzothiazole).
Thiofide (MBTS) [392], [602]

5-5′-Dithiobis-(2-nitrobenzoic acid).
Sequalog 38-9060-00 [172]

Dithiocarbamate.
Butyl Eight [894]
Rychem® 830 [756]

4.4′ Dithio dimorpholine.
Akrochem Accelerator 'R' [15]
Naugex SD-1 [881]
Sulfasan R [392], [602]

4,4′-Dithiomorpholine.
DSM [708]
Vanax A [894]

Di-o-tolyl guanidine.
Akrochem DOTG [15]
DOTG [50]
Mastermix DOTG 4163 MB [392]
Vanax DOTG [894]
Vulcafor DOTG [212], [910]
Vulkacit DOTG/C [597]

Ditridecyl adipate.
PlastHall DTDA [379]

Ditridecyl maleate.
Staflex DTDM [722]

Di-tridecyl phosphite.
Weston DTDP [129]

Ditridecyl phthalate.
Staflex DTDP [722]

Di-tridecyl sodium sulfosuccinate.
Monawet MT-70, MT-80 H2W [599]

Ditridecyl thiodipropionate.
Cyanox 711 [50]
DTDTDP [74]
Evanstab 13 [303]

Di-3,5,5-trimethylhexyl phthalate.
Vestinol NN [178]

Dm Dm hydantoin.
Glydant [358]

Docusate sodium products.
Disonate Products [510]

Docyl diethanolamide.
Nitrene C Extra [400]

(2)-9-Dodecenyl acetate.
No Mate Shootgard [21]
Orfamone [947]

Dodecyl alcohol with ethylene oxide, 9.5 mols, condensate.
Lubrol 12A-9 [442]

n-Dodecylamine.
Amine 12D [487]
Armeen 12D [79]
Lilamin 12D [487]

Dodecyl amine, primary.
Kemamine P-690 [429a]
Nissan Amine BB [636]

Dodecyl amine, primary, distilled.
Kemamine P-690 D [429a]

Dodecylammonium methanearsonate.
Super Dal-E-Rad AMA [904]

n-Dodecylbenzene, dodecylbenzene sulfonate sodium salt.
Elfan WA, WA Pulver [16]

n-Dodecylbenzene, dodecylbenzene sulfonate triethanolamine salt.
Elfan WAT [16]

Dodecyl benzene sodium sulfonates.
Nansa HS 80/S, HS 85/S, SL 30, SS 30, SS 60 [23]

Dodecyl benzene sodium sulfonate, straight-chain.
Conoco C-550, C-560 [206]

Dodecyl benzene, SO_3 sulfonated, isopropylamine salt.
Manro HCS [545]
Nansa YS 94 [26]

Dodecyl benzene, SO_3 sulfonated, triethanolamine salt.
Nansa TS 60 [26]

Dodecylbenzene, straight-chain.
Marlican [178]

Dodecyl benzene sulfonate, calcium salt, linear.
Chimipon CA [514]

Dodecyl benzene sulfonate, calcium salt, straighter chain.
Arylan CA [509]

Dodecyl benzene sulfonate flake, sodium salt, straight chain.
Arylan SX Flake [253], [509]

Dodecyl benzene sulphonate, linear.
Chimipon BAC [514]

Dodecyl benzene sulfonate liquid, sodium salt, straight chain.
Arylan SC30 [253], [509]

Dodecyl benzene, sulfonated salts.
Icowet® 3300 [109]

Dodecyl benzene sulfonate, sodium salt.
Maranil Paste A 50, Paste A 75, Powder A [401]

Dodecyl benzene sulfonate, sodium salt, straight chain.
Arylan SBC 25 [253], [509]

Dodecyl benzene sulfonate (TEA salt), straighter chain.
Arylan TE/C [509]

Dodecyl benzene sulfonate, triethanolamine salt, straight chain.
Arylan TE/C [253]

Dodecyl benzene sulfonate, triethanolamine sodium salt, linear.
Chimipon TSB [514]

Dodecyl benzene sulfonic acid.
Alkasurf LA Acid [33]
Arsul DDB, LAS [76]
Arylan SC Acid [509]
Bio Soft JN [825a]
Calsoft LAS-99 [697]
Conco AAS-98S [209]
DDBSA 99-b [602a]
Elfan WA Sulfosaure [16]
Elfan WA Sulphonic Acid [16], [764]
Emulsifier 99 [697]
Hetsulf Acid [407]
Manro BA, NA [545]
Marion AS 3 [178]
Mars SA-98S [554]
Nansa 1042, 1042/P, SBA [25]
Nansa SSA [548]
Nansa SSA/P [25]
Polyfac LAS-97 [918]
Polystep A-17 [825]
Reworyl K [725], [725a]
Reworyl-Sulfonic Acid K [274], [725]
Richonic Acid B [732]
Rueterg Sulfonic Acid [320]
Serdet DM [788]
Steinaryl-Sulfonic Acid K [274]
Tex-Wet 1197 [460]
Witco® 1298 Sulfonic Acid, Acid B [935]

Dodecylbenzene sulfonic acid, amine salt.
Chemsulf 75-IA [170]
Emulsifier IA, IAH [170]
Marlon AMX [178]
Tex-Wet 1006 [460]
Witconate™ 93S, P10-59, YLA [935]

Dodecyl benzene sulfonic acid (branched chain).
Arylan S Acid [253]
Chemsul B-555 [170]

Dodecyl benzene sulfonic acid (branched chain) *(cont'd)*.
Manro HA [545]
Polystep A-17 [825]
Ufacid TPB [874]

Dodecyl benzene sulfonic acid, calcium salt.
Arylan CA [253], [509]
Chemannate C-60B, C-70B [171]

Dodecyl benzene sulfonic acid, isopropylamine salt.
Cindet GE [193]

Dodecyl benzene sulfonic acid, linear.
Alkasurf LA Acid [33]
Ardet LAS [73]
Arsul LAS [76]
Chemsul L-555 [170]
Manro BA, NA [545]
Merpisap AS 98 [285]
Sterling LA Acid [145]
Surco DDBSA 97% [663]
Tairygent CA-1 [335]

Dodecyl benzene sulfonic acid, linear, sodium salt.
Geropon DBL [550]

Dodecyl benzene sulfonic acid, sodium salt.
Nansa 1106/P [25]
Tex-Wet 1047 [460]

Dodecyl benzene sulfonic acid, SO_3-sulfonated.
Nansa SBA, SSA [26]

Dodecyl benzene sulfonic acid, straight chain.
Arylan SBC Acid, SC Acid, SP Acid [253], [509]
Condasol Sulfonic Acid K [274]
Conoco SA-597 [206]
Marlon AS_3 [178]

Dodecylbenzyltrimonium chloride.
Rewoquat B 18 [725]

6-Dodecyl-1,2-dihydro-2,2,4-trimethylquinoline.
Santoflex® DD Antioxidant [392], [602]

Dodecyldimethylamine.
Adma™ 2 [302]
Armeen DM12D [16]
Nissan Tertiary Amine BB [636]
Onamine 12 [663]

Dodecyl dimethyl benzylammonium chloride.
Swanol CA-101 [631]

Dodecyldimethyl(2-phenoxy-ethyl) ammonium bromide.
Fungitex R [190]

Dodecyl diphenyl-ether disulfonic acid.
Dowfax 2A0 [261]

Dodecyl diphenyl ether sulfonate, sodium salt.
Conco Sulfate 2A1 [209]

Dodecyl diphenyl oxide disulfonic acid.
Conco Sulfate 2A1 Acid [209]
Dowfax 2A0 [261]

n-Dodecyl ether, ethoxylated.
Nikkol BL-1SY thru BL-8SY [631]

Dodecyl gallate.
Progallin LA [632]

n-Dodecyl mercaptan.
DDM-100 [682]
4P Mercaptan [682]
Sulfole 120 [695]

Dodecyl mercaptan ethoxylate.
Siponic 218, 260, SK [29]

Dodecyl mercaptan, polyethoxylated.
Seconic [825a]

4-Dodecyloxy-2-hydroxybenzophenone.
Eastman Inhibitor DOBP [278]

Dodecyl phenol + EO 5 moles.
Teric DD5 [439]

Dodecyl phenol + EO 9 moles.
Teric DD9 [439]

Dodecyl phenol + EO 14 moles.
Teric DD14 [439]

Dodecyl phenol ethoxylates.
Flo Mo 5D, 6D, 10D, 14D [247]

Dodecyl phenol ethoxylate 6 mole.
Sterox DF [602a]

Dodecyl-phenol (EO 10), ethoxylated.
Iconol DDP-10 [109]

Dodecyl phenol ethoxylate 10.5 mole.
Sterox DJ [602a]

Dodecylphenol ethylene oxide adduct, 5 mole.
T-DET DD-5 [859]

Dodecylphenol ethylene oxide adduct, 7 mole.
T-DET DD-7 [859]

Dodecylphenol ethylene oxide adduct, 9 mole.
T-DET DD-9 [859]

Dodecylphenoxypoly(ethyleneoxy)ethanol.
Igepal RC-520, RC-620, RC-630 [338]

Dodecyltrimethylammonium chloride.
Arquad 12-33; Arquad 12-50 [79]
Nissan Cation BB [636]

Dodoxynol-5.
T-DET DD-5 [859]

Dodoxynol-7.
T-DET DD-7 [859]

Dolomite, burned.
Mag-Li-Kote [606]

Dolomite, calcined.
Doloxide [606]

Dolomite, pulverized.
D220, D303, D306, D3002 [203]

Domiphen bromide.
Fungitex R [190]

Doxorubicin hydrochloride.
Adriamycin [7]

Doxycycline hyclate.
Doxy-Caps [281]
Vibramycin Capsules [510]

Drometrizole.
Tinuvin P [190]

DTPA.
SEQ80 [903]

DTPA, calcium salt.
PLEX PC [903]

EDTA.
Cheelox BF Acid [338]
Perma Kleer Acid [663]
Questex 4H [821]
SEQ 100 [903]
Sequestrene AA [190]
Versene Acid, Acid NF [261]

EDTA complex.
Keyval TN [893]

EDTA, sodium salt.
Conco SEQ [209]
Intraquest® TA Solution [225]

EDTA, tetra sodium salt.
Aquamollin Brands [416]
Cheelox® BF-12, BF-13, BF-78 [338]
Kemplex 100 [338]

Egg albumen.
Sol-U-Tein EA [308]

Eicosanyl-docosanyl primary amine.
Kemamine P-190 [429a]

Eicosanyl-docosanyl primary amine, distilled.
Kemamine P-190 D [429a]

Endo-cis-bi-cyclo-(2.21)-5-heptene-2,3-dicarboxylic anhydride.
Endic Anhydride [898]

Epiclorohydrin.
Herclor [406]
Hydrin [361]

Epinephrine.
Adrenalin [674]

(cis)-7,8-Epoxy-2-methyloctadecane.
Nomate Biogard [21]

Epoxy soya plasticizer.
Plastolein 9232 [289]

Erucamide.
Armid E [16]
Crodamide E, ER [223]
Kemamide E [429]
Petrac Eramide [688]

Erythromycin.
E-Mycin® [884]

Erythromycin ethylsuccinate.
E-Mycin E® [884]

Estertin mercaptide.
Stanclere T-208, T-222, T-233, T-250, T-250SD [459]

Estradiol cypionate.
ECP® [884]

Estrogens, esterified.
Menest® 0.3 mg, 0.625 mg, 1.25 mg, 2.5 mg [114]

1,1′-(1,2-Ethanediyl)bis (3,3,5,5-tetramethylpiperazinone)
Good-Rite UV3034 [361]

Ethanolamine alkyl benzene sulfonate.
Textol 80 (L) [946]

2,2′-(1,2-Ethenediyldi-4,1-phenylene)bisbenzoxazole.
Eastobrite OB-1 [278]

Ether sulfate.
Hostapon CAS [416]

Ether thioether.
Vulkanol 85 [597]
Vulkanol OT [112]

Ethinyl estradiol.
Feminone® [884]

N-(p-Ethoxycarbonyl-phenyl)-N′-ethyl-N′-phenylformamidine.
Givsorb UV-2 [353]

2-Ethoxy-2,3-dihydro-3,3-dimethyl-5-benzofuranyl methanesulphonate.
Nortron [118]

6-Ethoxy-1,2-dihydro-2,2,4-trimethylquinoline.
Santoflex AW Antioxidant [392], [602]

2-(2-Ethoxyethoxy)-ethyl acrylate; with MEHO inhibitor 100 ppm.
SR-256 [766]

2-ξ1-(Ethoxyimino) butylξ-5-ξ2-(ethylthio) propyl -3-hydroxy-2-cyclohexen-1-one.
Poast [638]

4-Ethoxy-m-phenylenediamine sulfate.
Rodol EOX [531]

Ethoxyquin.
Santoquin [602], [602a]

5, Ethoxy-3-trichloro-methyl-1,2,4-thiadiazole.
Koban [542]
Terrazole [659]
Truban [542]

Ethylacetoacetate chelate.
Tyzor DC [269]

Ethyl alcohol.
Neosol [791]

Ethyl-p-amino-benzoate.
Anaesthesin [932]

Ethyl-p-aminobenzoate, 2 mole propoxylate.
Amerscreen P [44]

2-(Ethylamino)-4-isopropylamino-6-methyl-thio-5-triazine.
Evik [190a]

Ethyl n-benzoyl-N-(3,4-dichloro-phenyl)-2-aminopropionate.
Suffix [794]

Ethyl bis (polyethoxy ethanol) alkyl ammonium ethyl sulfate.
Varstat 66 [796]

4-Ethyl-4-butyl-valerolactone.
Costaulon 121910 [406]

Ethyl 2[4(6-chloro-2-benzorazolyloxy-)-phenoxy-]-propanate.
Furore [416]

Ethyl 2[4(6-chloro-2-benzothiazolyloxy-)-phenoxy-]-propanoate.
Joker [416]

Ethyl 2-cyano-3,3-diphenylacrylate.
Uvinul N-35 [338]

S-Ethylcyclohexyl-thiocarbamate.
Ro-Neet [821]

Ethyl 4,4′dichlorobenzilate.
Chlorobenzilate [190]

Ethyl 3-(3,5-dichlorophenyl)-5-methyl-2,4-dioxo-5-carboxylate.
Serinal [310]

O-Ethyl-O-(2,4-dichlorophenyl)-phosphonothionate.
S-Seven [639]

O-Ethyl-O-(2,4-dichlorophenyl)-S-n-propyl-dithiophosphate.
Tokutkim [112]

3-Ethyl-2,4-dioxaspiro (5,5) undec-8-ene.
Spiroflor [399], [400]

O-Ethyl S,S-dipropyl phosphorodithioate.
Macap [730]

S-Ethyl dipropylthiocarbamate.
Eptam [821]
Genep EPTC [705]

Ethylene bisdithiocarbamate, agricultural fungicides salts.
Dithane [742]

N,N Ethlenebis monomer acid amide.
Kemamide W-5 [429]

Ethylene bis oleamide.
Kemamide W-20 [429]

Ethylene bis-stearamide.
Alkamide STEDA [33]
Kemamide W-40, W-45 [429], [430]
Lipowax C [525]
Markamide W-40 [74]

1-Ethylene bis (2-tallow, 1-methyl, imidazolinium-methyl sulfate).
Varisoft 6112 [796]

Ethylene-chlorotrifluoroethylene.
Liquinite LMH [527]

Ethylenediamine.
Kalex 100 [391]

Ethylene diamine dinonylnaphthalene sulfonate in kerosene.
Na-Sul EDS [894]

Ethylene diamine polyoxypropylene compounds, ethoxylated.
Alkatronic EDP 8-4, EDP 28-1, EDP 28-2, EDP 28-7, EDP 38-1, EDP 38-4, EDP 38-8 [33]

Ethylened, amine tetraacetate, tetrasodium salt.
Chelon 100 [821]

Ethylenediamine tetraacetic acid.
Hamp-ene Acid [364b]
Kalex Acids [390]
Questex 4H [821]
Tetralon Acid [37]
Trilon BS [106]

Ethylenediaminetetraacetic acid.
Versene Diammonium EDTA [261]

Ethylenediaminetetraacetic acid dihydrate, disodium salt.
Versene NA [261]

Ethylenediaminetetraacetic acid dihydrate, disodium salt, calcium chelate.
Versene CA [261]

Ethylene diamine tetraacetic acid, disodium salt.
Trilon BD [106]

Ethylenediaminetetraacetic acid, disodium salt, dihydrate.
Hamp-ene® Na_2 [364b]

Ethylenediaminetetraacetic acid, tetraammonium salt.
Versene Tetraammonium EDTA [261]

Ethylenediaminetetraacetic acid, tetrasodium salt.
Hamp-ene 100, 220 [364b]
Kalex Concentrate, Regular [391]
Marquest 100 [554]
Tetralon A [37]
Trilon B Liquid, B Powder [106]
Versene AG 100 [261]

Ethylenediaminetetraacetic acid, tetrasodium salt, dihydrate.
Hamp-ene Na_4 [364b]

Ethylenediaminetetraacetic acid, trisodium salt, trihydrate.
Hamp-ene Na_3 T [364b]

Ethylenediaminetetra(methylene phosphoric acid).
Dequest 2041 [602]

Ethylene diglycol stearate.
Tegin DGS [360]

Ethylene dioleamide.
Kemamide W-20 [429a]

Ethylene distearamide.
Kemamide W-35, W-40, W-42, W-45 [429a]

Ethylene dithiodiethanol 2.2.
Tegochrome® 22 [856]

Ethylene glycol dibutyl ether.
Dibutyl Cellosolve [878]

Ethylene glycol diethyl ether.
Ethyl Glyme [367]

Ethylene glycol dilaurate.
Cithol EGDL N/E, EGDL S/E [223]
CPH-108-N [379]

Ethylene glycol dimethacrylate.
SR-206 [766]

Ethylene glycol dimethyl ether dimethoxyethane.
Monoglyme [367]

Ethylene glycol dioleate.
Cithrol EGDO N/E, EGDO S/E [223]

Ethylene glycol distearate.
Cithrol EGDS N/E, EGDS S/E, EGML N/E [223]
CPH-360-N [379]
Cyclochem EGDS [234]
Drewmulse EGDS [266]
EGDS V.A. [360]
Elfan L310 [16], [764]
Emerest 2355 [289]
Glykoldistearat 90 [856]
Kemester EGDS [429], [430]
Kessco-Ethylen-glykol-distearat [16]
Kessco Glycol Esters [825]
Mapeg EGDS [564]
Nikkol Estepearl 10, 15 [631]
Pegosperse 50-DS [358]
Radiasurf 7269 [658]

Ethylene glycol dodecyl ether.
Nikkol BL-1SY [631]

Ethylene glycol hydroxystearate.
Naturechem EGHS [158]

Ethylene glycol mono butyl ether.
Butyl Cellosolve [878]
Butyl Oxitol [791]
Ektasolve EB Solvent [278]
Glycol Ether EB [87], [791]

Ethylene glycol monobutyl ether acetate.
Butyl Cellosolve Acetate [878]
Ektasolve EB Acetate [278]

Ethylene glycol mono-dioleate.
Glykololeat [856]

Ethylene glycol mono-distearate.
Tegin A 412, G [856]

Ethylene glycol monoethyl ether.
Ektasolve EE Solvent [278]
Oxitol [791]

Ethylene glycol monoethyl ether acetate.
Ektasolve EE Acetate [278]

Ethylene glycol monohexyl ether.
n-Hexyl Cellosolve [878]

Ethylene glycol monolaurate.
Cithrol EGML S/E [223]
CPH-113-N [379]

Ethylene glycol monomethyl ether.
Ektasolve EM Solvent [278]
Methyl Cellosolve [878]
Methyl Oxitol [791]

Ethylene glycol monomethyl ether acetate.
Methyl Cellosolve acetate [878]

Ethylene glycol monooleate.
Cithrol EGMO N/E, EGMO S/E [223]
CPH-61-N, CPH-228-N [379]
Tegin GO [360]

Ethylene glycol monopropyl ether.
Ektasolve EP [278]

Ethylene glycol monoricinoleate.
Cithrol EGMR N/E [223]
Flexricin 15 [642]

Ethylene glycol monostearate.
AMS-33 [397]
Cerasynt M, MN [895]
Chimipal SGE [514]
Cithrol EGMS N/E, EGMS S/E [223]
Clindrol SEG [197]
CPH-37-NA [379]
Cutina EGMS [223]
Cyclochem EGMS [234]
Drewmulse EGMS [265]
Emerest 2350 [289]
Empilan EGMS [23], [25], [26]
Hodag EGMS [415]
Kessco-Ethylen-glykol-monostearat [16]
Kessco Glycol Esters [825]
Lexemul EGMS [453]
Lipo EGMS [525]
Mapeg EGMS [564]
Mazol EGMS [564]
Pegosperse 50 MS [358]
Radiasurf 7270 [658]
Schercemol EGMS [771]

Ethylene glycol stearate.
Cithrol EGMS [223]
Cutina AGS [399], [401]
Monthybase [339]
Monthyle [339]
Stearic Monoethanolamine Ltd. [223]

Ethylene oxide.
Iberpal B.I.G. [386]
Iberscour P [386]

Ethylene oxide adduct.
Elfugin V [762]
Nilo VON [762]
Sandopan DTC, LF [762]
Sandozin NIT [762]

Ethylene oxide alcohol condensation product.
Atranonic Polymer 40+2 [95]

Ethylene oxide carboxylate.
Crestolan Conc. [222]

Ethylene oxide condensate.
Alkanol A-CN [269]
Crestonon, Crestonon H, Crestonon NI Conc. [222]
Dispersall [753]
Dispersol VL [442]
Empilan KL10, KL20 [25]
Intralan Salt N [225]
Intratex AN [225]
Intravon JF, JU [225]
Marvanol DOF [552]
Merpol OJS [269]
Protopan [710]
Protowet 100, E-4 [710]

Ethylene oxide condensate, fatty alcohol.
Dehydol 100 [401]
Empilan KA 3, KA 5, KA 8, KA 590, KA 880, KA 1080 [25]
Empilan KS 6, KS 690 [23], [25]
Matexil PN-PR [442]

Ethylene oxide condensate, fatty amine.
Ethylan TC, TF, TH-2, TN-10, TT-15 [253]

Ethylene oxide nonylphenol.
Iberwet W-100 [386]

Ethylene-propylene, fluorinated.
Liquinite FEP [527]

Ethylene thiourea.
ETU [402]

Ethylene thiourea in DOP.
Mastermix ETU 4407 PD [392]

Ethylene thiourea in EPR.
Mastermix ETU 4347 MB [392]

2-Ethyl-ethyl capronate.
Irotyl [399], [400]

S-Ethyl hexahydro-1 H-azepine-1-carbothinate.
Ordram [821]

2-Ethyl hexanol phosphoric oxide.
Victawet 35B [821]

2-Ethylhexanol sulfosuccinic acid diisooctylester sodium salt.
Elfanol 883 [16]

2-Ethylhexoic acid, lead salt.
L-26 [616]

2-Ethylhexoic acid, metallic salts.
Hexogen [151a]
Octasols [389]

2-Ethylhexyl adipate.
Crodamol DOA [223]

2-Ethylhexyl 2-cyano-3,3-diphenylacrylate.
Uvinul N-539 [338]

2-Ethylhexyl diphenyl phosphate.
Santicizer 141 [602]

Ethylhexyl diphenyl phosphite.
Weston EHDPP [129]

Ethylhexyl laurate.
Radia 7127 [658]

2-Ethylhexyl-p-methoxycinnamate.
Parsol MCX [353]

n-(2-Ethylhexyl)-5-norbonene-2,3-dicarboximide.
MGK 264 [568]

Ethylhexyl oleate.
Radia 7331 [658]
Rilanit EHO [401]

2-Ethylhexyl palmitate.
Crodamol OP [223]
Radia 7129 [658]

2-Ethylhexyl phosphate, acid form.
Servoxyl VPTZ 100 [788]

2-Ethylhexyl polyglycol ether phosphate, acid form.
Servoxyl VPTZ 3/100 [788]

2-Ethylhexyl stearate.
Rilanit EHS [401]

2-Ethylhexyl tallow fatty acid ester.
Rilanit EHTi [401]

Ethyl m-hydroxycarbanilate carbanilate (ester), or 3-ethoxycarbon-y-laminophenyl-n-phenylcarbamate.
Betanex [645]

Ethyl hydroxymethyl oleyl oxazoline.
Unamine OZ [529]

2,2′-Ethylidenebis (4,5-di-t-butylphenol).
Isonox 129 [770]
Vanox 1290 [894]

2-Ethyl-mercaptomethyl-phenyl-n-methyl-carbamate.
Croneton [112]

n-Ethyl-o-methyl-o-(2-chloro-4-methylmercaptaphenyl) phosphoramidothioate.
Mitemate [634]

Ethyl 3-methyl-4-(methylthio) phenyl (1-methylethyl) phosphoramidate.
Nemacur [112]

n-Ethyl-n-(2-methyl-2-proplnyl)-2,6-dinitro-4 (trifluro methyl) benzenamine.
Sonalan [282]

o-Ethyl o-ξ4-(methylthio) phenylξ s-propyl phosphorodithioate.
Bolstar [112]

Ethylmorphine hydrochloride.
Dionin [577]

Ethyl morrhuate.
Liponate EM [525]

Ethyloxide condensate.
Eccopan 600-L [277]

Ethyl paraben.
Lexgard E [453]
Nipagin A [632]
Protaben E [713]

Ethyl parathion.
Ethyl Parathion™ [602]

O-Ethyl-S,-phenylethylphosphonodithioate.
Dyfonate [821]

α -Ethyl- α -propyl-acrylaniline.
Phenex [379]

n-(1-Ethylpropyl)-3,4-dimethyl-2,6-dinitrobenzenamine.
Prowl [50]

2-(5-Ethyl-2-pyridyl) benzimidazole.
EPBZ [797]

Ethyl silicate.
Silbond [821]

S-ξ2-(Ethylsulfinyl)ethylξ 0,0-dimethyl phosphorothioate.
Metasystorf-R [112]

3-(5-ethyl-sulphonyl-1,3,4-thradiazol-2-yl) 1,3-dimethylurea.
Ustilan [112]

2-Ethylthio-4,6-bis (isopropylaminol-s-triazine.
Sancap [190]

S-Ethylthio trifluoroacetate.
Sequalog 43-7440-00 [172]

N-Ethyl o, p-toluene-sulfonamide.
Santicizer 8 [602]

o-Ethyl o-2,4,5-trichlorophenyl ethylphosphonothioate.
Agritox [112]

Ethyl urocanate.
Parasonarl Mark II [633]

Ethyl vanillin.
Ethavan® [602]

Etidronic Acid.
Turpinal SL [400]

Etocrylene.
Uvinul N-35 [109]

Europium(III) chloride.
Ultralog 44-0100-00 [172]

Europium(III) nitrate, 5H2O
Ultralog 44-0480-00 [172]

Europium(III) oxide.
Ultralog 44-0500-00 [172]

Fluocinolone acetonide *(cont'd)*.
Synalar Creams 0.025%, 0.01%, Ointment 0.025%, Topical Solution 0.01%, -HP Cream 0.2% [844]
Synemol Cream 0.025% [844]

Fluocinonide.
Lidex Cream 0.05%, Ointment 0.05%, -E Cream 0.05% [844]
Topsyn Gel 0.05% [844]

Fluorouracil.
Adrucil [7]
Efudex Topical Solutions and Cream [738]
Fluoroplex Topical Solution & Cream [403]
Fluorouracil Ampuls [738]

Fluorine & fluoride preparations.
Adeflor Chewable Tablets, Drops [884]
Fluoritab Tablets & Fluoritab Liquid [331]
Luride Drops, Lozi-Tabs Tablets [427]
Mulvidren-F Softab Tablets [831]
Pediaflor Drops [747]
Phos-Flur Oral Rinse/Supplement [427]
Point-Two Dental Rinse [427]
Poly-Vi-Flor 1.0 mg Vitamins w/ Fluoride Chewable Tablets, 0.5 mg Vitamins w/ Fluoride Drops, 1.0 mg Vitamins w/ Iron & Fluoride Chewable Tablets, 0.5 mg Vitamins w/ Iron & Fluoride Drops [571]
Thera-Flur Gel-Drops [427]
Tri-Vi-Flor 0.25 mg Vitamins w/ Iron & Fluoride Drops [571]
Vi-Daylin/F ADC Drops, ADC + Iron Drops, Chewable, Drops, + Iron Chewable, + Iron Drops [747]
Vi-Penta F Chewables, Infant Drops, Multivitamin Drops [738]

Fluorocarbon textile.
Zepel [270]

Fluorophosphoric acids.
FP Acids [670]

Fluorosalan.
Fluorophene [690a]

Fluorspar.
Pariflux [333]

Fluoxymesterone.
Android-F Tablets [136]
Fluoxymesterone Tablets [344]
Fluoxymesterone Tablets [723]
Fluoxymesterone Tablets [769]
Halotestin® [884]

Fluphenazine decanoate.
Prolixin Decanoate [817]

Fluphenazine hydrochloride.
Permitil Oral Concentrate, Tablets [772]
Prolixin Elixir & Injection, Tablets [817]

Flurandrenolide.
Cordran Ointment & Lotion, SP Cream, Tape, -N [254]

Flurazepam hydrochloride.
Dalmane Capsules [738]

Folic acid.
Added Protection III Multi-Fitamin & Multi-Mineral Supplement [589]
Al-Vite [267]
Cefol Filmtab Tablets [3]
Cevi-Fer Capsules (sustained release) [350]
Dayalets Filmtab, plus Iron Filmtab [3]
Eldec Kapseals [674]
Eldercaps [563]
Enviro-Stress with Zinc & Selenium [907]
Fero-Folic-500 [3[
Filibon F.A., Forte [516]
Folic Acid [523]
Folic Acid Tablets [240]
Folvite [516]
Folvron [516]
Hemocyte Injection, -F Tablets [888]
Hemo-Vite, Liquid [267]
Iberet-Folic-500 [3]
Ircon-FA [491]
Iromin-G [593]
Mega-B [72]
Megadose [72]
Mevanin-C Capsules [117]
Niferex-150 Forte Capsules [166]
Nu-Iron-Plus Elixir, Nu-Iron-V Tablets [563]
Orabex-TF [512]
Pramet FA [747]
Pramilet FA [747]
Prenate 90 Tablets [124]
Pronemia Capsules [516]
Stuartnatal 1+1 Tablets [831]
Trinsicon/Trinsicon M Capsules [354]
Vicon Forte Capsules [354]
Vio-Bec Forte [750]
Vitafol Tablets [304]
Zenate Tablets [723a

Formaldehyde.
Pedi-Dri Foot Powder [679]

Formaldehyde-methyl-cyclo-dodecylacetal.
Boisambreme [399], [400]

Formaldehyde sulfoxylate.
Cetalon SZ [643]

Fructose.
Fructose, No. 54016 [738]
Fru-Tabs [689]

Fumaric acid.
CWS [602a]

Furfuryl alcohol.
FA [716]

Furosemide.
Furosemide Injection [287]
Furosemide Injection, USP [674]
Furosemide Injection, USP [942]
Furosemide Tablets [344]
Furosemide Tablets [769]
Furosemide Tablets, USP [674]
Lasix Oral Solution [418]
Lasix Tablets and Injection [418], [510]
SK-Furosemide Tablets [805]

Gemfibrozil.
Lopid Capsules [674]

Gentamicin.
Gentamicin Ophthalmic Ointment & Solution [769]

Gentamicin sulfate.
Apogen® Pediatric Sterile Solution, 10 mg per ml, Apogen® Sterile Solution, 40 mg per ml, Vials, Syringe [114]
BayGent [111]
Garamycin Cream 0.1% and Ointment 0.1%, Injectable,Intrathecal Injection, I.V. Piggyback Injection, Ophthalmic Ointment-Sterile, Ophthalmic Solution-Sterile, Pediatric Injectable [772]
Gentamicin Cream 0.1% & Ointment 0.1% [769]
Gentamicin Sulfate Injection [287]
Ocu-mycin [655a]

Globulin.
Gammar, Immune Serum Globulin (Human) U.S.P. [80]
Hepatitis B Immune Globulin (Human) HyperHep [232]
Hu-Tet, Tetanus Immune Globulin (Human), U.S.P. [434]
Immuglobin [767]
Immune Globulin Intravenous, 5% (In 10% Maltose) Gamimune [232]
Immune Serum Globulin (Human), Immune Serum Globulin (Human) in Tubex [942]
Immune Serum Globulin (Human), U.S.P. Gamma Globulin [434]
Immune Serum Globulin (Human) Gamastan [232]
Imogam Rabies Immune Globulin (Human) [579]
Pertussis Immune Globulin (Human) Hypertussis [232]
Rabies Immune Globulin (Human) Hyperab [232]
Rabies Immune Globulin (Human), Imogam Rabies [579]
Tetanus Immune Globulin (Human) Hyper-Tet [232]
Tetanus Immune Globulin (Human), Tetanus Immune Globulin (Human) in Tubex [942]

Glucagon.
Glucagon for Injection Ampoules [523]

Glucoamylase (fungal).
Spezyme GA [313]

Glucose alkyl ether.
Triton BG10 [743]

Glucose glutamate.
Wickenol 545 [926]

Glucose isomerase.
Spezyme IGI [313]
Sweetzyme Type Q, Type A [654]

Glucose liquid.
Royal [216a]

Glucose oxidase.
DeeGee, Dee O [584]
Fermocozyme [313]

Glucoside, biodegradable.
Triton® BG-10 [742]

Glutamic acid hydrochloride.
Acidulin [523]
Kanulase [257]
Muripsin [649]

Glutaraldehyde.
Sepacid CE 5117, CE 5118, CE 5209 [106]
Sokalan® GDA [106]
UCAR Anti-Microbial [878]

Glutaraldehyde quaternary ammonium compounds.
Sepacid® CE 5225, CE 5265, CE 5273 [106]

Glutethimide.
Doriden Products [510]
Doriden Tablets [891]
Glutethimide Tablets [240]
Glutethimide Tablets [344]
Glutethimide Tablets [769]

Glyburide.
Micronase® [884]

Glycereth-12.
Mapeg ETG-12 [564]

Glycereth-26.
Liponic EG-1 [525]
Mapeg ETG-26 [564]

Gylceride ester, fatty.
Hodag PGO [415]

Glyceride, ethoxylated.
Syn Lube 6277 [588]

Glycerides, fatty acid, diacetyl tartaric acid ester.
Datamuls 42, 4820 [856]

Glycerides, fatty, sulfoacetate.
Emargol® L [935]

Glycerides, sulfated.
Finishing Oil O [710]

Glyceride sulfate, fatty.
Protowet FHL [710]
Sulfonated Castor Oil [710]

Glycerol monoricinoleate.
Aldo MR [358]
Alkamuls GMR [33], [34]
Drewmulse GMRO [266]
Hodag GMR, GMR-D [415]
Radiasurf 7153 [658]

Glycerol monosoyate.
Radiamuls 602, MG 2602 [658]

Glycerol monostearate.
Ahcovel Base N-15 [438]
Aldo HMS, MS, MSA, MSC, MSD, MSD FG, MS Industrial, MSLG [358]
Alkamuls GMS [33], [34]
Alkamuls GMS-45 [33]
Armostat 801 [653]
Cutina® GMS [401]
Drewmulse 20/200/V, 21, 30/900, 200, 900, GMS, TP, V, V-SE [715]
Emerest® 2400, 2401, 2403, 2407 [290]
Empilan GMS LSE40, GMS LSE 80, GMS MSE40, GMS NSE40, NSE90, GMS SE40, GMS SE70 [25]
Grocor 5500, 6000, 6000 E, 6000SE [374]
Hodag GMS [415]
Imwitor 960K [482]
Kessco-Glycerinmonostearate, rein, s.e. [16]
Kessco Glycerol Esters [825]
Lexemul 530 [453]
Mazol 165, GMS [564]
Nikkol MGS-F50, MGS-F75 [631]
Nissan Monogly-I, -M [636]
Radiamuls 141, 142, 341, 600, 601, 900, MG 2141, MG 2142, MG 2600, MG 2900 [658]
Radiasurf 7140, 7141, 7600 [658]
Rilanit GMO, GMS [401]
Tegacid Regular V.A., Special [360]
Tegin 515 V.A., V.A. [360]
Witconol™ MST [935]

Glycerol monostearate, ethoxylated.
Compound 9264B [715]

Glycerol monostearate, 12-OH.
Radiasurf 7146 [658]

Glycerol monostearate/POE 20 sorbitan tristearate, 60:40.
Aldosperse TS-40 [358]

Glycerol monostearate/polysorbate, 80:20.
Drewmulse 300 [715]

Glycerol monostearate/polysorbate 65, 67:33.
Drewmulse 365 [265]

Glycerol monostearate/polysorbate 80, 80:20.
Drewmulse 700 [715]

Glycerol myristate with ca. monoester, 40%.
GMM-33 [397]

Glycerol oleate with ca. monoester, 40%.
GMO-33 [397]

Glycerol, polyethoxylated.
Liponic EG-1 [525]

Glycerol-polyethylene tallow fatty acid ester.
Lactomul 463 [176]

Glycerol sorbitan fatty acid ester.
Arlacel 481, 986 [92]

Glycerol stearate.
Geleol [339]
Gelot 64 [339]
Kessco GMS-24, GMS-866, GMS-T [79]
Schercemol GMS [771]

Glycerol triacetate.
Kessco Triacetin [79]

Glycerol tri-fatty acid ester.
Rilanit GTC [401]

Glycerol triheptanoate.
Radianol 7376 [658]

Glycerol trioleate.
Aldo TO [359]
Alkamuls GTO [33]
Emerest® 2423 [290]
Grocor 1000, 1200 [374]
Hodag GTO [415]
Radia 7161, 7163, 7303, 7363 [658]
Rilanit GTO [401]

Glycerol trioleate, sulfated.
Actrasol EO [811]
Hipochem Dispersol GTO [410]

Glycerol tristearate.
GTS-33 [397]
Riianit GTS [401]

Glyceryl-p-amino-benzoate.
Escalol 106[10] [895]

Glyceryl p-amino-benzoic acid.
NIPA GMPA [632]

Glyceryl dilaurate.
Cithrol GDL S/E [223]
CPH-289-N [379]
Emulsynt GDL [895]

Glyceryl dioleate.
Cithrol GDO N/E, GDO S/E [223]
CPH-141-N, CPH-226-N, CPH-359-N, CPH-391-N [379]
Kessco Glycerol Dioleate [79]
Mazol GDO [564]
Nikkol DGO-80 [631]

Glyceryl distearate.
Cithrol GDS N/E, GDS S/E [223]
CPH-124-N [379]
Kessco GDS 386F [79]
Nikkol DGS-80 [631]

Glyceryl di-tri-behenate.
Compritol 888 [339]

Glyceryl di-tri-stearate.
Precirol WL 2155 [339]

Glyceryl esters.
Cithrol G Range [223]

Glyceryl hydroxystearate.
Naturechem GMHS [158]
Softigen 701 [482]

Glycerol isostearate.
Peceol Iso [339]
Schercemol GMIS [771]
Tegin ISO [360]

Glyceryl lactostearate.
GLS [223]

Glyceryl laurate.
Kessco 675, Glycerol Monolaurate, GML 54 SE, X-675 [79]
Lauricidin [574]
Monomuls 90-L 12 [375]
Protachem MLD [713]

Glyceryl monocaprylate.
Cithrol GMC [223]

Glyceryl monoisostearate.
Schercemol GMIS [771]

Glyceryl mono laurate.
Cithrol GML N/E, GML S/E [223]
CPH-12-A-SE, CPH-34-N, CPH-107-N [379]

Glyceryl monooleate.
Capmul GMO [829]
Cithrol GMO N/E, GMO S/E [223]
CPH-31-N, CPH-64-N, CPH-205-NX, CPH-362-N, CPH-385-N [379]
Cyclochem GMO [234]
Dur-Em 114, 204, GMO [272]
Grocor 2000 [374]
Nikkol MGO [631]
Tegin O Special [360]

Glyceryl monoricinoleate.
Cithrol GMR, GMR N/E, GMR S/E [223]
Flexricin® 13 [158], [642]

Glyceryl mono shortening.
Capumul GMVS-K [829]

Glyceryl monostearate.
Capumul GMS [829]
Cerasynt 945, Q, SD, WM [895]
Cithrol GMS Acid Stable, GMS A/S ES 0743, GMS N/E, GMS S/E [223]
CPH-53-N, CPH-144-N, CPH-250-SE, CPH-292-SE, CPH-381-SE [379]
Cyclochem GMS, GMS 21, GMS 165, GMS/SE [234]
Dur-Em 117 [272]
Empilan GMS Series [25]
Empilan GMS LSE 32, GMS NSE 32, GMS NSE 40, GMS SE 32, GMS SE 40 [26]
Grocor 5500 [374]
Lexemul 55G, 503, 530, 561, AR, AS, T [453]
Lipo GMS 410, GMS 450, GMS 470 [525]
Lipomulse GMS 165 [525]
Nikkol MGS-A, MGS-ASE, MGS-B, MGS-BSE, MGS-DEX, MGS-F20, MGS-F40 [631]
Protowax 90 [710]
Schercemol GMS [771]
Tegin, 90, 515, GRB, M, Special [360]
Teginacid Special [360]
Vykamol N/E, S/E [223a]

Glyceryl oleate.
CPH-353-SE [379]
Kessco Glycerol Monooleate, GMO-F [79]
Monomuls 90-O 18 [375]
Rilanit GMO [400]
Tegin O, O Spezial [360]

Glyceryl ricinoleate.
CPH-9-SE, CPH-35-N [379]
Rilanit GMRO [400]
Softigen 701 [275]
Tegin RZ [360]

Glyceryl starch.
Vulca 90 [619]

Glyceryl stearate.
Aldo HMS, MSLG [358]
Gelot 64 [339]
Kessco Glycerol Monostearate 860, Glycerol Monostearate DH-1, GMS-63-F, GMS-177-F, GMS-222-F, GMS-365, GMS-537, GMS-861, GMS-3325, Glycerol Monostearate Pure [79]
Lamecreme KSM [375]
Lexemul 55G, 503, 515, 530, 603, 615, 630, 672 SE, T, T55-SE [453]r
Lipo GMS 410, GMS 450, GMr 470, GMS 600 [525]
Mazol GMS, GMS-D [564]
Protachem 26, GMS [713]
Rewomul MG [725]
Rewopal MG [725]
Schercemol GMS [771]

Glyceryl stearate *(cont'd)*.
Tegin, Spezial, 90, 515, 4480, GRB, ISO, M, MAV [360]
Tegomuls 4100, 4101 [360]
Witconol CA, MS, MST, RHT [935]

Glyceryl triacetate.
Kodaflex Triacetin [278]

Glyceryl triacetyl recinoleate.
Naturechem GTR [158]

Glyceryl tribehenate.
Syncrowax HR-C [223]

Glyceryl tri laurate.
CPH-283-N [379]

Glyceryl trioctanoate.
Noncort TIO [640]

Glyceryl tri oleate.
CPH-72-N [379]
Grocor 1200 [374]

Glycidyl acrylate.
SR-378 [766]

Glycidyl methacrylate.
SR-379 [766]

Glycidyl tri-methyl ammonium chloride.
Ogtac-85 [182]

Glyco amido stearate.
Cerasynt IP [895]

Glycol bori-borate.
Aquaresin [358]

Glycol carbohydrate complex.
Glucarine B [358]

Glycol dibehenate.
Rewopol PG 340 [725]

Glycol distearate.
Felsapon-GDS [823]
Kessco Ethylene Glycol Distearate [79]
Lexemul EGDS [453]
Lipo EGDS [525]
Mapeg EGDS [564]
Nikkol PEARL-1222 [631]
Pearl Luster Agent MS [55]
Pearlescing Agent MS [55]
Pegosperse 50 DS [358]
Rewopal PG 280 [725]

Glycol ester.
Atlas EM-2, EM-3, EM-13 [94]
EGMS (Ethylene Glycol Monostearate) [453]
EM-40, EM-400, EM-600 [483]
Lexolube 2J-237, 2N-212, 2N-237, 2T-237 [453]
Lexomul PEG 200 DL, PEG 400 DL, PEG 400 DO, PEG 400 ML, PEG 400 MO [453]
PL-2N-204, PL-2N-204 RG, PL-2T-237 [453]
Witbreak DGE-60, DGE-75, DGE-85, DGE-128A, DGE-169, DGE-182 [934], [935], [937a], [938]
Witbreak DRA-21, DRA-50 [935], [937a], [938]

Glycol ester, ethoxylated.
Lipal [715]

Glycol ester, phosphated.
ESI-Terge 330 [294]
ESI-Terge P-330 [892]

Glycol ethers.
Gafcol [338]
Jaysolve [296]

Glycol, fatty acid ester.
Keripon NC [77]
Witconol™ F26-46 [935]

Glycol fatty ester.
Emulser DG [77]

Glycol hydroxystearate.
Naturechem EGHS [158]

Glycol mono di esters.
Mackester Series [567]

Glycol palmitate, C14-16.
Mexanyl GR [185]

Glycol stearate.
Kessco Ethylene Glycol Monostearate [79]
Lexemul EGMS [453]
Lipo EGMS [525]
Mapeg EGMS [564]
Monthybase [339]
Monthyle [339]
Nikkol PEARL-1218 [631]
Pegosperse 50 MS [358]
Protachem EGMS [713]
Schercemol EGMS [771]
Tegin G [360]

Glyco monostearate.
Felsapon-GMS [823]

Glycopyrrolate.
Glycopyrrolate Tablets [240]
Robinul Forte Tablets, Injectable, Tablets [737]

Glycylglycine.
Chemalog Grade 47-4880-00 [172]

Glycyrrhetinyl stearate.
SGS [557]

Glyphosate, isopropylamine salt.
Roundup® Glyphosate [602]

Gonadorelin hydrochloride.
Factrel [97]

Grape seed oil.
Lipovol G [525]

Graphite.
Asbury 230-U [85]
Grafoil [411]
National Boronated Graphites [878]
Poco [699]
Prodag [71]

Graphite, amorphous.
Asbury 505 [85]

Graphite, artificial.
Ashbury 4078 [85]
Asbury A-98 [85]

Graphite fiber.
Magnamite® [406]

Graphite yarn.
Thornel [878]

Griseofulvin.
Fulvicin P/G Tablets, 165 & 330 Tablets, -U/F Tablets [772]
Grifulvin V [666]
Grisactin, Ultra [97]
Gris-PEG Tablets [258]

Groundnut fatty acid.
Loramine DO 280/S [274]

Guaifenesin.
Actifed-C Expectorant [139]
Adatuss DC Expectorant Cough Syrup [560]
Ambenyl-D Decongestant Cough Formula [551]
Anatuss Tablets & Syrup, with Codeine Syrup [563]
Asbron G Inlay-Tabs/Elixir [258]
Breonesin [133]
Brexin Capsules & Liquid [767]
Bromphen DC Expectorant, Expectorant [769]
Brondecon [674]
Bronkolixir [133]
Bronkotabs [133]
Codimal Expectorant [166]
Conex, with Codeine [661]
Congess Jr. & Sr. T.D. Capsules [326]
Coryban-D Cough Syrup [690]
Cremacoat 2, 3 [902]
Decongestant Expectorant [769]
Deproist Expectorant w/ Codeine [344]
Detussin Expectorant [769]
Dilaudid Cough Syrup [495]
Dilor-G Tablets & Liquid [767]
Donatussin DC Syrup, Drops [513]
Dorcol Pediatric Cough Syrup [257]
Dura-Vent [271]
Elixophyllin-GG [115]
Emfaseem Capsules [765]
Entex Capsules, LA Tablets, Liquid [652]
Ex-Span Tablets [748]
Fedahist Expectorant [746]
Guaifed Capsules (Timed Release) [612]
Guaifenesin Syrup [751]
Guaituss Syrup, A-C Syrup, D-M Syrup [769]
Head & Chest [709]
Histalet X Syrup & Tablets [723a]
Hycotuss Expectorant [269]
Hytuss Tablets and Hytuss-2X Capsules [436]
Lufyllin-GG [911a]
Mudrane GG Elixir, GG Tablets, GG-2 Tablets [703]
Naldecon-CX Suspension, -DX Pediatric Syrup, -EX Pediatric Drops [134]
Novahistine Expectorant [580]
Nucofed Expectorant, Pediatric Expectorant [114]
Poly-Histine Expectorant Plain, Expectorant with Codeine [124]
Pseudo-Bid Capsules, -Hist Expectorant [420]
P-V-Tussin Tablets [723a]
Quibron & Quibron-300, Quibron Plus [572]
Respaire-SR Capsules [513]
Rhindecon-G Capsules [566]
Robitussin A-C, -DAC [737]
Robitussin Syrup [510]
Rymed Capsules, -JR Capsules, Liquid, -TR Capsules [281]
Ryna-CX Liquid [911a]
Slo-Phyllin GG Capsules, Syrup [746]
Sorbutuss [239]
S-T Forte Syrup & Sugar-Free [779]
Synophylate-GG Tablets/Syrup [166]
Tedral Expectorant [674]
Theolair-Plus Tablets & Liquid [734]
T-Moist Tablets [930]
Triafed-C Expectorant [769]
Triaminic Expectorant, Expectorant w/ Codeine [257]
Trinex Tablets [560]
Tussar SF, -2 [891]
Tussend Expectorant [580]
Unproco Capsules [723]
Zephrex Tablets, -LA Tablets [124]

Guanadrel sulfate.
Hylorel Tablets [682]

Guanethidine sulfate.
Ismelin [189]

Guanidine acetate 99.99+%.
Ultralog Grade 47-7800-00 [172]

Guanidine carbonate 99.99+%.
Ultralog Grade 47-7860-00 [172]

Guanidine hydrochloride 99.99+%.
Ultralog Grade 47-7920-00 [172]

Guanidine nitrate 99.99+%.
Ultralog Grade 47-7960-00 [172]

Guanidine phosphate dibasic 99.99+%.
Ultralog Grade 47-8040-00 [172]

Guanidine phosphate monobasic 99.99+%.
Ultralog Grade 47-8060-00 [172]

Guanidine sulfate 99.99+%.
Ultralog Grade 47-8100-00 [172]

Guar gum.
Jaguar Gum #124, Gum A-20-D, Plus [163]

Guar gum, refined.
Guar Gum [406]

Guar hydroxypropyl trimonium chloride.
Cationic Guar C-261 [399], [400]
Jaguar C-13, C-13S, C-13SD, C-15, C-17 [163]

Gum arabic.
Acasol [199a]
Emulgum [199a]
Whipgum [199a]

Hectorite.
Ben-A-Gel [640a]
Bentone E-W Rheological Additive [641a]
Macaloid [640a]

HEDTA.
SEQ 120 [903]

HEEDTA, trisodium salt.
Cheelox® HE-24 [338]

Heliotropin, sodium bisulfite complex.
Helopex [800]

Hematinic iron products.
Jefron [261]

Hemicellulose extract.
Masonex [559]

Hemisulfur mustard seed.
HSM [128]

1,4,5,6,7,8,8-Heptachloro-3a.4.7.7a-tetrahydro-4,7-methanoindene.
Drirox H-34 [898]

2-Heptadecyl-2-imidazoline acetate.
Crag Fruit Fungicide 341 [12a]

Heptadecyl imidazoline, quaternized.
Monaquat ISIES [599]

Heptadecyl sodium sulfate.
Rexowet 77 [292]

Heptakis (dipropylene-glycol) triphosphite.
Weston PTP [129]

Heptane 99+%.
Sequalog Grade 48-7100-00 [172]

Heptyl acrylate.
Norsocryl [651]

2-n-Heptylcyclopentanone.
Frutalone -112730 [406]

Hexabromocyclododecane.
Great Lakes CD-75P [369]
Saytex HBCD [302]

Hexachloroacetone.
HCA [35]

1,2,3,4,5,6-Hexachlorocyclohexane.
HexaFlor [730]

1,2,3,4,5,6- Hexachlorocyclohexane, α isomer.
Exagama [730]

Hexachlorocyclopentadiene.
PCL [898]

Hexachloroepoxyoctahydro-endo, endodimethanonaphthalene.
Endrex [794], [898]

Hexachloro-epoxy-octahydroendo, exo-dimethanonaphthalene.
Dieldrex [794]

1,2,3,4,10-Hexachloro-1,4,4a,5,8a-hexahydro-exo-1,4-endo-5,8-dimethanonaphthalene.
Aldrex [794]

6,7,8,9,10,10-Hexachloro-1,5,5a,6,9,9a-hexahydro-6,9-methano-2,4,3-benzodioxa-thiepin-3-oxide.
Thiodan [331a]

Hexachlorophene.
G-11 [353]

(Z,Z)-7,11-Hexadecadien-L-ol-acetate (Z,E)-7,11-hexadecadien-L-ol-acetate.
No Mate PBW [21]

n-Hexadeconal.
Cyclogol Cetyl Alcohol NF [234]
Rilanit G 16 [401]

n-Hexadecanol, n-octadecanol.
Cyclogol Cetyl-Stearyl Alcohol [234]

z-11-Hexadecenal.
No Mate Chokegard [21]

n-Hexadecylamine.
Amine 16D [487]
Lilamin 16D [487]

Hexadecylamine, primary.
Armeen 16D [79]
Kemamine P-880 [429a]
Nissan Amine PB [636]

2-Hexadecyl stearate.
Grocor 5820 G [374]

Hexadecyl amine primary, distilled.
Kemamine P-880 D [429a]

n-Hexadecyl 3,5-di-t-butyl-4-hydroxybenzoate.
Cyasorb UV 2908 [50]

Hexadecyldimethylamine.
Adma™ 6 [302]
Armeen DM16D [16]
Nissan Tertiary Amine PB [636]
Onamine 16 [663]

Hexadecyldimethylamine oxide.
Aromox DM16D-W [16]

Hexadecyl diphenyl ether disulfonate, sodium salt.
Conco Sulfate 4C3 [209]

n-Hexadecyl ether, ethoxylated.
Nikkol BC-1SY thru BC-8SY [631]

n-Hexadecyl 2-ethylhexanoate.
Excepar1 HO [478]

Hexadecyl trimethyl ammonium chloride.
Nissan Cation PB-40, PB-300 [636]

N-hexadecyl-N,N,N-trimethyl ammonium-p-toluene sulfonate.
Cetats [408]

Hexaethylene glycol dodecyl ether.
Nikkol BL-6SY [631]

1-1-1-3-3-3-Hexafluoroisopropanol 99+%.
Sequalog 49-3800-00 [172]

Hexaglycerol dioleate.
Caprol 6G20 [829]

Hexaglycerol distearate.
Aldo HGDS [358]
Caprol 6G2S [829]
Drewmulse 6-2-S [715]
Drewpol 6-2-S [715]
Hodag SVO-629 [415]
Polyaldo HGDS [358]

Hexahydrotriazine.
Soi Bact G [807]

Hexahydro-1,3,5-triethyl-S-triagine.
Vancide TH [894]

Hexahydro-1,3,5-tris (2-hydroxethyl)-s-triazine.
Onyxide 200 [663]

Hexakis (2-methyl-2-phenylpropyl)-distannoxane.
Torque [794]
Vendex [791]

Hexamethoxymethylmelamine.
Cymel 303 [52a]
Cyrez 963 Resin, 965 [50]

Hexamethyldisiloxane.
Abil K 520 [856]

N,N i-Hexamethylene bis (3,5-di-t-butyl-4-hydroxy-hydro-cinnamamide).
Irganox 1098 [190]

1,6-Hexamethylene bis-(3,5-di-t-butyl-4-hydroxyhydrocinnamate).
Irganox 259 [190]

Hexamethylene diamine carbamate.
Diak No. 1 [269]

Hexamethylene diamine tetra (methylene phosphonate).
Dequest 2051 [602]

Hexamethyleneimine.
HMI [163]

Hexamethylenetetramine masterbatch.
Mastermix Hexa 400 MB [392]

Hexamethylenetetramine paste.
Mastermix Hexa 402 PD [392]

Hexamethyl tetracosane.
Squalane [223]

1,6-Hexanediol diacrylate/hydroquinone inhibitor, 100 ppm.
SR-238 [766]

1,6-Hexanediol methacrylate/hydroquinone inhibitor, 100+- 25 ppm.
SR-239 [766]

Hexapotassium hexamethylene diamine tetra (methylene phosphonic acid).
Dequest® 2054 [602]

Hexyl acrylate.
Norsocryl [651]

Hexylcaine hydrochloride.
Cyclaine [577]

2-Hexyl decanol-1.
Primarol 1006 [400]
Standamul G-16 [400]

2-Hexyldecyl stearate.
Grocor 5820 G [374]

Hexyl laurate.
Cetiol A [399], [400]
Standamul CTA [400]

n-Hexyl methacrylate.
SR-211 [766]

Hexyloxypropylamine.
Tomah PA-10 [866]

Hexyl polyglycol ether phosphate, acid form (EO 1).
Servoxyl VPWZ 1/100 [788]

DL-Homoalanin-4-yl(methyl)phosphinic acid.
Basta [416]

Homomenthyl salicylate.
Kemester HMS [429]

Honey.
Lipo HON [525]

Hyaluronic Acid.
Protein Extract Q [424]

Hyaluronidase.
Alidase [782]

Hydralazine HCL.
Apresoline Tablets [510]

Hydrazine.
Deoxy-Sol [307]

Hydrazine anhydrous.
Sequalog Grade 50-6619-00 [172]

Hydroabietyl alcohol.
Abitol® [406]

Hydrobromic acid.
Hydrobromic Acid [904]

Hydrochloric acid.
Acculog 50-8000-00, 50-8000-10, 50-8000-20 [172]

Hydrochlorothiazide.
Hydrochlorulan Tablets [510]

Hydrocodone bitartrate.
Hydrogesic Tablets CIII [281]

Hydrocodone bitartrate + aspirin, 7 mg:325 mg.
Anexsia® -D [114]

Hydrocodone bitartrate, pseudoephedrine HCL, guaifenesin, 5 mg:60 mg:per 200 mg:5 ml.
SRC Expectorant CIII [281]

Hydrocortisone.
Cortef® Sterile Suspension IM and Tablets [884]
Dermacort [750]
Proctocort [750]

Hydrocortisone acetate.
Cortaid® [884]
Cortef®, Cortef Acetate® [884]

Hydrocortisone cream.
Theracort [377a]

Hydrocortisone cypionate.
Cortef® [884]

Hydrocortisone sodium succinate.
Solu-Cortef® [884]

Hydrogen peroxide.
Super D [332]

Hydrogen phosphate, C_{12}-C_{14} alcohol.
Merpophal SP [285]

Hydrogenphosphate, $C_{16/18}$ alcohol.
Merpiphos B 52 [285]

Hydroquinone.
Rodol HQ [531]
Tecquinol [278]

Hydroquinone, alkylated.
Antioxidant 451 [881]

Hydroquinone monomethyl ether.
Eastman HQMME [278]

Hydrolysate.
Sterling CHP [145]

2-Hydroxy-4-acryloxyethoxy-benzophenone.
Cyasorb UV2126 [50]

Hydroxyamine phosphate ester.
Fostex T [400]

Hydroxy anisole, butylated.
Sustane BHA 1-F [883]
Tenox BHA [278]

p-Hydroxy benzoates.
Aseptoform [370]

p-Hydroxy-benzoic acid, esters.
Parasept [850]

1-Hydroxybenzotriazole monohydrate.
Sequalog Grade 51-1050-00 [172]

Hydroxybutyl methylcellulose.
Methocel HB [261]

Hydroxy-citronellal-methyl anthranilate (methyl N-3,7-dimethyl-7-hydroxyoctylidene-anthranilate).
Aurantiol [353]

2(2′-Hydroxy-3′,5′-di-t-amylphenyl) benzotriazole.
Tinuvin 328 [190]

2(2′-Hydroxy-3′,5′-di-t-butylphenyl) benzotriazole.
Tinuvin 320 [190]

2-[2-Hydroxy-3-5,-di-(1,1-dimethylbenzyl)phenyl)-2H-benzotriazole.
Tinuvin 900 [190]

4-Hydroxy-3,5 diiodobezonitrile.
Iotril [540]

Hydroxydimethylarsine oxide.
Rad-E-Cate 25 [904]

3-Hydroxy-N,N-dimethyl-cis-crotonamide, dimethyl phosphate.
Bidrin [791]

Hydroxyethane diphosphonic acid.
Fostex P [399], [400]

1-Hydroxyethyl-2-alkyl-imidazoline.
Rewopon IM-CA, IM-OA [725a]

N-Hydroxyethyl-2-amino-4-hydroxytoluene sulfate.
Rodol EACS [531]

1-Hydroxyethyl-1-benzyl-2-alkyl imidazolinium chloride.
Uniquat CB-50 [529]

Hydroxyethyl-cellulose.
Natrosol [404a], [406]
Natrosol 150, 250 [406]
Tylose H Series [416]
Viscontran HEC [400]

Hydroxy ethyl coco imidazoline.
Servamine KOD 360 [788]
Unamine C [529]

N-(Hydroxyethyl)ethylenediamine-triacetate trisodium salt.
Cheelox HE-24 [338]
Chelon 120 [821]

N-Hydroxyethylethylenediaminetriacetic acid.
Hamp-ol Acid [364b]

Hydroxyethylethylenediaminetriacetic acid, trisodium salt.
Chel DM-41 [190]
Hamp-ol® 120 [364b]
Trilon D Liquid [106]
Versenol 120 [261]

N-Hydroxyethylethylenediaminetriacetic acid, trisodium salt dihydrate.
Hamp-ol® Crystals [364b]

1-Hydroxyethyl-2-heptadecenyl imidazoline.
Loramine IM-CA, IM-OA [274]
Onamine RO [663]
Rewomine IM-CA, IM-OA [274]
Steinamin IM-CA, IM-OA [274]

1-(2-Hydroxyethyl)-2-[(Z)-8-heptadecyl]-2-imidazoline.
Miramine OC [592]

Hydroxyethyl hydroxypropyl C12-15 alkoxypropylamine oxide.
Varox 185 PH [796]

Hydroxy-ethylidene diphosphonic acid.
Dequest 2010 [602]

Hydroxyethylimidazoline.
Servamine KOO 360 [788]

Hydroxyethyl imidazoline; oleic hydrophobe.
Alkazine O [33]

1-(2-Hydroxyethyl)-2-norcoco-2-imidazoline.
Miramine CC [592]

1-(2-Hydroxyethyl)-2-norsoya-2-imidazoline.
Miramine SC [592]

2-Hydroxyethyl-n-octyl sulfide.
MGK Repellent 874 [568]

1-Hydroxyethyl-2-oleyl-imidazoline.
Unamine O [529]

Hydroxyethyl stearyl amide.
Cerasynt D [895]

1-Hydroxyethyl-2-stearyl imidazoline.
Unamine S [529]

1-Hydroxyethyl-2-tall oil imidazoline.
Unamine T [529]

2-Hydroxy fatty alcohol-alkoxylate.
Eumulgin L [401]

2-Hydroxy-4-isooctoxy-benzophenone.
Carstab 701, 702 [192]

2-Hydroxy-4-methoxy-benzophenone.
Acetorb A [5]
Cyasorb UV9 [50]
Syntase 62 [624]
UV-Absorber 325 [597]
Uvasorb MET [862a]
Uvinol M-40 [109], [338]

2-Hydroxy-4-methoxy-benzophenone-5-sulfonic acid.
Uvinul MS-40 [338]

2[(Hydroxymethyl)-amino] ethanol.
Troysan 174 [869]

2[(Hydroxymethyl)-amino]-2-methylpropanol.
Troysan 192 [869]

3-Hydroxy-N-methyl-(cis)-crotonamide, dimethyl phosphate.
Azodrin [791]

3-Hydroxy-5-methylisoxazole.
Tachigaren [763]

2(2′-Hydroxy-5′-methylphenyl) benzotriazole.
AM-600 [315]
Mixxim LS-14 [307a]
Tinuvin P [190]
Uvasorb SV [862a]
Vanox UV-1 [894]
Viosorb 520 [5]

N-((1S)-(1,4,6/5)-3-Hydroxymethyl-4,5,6-trihydroxy-2-cyclohexenyl) (O-beta-D-glucopyrano-syl-(1⇒3)-(1S)-(1,2,4/3,5)-2,3,4-trihydroxy-5-hydroxy-methylcyclohexyl)amine.
Validacin [848]

Hydroxy octadecyl methyl glucamine.
Carsamine G-18 [529]

2-Hydroxy-4-n-octoxy-benzophenone.
Carstab 700, 705 [192]
Mark 1413 [75]
Syntase 800 [624]
Uvinul 408 [109]

2-(2-Hydroxy-5-t-octyl-phenyl) benzotriazole.
Cyasorb UV-5411 [50]

Hydroxypropylcellulose.
Klucel®, 6, H, HF, HP, M [406]

Hydroxypropyl guar.
Gelcharg HP4 [400]
Jaguar HP-60 [163]

Hydroxypropyl methylcellulose.
Methocel E, F, J, K [261]
Viscontran MHPC [400]

6-Hydroxy-3 (2H) pyridazinone.
Royal MH-30 [881]

5-(α -Hydroxy- α -2-pyridylbenzyl)-7-(α -2-pyridylbenzylidene)-5-norbornene-2,3-dicarboximide.
Raticate [565]

12-Hydroxystearic acid.
Radiacid 200 [658]
12-HSA [158]

N-Hydroxysuccinimide.
Sequalog Grade 52-6440-00 [172]

4-Hydroxy-3-(1,2,3,4-tetrahydro-1-naphthyl) coumarin.
Racumin [112]

Hydroxy toluene, butylated.
BHT [5]
Butylated Hydroxytoluene [882]
CAO® -1, -3 Food Grade [797]
Sustane BHT [883]

Hydroxy urea.
Hydrea [817]

Hydroxyzine HCI.
Atarax Tablets [510]

Hydroxyzine pamoate.
Vistaril Capsules [510]

Hygromycin B.
Hygromix [282]

Hyoscyamine compound.
Donnatol Products [510]

Iron preparations *(cont'd).*
Hemocyte Plus Tabules, -F Tablets [888]
Hemo-Vite, Hemo-Vite Liquid [267]
Heptuna Plus [739]
Hytinic Capsules & Elixir [436]
I.L.X. B_{12} Elixir Crystalline, Tablets [489]
Iberet, -500, -500 Liquid, -Folic-500, Liquid [3]
Imferon [580]
Iromin-G [593]
Isomil, SF, SF 20 [747]
Mevanin-C Capsules [117]
Niferex Forte Elixir, Tablets/Elixir, w/ Vitamin C, -150 Capsules, -150 Forte Capsules, -PN [166]
Nu-Iron Elixir, 150 Caps, -Plus Elixir, -V Tablets [563]
Perihemin [516]
Peritinic Tablets [516]
Poly-Vi-Flor 1.0 mg Vitamins w/ Iron & Fluoride Chewable Tablets, 0.5 mg Vitamins w/ Iron & Fluoride Drops [571]
Prenate 90 Tablets [124]
Pronemia Capsules [516]
Similac With Iron, With Iron 13, With Iron 20, With Iron 24 [747]
Stuartinic Tablets [831]
Tabron Filmseal [674]
TriHemic 600 [516]
Tri-Vi-Flor 0.25 mg Vitamins w/ Iron & Fluoride Drops [571]
Vi-Daylin Plus Iron ADC Drops, Plus Iron Chewable, Plus Iron Drops, Plus Iron Liquid, /F ADC + Iron Drops, /F + Iron Chewable, /F + Iron Drops [747]

Isethionate, sodium salt, fatty acid.
Hostapon KA Special [416]

Iso-C_{13} alcohol 4 polyglycol ether.
Merpoxen TR 40 [285]

Iso-C_{13} alcohol 7 polyglycol ether.
Merpoxen TR 70 [285]

Iso-C_{13} alcohol 10 polyglycol ether.
Merpoxen TR 100 [285]

Iso-C_{13} alcohol 4-polyglycolether hydrogen phosphate.
Merpiphos PP [285]

Iso-C_{13} alcohol 6.5 polyglycolether hydrogen phosphate.
Merpiphos UP [285]

2-Isoalkyl (C_{14}-C_{20}), 1-hydroxyethyl, 1-ethyl imidazolinium ethyl sulfate.
Schercoquat IIS-RD [771]

Isoalkylsulfate, sodium salt.
Sulfetal 4105 [948]

Isobornyl thiocyanoacetate.
Thanite [406]

Isobutane.
A-31 [9], [695]
Hydrocarbon Propellant A-31 [695]

Isobutoxypropanol.
Dowanol PIB-T [261]

Isobutyl acrylate.
Norsocryl [651]

1-Isobutylcarbomoyl imidazolin-2-one.
Merpelan AZ [112]

Isobutylene.
Amcycle Butyl [61], [422], [611]

Isobutyl isobutyrate.
Eastman IBIB [278]

Isobutyl oleate.
Radia 7230 [658]

Isobutyl stearate.
Ibulate [507]
Kessco Isobutyl Stearate [79]
Radia 7240, 7241 [658]
Uniflex IBYS [876]

Isobutyraldehyde oxime/mineral spirits, 50:50.
Isobutyraldehyde Oxime [63]

Isobutyric acid and propionic acid, 60:40%.
Tenox IBP-2 [278]

Isocarboxazid.
Marplan Tablets [738]

Isoceteareth-8 stearate.
Isoxal E [901]

Isoceteth-10.
Nikkol BEG-1610 [631]

Isoceteth-20.
Arlasolve 200 [438]
Nikkol BEG-1620 [631]

Isoceteth-30.
Nikkol BEG-1630, BH-30 [631]

Isocetyl alcohol.
Cyclal G16 [939]
Michel XO-144 [581]
Standamul G-16 [400]

Isocetyl alcohol hexadecanol.
Standamul G-16 [399], [400]

Isocetyl isostearate.
Nikkol ICIS [631]

Isocetyl stearate.
Ceraphyl 494 [895]
Crodamol ICS [223]

Isocetyl stearate *(cont'd).*
Lexol HDS [453]
Nikkol ICS-R [631]
Schercemol ICS [771]
Standamul 7061 [399], [400]

Isocetyl stearoyl stearate.
Ceraphyl 791 [895]

Isocyanate, blocked.
Chemiflex 315XA(80) [764]

Isocyanate foam systems.
Isomate [884]

Isocyanate, polyfunctional aliphatic.
Desmodur N [112]

Isocyanato-urethane.
Desmodur KN [112]

Isodeceth-5.
Carsonon D-5 [529]
Macol W-5 [564]

Isodecyl acrylate.
SR-395 [766]

Isodecyl diphenyl phosphate.
Santicizer 148 [602]

Isodecyl isononanoate.
Wickenol 152 [926]

Isodecyl methacrylate.
SR-242 [766]

Isodecyl oleate.
Ceraphyl 140A [895]

Isodecylparaben.
Keratoplast Codex 01314 [901]

d-Isoephedrine hydrochloride.
Isoetharine Hydrochloride Inhalation [751]

Isoetharine.
Bronkometer [133]
Bronkosol [133]

Isoetharine hydrochloride.
Arm-A-Med™ [80]

Isoflupredone acetate.
Predef® [884]

2-Isoheptadecyl, 1-hydroxyethyl, 1-benzyl imidazolinium chloride.
Schercoquat IIB [771]

2-Isoheptadecyl, 1-hydroxyethyl, 1-ethyl imidazolinium ethyl sulfate.
Schercoquat IIS [771]

Isohexacosanol.
Standamul GTO-26 [400]

Isohexyl laurate.
Kessco X-650 [79]

Isolaureth-6.
Tergitol TMN-6 [878]

Isometheptene mucate.
Midrin Capsules [153]
Migralam Capsules [505]

Isoniazid.
INH Tablets [189]
Isoniazid Tablets [240]
Laniazid Tablets [510]
Rifamate [580]

Isononyl isononanoate.
Wickenol 151 [926]

Isononyl stearate.
Radia 7510 [658]

Iso-octyl iso-decyl phthalate.
Staflex ODP [722]

Isocetyl oleate.
Radia 7330 [658]

Isooctyl polyglycolester phosphate, sodium salt.
Merpiphos A 30/50 [285]

Isooctyl acrylate.
Norsocryl [651]

Iso-octyl stearate.
Cetiol 868 [401]
Radia 7130, 7131 [658]

Isooctyl tallate.
PlastHall 100 [379]

Isoparaffin, C7-8.
Isopar C [306]

Isoparaffin, C8-9.
Isopar E [306]

Isoparaffin, C9-11.
Soltrol 100 [695]

Isoparaffin, C9-14.
Soltrol 145 [695]

Isoparaffin, C10-11.
Isopar G [306]

Isoparaffin, C10-13.
Soltrol 130 [695]

Isoparaffin, C11-12.
Isopar H, K [306]

Isoparaffin, C11-13.
Isopar L [306]

Isoparaffin, C13-14.
Isopar M [306]

Isoparaffin, C13-16.
Soltrol 220 [695]

Isopropamide iodide.
Combid Spansule Capsules [805]
Darbid [805]

Isopropamide iodide *(cont'd)*.
Ornade Spansule Capsules [805]
Prochlor-Iso Timed Release Capsules [769]
Pro-Iso Capsules [344]

Isopropanolamine lauryl sulfate.
Carsonol ILS [529]
Incronol ILS [223b]

3-Isopropoxycarbonyl-aminophenyl-n-ethyl carbamate.
Verdinal [772]

(E)-0-2-Isopropoxy-carbonyl-1-methylvinyl O-methyl ethylphosphoramidothioate.
Safrotin [760a]

o-Isopropoxyphenyl methylcarbamate.
Baygon [184]
Isocarb [530]

Isopropyl alcohol.
Cetylcide Solution [168]
Komed Acne Lotion, HC Lotion [103]
Tinver Lotion [103]

Isopropyl alkyl benzene sulfonate, amine branched.
Polystep A-11 [825]

Isopropylamine.
Mono Isopropylamine (MIPA) [906]

Isopropylamine alkylbenzene sulphonate.
Gardilene IPA [23], [24]
Pentine 1185 [198]

Isopropylamine dodecylbenzenesulfonate.
Alkasurf IPAM [33]
Calimulse, Calimulse PRS [697]
Conco AAS Special 3 [209]
Nansa YS 94 [23], [25], [548]
Richonate YLA [732]
Rueterg IPA [320]

Isopropylamine sulfonate.
Calimulse PRS [697]

Isopropyl 4-aminobenzene sulfonyl, di(dodecyl benzenesulfonyl) titanate.
Ken-React KR-26S [488]

2-Isopropylamino-4-(3-methoxypropylamino)-6-methylthio-1,3,5-triazine.
Gesaran [190]

2-Isopropylamino-4-methyl=amino-6-methylthio-s-triazine.
Semeron [190]

R-(-) Isopropyl n-benzoyl-n-(3-chloro-4-fluorophenyl)-2-aminopropionate.
Suffix BW [794]

Isopropylbenzylsalicylate.
Megasol [901]

Isopropyl carbanilate.
Beet-Kleen [792]
Chem Hoe [561a], [705]
Premalox [561a]
Triherbide IPC [683]

Isoprpyl m-chlorocarbanilate.
Beet Kleen [792]
Chloro IPC [683], [705]
Furloe [705]
Sprout Nip [705]

Isopropyl di(4-amino benzoyl) isostearoyl titanate.
Ken-React KR-37S [488]

Isopropyl 4,4′-dibromobenzilate.
Acarol [190a]
Folbex VA [190]
Neoron [190]

4-Isopropyl-5,5-dimethyl-1,3-dioxane.
Anthoxan [399], [400]

Isopropyl ester dimethylammonium methylsulfate, difatty acid.
Rewoquat CR 3099 [725a]

4,4′-Isopropylidenediphenol.
Bisphenol A [261]

Isopropyl isostearate.
Emerset 2310 [290], [291]
Lan-O-Derm [197]
Schercemol 318 [771]
Wickenol 131 [926]

Isopropyl lanolate.
Amerlate P [44]
Crodalan IPL [223]
Lanisolate [507]
Ritasol [735]

Isopropyl laurate.
Crodamol IPL [223]

Isopropyl linoleate.
Ceraphyl IPL [895]
Wickenol 129 [926]

Isopropyl myristate.
Crodamol IPM [223]
Deltyl Extra [353]
Domol Bath & Shower Oil [585]
Emerest 2314 [289], [291]
Estol 1512, 1513, 1514 [875]
I.P.M. [18], [401]
Isopropyl Myristate 2/044111 [263]
JA-FA IPM [465]
Kessco-Isopropylmyristat [16]
Kessco Isopropyl Myristat [79]
Kesscomir [79]
Lanesta-31 [507]
Lexol IPM [453]
Liponate IPM [525]
Promyr [44]

Isopropyl myristate *(cont'd).*
Radia 7190 [658]
Schercemol IPM [771]
Starfol IPM [87]
Stepan D-50 [825]
Unimate IPM [876]
Wickenol 101 [926]

Isopropyl myristo-palmitate.
Radia 7220 [658]

Isopropyl oleate.
JA-FA IPO [465]
Radia 7231 [658]
Schercemol IPO [771]

Isopropyl palmitate.
Crodamol IPP [223]
Deltyl Prime [353]
Emerest 2316 [289], [291]
Estol 103, 1517 [875]
IPP [401]
Isopropyl Palmitate 2/044121 [263]
JA-FA IPP [465]
Kessco-Isopropylpalmitat [16]
Kessco Isopropyl Palmitate [79]
Lexol IPP [453]
Liponate IPP [525]
Propal [44]
Radia 7200 [658]
Starfol IPP [87]
Stepan D-70 [825]
Unimate IPP [876]
Wickenol 111 [926]

(o)-isopropyl-phenol.
Prodox® 131, 133 [315]

Isopropyl N-phenyl-carbamate.
Chem-Hoe [705]

N-(4-Isopropylphenyl)-N',N'-dimethylurea.
Arelon, Arelon P [416]
Belgran [730]
Doublet [730]
Graminon [190]
IP 50, IP Flo [730]
Prodix [730]
Tolkan [730]
Twin-Tak [730]

N-isopropyl-N'-phenyl-p-phenylene diamine.
Akrochem Antioxidant PD-1 [15]
Flexzone 3-C [881]
Permanax IPPD [212], [910]
Santoflex IP [392], [602]
Vulkanox 4010 NA [597]

Isopropyl stearate.
IPS [401]
JA-FA IPS [465]
Kessco-Isopropylstearat [16]
Radia 7195 [658]
Unimate IPS [876]
Wickenol 127 [926]

Isopropyl tri(dioctyl phosphate) titanate.
Ken-React KR-12 [488]

Isopropyl tri(dioctyl pyrophosphate) titanate.
Ken-React KR-38S [488]

Isoproterenol hydrochloride.
Arm-A-Med™ [80]
Isoproterenol HCI Injection [287]
Mucomyst with Isoproterenol [572]
Vapo-Iso Solution [324]

Isoproterenol preparations.
Aerolone Solution [523]
Duo-Medihaler [734]
Isuprel Hydrochloride Compound Elixir, Glossets, Solution 1:200 & 1:100, Injection 1:5000, Mistometer [133]
Medihaler-Iso [734]
Norisodrine Aerotrol, Sulfate Aerohalor, w/ Calcium Iodide Syrup [3]

Isosorbide dinitrate.
Angicon [512]
Dilatrate-SR [719]
Iso-Bid Capsules [350]
Isordil Chewable (10 mg.), Oral Titradose (5 mg.), Oral Titradose (10 mg.), Oral Titradose (20 mg.), Oral Titradose (30 mg.), Sublingual 2.5 mg., 5 mg. & 10 mg. [463]
Isordil Tablets [510]
Isordil Tembids Capsules & Tablets (40 mg.), (10 mg.) w/ Phenobarbital (15 mg.) [463]
Isosorbide Dinitrate Oral Tablets, Dinitrate Sublingual Tablets [240]
Isosorbide Dinitrate T.D. Capsules & Tablets [344]
Isosorbide Dinitrate Tablets [344]
Isosorbide Dinitrate Tablets - Oral & Sublingual [769]
Isosorbide Dinitrate Timed Capsules & Tablets [769]
Isotrate Timecelles [394]
Sorate-5 Chewable Tablets, -2.5 Sublingual Tablets, -5 Sublingual Tablets, -10 Chewable Tablets, -40 Capsules [868]
Sorbide [563]
Sorbitrate Tablets [831]

Isostearamide DEA.
Hetamide IS [407]
Lamesoft ID [375]
Mackamide ISA [567]
Monamid 150-IS [599]
Nitrene 300 IS [399], [400]

Isostearamide DEA *(cont'd)*.
Schercomid ID, SI, SI-M [771]
Standamid ID [400]
Witcamide 5118S [935]

Isostearamide MEA.
Schercomid IME [771]

Isostearmide MIPA.
Schercomid IMI [771]

Isostearamido amine.
Mackine 401 [567]

Isostearamido betaine.
Mackam ISA [567]

Isostearamido morpholine.
Mackine 421 [567]

Isostearamidopropylkonium chloride.
Schercoquat IB [771]

Isostearamidopropyl amine oxide.
Incromine Oxide I [223b]

Isostearamido-propyl betaine.
Cycloteric BET I-30 [234]
Incronam I-30 [223b]
Mackam ISA [567]
Schercotaine IAB [771]

Isostearamidopropyl dimethylamine.
Cyclomide IODI [234]
Incromine IB [223b]
Mackine 401 [567]
Schercodine I [771]

Isostearamidopropyl dimethylamine lactate.
Incromate IDL [223b]
Mackalene 416 [567]
Richamate 6613 [732]

Isostearamido-propyl dimethyl-amino gluconate.
Katemul IGU, IGU-70 [771]

Isostearamido-propyl dimethylamino glycolate.
Katemul IG, IG-70 [771]

Isostearamidopropyl ethyldimonium ethosulfate.
Schercoquat IAS, IAS-LC [771]

Isostearamidopropyl morpholine.
Incromine ISM [223b]

Isostearamidopropyl morpholine betaine.
Incronam ISM-30 [223b]

Isostearamidopropyl morpholine lactate.
Emcol ISML [934], [935]
Incromate ISML [223b]
Richamate ISML [732]

Isostearamidopropyl morpholine oxide.
Incromine Oxide ISMO [223b]

Isosteareth-2.
Arosurf 66-E2 [796]
Hetoxol IS-2 [407]

Isosteareth-3.
Hetoxol IS-3 [407]

Isosteareth-6 carboxylic acid.
Sandopan TA-10 [762]

Isosteareth-10.
Arosurf 66-E10 [796]
Hetoxol IS-10 [407]

Isosteareth-12.
Ethlana 12 [507]

Isosteareth-20.
Arosurf 66-E20 [796]
Hetoxol IS-20 [407]

Isosteareth-22.
Ethlana 22 [507]

Isosteareth-50.
Ethlana 50, 50-M [507]

Isosteareth acid.
Century 1105 [876]
Emersol [288a]
Emersol 871, 875 [288]

Iso-stearic alkanolamide.
Monamid 150-IS [599]
Witcamide 823/10 [937]

Isostearic, amphoteric.
Monateric ISA 35% [599]

Isostearic diethanolamide.
Mackamide ISA [567]
Standamid ID [400]

Isostearic glycerides, ethoxylated.
Labrafil ISO [339]

Isostearic imidazoline.
Schercozoline I [771]

Isostearic monoester.
Imwitor 780K [482]

Isostearoamphocarboxypropionate.
Schercoteric IS-SF-2 [771]

Isostearoamphoglycinate.
Rewoteric AM-21S [725]

Isostearoamphopropionate.
Miranol ISM Conc. [592]
Monateric ISA-35, ISA-CI [599]
Schercoteric I-AA [771]

Isostearoyl lactate.
Patlac IL [674a]

Isostearyl alcohol.
18G [594]
Adol 66 [796]
Michel XO-146 [581]

Jojoba butter.
Jojoba Butter [473]

Jojoba, isomerized.
Jojo Butter-31 [473]

Jojoba meal.
APMC Jojoba Meal [11a]

Jojoba oil.
Cojoba [218a]
Golden Jojoba Oil [11a]
Jojoba Liquid Wax [473]
Lipovol J [525]

Jojoba oil, hydrogenated.
Jojoba Wax [473]

Jojoba oil, partially bleached.
Sonora Grade Jojoba Oil [472a]

Jojoba oil, trans-isomerized.
Jojobutter 31 Group [472a]

Jojoba seeds.
APMC Jojoba Seeds [11a]

Jojoba wax, hydrogenated.
Hydroba 70 [472a]

Kanamycin sulfate injection.
Klebcil® Injection 1 Gm, Pediatric Injection 75 mg, Injection 500 mg [114]

Kaolin.
Buca [336]
Clay [857]
Kaopaque 10, 20 [347]
Lion English Kaolin [188]
Medicinal Kaolin NF 825 [924]
'SIM' Colloidal Kaolin [188]
Stockalite [381]
Supreme [381]
Vanclay [894]
Wilklay [928]

Kaolin (aluminum silicate).
Al-Sil-Ate LO [336]
Catalpo [336]
EPK [311]

Kaolin, calcined (dehydroxylated aluminum silicate).
SP-33 [336]
Whitetex [336]

Kaolin (china clay).
Kaolin [347], [381], [804]

Kaolin clay, air-floated.
Bilt-Cote H-1, H-5, S-1, S-5 [894]

Kaolin clay, hard.
Crown Clay [379], [810]
Dixie Clay [894]
Dover Clay [154]
Recco Clay [154]
Secco Clay [379], [810]
Type 80 [810]
White Crown Clay [810]

Kaolin clay, soft.
Langford Clay [894]
McNamee Clay [894]
Swanee Clay [154]

Kaolinite.
Hydrite 121-S, Flat D, PX, PXS, R, RS, UF [347]
Nopok [428]
Nuflo [428]
Suprex [428]
Tako [857a]

Kaolin, purified.
Electros [188]

Kaolin, surface-modified, calcined.
Translink 37 [336]

Kelp.
Pacific Kelp [576]

Keratin.
Kerapro [424]
Secol Keratin [783a]

Ketone formaldehyde condensate.
Cardosol Brand Resin [862]

Ketone peroxide in dimethyl phthalate diluent.
Lupersol DDM, DDM-30, Delta-X [536]

Ketone peroxide in proprietary diluent.
Lupersol DNF, DSW [536]

Kraft lignin.
Reax 100M [917a]

Kraft lignin product, sulfonated.
Reax 92 [917a]

Kraft lignin, sodium salt.
Reax 85A [917a]

Kraft lignin, sulfonated alkoxylated, sodium salt.
Reax 90P [917a]

Kraft lignin, sulfonated, sodium salt.
Reax 80A, 80C, 81A, 82, 83A, 95A [917a]

Kraft pine lignin.
Indulin AT [917a]

Kraft pine lignin, sodium salt.
Indulin C [917a]

Labetalol hydrochloride.
Trandate® [354]

Lactase.
Amerzyme Fungal Lactase [54]

Lactate ester.
Patlac IL [676]

Lactic acid.
Patlac LA 88% [735]

Lactic acid chelate, ammonium salt.
Tyzor LA [269]

Lactic acid esters.
Tegin L Series [360]

Lactic acid fatty acid glyceride.
Axol L 61, L 62 [856]

Lacto-glycerides.
Lamegin GLP 10, GLP 20 [176]

Lactylic acid & its salt, caprylic fatty acid ester.
Grindtek FAL4 [372]

Laneth-5.
Polychol 5 [223]
Ritawax 5 [735]

Laneth-10.
Polychol 10 [223]

Laneth-10 acetate.
Ritawax AEO [735]

Laneth-15.
Polychol 15 [223]
Ritawax 15 [735]

Laneth-16.
Polychol 16 [223]

Laneth-20.
Polychol 20 [223]

Laneth-40.
Polychol 40 [223]
Ritawax 40 [735]

Lanthanum chloride.
Ultralog 54-8120-00 [172]

Lanthanum oxide.
Ultralog 54-8220-00 [172]

Lapyrium chloride.
Emcol E-607L [934], [935]

Lard glyceride.
Myverol 18-40 Distilled Monoglycerides [278]

Lard glyceride, acetylated.
Tegin E66 [360]

Lard glyceride, acetylated hydrogenated.
Tegin E61 [360]

Lard monodiglyceride.
Tegomuls 19 [856]

Lauramide DEA.
Alkamide L9DE [33]
Aminol LA-12, LM-5, LM-30, LM-30-C Special [320]
Carsamide SAL-7, SAL-9, SAL-82 [529]
Clindrol Superamide 100L, Superamide 100LM, 200-L [197]
Comperlan LD [399], [400]
Comperlan LDO, LDS [399]
Comperlan LMD [400]
Condensate L-90, L-90A, L-5560, LM11D, LM-5560 [174]
Condensate PE, PL [209]
Cyclomide DL 203, DL 203/S, DL 203/SL, DL 207/S, DL 207/SL [234]
Cyclomide LD [939]
Cyclomide LE [234]
Emery 6590 [291]
Emid 6510, 6511, 6513, 6519, 6541 [291]
Empilan LDE, LDE/FF, LDX [548]
Hetamide ML [407]
Hyamide 90-C [174]
Lipamide LMWC [525]
Mackamide L-95, LMD [567]
Mazamide LM-20, LS-173, LS-196 [564]
Monamid 150-GLT, 150 LMWC, 150 LW, 150 LWA, 716, 770 [599]
Monamine ACO-100, LM-100 [599]
Ninol 52-LL, 4821, AA 62, AA-62 Extra, P-616, P-621 [825]
Nitrene 125 LS [399], [400]
Nitrene L-90 [400]
Onyxol 336, 345, SD [663]
Product CPI-24-2, LT-18-34, LT 18-48 [197]
Protamide L-5-560, L-80M, L90, L90A, LM 73, LM 5560 [713]
Rewo-Amid DL 203, DLMS [725]
Richamide 6310, LM-46, STD, STD-HP [732]
Schercomid 1214 [771]
Schercomid LD [732]
Schercomid SL-Extra, SLM, SLM-C, SLMC-75, SL-ML, SLM-LC, SL-ML-LC, SLM-S [771]
Sipomide 843 [29]
Standamid LD, LDO, LDS [399], [400]
Sterling LDEA-90 [145]
Super Amide, L-9, L-9A, L-9C, LL, LM [663]
Surco Super Amide L9 [663]
Synotol L 60, LM-60 [715]
Unamide J-56, LDX, LMDX [529]
Varamide L-1, ML-4 [796]
Witcamide 5195 [935]
Witcamide 6310 [934], [935]

Lauramide DEA *(cont'd).*
Witcamide STD-HP [934]

Lauramide diethanolamide.
Incromide L-90 [223b]

Lauramide MEA.
Alkamide L9ME [33]
Carsamide SAL-7M [529]
Comperlan LM [399]
Comperlan LMM [400]
Cyclomide L203 [234]
Cyclomide LM [939]
Empilan LME [548]
Hetamide MML [407]
Kessco X-159 [79]
Mackamide LMM [567]
Monamid LMA, LM-MA [599]
Rewo-Amid L 203 [725]
Standamid LM [399], [400]

Lauramide MIPA.
Clindrol 101LI [197]
Comperlan LP [400]
Cyclomide LIPA [234]
Cyclomide LP [939]
Empilan LIS [548]
Monamid 692, LIPA, LMIPA [599]
Nidaba 3 [901]
Onyxol 368 [663]
Rewo-Amid IPL 203 [725]

Lauramidopropylamine oxide.
Mackamine LAO [567]

Lauramidopropyl betaine.
Mackam LA [567]
Mirataine BB [592]
Monateric LMAB [599]

Lauramidopropyl dimethylamine.
Lexamine L13 [453]
Schercodine L [771]

Lauramidopropyl PG-dimonium chloride phosphate.
Monaquat P-TD [599]

(3-Lauramidopropyl) trimethylammonium methyl sulfate.
Cyastat® LS [52]

Lauramine.
Armeen 12, 12D [79]
Kemamine P-690 [429a]

Lauramine oxide.
Alkamox LO [33]
Ammonyx DMCD-40, LMO, LO [663]
Aromox DMMC-W [79]
Conco XAL [209]
Emcol L [934], [935]
Empigen OB [548]
Jordamox LDA [474]
Mackamine LO [567]
Mazox LO [564]
Monalux LO [599]
Ninox L [825]
Rewominoxid L 408 [725]
Schercamox DML [771]

Lauraminopropionic acid.
Deriphat 170C [399]
Velvetex 710L [399], [400]

Lauramphо PG-glycinate phosphate.
Monaquat P-TL [599]

Laureth-1.
Alkasurf LAN-1 [33]
Hetoxol L-1 [407]
Lipocol L-1 [525]
Procol LA-1 [713]
Siponic L-1 [29]

Laureth (1) sulfate.
Texapon ES-1

Laureth-2.
Carsonon L-2 [529]
Dehydol LS 2 [400]
Empilan KB2 [548]
Hetoxol L-2 [407]
Incropol L-2 [223b]
Procol LA-2 [713]

Laureth-3.
Carsonon L-3 [529]
Dehydol LS 3 [400]
Emalex 703 [628]
Empilan KC3 [548]
Hetoxol L-3N [407]
Procol LA-3 [713]
Siponic L-3 [29]

Laureth-3 phosphate.
Cyclophos PL3 [234]

Laureth-4.
Brig 30, 30 Special [438]
Cyclogol ZBL4 [939]
Dehydol LS 4 [400]
Emthox 5875, 5882 [291]
Ethosperse LA-4 [358]
Hetoxol L-4N [407]
Lanycol-30 [507]
Lipal 4LA [715]
Lipocol L-4 [525]
Macol LA-4 [564]
Procol LA-4 [713]
Simulsol P 4 [785]
Siponic L-4 [29]

Laureth-5.
Carsonon L-5 [529]
Procol LA-5 [713]

Laureth-5 carboxylic acid.
Akypo 23Q38 [182]

Laureth-6.
Cyclogol ZBL6 [939]

Laureth-6 citrate.
Citrest LSU [234]

Laureth-7.
Incropol L-7, L-7-90 [223b]
Marlipal MG [178]
Procol LA-7 [713]

Laureth-8 phosphate.
Protaphos SDA [713]

Laureth-9.
Carsonon L-9, L-985 [529]
Cyclogol ZBL9 [939]
Hetoxol L-9N, LS-9 [407]
Lipal 9LA [715]
Macol LA-9 [564]
Procol LA-9 [713]

Laureth-10.
Dehydol 100 [400]
Procol LA-10 [713]

Laureth-11.
Cyclogol ZBL11 [939]
Procol LA-11 [713]

Laureth-12.
Alkasurf LAN-12 [33]
Carsonon L-12 [529]
Cyclogol EL [234]
Ethosperse LA-12 [358]
Incropol L-12 [223b]
Lipal 12LA [715]
Lipocol L-12 [525]
Macol LA-12 [564]
Procol LA-12 [713]
Siponic L-12 [29]

Laureth-13.
Cyclogol ZBL13 [939]
Procol LA-13 [713]

Laureth-15.
Alkasurf LAN-15 [33]
Cyclogol ZBL15 [939]
Procol LA-15 [713]

Laureth-20.
Emalex 720 [628]
Procol LA-20 [713]

Laureth-23.
Alkasurf LAN-23 [33]
Brij 35, 35SP [438]
Emthox 5877, 5964 [291]
Hetoxol L-23N [407]
Hodag Nonionic L-23 [415]
Incropol L-23 [223b]
Lanycol-35 [507]
Lipal 23LA [715]
Lipocol L-23 [525]
Macol LA-23 [564]
Procol LA-23 [713]
Simulsol P 23 [785]
Siponic L-25 [29]

Laureth-25.
Cyclogol ZBL25 [939]
Procol LA-25 [713]

Laureth-30.
Cyclogol ZBL30 [939]
Incropol L-30 [223b]
Procol LA-30 [713]

Laureth-40.
Procol LA-40 [713]

Lauric acid.
Acme 760, 765 [6]
Emery 652 [289]
Hetamide LA, ML [407]
Hystrene 9512 [429a]
Neo-Fat 12, 12-43 [79]
PRIFAC 2920 [875]

Lauric acid diethanolamide.
Clindrol 200L [197]
Comperlan LD [401]
Loramine DL 203/S [274]
Loropan LD [858]
Profan AA62 Extra [764], [948]
Rewomid 203/S [274]
Rewomid DL 203 [725]
Rewomid DL 203/S [725], [725a]
Schercomid 1214 [771]
Stafoam DL [636]
Standamid LD [400]
Steinamid 203/S [274]
Synotol L-60, L-90 [266]

Lauric acid diethanolamine condensate.
Onyxol 336, 345 [663]

Lauric acid hexyl ester.
Cetiol A [401]

Lauric acid isopropanolamide.
Comperlan LP [401]
Profan AD31 [764], [948]
Standamid LP [398]

Lauric acid monoethanolamide.
Comperlan LM [401]
Lauridit LM [16], [17]
Loropan LM [858]

Lauric acid monoethanolamide, sulfosuccinate.
Cyclopol SBL203 [234]

Lauric acid monoethanolamide sulfosuccinate, sodium salt.
Marlinat HA 12 [178]

Lauric acid monoisopropanolamide.
Rewomid IPL 203 [725]

Lauric acid, substituted imidazoline.
Schercozoline L [771]

Lauric acid POE 4.
Crodet L4 [223]

Lauric acid POE 8.
Crodet L8 [223]

Lauric acid POE 12.
Crodet L12 [223]

Lauric acid POE 24.
Crodet L24 [224]

Lauric acid POE 40.
Crodet L40 [223]

Lauric acid POE 100.
Crodet L100 [223]

Lauric acid polydialkanolamide.
Loramine DL 203 [274]
Rewomid DL 203 [274]

Lauric acid, zinc salts.
Laurex [881]

Lauric alkanolamide.
Alkamide CL-55 [33]
Alkamide L9DE [33], [34]
Hartamide LE-90 [391]
Mazamide LS-196 [564]
Monamid 150-LW, 716 [599]
Monamine ACO-100 [599]
Unamide J-56 [529]

Lauric alkanolamide, ethoxylated.
Unamide L-2, L-5 [529]

Lauric alkylolamine condensate.
Nopcogen 14-L [252], [253]

Lauric amido alkyl dimethyl amine.
Schercodine L [771]

Lauric-amido-propyldimethyl-amine oxide.
Felsoxid-KP-100 [823]

Lauric amphoglycinate.
Sipoteric 1127 [29]

Lauric DEA.
Emid® 6519 [290]

Lauric diethanolamide.
Alkamide 2124 [33]
Ardet DLA, LDA, LMC [73]
Carsamide LLDX [529]
Carsamide SAL-7 [155]
Carsamide SAL-9 [529]
Cedemide A Extra [256]
Chimipal LDA [514]
Clindrol Superamide 100L [197]
Condensate PL [209]
Crillon LDE [223]
Cyclomide DL203, DL203/S [234]
Emid 6541 [289]
Empilan LDE [23], [25], [26]
Empilan LDE/FF [23]
Empilan LDX [23], [25]
Ethylan MLD [253], [509]
Hartamide LDA 70, LDA 90, LDA Extra [390]
Incromide LCL, LR [223b]
Intermediate 512 [429]
Monamid 150-LW, 150-LWA, 716, 982, 1034 [599]
Monamine ACO-100 [599]
Product LT 18-48 [197]
Standamid LDM, LDS [400]
Sterling LDEA-90 [145]
Unamide J56, LDX [529]
Varamide L-1. ML-1, SL-9 [796]
Witcamide® 6310, STD-HP [935]

Lauric diethanolamide, sodium sulfosuccinate ester.
Alkasurf SS-L9DE [33], [34]

Lauric diethanolamide, super.
Aramide LDS [76]
Emid 6510, 6511 [289]

Lauric diethanolamine.
Alkamide L9DE [33]
Carsamide SAL-9 [155]

Lauric diglycolamide.
Incromide LMD [223b]

Lauric fatty acid, 95%.
Hystrene 9512 [430]

Lauric fatty acid alkanolamide.
Condensate PE [209]
Super Amide L9, L9C [663]

Lauric fatty acid DEA.
Super Amide LL [663]

Lauric fatty acid diethanolamide.
Merpinamid LD/E [285]
Onyxol 336 [663]

Lauric fatty acid ethoxylate.
Alkasurf L-2, L-5, L-9 [33]

Lauric fatty acid monoisopropanolamide.
Merpinamid LMIPA [285]

Lauric imidazoline.
Schercozoline L [771]

Lauricimidazoline amphoteric.
Amphoterge L [529]

Lauric imidazoline betaine.
Empigen CDL 10, CDL 30 [23], [25]

Lauricimidazoline dicarboxy amphoteric.
Amphoterge L-2 [529]

Lauric imidazoline dicarboxylate.
Mackam 2 L [567]

Lauric imidazoline mono carboxylate.
Mackam 1 L [567]

Lauric isopropanolamide.
Clindrol 101L1 [197]
Comperlan LP [399]
Empilan LIS [23], [25]

Lauric MEA ethoxylate.
Alkamidox L-2, L-5 [33]

Lauric MIPA.
Alkamide LIPA [33]
Incromide LI [223b]

Lauric/MIPA alkanolamide.
Alkamide LIPA [33]

Lauric monoethanolamide.
Alkamide L9ME [33]
Ardet LEA [73]
Cedemide M Extra [256]
Crillon LME [223]
Cyclomide L203 [234]
Empilan LME [23], [25]
Hartamide LMEA-90 [390]
Loramine L203 [274]
Monamid LMA [599]
Rewomid L 203 [274], [725]
Steinamid L203 [274]

Lauric monoethanolamide, ethoxylated.
Alkamidox L-2, L-5 [33]

Lauric monoethanolamide, ethylene oxide condensate.
Amidox L-2, L-5 [825]

Lauric monoethanolamide, sodium sulfosuccinate ester.
Alkasurf SS-L9ME [33]

Lauric monoethanolamide sulfosuccinate, sodium salt.
Emery 5310 [290]

Lauric monoisopropanolamide.
Alkamide LIPA [33], [34]
Ardet LIPA [73]
Cyclomide LIPA [234]
Empilan LIS [26]
Loramine IPL 203 [274]
Monamid LIPA [599]
Rewomid IPL 203 [274]
Steinamid IPL 203 [274]

Lauric myristic acid, diethanolamide, 70:30.
Clindrol Superamide 100LM [197]
Synotol LM-90 [266]

Lauric myristic acid monoethanolamide.
Comperlan LMM [401]

Lauric-myristic alkanolamide.
Alkamide L7DE [33], [34]
Alkamide L7DE-BT [33]
Aramide LMDS [76]
Hartamide LM 11D [391]
Mazamide LS-173 [564]
Monamid 150-LMW-C, LM-100 [599]

Lauric-myristic amide, super.
Hartamide CE80 [391]

Lauric-myristic betaine.
Mackam LMB [567]

Lauric myristic DEA.
Alkamide L7DE, L7DE-BT [33]

Lauric/myristic diethanolamide.
Ardet LMD [73]
Comperlan LD, LMD [399]
Cyclomide DL207/S, DL207/SL [234]
Nitrene L-76 [400]
Unamide LMDX [529]

Lauric/myristic diethanolamide, 70:30.
Monamid 150-LMW-C, R 31-42 [599]
Monamine LM-100 [599]
Schercomid SLM-S [771]

Lauric-myristic diethanol superamide.
Mazamide LS-173 [564]

Lauric myristic fatty acid alkanolamide.
Super Amide LM [663]

Lauric-myristic monoethanolamide.
Alkamide L7ME [33]
Hartamide LMEA-70 [390]
Monamid LMMA [599]

Lauric/myristic monoisopropanolamide.
Monamid LMIPA [599]

Lauric myristyl diethanol superamide.
Unamide LMDX [529]

Lauric polyamine condensate.
Nopcogen 16-L [253]

Lauric sulfosuccinate.
Mackanate LM-40 [567]

Lauroamphocarboxyglycinate.
L/M Analog of Miranol HM Special [592]
Miranol HM Special, HM Special Conc., H2M Conc. [592]
Monateric 951A [599]
Rewoteric AM-2H [725]
Schercoteric LS-2 [771]
Stripped Coco Analog of Miranol HM Special [592]

Lauroamphocarboxypropionate.
Miranol H2M-SF 70%, H2M-SF Conc., H2M-SFE Conc. [592]
Rewoteric AM-2LSF [725]

Lauroamphocarboxypropionic acid.
Miranol H2M Anhydrous Acid [592]

Lauroamphoglycinate.
Miranol HM Conc. [592]
Monateric LM-24, LM-M24, LMM-30, LM-M42 [599]
Rewoteric AM-2[illegible] [725]
Schercoteric LS[illegible] [illegible]1]
Sipoteric 1127 [2[illegible]]
Velvetex G-20 [400]

Lauroamphopropionate.
Miranol HM-SF Conc., HM-SFE Conc. [592]
Rewoteric AM-LSF [725]
Schercoteric LS-SF [771]

Lauroamphopropylsulfonate.
Miranol HS Conc. [592]
Schercoteric LS-EP [771]

Lauroyl colamino formyl methyl pyridinium chloride.
Emcol E 607 L [934], [935], [937a], [938]

Lauroyl diethanolamide.
Nissan Stafoam DL [636]
Nitrene L-90 [399], [400]

Lauroyl/myristoyl diethanolamide, 70:30.
Nitrene L-76 [399], [400]

Lauroyl peroxide.
Alperox C [536]
Alperox-F (flake) [536], [536a]
Laurox, Laurox-W40 [16], [653]

Lauroyl sarcosinate, sodium salt.
Nikkol Sarcosinate LN [631]
Secosyl [825a]

Lauroyl sarcosine.
Crodasinic L [223]
Hamposyl L [363], [364b]
Maprosyl L [663]
Sarkosyl L, NL-30 [190]

Laurtrimonium chloride.
Dehyquart LT [400]
Kemamine Q-6903B [429a]

Lauryl acrylate.
SR-335 [766]

Lauryl alcohol.
Alfol 12 [206]
Cachalot L-50, L-90 [581]
CO-1214 [709]
Kalcohl 20 [478]
Laurex L1, NC, 12 [23], [548]

Lauryl alcohol ether sulfate, alkanolamine salt.
Zetesol 856, 856 D, 2210 [948]

Lauryl alcohol ether sulfate, sodium salt.
Zetesol NL [948]

Lauryl alcohol, ethoxylated.
Chemal LA-4, LA-9, LA-12, LA-23 [170]
Siponic L1, L4, L7, L-7-90, L12, L16, L25 [29]

Lauryl alcohol ethoxylate (EO 12 moles).
Polystep F-12 [825]

Lauryl alcohol ethoxylate (EO 30 moles).
Polystep F-13 [825]

Lauryl alcohol, ethoxylated, sodium sulfosuccinate ester.
Alkasurf SS-LA-3 [33], [34]

Lauryl alcohol, ethoxylated, sulfosuccinate.
Cyclopol SBFA30 [234]

Lauryl alcohol ethoxylate, sulfosuccinate half ester, sodium salt.
Alconate L3 [29]

Lauryl alcohol polyglycolether sulfosuccinate, sodium salt.
Emery 5320 [290]
Setacin 103 Spezial [948]

Lauryl alcohol sulfate.
Emkasol WAS [292]
Empicol LM 45, LXV [23]

Lauryl alcohol sulfate, alkylolamine salt.
Sulfetal C JOT 38, C JOT 60 [948]

Lauryl alcohol, sulfated, sodium salt.
Elfan 200 [16], [764]
Sulfetal K 35, K 90, L 95 [948]

Lauryl alcohol sulfate, triethanolamine salt.
Sulfetal KT 400 [948]

Lauryl alkanolamide.
Sandoz Amide PE, PL [761]

Lauryl allyl sulfosuccinate, sodium salt.
Trem LF 40 [252]

Laurylamine.
Radiamine 6163 [658]

Laurylamine acetate.
Crodamac 1.12A [223]

Laurylamine (distilled).
Radiamine 6164 [658]

Lauryl amine oxide.
Mackamine LO [567]
Sandoz Amine Oxide XA-L [761]

Lauryl amine, primary.
Lilamin 163 [522]

Lauryl betaine.
Empigen BB [548]
Lonzaine 14 [529]
Mafo LOB [564]

Lauryl betaine *(cont'd)*.
Mirataine LDMB [592]
Product DDN [269]
Rewoteric AM-DML [725]
Swanol AM-301 [631]
Varion CDG [796]

Lauryl chloride.
Barchlor 125 [529]

Lauryl diethanolamide.
Merpinamid KD11, LD/E [285]
Nissan Stafoam L [636]
Tohol N-230, N-230X [864]

Lauryl dimethyl amine.
Barlene 12S [529]

Lauryl dimethylamine C21-dicarboxylate.
Armocare 5098 [79]

Lauryl dimethyl t-amine (distilled).
Lilamin 312 D [522]

Lauryl dimethyl amine oxide.
Alkamox LO [33], [34]
Ammonyx DMCD-40, LO [663]
Barlox 12 [529]
Conco XA-L [209]
Incromine Oxide L [223b]
Loramine L 408 [274]
Mazox LO [564]
Ninox L [825]
Oxamin LO [439]
Rewominox L 408 [725]
Varox 365 [796]

Lauryl dimethylamino acetic acid betaine.
Swanol AM-301 [631]

Lauryl dimethyl-ammonium betaine.
Rewoteric AM-DML [725]

Lauryl dimethylbenzyl-ammonium bromide.
Amonyl BR 1244 [711]

Lauryl dimethyl benzyl ammonium chloride.
Catinal CB-50 [864]
Dehyquart LDB [401]
Hartex San Q 50 [390]
Loraquat B-50 [274]
Rewoquat B 50 [725]
Tequat PAN [514]
Vantoc CL [442]

Lauryl ether phosphate, acid form (EO 3).
Servoxyl VPAZ 3/100 [788]

Lauryl ether sulfate.
Akyposal 100 DAL, 100 DE, 100 DEG [182]
Elfan NS [16]
Steol 4N [825], [825a]

Lauryl ether sulfate, sodium salt.
Calfoam ES-30 [697]
Steol 4N [825]

Lauryl ether sulfosuccinate.
Mackanate EL [567]

Lauryl ethoxylate.
Ethosperse LA-23 [358]

Lauryl ethoxy sulfate (EO 3), ammonium salt.
Sandoz Sulfate 216 [761]

Lauryl ethoxy sulfate (EO 3), sodium salt.
Sandoz Sulfate 219, ES-3 [761]

Lauryl ethoxy sulfate (EO 3.5), sodium salt.
Sandoz Sulfate WE [761]

Lauryl imidazoline.
Schercozoline L [771]

Lauryl imidazoline betaine.
Carsonam L [155]

Lauryl isopropanolamide.
Clindrol 102-LI [197]

Lauryl isoquinolinium bromide.
Isothan Q-15, Q-75 [663]

Lauryl isostearate.
Isostearene L [901]

Lauryl lactate.
Ceraphyl 31 [895]
Crodamol LL [223]
Cyclochem LVL [234]
Schercemol LL [771]

Lauryl methacrylate.
Empicryl LM [23], [25]
SR-313 [766]

Lauryl monoethanolamide.
Merpinamid KM, LM [285]

Lauryl-myristyl alcohol.
Kalcohl 5-24, 6-24, 7-24 [478]

Lauryl myristyl dimethyl amine oxides.
Bio-Surf PBC-460 [120]

Lauryl phosphate, acid form.
Servoxyl VPAZ 100 [788]

Lauryl pyridinium bisulfate.
Dehyquart D [401]

Lauryl pyridinium chloride.
Catigene CLP/50 [116], [825a]
Charlab LPC [159]
Dehyquart C, C Crystals [400], [401]

Lauryl sarcosine.
Sarkosyl L [190]

Lauryl sulfate.
Cycloryl XL [234]
Maprofix [663]

Lauryl sulfate *(cont'd)*.
Stepanol [825]

Lauryl sulfate, amine salt.
Melanol LP 20 [711]

Lauryl sulfate, ammonium ethoxylated.
Empicol EAB [25]

Lauryl sulfate, ammonium salt.
Melanol LP 20 [711]
Sandoz Sulfate A [761]
Sipon L-22 [29]

Lauryl sulfate, diethanolamine salt.
Melanol LP 20 D [711]
Sandoz Sulfate EP [761]

Lauryl sulfate, isopropanolamine salt.
Melanol LP 1 [711]

Lauryl sulfate monoethanolamine salt.
Elfan 240M [16], [764]
Melanol LP 20 M [711]

Lauryl sulfate, sodium salt.
Melanol CL 30 [711]
Sandoz Sulfate WA Dry, WAG, WAS, WA Special [761]

Lauryl sulfate TEA.
Sunnol LST [524]

Lauryl sulfate triethanolamine salt.
Elfan 240T [16], [764]
Melanol LP 20 T [711]
Perlankrol ATL-40 [253]
Sandoz Sulfate TL [761]

Lauryl sulfoacetates.
Lathanol [825]

Lauryl superamide, 92-95%.
Aminol LM-5C [320]

Lauryl trimethyl ammonium chloride.
Chemquat 12-33, 12-50 [170]
Dehyquart LT [401]

Lavender extract.
Lavender HS, LS [20]

Lead(II) acetate.
Ultralog 55-0780-00 [172]

Lead(II) chloride.
Ultralog 55-0920-00 [172]

Lead chlorosilicate complex.
Lectro 60, 60 XL [88], [615a]

Lead chlorosilicate-sulfate complex, basic.
Lectro 80 [88]

Lead chromate.
Chrome Yellow Light Y-433-D [269]

Lead diamyldithiocarbamate.
Amyl Ledate [894]
Vanlube 71 [894]

Lead dimethyl dithiocarbamate.
LDMC [402]
Ledate [894]
Methyl Ledate [894]

Lead dinonylnaphthalene sulfonate in light min. oil.
Na-Sul LS [894]

Lead fumarate, tetrabasic.
Lectro 78 [88]

Lead maleate, tribasic.
Tri-Mal [88]

Lead(II) oxide.
Ultralog 55-1280-00 [172]

Lead phosphite, dibasic.
Dyphos, Dyphos XL [88], [615a]

Lead phthalate, dibasic.
Dythal, Dythal XL [88], [615a]

Lead silicate sulfate, basic.
Tribase E, E XL [88]

Lead silico chromate.
Oncor F31, Y47A [641a]

Lead silico chromate, basic.
Oncor M50 [641a]

Lead stearate.
Leadstar [88]

Lead stearate, dibasic.
DS-207 [88]

Lead sulfo phthalate complex.
Lectro 90 [88]

Lecithin.
Actiflo® Series [167]
Alcolec, 619B, 621, 628G, Granules & Powder, RCX-1, S, Z-3 [57]
Centrocap® Series [167]
Centrol® Series [167]
Centrophase® Series [167]
Centrophil® Series [167]
Crolec 4135 [223]
Kelecin [813]
Lecicap Series [731]
Lecigran Series [731]
Lecikote 8555 [731]
Leciprime Series [731]
Lecisoy Series [731]
Lecisperse Series [731]
Lexin CSO, K [57]
Lexinol [57]
M-C-Thin 45 [534]
Sta-Sol [818]

Lecithinamide DEA.
Nidaba 318 [901]

Lecithin compound.
Metarin F [534]

Lecithin, fluid.
Centrophase 31 [166a]

Lecithin fraction.
Asol [533], [534]
Chocothin [533], [534]
Emulpur [534]
Metarin, Metarin P [534]

Lecithin fraction, oil-free.
Emulpur N [534]

Lecithin, heat-resistant.
Centrophase HR [166a]

Lecithin, hydrophyllic.
Emulbesto [534]
Emulfluid [534]

Lecithin, hydroxylated.
Alcolec Z-3 [57]
Centrolene® Series [167]

Lecithin natural.
Actiflo 68, 70 [167]
Canasperse U.B.F. [145]
M-C-Thin AF-1 [534]
Soyalec DBF, DBP, SBF, SBP, UBF, WDF-FG [145]

Lecithin product, water soluble.
Alcolec 634, 638, HS-3 [57]

Lecithin, spray-dried.
Emulthin M-35, M-501 [534]

Lecithin, standardized.
M-C-Thin [533], [534]

Lecithin, water-dispersible.
Alcolec 439-C [57]
Crolec 4390 [223]
Troykyd Lecithin W.D. [869]

Lemon extract.
Lemon HS [20]

Leucyl aminopeptidase substrate.
LNA [128]

Lidocaine HCL.
Zylocaine [510]

Lignin.
Indulin [917a]

Lignin amine.
Indulin W-3 [917a]

Lignocellulose.
Furafil [716]
Industrial Shell Flour [11]

Lignosulfonate.
Kelig 32 [47]
Lignosite, Lignosite 260 [349]
Marabond 21 [47]
Maracarb [47]
Maraçell C [719a]
Marasperse [47], [719a]
Marasperse 52 CP [47]
Norlig [47], [719a]
Polyfons [918]
Reax [918]
Rychem® 824 [756]

Lignosulfonic acid, salts.
Lignosite [349]

Limestone, calcitic.
C-99½%, C-99½% Electro [428]
CCC Natural Calcium Carbonate [428]
Q-1, Q-1 Electro, Q-White [428]

Limestone, dolomitic.
Stonelite [656]

Lincomycin hydrochloride.
Lincocin® [602]
Lincomix® [602]

Lindane.
HGI [421]
Isotox [183]

Linen-phenolic laminate.
Insurok T-733 [812]

Linolea-erucic (colza) glycerides, ethoxylated.
Labrafil WL 1905 CS [339]

Linoleamide.
Nitrene 150 LD [399], [400]
Schercomid SLS [771]

Linoleamide DEA.
Aminol LNO [320]
Clindrol 15-73 [197]
Comperlan F [400]
Comperlan SOMD [399]
Condensate LNO [174]
Cyclomide DIN 278, DOTS [234]
Cyclomide ND [939]
Emid 6540 [291]
Foamole A, L [895]
Hetamide LL [407]
Monamid 15-70W, 150-ADY [599]
Monamine ADY-100 [599]
Nitrene 100 SD [399], [400]
Product LT 15-73-1 [197]
Protamide LNO [713]
Rewo-Amid F [725]
Schercomid SLE, SLS [771]
Standamid SOMD [399], [400]

Linoleamide MEA.
Cyclomide NM [939]
Hetamide MLL [407]

Linoleamide MIPA.
Cyclomide NP [939]
Monamid 835 [599]

Linoleamidopropyl dimethylamine.
Incrome LSB [223b]

Linoleamidopropyl dimethylamine lactate.
Incromate LDL [223b]

Linoleic acid.
Emersol 315 [288]
Polylin 155 [494]

Linoleic acid alkanolamide.
Comperlan F [401]

Linoleic alkanolamide.
Monamid 15-70W [599]

Linoleic DEA.
Emid® 6540 [290]

Linoleic diethanolamide.
Clindrol LT 15-73-1 [197]
Cyclomide DIN-295/S, DOTS [234]
Empilan 2125/A [23]
Incromide LA [223b]
Monamid 15-70W [599]
Rewomid F [725a]
Standamid SOD [400]

Linoleic-oleic diethanolamide.
Product LT 15-73-1 [197]

Linseed acid.
L-310 [709]

Linseed oil.
Zymol [70]

Linseed oil, epoxidized.
Admex ELO [796]
Drapex 10.4 [74]
Epoxol 9-5 [840]
Flexol Plasticizer LOE [878]

Linuron.
Lorox [270]

Lipase.
Fermlipase, Fermlipase™ PL [313]
Lipase-AN 5000 [54]

Litharge.
Neolith [617a]

Lithium carbonate.
Lithobid [750]
Lithonate [750]
Lithotabs [750]
Ultralog 55-8100-00 [172]

Lithium citrate.
Lithonate-S [750]

Lithium dinonylnaphthalene sulfonate in light min. oil.
Na-Sul 707 [894]

Lithium lauryl sulfate.
Texapon LLS [401]

Lithium stearate.
Witco Lithium Stearate 306 [935]

Lysine.
Lyamine [577]

Lysozyme.
Lysoferm™ [313]

Magnesia brick.
Magnesite H-W [384]

Magnesia catalyst.
Mg-0601 T 1/8″ [389]

Magnesia-chrome refractory.
Guidon [384]

Magnesium.
Cycloryl MG [234]

Magnesium alkyl benzene sulphonate.
Nansa MS45 [23]

Magnesium aluminum silicate.
Magnabrite [47a]
Van Gel, Van Gel B [894]
Veegum, F, HS, HV, K, T [894]

Magnesium aspartate.
Carbossalina Codex 01107 [501a]
Carboxaline [901]

Magnesium-calcium carbonate.
T-R Camadil, T-R Micro Fine [466]

Magnesium calcium silicate (industrial talc).
I.T. 3X, 5X, 325, FT, X [894]

Magnesium carbonate.
Magcarb L [379], [577]
Magnesium Carbonate, Heavy NF 320, Light NF 309 [924]
Marinco CL [577]
Stan-Mag Magnesium Carbonate [392]

Magnesium dodecyl benzene sulphonate.
Nansa MS 45 [23], [25]

Magnesium hydroxide.
Magnesium Hydroxide, Paste 314, Powder NF 370 [924]

Magnesium laureth sulfate.
Cycloryl GD [939]
Empicol EGB [548]
Montelane MG [785]
Rewopol MGL 45 [725]
Texapon MG [400]

Magnesium lauryl ether sulfate.
Drewpon ESG [265]
Empicol EGB [25]
Sactipon 2 OMG [521]
Sactol 2 OMG [521]
Texapon MG [401]
Tylorol MG [858]

Magnesium lauryl sulfate.
Carsonol MLS [155], [529]
Conco Sulfate M [209]
Cycloryl GA [939]
Cycloryl MG [234]
Drewpon MG [265], [266]
Empicol ML26 [23], [25], [548]
Empicol MLV [23]
Maprofix MG [663]
Norfox MLS [650]
Polystep B-9 [825]
Richonol Mg [732]
Sipon LM, MLS [29]
Standapol MG [399], [400]
Stepanol MG [825]
Sulfotex MG [399]
Surco MG-LS [663]

Magnesium montmorillonite.
Macaloid [616]
Propaloid-T [616]
Tansul 7 [616]

Magnesium myristate.
Satinex [140]

Magnesium oxide.
Elastomag [607]
Elastomag 30, 60, 90, 100, 100R, 170 [15], [607], [752]
Mag Chem 900 [555]
Maglite A [379]
Maglite D, K, Y [379], [577]
Magnesium Oxide, Heavy USP 310, Light USP 311 [924]
MAGOX® [161], [379], [392], [804]
Stan Mag Bars, MLW [392]
Ultralog 56-6905-00 [172]

Magnesium oxide, bar form.
Ken-Mag [488]

Magnesium oxide beads.
Stan-Mag Beads and Mini Beads [392]

Magnesium oxide, calcined.
Stan Mag 112 [392]

Magnesium oxide dispersion.
Akro-Mag Blue Label Bar [15]

Magnesium perchlorate, anhydrous granular.
Dehydrite [83]

Magnesium phosphide.
Degesch Fumi-Cel, Fumi-Strip [243]

Magnesium ribbon.
Galvoline [261]

Magnesium rod.
Galvarod [261]

Magnesium salicylate.
Magaǹ [7]

Magnesium silicate.
Alberene Stone [22]
Brite Sorb [43]
Britesorb 90 [706]
Ceramitalc 10AC, HDT [894]
C-400 Talc [236]
Emtal 41 Talc, 42 Talc, 43 Talc, 44 Talc, 500 Talc, 549 Talc [379]

Magnesium silicate *(cont'd)*.
Florisil [330]
Magnesol [332]
Mistron Vapor Compacted, Densified [236]
Nytal 100HR [894]
RWS Talc [752]
Vantalc 6H [894]
Vertal 5 Talc, 10 Talc, 15 Talc [752]

Magnesium silicate, hydrous.
Ben-A-Gel [615a]
Nytal 99, 100, 200, 300, 400 [894]
Omyatalc PL 2, PL 2 CS [660]
Vantalc 500, 700, 900, 1350, 1500 [894]

Magnesium silicate, non-hygroscopic.
Talc [379]

Magnesium silicate (talc).
MP 12-50, 15-38 [691]
OMYA Vertal 5, XXX [660]
Vantalc 6H [894]

Magnesium stearate.
Magnesium Stearate [392], [804]
Petrac MG-20 NF [688]
Radiastar 1100 [658]
Witco Magnesium Stearate N.F. [935]

Magnesium sulfate.
Ultralog 56-7525-00 [172]

Magnesium sulfate heptahydrate.
Epsom Salt [261]

Magnesium sulfonate.
Hybase M-400 [936]

Magnesium sulfonate, overbased.
Hybase M-300, M-400 [936]

Magnesium xylene sulfonate.
Eltesol MGX [23], [25]

Maleate microcrystalline.
CPM [19a]

Malt extract.
Malt Extract ND-201 [707]

Maltitol.
Mabit [396]

Maltodextrin.
Mor-rex 1908, 1910, 1918 [366]

Maltol.
Veltol [691]

D-(+)-Maltose monohydrate.
Chemalog 56-9600-00 [172]

Malt syrup, diastatic.
Fermentase [707]

Manganese chloride.
Mangatrace [80]

Manganese (II) chloride.
Ultralog 56-9990-00 [172]

Manganese dioxide.
KM® [490]
Multisorb A.R. [542]

Manganese (IV) dioxide.
Ultralog 57-1005-00 [172]

Manganese ethylenebisdithio-carbamate.
Dithane M-22, M-22 Special [725]
Maneb 80 [230a]
Manzate, Manzate D [270]
Nespor [310]
Polyram M [107]
Tersan LSR [270]
Trimangol [683]
Tubothane [561a]

Manganese metal, electrolytic.
Tronamang® [490]

Manganese, nitrided.
Nitrelmang [333]

Manganese oxide, brown.
Manganox [333]

Manganese phosphate coatings.
Irco Lube [457]

Manganese powder, electrolytic.
Plast-Manganese [357]

Manganese sulfate.
Tecmangam [278]

Manganese sulfide.
Mansulox [333]

Manganese violet.
Manganese Violet 7101 [924]
Mango Violet C43-001, C43-3001 [835]
Pur-Man Violet B.C. 34-3512-1 [414]

Manganous oxide.
Nu-Manese [852]

Mannide monooleate.
Arlacel A [92]

Manoxol OT/sodium benzoate, 85:15%.
Monoxol OT/B [544]

MEA-laureth sulfate.
Alkposal MLES-35 [182]

MEA-lauryl sulfate.
Afron 22 [901]
Cycloryl SA [939]
Empicol EL, LQ27, LQ33, LQ77, XT45 [548]
Melanol LP 20 M [785]
Rewopol MLS 30 [725]
Serdet DFM 30 [175]
Standapol MLS [400]
Texapon MLS [400]

Meclizine HCL.
Antivert Tablets [510]

Medroxyprogesterone acetate.
Depo-Provera® [602]
Provera® [602]

MEK peroxide in dimethyl phthalate.
Hi-Point 90 Red, PD-1 [890]

Melamine adhesives.
Diaron [722]

Melengestrol acetate.
MGA® [602]

p-Menthane hydroperoxide.
PMHP [414a]

Menthyl salicylate.
Filtrol [205]

Meperidine hydrochloride.
Demerol Hydrochloride[162] [932]

Meprobamate.
Equanil Tablets [510]

Methionine hydroxy analogue calcium 90%.
Hydan [270]

Merbromin.
Mercurochrome [435]

2-Mercaptobenzothiazole.
Accelerator M [5]
Akrochem MBT [15]
Captax [362], [894]
MBT [15], [708], [881]
Pennac MBT-O [682]
Rokon [894]
Rotax [278], [894]
Royal MBT [752]
Thiotax (MBT) [392], [602]
Vulcafor MBT [212], [910]

Mercaptobenzothiazole disulfide.
MBTS [708]

2-Mercaptobenzothiazole, zinc salt.
Akrochem ZMBT [15]
OXAF [881]
Vulkacit Merkapto/MGC, ZM [597]
Zenite [270]

n-(2-Mercaptoethyl) benzenesulfonamide, s-(0,0-diisopropyl phosophorodithioate) ester.
Betasan [677]
Pre-San [542], [821]

2-Mercaptoimidazoline.
NA-22 [270]

2-Mercaptoimidazoline (ethylene thiourea).
Vulkacit NPV/C2 [112]

N-(Mercaptomethyl) phthalimide S-(O,O-dimethyl phosphoro-dithioate).
Imidan [821]
Prolate [821]

2-Mercaptotolylimidazole.
Vanox MTI [894]

Mercuric chloride.
Fungchex [691]

Mercuric oxide.
Santar [762a], [799a]

Mercuric oxide, yellow.
Ocu-Merox [655a]

Mercurous chloride/mercuric chloride/inert ingredients, 60:30:10.
Calo-Clor [542]

Mercury(II) chloride.
Ultralog 57-8260-00 [172]

Mercury(II) oxide.
Ultralog 57-8310-00 [172]

Meroxapol 105.
Pluronic 10 R5 [109]

Meroxapol 108.
Pluronic 10 R8 [109]

Meroxapol 171.
Pluronic 17 R1 [109]

Meroxapol 172.
Pluronic 17 R2 [109]

Meroxapol 174.
Pluronic 17 R4 [109]

Meroxapol 178.
Pluronic 17 R8 [109]

Meroxapol 251.
Pluronic 25 R1 [109]

Meroxapol 252.
Pluronic 25 R2 [109]

Meroxapol 254.
Pluronic 25 R4 [109]

Meroxapol 255.
Pluronic 25 R5 [109]

Meroxapol 258.
Pluronic 25 R8 [109]

Meroxapol 311.
Pluronic 31 R1 [109]

Meroxapol 312.
Pluronic 31 R2 [109]

Meroxapol 314.
Pluronic 31 R4 [109]

Methacrylate.
Crylcon 3000 Series, 7000 Series [269]

Methacrylato chromic chloride.
Volan, Volan L [269]

Methicone.
Dow Corning 1107 Fluid [262]
Silicone 1107 Fluid [262]

Methionine hydroxy analog, calcium salt.
MHA [602a]

Methocarbamol.
Robaxin Tablets [510]

2-(Methoxycarbonylamino)-benzimidazole.
Bavistin [107], [108]
Delsene [270]
Derosal [416]
Pillarstin [696a], [945a]

2-Methoxy-3,6-dichloro-benzene.
Banair [439]

2-Methoxy-3,6-dichlorobenzoic acid.
Banex [898]
Banvel 4S, 4WS, CST, D, Herbicide, II Herbicide [898]

Methoxydiglycol.
Dowanol DM [261]

Methoxyethanol.
Dowanol EM [261]

Methoxy-ethoxypropylamine.
Surfam P-MEPA [554]

Methoxy ethyl mercury acetate.
Panogen, Panogen M [792]

Methoxyethyl mercury chloride chloro (-2-methoxyethyl mercury.
Agallol [112]
Emisan 6 [304a]
Tayssato [485a]

1-Methoxy indane.
Phloralid-141110 [406]

4-Methoxy-4-methyl-pentanone-2.
PentoXone [791]

p-Methoxy phenol.
Hydroquinone Monomethyl Ether [278]

N-4-(p-Methoxy-phenoxy)-phenyl-N′,N′-dimethylurea.
Lironion [190]

4-Methoxy-m-phenylenediamine.
Rodol BA [531]

2-Methoxy-p-phenylenediamine sulfate.
Rodol PDAS [531]

Methoxy polyethylene glycol 400 monooleate.
Emery 6779 [289]

Methoxypropanol.
Dowanol PM [261]

Methoxypropylamine.
Surfam P5 Distilled, P5 Technical [554]

Methscopolamine bromide.
Pamine® [884]

Methyl acetyl ricinoleate.
Flexricin P-4 [640a]

Methylacrylate.
Norsocryl [651]

1-Methyl-2 alkyl-3 alkylamidoethyl-imidazolinium-methosulfate.
Loraquat M 5040 [274]

Methyl (1) alkylamidoethyl (2) alkyl imidazolinium.
Amasoft 16-7 [190]

1-Methyl-1-alkylamidoethyl-2-alkylimidazolinium methosulfate.
Carsosoft S-75, S-90 [529]
Varisoft 475 [796]

Methyl alkyl polysiloxane.
SF 1134, 1147 [341]

Methyl alkyl polysiloxane silicone fluid.
SF 1080, 1091 [341]

Methyl-p-aminophenol sulfate.
Metol [389]

p-Methylaminophenol sulfate.
Rodol PM [531]

Methyl anthranilate.
MA [797]

Methyl behenate ester.
Kemester 9022 [429]

Methyl N-benzoyl-N-(3-chloro-4-flurophenyl)-2-aminopropioate.
Lancer [444]

α-Methylbenzyl 3-hydroxy-cis-crotonate, dimethyl phosphate.
Ciodrin [794]
Cypona E.C. [794]
Duo-Kill [794]

Methyl-bicyclo[2.2.1]heptene-2,3-dicarboxylic anhydride isomers.
Nadic Methyl Anhydride[175] [35]

Methyl bis (hydr. tallowamidoethyl) 2-hydroxyethyl ammonium methyl sulfate.
Varisoft 110 [796]

Methylbis (2-hydroxy-ethyl) coco ammonium chloride.
Ethoquad C/12 [79]
Variquat 638 [796]

Methylbis (2-hydroxy-ethyl) octadecyl ammonium chloride.
Ethoquad 18/12 [79]

Methylbis (2-hydroxy-ethyl) oleyl ammonium chloride.
Ethoquad O/12 [79]

Methyl bis (oleylamido-ethyl) 2-hydroxyethyl ammonium methyl sulfate.
Varisoft 222-LT [796]

Methyl bis (tallow amidoethyl) 2-hydroxyethyl ammonium methyl sulfate.
Varisoft 222, 222-90%, 222-HV [796]

Methyl bis (tallow amidoethyl) 2-hydroxypropyl ammonium methyl sulfate.
Varisoft 238, 238-90% [796]

Methyl bromide.
Profume [261]

Methyl bromide/chloropicrin, 98:2%.
Pestmaster Soil Fumigant-1 [582]

Methyl 1-(butylcarbamoyl)-2-benzimidazolecarbamate.
Benlate [270]

Methylcellulose.
Methocel, Methocel A, A4C, A4M, A15-LV [261]
Viscontran MC [400]

Methyl chloride.
Artic [270]

Methyl 2-chloro-9-hydroxyfluorene-9-carboxylate.
CF 125 [162a]
Curbiset [162a], [291a]
Maintain CF 125 [881]
Multiprop [162a]

2-Methyl-4-chlorophenoxyacetic acid.
Vacate [780]

2-Methyl-4 chlorophenoxyacetic acid, dimethylamine salt.
Vacate [780]

2-Methyl-4-chlorophenoxy-acetic acid, isooctyl ester.
Shamrox [780]

4-(2-Methyl-4-chlorophenoxy) butyric acid.
Can-Trol [730], [730a]

2-(2-Methyl-40-chlorophenoxy) propionic acid.
Methoxone M [444]

2-(2-Methyl-40-chlorophenoxy) propionic acid, potassium salt.
Mecopex [195]

Methyl cocoate.
CE-618 [709]
Kemester 325 [430]

Methyl coconate.
Emery Methyl Esters [290]
Kemester 325 [429]

1-(2-Methylcyclohexyl)-3-phenylurea.
Tupersan [270]

Methyl-cyclo-octyl-carbonate.
Jasmacylat [399], [400]

Methylcyclopentenolone.
Cyclotene [261]

Methyl 5-(2,4-dichloro-phenoxy)-2-nitrobenzoate.
Modown [730], [730a]

Methyl 2-ξ4-(2′,4′-dichorophenoxy)-phenoxyξ propanoate.
Hoe-Grass [55], [416]
Hoelon 3EC [55], [416]
Hoelon [55], [416]
Illoxan [55], [416]

Methyl di-coconut-t-amine.
Kemamine T-6501 [429], [430]

4-Methyl-7-diethyl-amino coumarin.
MDAC [192]

Methyl di-hydrogenated tallow-t-amine.
Kemamine T-9701 [429], [430]

2-Methyl-5, 6-dihydro-4-H-pyran-3-carboxylic acid anilide.
Sicarol [416]

Methyl N′,N′-dimethyl-N-ξ(methyl carbamoyl) oxyξ -1-thiooxamimidate.
Vydate L Insecticide 1 Nematicide [270]

Methyl N-2,6-dimethyl phenyl-N-furoyl (2)-alaninate.
Fongarid [190]

Methyl 2 ξξξξ(4,6-dimethyl-2-pyrimidinyl)aminoξ-carbonylξ aminoξ sulfonylξ benzoate.
Oust Weed Killer [270]

6-Methyl-1,3-dithiolo ξ4,5-bξ quinoxalin-2-one.
Morestan [112], [597a]

Methyldodecylbenzyl trimethyl ammonium chloride/methyldodecylxylene bis (trimethylammonium chloride), 80:20%.
Hyamine 2389 [742]

Methylene bis-o-chloroaniline.
Cyanaset M [50]

2,2′-Methylene bis(4-chlorophenol).
Fungicide FX, GM [77]

Methylenebisdiamylphenoxypolyethoxyethanol
Triton X-155 [742]

4,4′-Methylene bis(di-butyldithiocarbamate).
Vanlube 7723 [894]

4,4′-Methylenebis(2,6-di-t-butylphenol).
Ethanox 702 [302]

2,2′-Methylene-bis-(4-ethyl-6-t-butyl phenol).
Antioxidant 425 [50]
Cyanox® 425 [52]

2,2′ Methylene-bis (4-methyl-6-t-butyl) phenol.
Akrochem Antioxidant 235 [15]
Antioxidant 2246 [50]
CAO-5, -14 [796]
CAO®-R [797]
Santowhite PC Antioxidant [392], [602]
Vanox 2246 [894]
Vulkanox BKF [597]

2,2′-Methylene-bis (6-(1-methylcyclohexyl)-p-cresol).
Permanax WSP [212], [910]

2,2′-Methylene-bis(4-methyl-6-cyclohexyl phenol).
Vulkanox ZKF [597]

2,2′-Methylene-bis [4-methyl-6-(1-methylcyclohexyl) phenol].
Nonox® WSP Antioxidant [438]

Methylene bis (naphthalene sulfonic acid), disodium salt.
Dispersol T [442]

Methylenebis (4-phenyl isocyanate), bisphenol adduct.
Hylene MP [270]

Methylenebis (thiocyanate).
Amerstat 282 [264]
Count-Down [904]
Methylene Bis (Thiocyanate) Technical [904]
Microbi-Cide 30 [904]
Rychem 810 [756]
Vineland MBT 10%, 30% [904]

2,2-Methylenebis (3,4,6-trichlorophenol).
Hexide [477]

Methylene chloride.
Aerothene MM [261]
Stauffer MC+ [821]

Methylene glutaronitrile.
MGN [889]

Methylene iodide.
Mi-Gee [346]

2-(1-Methylethoxy) phenol methylcarbamate.
Baygon [112], [597a]
Blattanex [112]
Hercon Insectape with Baygon, Insectape Professional Strength with Propoxur, Insectape with Propoxur [404]
Pillargon [696a], [945a]
Roach-Chek, Roach Tape [404]
Suncide [112]
Tugon Fliegenkugel [112]
Unden [112]

3-(1-Methylethyl)-1H-2,1,3-benzothiadiazin-4(3H)-one 2,2-dioxide.
Basagran [107]

Methyl ethyl cellulose.
Edifas [173]

1-Methylethyl 2-{{ethoxy}(1-methylethyl)-amino}phosphinothioyl} oxy}-benzoate.
Amaze [106], [597a]
Oftanol [106], [597a]

Methyl ethyl ketone peroxide.
Cadox F.85, M-30, Cadox® M-50, Cadox M-105, MDA-30 [16]
Esperfoam FR [890]
Lupersol DFR, DSW [536a]
Norox FS-10019, W-60 [644]
Superox 701, 703 [722]
Thermacure FR [335a]

Methyl ethyl ketone peroxide in a plasticizer sol'n.
Superox 702, 732 [722]

Methyl ethyl ketoxime.
Skino #2 [605]
Troykyd Anti-Skin B [869]

2-Methyl-furan-3-carboxanilide.
Pano-ram [486]

Methyl gallate.
Progallin M [632]

Methyl gluceth-10.
Glucam E-10 [44]

Methyl gluceth-20.
Glucam E-20 [44]

Methyl gluceth-20 sesquistearate.
Glucamate SSE-20 [44]

Methyl glucose sesquioleate.
Glucate SO [44]

Methyl glucose sesquistearate.
Glucate SS [44]

Methyl glucoside.
Sta-Meg [818]

Methyl glucoside, alkoxylated.
Glucam Derivatives [44]

Methyl glucoside dioleate, ethoxylated (120).
Glucamate® DOE-120 [44], [45]

Methyl glucoside sesquistearate.
Glucate SS [44], [45]

Methyl glucoside sesquistearate, ethoxylated (20).
Glucamate® SSE-20 [44], [45]

Methyl-2-n-hexyl-3-oxo-cyclopentanecarboxylate.
Dihydro Isojasmonate-121440 [406]

Methyl-1-hydr. tallow amido ethyl-2-hydr. tallow imidazolinium-methyl sulfate.
Varisoft 445 [796]

Methyl-m-hydroxycarbanilate-m-methylcarbanilate.
Betanal [772]
Spinaid [645]

Methyl m-hydroxycarbanilate-m-methylcarbanilate/ethyl m-hydroxycarbanilate carbanilate, 1:1.
Betamix [645]

Methyl hydroxycellulose.
Tylose P, P-x, PS-x, P-Z Series [416]

Methyl hydroxyethylcellulose.
Tylose MB, MH, MHB, MHB-y, MHB-yp, MH-K, MH-xp [416]

2-Methyl-4-hydroxypyrrolidine.
Tiazamine [901]

Methyl 12-hydroxy-stearate.
Cenwax ME [876]

3-Methyl-5-isopropylphenyl-N-methylcarbamate.
Carbamult [772]

3-Methyl-4,5-isoxazoledione-4-ξ(2-chlorophenyl)hydrazoneξ.
Mil-Col [444]
SAIsan [444]

Methyl laurate.
CE-1218, -1270, -1280, -1290, -1295 [709]
Emery Methyl Esters [290]
Kemester 9012 [429a]
Radia 7118 [658]

Methyl linoleate.
Emery Methyl Esters [290]

4- and 5-Methylmercaptobenzimidazole.
Vulkanox MB-2/MGC [597]

4- and 5-Methylmercaptobenzimidazole, zinc salt.
Vulkanox ZMB-2/G [597]

S-Methyl-N((methylcarbamoyl) oxy)-thioacetimidate.
Lannate [270]
Nudrin [678a], [791]

7-Methyl-3-methylene-1,6-octadiene, min purity 75%.
Myrcene 85 [357]

2-Methyl-2-(methylthio)propionaldehyde O-(methylcarbamoyl) oxime.
Temik [877]

N-Methylmorpholine 99%.
Sequalog Grade 62-0020-00 [172]

Methyl myristate.
Emery Methyl Esters [290]
Kemester 9014 [429a]

Methyl naphthalene compound.
Carrier TW-200 [630]

Methyl naphthalene, emulsified.
Marvanol Carrier J-28 [552]

N-Methyl-N-nitroso-p-toluenesulfonamide.
Diazald [32]

Methyl nonyl ketone.
MGK Dog & Cat Repellent [568]

Methylol amide surfactant.
Neoprotex KM [630]

Methyl oleate.
Emerest 2301 [289]
Estol 1404 [875]
Grocor 8002, 8008 [374]
Kemester 105 [429], [429a], [430]
Kemester 115, 205 [429]
Radia 7060 [658]
Stepan C-68 [825]

Methyl oleate (white).
Kemester 205 [429], [430]

N-Methyl oleoyl taurate, sodium salt.
Nissan Diapon TO [636]

1-Methyl-L-oleylamido-ethyl-2-oleyl-imidazolinium methosulfate.
Incrosoft CFI [223b]
Varisoft 3690 [796]

Methylol melamine compound.
Imperon Fixing Agent HWA [416]

Methyl palmitate.
Emery Methyl Esters [290]
Radia 7120 [658]

Methyl palmitate oleate.
Petrosan 102 [724]

Methylparaben.
Aseptoform [370]
Lexgard M [453]
Methyl Parasept [476a], [850]
Nipagin M [632]
Preservaben M [867]
Protaben M [713]

Methyl parathion.
Methyl Parathion™ [602]

Methyl pentachlorostearate.
CMS-32 [483]

Methyl phenyl carbinyl acetate.
Gardenol [353]

1-Methyl-3-phenyl-5-{3-(trifluoromethyl) phenyl}-4-(1H)-pyridinone.
Sonar [282]

3-(2-Methylpiperidino)propyl 3,4-dichlorobenzoate.
Pipron [282]

Methyl polyethanol quaternary amine.
Peregal OK [338]

Methylpolyoxyethylene (15) coco ammonium chloride.
Ethoquad C/25 [79]

Methylpolyoxyethylene (15) octadecyl ammonium chloride.
Ethoquad 18/25 [79]

Methylpolyoxyethylene (15) oleyl ammonium chloride.
Ethoquad O/25 [79]

Methyl polysiloxane.
Abil-Methyl-Silcione Oils (20-100.000 mm^2s^{-1}) [856]

Methylprednisolone.
Medrol® [510], [884]

Methylprednisolone acetate.
Depo-Medrol® [884]
Medrol Enpak® [884]

Methylprednisolone sodium succinate.
Solu-Medrol® [884]

2-(1-Methyl-2-propyl)-4,6-dinitrophenyl isopropylcarbanate.
Acrex [486]

N-Methyl-2-pyrrolidone.
M-Pyrol [338]

3-(1-Methyl-2-pyrrolidyl) pyridine.
Black Leaf 40 [122a]

6-Methyl-2,3-quinoxalinedithiol cyclic carbonate.
Morestan [184]

Methyl ricinoleate.
Estrasan 1-A [724]
Flexricin P-1 [640a]

Methyl siliconates.
Tegosivin® [856]

Methyl soyate.
Kemester 226 [429], [430]

Methyl stearate.
Grocor 8058 [374]
Kemester 9718 [429a]
Radia 7110 [658]

Methyl stearate ester.
Kemester 9018 [429]

4- α -Methyl-styrylphenol.
Prodox® 121 [315]

Methyl sulfanilylcarbamate.
Asulox, 40, F [561a]

4-Methylsulphonyl-2,6-dinitro-N,N-dipropylaniline.
Planavin [794]

N-Methyl tallow-alkyl taurate, sodium salt.
Nissan Diapon T [636]

1-Methyl-1-tallow amido-ethyl-2-heptadecyl imidazolinium methosulfate.
Ammonyx 4080 [663]

Methyl-1-tallow amido ethyl-2-tallow imidazolinium-methyl sulfate.
Varisoft 475 [796]

Methyl tallowate.
Emery Methyl Esters [290]
Grocor 8041 [374]

Methyl taurate esters.
Protapons [713]

Methyl tauride, sodium salt, fatty acid.
Hostapon CT Paste [55], [416]
Hostapon STT Paste, T [416]

(cis)-3-Methyl- Δ-tetrahydrophthalic anhydride.
Curacid 400 [238]

3-Methyl-thiazolidine-thione-2.
Vulkacit CRV/LG [597]

3-(Methylthio)-O-{(methyl-amino) carbonyl} oxime-2-butanone.
Afiline [910a]
Drawin 755 [910a]

Methyltin mercaptide.
Advastab TM-181, TM-185, TM-692 [192]

Methyl tri ammonium chloride (C_8-C_{10}).
Adogen 464 [796]

5-methyl-1,2,4-triazolo { 3,4-b}-benzothiazole.
Beam [282]

Methyl tuads/ethyl tuads, 60:40.
Methyl-Ethyl Tuads [894]

Methyl violet-phosphomolybdic acid salt.
Atlas 9673 Paragon Violet Toner [496]

Metronidazole.
Flagyl Tablets [510]

Mibolerone.
Cheque® [884]

Mica.
Mica, Cosmetic, Bacteria Controlled 280 [924]

Mica *(cont'd).*
Micatex [616]

Mineral oil.
Bayol 72, 90, 92 [306]
Benol White Mineral Oil [936]
Kaydol White Mineral Oil [936]
Klearol White Mineral Oil [936]
Liteteck [684]
Marcol 52, 62, 70, 72, 82, 87, 90, 97, 130, 145 [306]

Mineral oil, emulsifiable.
Prosol C70 [390]
Spinolene SO, SOA, SPK [788]

Mineral oil, sulfonated.
Twitchell 6805 Oil, 8262, 8266, 8905, 8910, 8955 [290]

Mineral oil, white.
Blandol [934], [936]
Carnation [936]
Drakeol [684]
Freezene [934]
Klearol [934]

Minkamidopropyl dimethylamine.
Foamole B [895]

Mink-amido-propyl dimethyl hydroxyethyl ammonium chloride.
Ceraphyl 65 [895]

Mink oils.
Emulan [293]
Naturol [507]
Olio Di Visone [901]

Minoxidil.
Loniten® [884]

MIPA-dodecylbenzenesulfonate.
Hetsulf IPA [407]

MIPA-lauryl sulfate.
Melanol LPI [785]

Molybdenum complex.
KemGard 425, 911A, 911B, 911C [797]

Molybdenum disulfide.
Molysulfide [146]

Molybdenum(VI) oxide.
Ultralog 64-2165-00 [172]

Molybdenum oxysulfide dithiocarbamate.
Molyvan A [894]

Molybdic oxide.
U-Pol [196]

Monoalkyl sulfosuccinate.
Lankrapol KNB 22 [253], [509]

Monoammonium phosphate-based formulation.
Foray [66]

Mono-t-butyl hydroquinone.
Eastman MTBHQ [278]
Sustane TBHQ [883]

Monocalcium phosphate.
H.T. [602a]

Monocalcium phosphate, anhydrous.
Py-Ran [602]

Monocalcium phosphate monohydrate.
H.T.® [602]

Monocarboxylate amphoteric.
Cycloteric MV-SF [234]

Monodecyl diphenyl phosphite.
Mark MDDPP [75]

Mono-diglycerides.
Admul MG 4103, MG 4123, MG 4143 [704]
Emuldan [373]

Mono-diglycerides, acetic acid ester.
Axol E 41, E 66 [856]
Tegin E 41, E 61, E 66 [856]

Mono-diglycerides, citric acid ester.
Axol C 64 [856]
Lamegin ZE 30, ZE 60 [176]
Tegin C 1 R, C 61, C 62, C 64 [856]

Monodiglycerides, diacetyl tartaric acid ester.
Lamegin DWF, DWH, DWP [176]

Mono-diglycerides, dispersed.
Lamesoft 156 [176]

Mono-diglycerides, lactic acid ester.
Axol L 624, L 626 [856]
Tegin L 61, L 62 [856]

Monoethanolamide, higher fatty acid, sulfosuccinate.
Cyclopol SBR3 [234]

Monoethanolamine alkyl sulfate.
Manro ML 33S [545]

Monoethanolamine lauryl ether sulfate.
Akyposal MLES 35, MLS 30 BOD, MLS 55 [182]
Drewpon ECM [265], [266]
Sactipon 2 OM [521]
Tylorol LM [858]

Monoethanolamine lauryl ethoxy (1) sulphate.
Empicol 0285 [23]

Monoethanolamine lauryl sulfate.
Alkasurf MLS [33]
Condanol MLS 35 [274]
Cycloryl SA [234]
Drewpon MEA [265], [266]
Empicol EL [25]
Empicol LQ33 [23], [24]

Monoethanolamine lauryl sulfate *(cont'd)*.
Empicol LQ 33/T, LQ 70 [23], [25]
Manro ML 33 [545]
Rewopol MLS 35 [725]
Sactipon 2 M [521]
Sactol 2 M [521]
Sulfatol 33 MO [1]
Sulphonated Lorol Liquid MA, Liquid MR [745]
Texapon MLS [401]
Zoharpon LAM [946]

Monoethanolammonium lauryl sulfate.
Rewopol MLS 30 [725a]

Monoglycerides.
Admul MG 4163, MG 4203, MG 4223, MG 4304, MG 4404, MG 5103, MG 6103, MG 6404, MG 6504 [704]
Dimodan CP, LS, P, PM, PV, PVP, S, TH [373]
Myvatex Mighty Soft (K) [278]
Myverol [278]
Normulgen M, O [150]
Radiamuls MG [658]

Monoglycerides (40%).
G. 38 Emulsifier [339]
Maisine [339]
Olicine [339]

Monoglycerides, acetic acid esters.
Cetodan, Cetodan 50-00A [373]

Monoglycerides, acetylated.
Axol E 67 [856]
Cetodan [373]
Dynacet [275], [482]
Radiamuls AMG [658]
Vykacet L, T [223]

Monoglycerides, citric acid esters.
Acidan [373]
Imwitor 369, 370 [275]

Monoglycerides, citrylated.
Radiamuls CMG [658]

Monoglycerides, diacetylated tartaric acid esters.
Admul Data [704]
Datagel [223]
Datem Esters (various grades) [223]
Imwitor 1330, 2020, 2320 [275]
Panodan, Panodan 235 [373]

Monoglycerides, distilled.
Alphadim 90 AB [675], [676]
Alphadim 90 LC [676]
Amidan [373]
Dimodan [373]
Dri'N Soft [672]
Excel O-95N, T-95, T-95P, VS-95, VS-95S [478a]
Hymono 1103, 8803 [704]
Imwitor 191 [275]
Kureton A [478a]
Starplex 90 [675]

Monoglycerides, ethoxylated.
Chimipal FV [514]

Monoglyceride, fatty acid.
Rewomul MG [725]

Monoglycerides, hydrated distilled.
HSH [675]
Sof-plus [675]

Monoglycerides, kosher distilled.
Alphadim 90 SBK [675]

Monoglycerides, lactic acid esters.
Admul GLP, GLS [704]
Imwitor 333 [275]
Lactodan [373]

Monoglycerides, lactylated.
Radiamuls LMG [658]

Monoglycerides polyglycol ether, fatty acid.
Rewoderm ES 90 [725a]

Monoglycol ester, fatty acid.
Genapol PMS [416]

Monoisopropanol amine lauryl sulfate.
Empicol MIPA [23], [24]

Monomethyl p-amino-phenol sulfate.
Pictol [542]

Monomethyl dicoconut t-amine.
Kemamine T 6501 [429a]

Monomethyl dieicosanyl-docosanyl t-amine.
Kemamine T 1901 [429a]

Monomethyl dihydrogenated tallow t-amine.
Kemamine T 9701 [429a]

Monomethylol dimethylhydantoin.
Dantoin® MDMH [358]

Mono naphthalene sulfonic acid, condensed, ammonium salt.
Lomar PWA [253]

Mono naphthalene sulfonic acid, condensed, sodium salt.
Lomar PW [253]

Monooleate.
Grocor PEG 300 [374]

Monopentaerythritol.
Monopentek [850]

Monophenyl glycol.
Rewopal MPG 10 [725a]

Monosodium acid methanearsonate.
Check-Mate [904]

Monosodium acid methanearsonate *(cont'd).*
Dal-E-Rad 70 + W, 120 [904]
Weed-E-Rad + W [904]
Weed Hoe-2X [904]
Weed-Hoe 108, 120 [904]

Monosodium N-cocoyl-1-glutamate.
Acylglutamate CS-11 [14]
Amisoft GS-11 [14]

Monosodium glutathione, lypophilized.
Triptide [776]

Monosodium N-hydrogenated tallowyl-L-glutamate.
Acylglutamate HS-11 [14]
Amisoft HS-11 [14]

Monosodium N-lauroyl-L-glutamate.
Acylglutamate LS-11 [14]
Amisoft LS11 [14]

Monosodium-N-lauryl β-iminodipropionic acid.
Deriphat 160C [400]

Monosodium methanearsonate.
Ansar 170 H.C., 529 H.C. [780]
Arsonate Liquid [780]
Bueno 6 [780]
Daconate 6 [780]
Dal-E-Rad [904]
Herb All [671a]
Merge 823 [859]
Weed-E-Rad [904]
Weed Hoe [904]

Monosodium N-mixed fatty acid acyl glutamate.
Acylglutamate GS-11 [14]

Monosodium N-octyl- β-iminodipropionic acid.
Deriphat 130-C [400]

Monotallate ester polyethylene glycol.
Ethoxy #4 [916]

(Monotrichloro) tetra-(monopotassium dichloro)-penta-isocyanurate.
ACL 66 [602]

Monotriethanolamine N-cocoyl-1-glutamate.
Acylglutamate CT-12 [14]
Amisoft C-T-12 [14]

Monotriethanolamine n-hydrogenated-tallowyl-l-glutamate.
Amisoft HT 12 [14]

Monotriethanolamine n-lauroyl-l-glutamate.
Amisoft LT-12 [14]

Montan acid wax.
BASF Wax L, LS, S [106]

Montan ester wax.
BASF Wax DG, E, ES, LCP, LG, LK 3, SG [106]
Cereplast® F, MT [106]

Montan ester wax, partially saponified.
BASF Wax OP [106]

Montan wax.
Montan Wax - STRALPITZ [830]

Montmurillonite clay.
Florco [330]

Montmorillonite clay, white.
Gelwhite [347]

N-Morpholin-2-benzo thiazole-sulfenamide.
Pennac MBS [682]

2-(N-Morpholino)ethanesulfonic acid.
Chemalog 64-3560-00 [172]

Morpholinopropanesulfonic acid 99%.
Chemalog 64-3740-00 [172]

2-(Morpholinothio) benzothiazole.
Santocure Mor Vulcanization Accelerator [392], [602]

4-Morpholinyl-2-benzothiazole disulfide.
Morfax [894]

2-(4-Morpholinyldithio) benzothiazole.
Vulcuren 2 [112]

2-(4-Morpholinylmercapto) benzothiazole.
Delac MOR [881]

2-(4-Morpholinyl) mercaptobenzothiazole sulfenamide.
Sulfene 2 [708]

Morpholyl benzthiazyl sulphenamide.
Vulcafor MBS (G) [212], [910]

Mullite.
Tamul [169]

Mullite firebrick.
Kaomul [98]

Mullite refractory.
MV33 [565]
Tamax [169]

Myreth-3.
Emthox 5969 [291]
Hetoxol M-3 [407]

Myreth-3 caprate.
Standamul 1410E [400]

Myreth-3 laurate.
Schercemol MEL-3 [771]

Myreth-3 myristate.
Liponate 143M [525]
Schercemol MEM-3 [771]
Standamul 1414E [400]

Myreth-3 palmitate.
Schercemol MEP-3 [771]

Myreth-4.
Lipocol M-4 [525]
Procol MA-4 [713]

Myristalkonium chloride.
Barquat MS-100, MX-50, MX-80 [529]
Kemamine Q-7903B [429a]

Myristalkonium saccharinate.
Onyxide 3300 [663]
Rewoquat QA 100 [725]

Myristamide DEA.
Aminol MRC-A [320]
Carsamide MDEA [529]
Condensate M-100A, MRCA, PM [174]
Hetamide M [407]
Monamid 150 MW [599]
Protamide MRCA [713]
Rewo-Amid DLMSE [725]
Schercomid MD-Extra, SM [771]

Myristamide MEA.
Carsamide MMEA [529]
Hetamide MM [407]
Schercomid MME [771]
Witcamide MM [934]

Myristamidopropyl betaine.
Schercotaine MAB [771]

Myristamidopropyl dimethylamine.
Lexamine M-13 [453]
Schercodine M [771]

Myristamine oxide.
Ammonyx MCO, MO [663]
Barlox 14 [529]
Conco XAM [209]
Emcol M [934], [935]
Jordamox MDA [474]
Ninox M [825]
Schercamox DMA, DMM [771]

Myristic acid.
Crodacid [223]
Emery 655 [288]
Hystrene 9014 [429], [429a]
Neo-Fat 14 [79]
PRIFAC 2940 [875]

Myristic alkanolamide.
Monamid 150-MW [599]

Myristic diethanolamide.
Monamid 150-MW [599]

Myristic fatty acid, 90%.
Hystrene 9014 [430]

Myristic fatty acid alkanolamide.
Condensate PM [209]

Myristoamphoglycinate.
Miranol MM Conc. [592]

Myristoyl hydrolyzed animal protein.
Lexein A-200, A-210 [453]

Myristoyl sarcosinate, sodium salt.
Nikkol Sarcosinate MN [631]

Myristoyl sarcosine.
Hamposyl M [363]

Myristyl alcohol.
Alfol 14, 1412 [206]
Cachalot M-43 [581]
Dehydag Wax 14 [400]
Kalcohl 40 [478]

Myristyl alcohol, polypropoxylated.
Witconol APEM, APM [937a], [938]

Myristyl amido betaine.
Schercotaine MAB [771]

Myristyl amine oxide.
Sandoz Amine Oxide XA-M [761]

Myristyl-cetyl dimethyl amine oxide.
Ammonyx MCO [663]

Myristyl chloride.
Barchlor 14S [529]

Myristyl dimethyl amine.
Barlene 14S [529]

Myristyl dimethyl t-amine (distilled).
Lilamin 314 D [522]

Myristyl dimethyl amine oxide.
Ammonyx MO [663]
Conco XA-M [209]
Empigen OH [25]
Incromine Oxide M [223b]
Ninox M [825]

Myristyl dimethyl benzyl ammonium chloride.
Dibactol [319], [408a]

Myristyl dimethyl benzyl ammonium chloride dihydrate.
Arquad DM14B-90 [16]
Barquat MS-100 [529]
Dibactol [408]

Myristyl eicosanol.
Standamul G 32/36 [400]

Myristyleicosyl stearate.
Standamul 7115, G-3236 Stearate [400]

Myristyl lactate.
Cegesoft C17 [375]
Ceraphyl 50 [895]
Crodamol ML [223]
Cyclochem ML [234]
Liponate ML [525]
Nikkol Myristyl Lactate [631]
Schercemol ML [771]
Wickenol 506 [926]

Myristyl myristate.
Cegesoft C28 [375]
Ceraphyl 424 [895]
Crodamol MM [223]
Cyclochem MM/M [234]
Liponate MM [525]
MYM-33 [397]
Nikkol MM [631]
Schercemol MM [771]
Waxenol 810 [926]

Myristyl propionate.
Lonzest, Lonzest 143S [529]
Schercemol MP [771]

Myristyl stearate.
Cyclochem MST [234]

Myristyl trimethyl ammonium bromide.
Mytab [408], [408a]

Myrtrimonium bromide.
Mytab [408]

Nadolol.
Corgard [817]
Corzide [817]

Nafcillin sodium.
Nafcil [134]
Nallpen® for Injection—Buffered, 1 Gm., 2 Gm., 10 Gm., 500 mg. [114]
Unipen Injection, Capsules, Powder for Oral Solution, & Tablets [942]

Nalbuphine hydrochloride.
Nubain [269]

Nalidixic acid.
NegGram Caplets, Suspension [932]

Naloxone hydrochloride.
Narcan and Narcan Neonatal [269]
Talwin Nx [932]

Nandrolone decanoate.
Deca-Durabolin [664]
Kabolin [518]

Nandrolone phenpropionate.
Durabolin [664]

Naphazoline hydrochloride.
Clear Eyes Eye Drops [3]
4-Way Nasal Spray [134]
Ocu-Zoline [655a]
Privine Hycrochloride .05% Nasal Solution, .05% Nasal Spray [189]
Vasolin Opthalmic Solution [498]

Naphthalene acetamide.
Amid-Thin W [370a], [877]

1-Naphthaleneacetic acid.
Fruitone N [877]
Hormofix [376a]
NAA 800 [877]
Plucker [445]
Rootone [877]
Stik [331a]
Tip-Off [583]

Naphthalene condensate.
Intralan Salt HA [225]

Naphthalene, condensated, salt.
Demol RN [478]

Naphthalene, condensed, sodium salt.
Lomar D Liquid [253]

1,5-Naphthalenediol.
Rodol 15N [531]

2,3-Naphthalenediol.
Rodol 23N [531]

2,7-Naphthalenediol.
Rodol 27N [531]

Naphthalene-formaldehyde condensate, sulfonated, sodium salt.
Blancol Dispersant [338]
Blancol® N [338]
Tamol N Micro [725]

Naphthalene formaldehyde sulfonate.
Dyesperse DC [390]

Naphthalene, sulfated.
Nopcosant L [252]

Naphthalene sulfonate.
Harol D [365]
Lomar D [252]

Naphthalene sulfonate, condensed.
Lomar [643]
Stepantan A, NP 80 [825a]

Naphthalene sulfonate, condensed, sodium salt.
Synthrapol DA-AC [438]

Naphthalene sulfonate, sodium salt.
Protodye QNF [710]

Naphthalene sulfonic acid, condensed, ammonium salt.
Tamol® NIB [106]

Naphthalene sulfonic acid, condensated, salt.
Demol N [478]
Runox 1000 [864]

Naphthalene sulfonic acid, condensed, sodium salt.
Lomar D, LS [253]
Tamol NH, NN, NNO, NNOK, NNOK-SA, NOC [106]
Vultamol®, Vultamol SA Liquid [106]
Wettol D 2 [106]

Naphthalene sulfonic acid formaldehyde condensate, ammonium salt.
Synopen N [77]

Naphthalene-sulfonic acids, sodium salt.
Intraphor AC [253]
Tamol NNO, NNOK [106]

1,8-Naphthalic anhydride.
Protect [377]

Naphthenic acids.
Sunaptic [835]

Naphthenic hydrocarbon oil.
Cyclolube 62, 85, 132, 4053 [940]

Naphthenic hydrocarbon oil, ASTM type 103.
Cyclolube 100, 213,, 2310, NN-1, NN-2 [940]

Naphthenic hydrocarbon oil, ASTM type 104A.
Cyclolube 413, 3710 [940]

Naphthenic oil.
Process Oil C-255-NS, C-293NS, SR-111 [379]

Naphthenic oil *(cont'd).*
Shellflex 3211, 3271, 3681 [791]
Sunthene 310 [154], [835]

Naphthenic oil, ASTM D2226, type 102.
Sunthene 380 [154], [835]

Naphthenic oil, ASTM D2226, type 103.
Circo Light Rubber Process Oil [154], [835]
Circosol 410, 450, 4130, 4240 [154], [835]
Sunthene 410, 450, 4130, 4240, 5600 [154], [835]

Naphthenic oil, ASTM D2226, type 104A.
Sunthene 235 [154], [835]

Naphthenic petroleum oil, ASTM type 103.
Flexon 580, 641, 650, 660, 680 [306]

Naphthenic petroleum oil, ASTM type 104A.
Flexon 765, 766, 785 [306]

1-Naphthol.
Rodol ERN [531]

Naphthol dyeing assistant.
Eunaphtol AS [338]

2-(α-Naphthoxy)-N,N-diethylpropionamide.
Devrinol [821]

α-Naphthylacetic acid.
Phyomone [444]

1-Naphthylacetic acid.
Tipoff [583]

1-Naphthyl-N-methylcarbamate.
Ravyon [540]
Sevin [877]

N-1-Naphthylphthalamic acid.
Alanap [881]

Naproxen.
Naprosyn [844]

Naproxen sodium.
Anaprox Tablets [844]

Neatsfoot oil, sulfated.
Actrasol FG [811]
Atlas Sul. L-2 [94]
Sulfonated Neatsfoot 3089 [198]

Neatsfoot oil, sulfonated.
Crestoil [222]

Neomycin sulfate.
Bay Sporin Otic Suspension [111]
Biosol® AG 100 Premix, 325 Soluble [884]
Coly-Mycin S Otic w/Neomycin & Hydrocortisone [674]
Cortisporin Cream, Ointment, Ophthalmic Ointment, Ophthalmic Suspension, Otic Solution, Otic Suspension [139]
Cortixin Otic Suspension [498]
Mycifradin® [884]
Myciguent® [884]
Mycolog Cream & Ointment [817]
Mytrex Cream & Ointment [767]
Neodecadron Sterile Ophthalmic Ointment, Solution [578]
Neomix® [884]
Neomycin Sulfate Tablets [119], [751]
Neo-Polycin [580]
Neosporin Aerosol, G. U. Irrigant, Ointment, Ophthalmic Ointment Sterile, Ophthalmic Solution Sterile, Powder, -G Cream [139]
Neo-Synalar Cream [844]
Nyst-olone Cream & Ointment [771]
Tri-Thalmic Ophthalmic Solution [771]

Neo pentyl fatty esters.
Nissan Unister [636]

Neopentyl glycol diacrylate.
SR-247 [766]

Neopentyl glycol dicaprate.
Estemol N-01 [640]

Neopentyl glycol dimethacrylate.
SR-248 [766]

Neoprene.
Neoprene Latex Series [269]

Neoprene liquid.
NeoCoat, NeoLine [682]

Neostigmine bromide.
Prostigmin Tablets [738]

Neostigmine methylsulfate.
Neostigmine Methylsulfate Injection [287]
Prostigmin Injectable [738]

Netilmicin sulfate.
Netromycin Injection [772]

Niacin.
Cardioguard Natural Lipotropic Dietary Supplement [589]
Niacin, No. 69902 [738]
Niacin Tablets [751]
Nicolar Tablets [891]
Nico-Vert Capsules [281]
Rovite Tonic [748]

Niacinamide.
A.C.N. Tablets [686]
Albafort Injectable [105]
Al-Vite [267]
B-C-Bid Capsules [350]
Besta Capsules [394]
Eldercaps [563]

Niacinamide *(cont'd).*
Eldertonic [563]
Geravite Elixir [394]
Glutofac Tablets [489]
Hemo-Vite, Hemo-Vite Liquid [267]
Mega-B [72]
Megadose [72]
Natalins Rx, Tablets [572]
Niacinamide, No. 69905 [738]
Nu-dron-V Tablets [563]
Prenate 90 Tablets [124]
Therabid [593]

Niacinamide hydroiodide.
Iodo-Niaacin Tablets [661]

Nickel, 98.6%.
Permanickel 300 [432]

Nickel/aluminum, 94.5:4.5%.
Duranickel 301 [433]

Nickel azo pigment, yellow.
Permagreen Gold [797]

Nickel bis [O-ethyl (3,5 di-t-butyl-4-hydroxybenzyl)] phosphonate.
C-PPR-8006, -8007 [720]
Irgastab 2002 [190]

Nickel bis (octyl phenol sulfide).
UV Chek AM-101, AM-105 [196]

Nickel, carbonized.
Duocarb [927]

Nickel catalyst.
Ni-0104 P, -0104 T 1/8″, -0301 T 1/8″, -0707 T 1/8″, -0901 S 1/2″, -1404 T 3/16″, -2002 C 1″, -3210 T 3/16″, -3250 T 3/16″, -3266 E 1/16″, -5124 T 3/16″, -5132 P, -5333 T 3/16″ [389]
Nysel CN-14 (Ni 3611 L), HK-4 (Ni-3609 F), HK-12 (Ni-5708 L), Nysel (Ni-3201 F), SP-7 (Ni-5169 F) [389]

Nickel chloride.
Ultralog 65-5500-00 [172]

Nickel, chromium, aluminum and copper, 75:20:2.5:2.5%.
Evanohm [927]

Nickel/cobalt, 80:20%.
Emisaloy [927]

Nickel, cobalt, iron, minor ingredients, 29:17:53:1%.
Kovar [917]

Nickel dibutyldithiocarbamate.
Akrochem NiBud [15]
C-PPR-8002, -8003 [720]
Naugard NBC [881]
NBC [269], [270]
NDBC [708]
Robac Nibud [386a]
Rylex NBC [269]
UV Chek Am-104 [315]
Vanox NBC [894]

Nickel diisobutyldithiocarbamate.
ISO Butyl Niclate [894]

Nickel dimethyldithiocarbamate.
Methyl Niclate [894]

Nickel/manganese, 95:5%.
Mangrid [927]

Nickel molybdate.
HT-500 E 1/8″ [389]

Nickel(II) oxide.
Ultralog 65-7130-00 [172]

Nickel tungsten.
Ni-4301 E 1/16″ [389]

Niclosamide.
Niclocide Chewable Tablets [585]

Nicotinamide.
I.L.X. B_{12} Elixir Crystalline, Tablets [489]

b-Nicotinamide adenine dinucleotide.
Chemalog 65-8260-00 [172]

Nicotinic acid.
Lipo-Nicin [136]
Nico-400 [551]
Nicolar Tablets [891]
Verstat Capsules [765]

Nicotinyl alcohol.
Roniacol Elixir, Tablets, Timespan Tablets [738]

Nifedipine.
Procardia Capsules [691]

Nikethamide.
Coramine [189]

Ninhydrin reagent.
Sequalog 65-9540-00 [172]

Nithiazide.
Hepzide [577]

Nitric acid.
Acculog 66-0520-10 [172]

Nitrile-butadiene rubber.
Perbunan N 3807 NS, N 3810 [112]

Nitrile latex.
Darex 110L [364b]

Nitriloacetate trisodium salt.
Cheelox NTA-14, NTA-Na_3 [338]

Nitrilotriacetic acid.
Hampshire® NTA Acid [364b]
Trilon AS [106]

Nitrilotriacetic acid, trisodium salt.
Hampshire® NTA 150 [364b]

Nitrilotriacetic acid, trisodium salt *(cont'd).*
Trilon A 92, A Liquid [106]

Nitrilotriacetic acid trisodium salt, mono hydrate.
Hampshire® NTA Na_3 Crystals [364b]

2-Nitro-p-anisidine⇒o-acetoacetotoluidide.
Atlas A8716 Monofast Orange Toner [496]

m-Nitrobenzene sulfonic acid, sodium salt.
Ludigol F, 60 [338]

Nitrocarbonitrate blasting agents.
Hercomix 1 [406]

Nitrocellulose.
AS™ [406]
Herculoid [406]

Nitrocellulose laquer emulsions.
Hydrholac [742]

Nitrofurantoin.
Furadantin Oral Suspension, Tablets [652]
Nitrofurantoin Capsules & Tablets [769]

Nitrofurantoin macrocrystals.
Macrodantin Capsules [652]

Nitrofurantoin sodium.
Ivadantin [652]

Nitrofurazone.
Furacin Preparations, Soluble Dressing, Topical Cream [652]

Nitroglycerin.
Cardabid [765]
Nitro-Bid IV, Ointment, 2.5 Plateau Caps, 6.5 Plateau Caps, 9 Plateau Caps [551]
Nitrodisc [783]
Nitro-Dur Transdermal Infusion System [491]
Nitroglycerin Ointment 2% [769]
Nitroglycerin S.R. Capsules [344]
Nitroglyn [491]
Nitrol Ointment [499]
Nitrolin Timed Capsules [769]
Nitrong Ointment, 2.6 mg. Tablets, 6.5 mg. Tablets, 9 mg. Tablets [921]
Nitrospan Capsules [891]
Nitrostat Ointment, Tablets, IV [674]
Susadrin Transmucosal Tablets [580]
Transderm-Nitro Transdermal Therapeutic System [189]
Tridil [49]

2-Nitro-p-phenylenediamine.
Rodol 2R, Brown 2R [531]

4-Nitro-m-phenylenediamine.
Rodol LY [531]

4-Nitro-o-phenylenediamine.
Rodol 4J [531]

4-Nitro-o-phenylenediamine hydrochloride.
Rodol 4GP [531]

4-Nitro-o-phenylenediamine sulfate.
Rodol 4GSP [531]

2-Nitropropane.
Nipar S-20 [449]

p-Nitrosodimethylaniline.
Palldon [506]

N-Nitrosodiphenylamine.
Redax [362], [894]
Vantard R [894]
Vulcatard A [212], [910]

2-Nitroso-1-naphthol.
Cobon [506]

2-Nitro-p-toluidine⇒acetoacetanilide.
Atlas A8069 Monofast Yellow G Toner [496]

2-Nitro-p-toluidine⇒2-naphthol.
Atlas B3358 Toluidine Toner, B3483 Toluidine Toner [496]

Nonane-1,3-diol monoacetates.
Jasmonyl [353]

Nonoxynol-1.
Alkasurf NP-1 [33]
Surfonic N-10 [854]

Nonoxynol-2.
Carsonon N-2 [529]
Conco NI-21 [209]
Igepal CO-210 [338]
Macol NP-2 [564]

Nonoxynol-4.
Alkasurf NP-4 [33]
Arkopal N-040 [55]
Carsonon N-4 [529]
Conco NI-43 [209]
Hetoxide NP-4 [407]
Igepal CO-430 [338]
Macol NP-4 [564]
Makon 4 [825]
Neutronyx 622 [663]
Surfonic N-40 [854]
T-DET N-4 [859]
Tergitol NP-4 [878]

Nonoxynol-5.
Alkasurf NP-5 [33]
Igepal CO-520 [338]
Macol NP-5 [564]
Nikkol NP-5 [631]
Triton N-57 [742]

Nonoxynol-6.
Alkasurf NP-6 [33]
Arkopal N-060 [55]
Carsonon N-6 [529]
Conco NI-60 [209]
Igepal CO-530 [338]
Macol NP-6 [564]
Makon 6 [825]
Neutronyx 626 [663]
Surfonic N-60 [854]
T-DET N-6 [859]

Nonoxynol-7.
Alkasurf NP-7 [33]
Macol NP-7 [564]
Tergitol NP-7 [878]

Nonoxynol-8.
Alkasurf NP-8 [33]
Arkopal N-080 [55]
Carsonon N-8, N-2088 [529]
Igepal CO-610 [338]
Macol NP-8 [564]
Makon 8 [825]
T-DET N-8 [859]

Nonoxynol-9.
Alkasurf NP-9 [33]
Arkopal N-090 [55]
Carsonon N-9 [529]
Conco NI-90 [209]
Empilan NP9 [548]
Emthox 6964 [291]
Emulan [293]
Hetoxide NP-9 [407]
Igepal CO-630 [338]
Macol NP-9 [564]
Neutronyx 600 [663]
Protachem 630 [713]
Ramses Contraceptive Vaginal Jelly [773]
Semicid Vaginal Contraceptive Suppositories [922]
Sterling NPX [145]
Tergitol NP-9 [878]

Nonoxynol-9 iodine.
Biopal VRO-20 [338]

Nonoxynol-9 phosphate.
Gafac RE-610 [338]

Nonoxynol-10.
Alkasurf NP-10 [33]
Arkopal N-100 [55]
Carsonon N-10 [529]
Conco NI-100 [209]
Eumulgin 286 [400]
Hyonic PE-100 [252]
Igepal CO-660, CO-710 [338]
Macol NP-10 [564]
Makon 10 [825]
Simulson 1030 NP [785]
Surfonic N-95, N-102 [854]
T-DET N-9.5 [859]
Tergitol NP-10 [878]
Triton N-101 [742]

Nonoxynol-10 phosphate.
Protaphos P-610 [713]

Nonoxynol-11.
Alkasurf NP-11 [33]
Arkopal N-110 [55]
Carsonon N-11 [529]
T-DET N-10.5 [859]

Nonoxynol-12.
Alkasurf NP-12 [33]
Carsonon N-12 [529]
Igepal CO-720 [338]
Makon 12 [825]
Surfonic N-120 [854]
T-DET N-12 [859]

Nonoxynol-12 iodine.
Biopal NR-20 [338]

Nonoxynol-13.
Arkopal N-130 [55]
Tergitol NP-13 [878]

Nonoxynol-14.
Lamacit 877 [375]
Makon 14 [825]

Nonoxynol-15.
Alkasurf NP-15 [33]
Arkopal N-150 [55]
Conco NI-150 [209]
Igepal CO-730 [338]
Neutronyx 640 [663]
Surfonic N-150 [854]
Tergitol NP-15 [878]

Nonoxynol-18.
Lamacit KW 80-18 [375]

Nonoxynol-20.
Alkasurf NP-20 [33]
Conco NI-185 [209]
Igepal CO-850 [338]
Macol NP-20 [564]
Makon 20 [825]
Surfonic N-200 [854]
Tergitol NP-20 [878]

Nonoxynol-23.
Arkopal N-230 [55]

Nonoxynol-30.
Arkopal N-300 [55]
Carsonon N-30 [529]
Conco NI-187 [209]
Igepal CO-880 [338]
Macol NP-30 [564]
Makon 30 [825]
Surfonic N-300, NB-5 [854]

Nonoxynol-30 *(cont'd).*
T-DET N-30, N-307 [859]

Nonoxynol-40.
Alkasurf NP-40 [33]
Carsonon N-40 [529]
Conco NI-1900, NI-197 [209]
Hetoxide NP-40 [407]
Igepal CO-890 [338]
Macol NP-40 [564]
Surfonic N-400, NB-14 [854]
Tergitol NP-44 [878]

Nonoxynol-50.
Alkasurf NP-50 [33]
Carsonon N-50 [529]
Conco NI-2000 [209]
Igepal CO-970 [338]
Macol NP-50 [564]
T-DET N-50 [859]

Nonoxynol-100.
Carsonon N-100 [529]
Macol NP-100 [564]
T-DET N-100, N-1007 [859]

Nonyl naphthalene sodium sulfonate.
Emkal NNS [292]

Nonyl naphthalene sulfonic acid.
Emkal NNS Acid [292]

Nonyl nonoxynol-5.
Emalex DNP-5 [628]
Hetoxide DNP-5 [407]

Nonyl nonoxynol-7 phosphate.
GAFAC RM-410 [338]

Nonyl nonoxynol-10.
Hetoxide DNP-10 [407]
Igepal DM-530 [338]

Nonyl nonoxynol-10 phosphate.
Gafac RM-510 [338]
Monafax L10 [599]

Nonyl nonoxynol-15 phosphate.
Gafac RM-710 [338]

Nonyl nonoxynol-49.
Igepal DM-880 [338]

Nonyl nonoxynol-100.
Carsonon DM-276 [529]

Nonyl nonoxynol-150.
Igepal DM-970 [338]

Nonylphenol.
Berol Nonylphenol [116]
Hyonic PE-40, PE-90, PE-100, PE-120 [252]
Marlican [178]

Nonyl phenol, alkoxylated.
Serdox NFR [788]

Nonylphenol EO, 1.5 mol.
Merpoxen NO 15 [285]

Nonylphenol + EO, 2 moles.
Teric N2 [439]

Nonylphenol + EO, 3.5 moles.
Teric N3 [439]

Nonylphenol + EO, 4 moles.
Teric N4 [439]

Nonylphenol + EO, 5.5 moles.
Teric N5 [439]

Nonylphenol + EO, 8.5 moles.
Teric N8 [439]

Nonylphenol + EO, 9 moles.
Teric N9 [439]

Nonylphenol + EO, 10 moles.
Teric N10 [439]

Nonylphenol + EO, 11 moles.
Teric N11 [439]

Nonylphenol + EO, 12 moles.
Teric N12 [439]

Nonylphenol + EO, 13 moles.
Teric N13 [439]

Nonylphenol + EO, 15 moles.
Teric N15 [439]

Nonylphenol + EO, 20 moles.
Teric N20 [439]

Nonylphenol + EO, 30 moles.
Teric N30 [439]

Nonylphenol + EO, 40 moles.
Teric N40 [439]

Nonyl phenol (EO 50 mol).
Nopco RDY [252]

Nonylphenol + EO, 100 moles.
Teric N100 [439]

Nonylphenol ether phosphate, acid form (EO 4).
Servoxyl VPNZ 4/100 [788]

Nonylphenol ether phosphate, acid form (EO 5).
Servoxyl VPNZ 5/100 [788]

Nonylphenol ether phosphate, acid form (EO 10).
Servoxyl VPNZ 10/100 [788]

Nonylphenol ether phosphate, acid form (EO 15).
Servoxyl VPNZ 15/100 [788]

Nonyl phenol ethoxylate.
Alkasurf DNP-10, NP-1, NP-4, NP-5, NP-6, NP-8, NP-9, NP-10, NP-11, NP-15, NP-15 80%, NP-20, NP-20 70%, NP-30, NP-30 70%, NP-35, NP-35 70%, NP-40, NP-40 70%, NP-50, NP-50 70%, NP-100 70% [33]
Berol 02, 09, 259, 267, 292, WASC [116]
Cedepal CO-210, CO-430, CO-500, CO-530, CO-610, CO-630, CO-710, CO-730, CO-880, CO-887, CO-890, CO-977, CO-990, CO-997 [256]
Chemax NP-4, NP-6, NP-9, NP-10, NP-15, NP-30, NP-40 [170]
Elfapur N 50, N 70, N 90, N 120, N 150 [16]
Empilan NP4, NP6 [23], [26]
Empilan NP9 [26]
Empilan NP12, NP30 [23], [26]
Ethylan 20, 44, 55, 77, BCP, BV, DP, HA, HP, KEO, N30, N50, N92 [509]
Ethylan NP-1 [253], [509]
Ethylan TU [509]
Flo Mo 4N, 6N, 7N, 9N, 11N, 13N, 20N [247]
Levelan P208 [509]
Nonal 206, 208, 210 [864]
Peganol NP 1.5, NP 4, NP 5, NP 6, NP 9, NP 10, NP 15, NP 20, NP 30, NP 40, NP 50, NP 100 [129]
Polyfac NP-40 [918]
Polyfac NP-40-70 [917a], [918]
Rexol 25/1, 25/20, 25/30, 25/40, 25/50, 25/100-70%, 25/307, 25/507 [390]
Serdox NNP 1½ [788]
Solar NP [840]
Soprofor NP/Series [550]
Synperonic NP4, NP5, NP6, NP8, NP9, NP10, NP12, NP13, NP15, NP20, NP30 [443]
Syntopon A, A 100 [937a], [938]
Syntopon B [934], [937a], [938]
Syntopon B 300 [937a], [938]
Syntopon C [934], [937a], [938]
Syntopon D, E, F, G, H [937a], [938]

Nonylphenol (EO 1.5), ethoxylated.
Iconol NP-1.5 [109]
Norfox NP-1 [650]
T-Det N-1.5 [859]

Nonylphenol, ethoxylated 4-mole.
Carsonon N-4 [155]
Iconol NP-4 [109]
Norfox NP-4 [650]
Polystep F-1 [825]
Sterox ND [602a]
Tergitol NP-4 [878]

Nonylphenol (EO 5), ethoxylated.
Iconol NP-5 [109]
Sterox NE [602a]

Nonylphenol, ethoxylated 6-mole.
Carsonon N-6 [155]
Iconol NP-6 [109]
Norfox NP-6 [650]
Polystep F-2 [825]
Sterox NF [602a]
Tergitol NP-6 [878]

Nonyl phenol ethoxylate 6.5 mole.
Sterox NG [602a]

Nonylphenol (EO 7), ethoxylated.
Iconol NP-7 [109]
Tergitol NP-7 [878]

Nonyl phenol, ethoxylated 7.5 mol.
Norfox NP-7 [650]

Nonyl phenol ethoxylate (EO 8 moles).
Polystep F-3 [825]
Tergitol NP-8 [878]

Nonylphenol, ethoxylated, 8-8½ mole.
Carsonon N-8 [155]

Nonylphenol (EO 9), ethoxylated.
Iconol NP-9 [109]
Norfox NP-9 [650]
Tergitol NP-9, TP-9 [878]

Nonylphenol, ethoxylated, 9-10 mole.
Carsonon N-9 [155]

Nonyl phenol ethoxylate 9.5 mole.
Sterox NJ [602a]

Nonylphenol (EO 10), ethoxylated.
Iconol NP-10 [109]
Polystep F-4 [825]

Nonylphenol, ethoxylated, 10-11 mole.
Carsonon N-10 [155]

Nonyl phenol ethoxylate 10.5 mole.
Sterox NK [602a]
Tergitol NP-10, NPX [878]

Nonyl phenol, ethoxylated 11 mole.
Norfox NP-11 [650]
Sterox NL [602a]

Nonyl phenol ethoxylate (EO 12 moles).
Polystep F-5 [825]

Nonyl phenol ethoxylate 12.5 mole.
Sterox NM [602a]

Nonylphenol ethoxylate (EO 13 moles).
Tergitol NP-13 [878]

Nonyl phenol ethoxylate (EO 14 moles).
Polystep F-6 [825]

Nonyl phenol ethoxylate (EO 15 moles).
Polystep F-7 [825]

Nonylphenol (EO 20), ethoxylated.
Iconol NP-20 [109]
Polystep F-8 [825]
T-Det N-20 [859]

Nonylphenol, ethoxylated, 30 mole.
Carsonon N-30 [155]
Iconol NP-30, NP-30-70% [109]
Polystep F-9 [825]

Nonylphenol (EO 40), ethoxylated.
Alkasurf NP-40 [33]
Iconol NP-40, NP-40-70% [109]
Polystep F-10 [825]
Tergitol NP-40 [878]

Nonyl phenol ethoxylate (ethylene oxide, 50 moles).
Ethylan N50 [253]
Iconol NP-50 [109]

Nonylphenol (EO 70), ethoxylated.
Iconol NP-70, NP-70-70% [109]

Nonylphenol (EO 100), ethoxylated.
Iconol NP-100 [109]

Nonylphenol ethoxylate, sulfated, aluminum salt.
Carsonol ANS [529]

Nonyl phenol ethoxylate, sulfosuccinate, sodium salt.
Protowet 4196 [710]

Nonyl phenol with ethylene oxide, 4 moles.
Chemcol NPE-40 [171]
T-Det N-4 [859]

Nonyl phenol with ethylene oxide, 5.5 mol.
Lubrol N5 [478a]

Nonyl phenol with ethylene oxide, 6 moles.
Chemcol NPE-60 [171]

Nonyl phenol plus ethylene oxide, 9.5 moles.
Gradonic N-95 [365]

Nonyl phenol with ethylene oxide, 10 moles.
Chemcol NPE-100 [171]

Nonyl phenol with ethylene oxide, 13 mols.
Lubrol N13 [478a]

Nonyl phenol with ethylene oxide, 14 moles.
Chemcol NPE-140 [171]

Nonyl phenol with ethylene oxide, 20 moles.
Chemcol NPE-200 [171]

Nonyl phenol with ethylene oxide, 40 moles.
Chemcol NPE-400 [171]

Nonylphenol with ethylene oxide, 70 moles.
T-Det N-70 [859]

Nonylphenol-30 mole ethylene oxide adduct.
T-DET N-307 [859]

Nonylphenol-40 mole ethylene oxide adduct.
T-DET N-407 [859]

Nonylphenol-70 mole ethylene oxide adduct.
T-DET N-707 [859]

Nonylphenol-100 mole ethylene oxide adduct.
T-DET N-100, N-1007 [859]

Nonyl phenol ethylene oxide condensate.
Agral 90 [444]
Ethylan 44, 55, 77, BCP, BV, DP, KEO, TU [253], [509]
Hyonic PE-90, PE-100 [252]
Levelan P148, P357 [509]
Lissapol N, NX, NXP [441]
Sterox® ND Surfactant, NE Surfactant, NJ Surfactant, NK Surfactant [602]
Synperonic N, NP4, NP5, NP6, NP8, NP9, NP10, NP13, NP20, NP35, NPE1800, NX, NXP [441]
Valdet 561 [892]

Nonyl phenol ethylene oxide condensate (ethylene oxide, 20 moles).
Ethylan 20 [253], [509]

Nonyl phenol ethylene oxide condensate (ethylene oxide, 25 moles).
Ethylan HP [253], [509]

Nonyl phenol ethylene oxide condensate (ethylene oxide, 35 moles).
Ethylan HA [253], [509]

Nonylphenol fatty esters, ethoxylated.
Iconol CNP-1, CNP-10 [109]

Nonylphenol phosphate, ethoxylated.
Polyfac PN-209 [917a]

Nonylphenol polyethoxyethanol.
Triton N-40, N-42, N-60, N-150 [743]

Nonylphenol, polyethoxylated.
Carsonon N-4, N-6, N-9, N-10, N-11, N-12, N-30, N-100 [529]
Carsonon N-100, 70% [155]
Carsonon ND-317 [529]

Nonylphenol, polyethoxylated, (EO 4).
Hyonic NP-40 [252]

Nonylphenol, polyethoxylated, (EO 6).
Hyonic NP-60 [252]

Nonylphenol, polyethoxylated, (EO 9).
Hyonic NP-90 [252]

Nonylphenol, polyethoxylated, (EO 10).
Hyonic NP-100 [252]

Nonylphenol, polyethoxylated, (EO 11).
Hyonic NP-110 [252]

Nonylphenol, polyethoxylated, (EO 12).
Hyonic NP-120 [252]

Nonylphenol, polyethoxylated, (EO 40).
Hyonic NP-407 [252]

Nonylphenol, polyethoxylated, (EO 50).
Hyonic NP-500 [252]

Nonylphenol polyethylene glycol.
Tergitol NP-44 [878]

Nonyl phenol polyethylene glycol ether.
Arnox 930, 940, 950 [76]
Lamacit 877 [176]
Tergitol® NP-4, NP-7, NP-8, NP-9, NP-10, NP-40 [879]

Nonylphenol polyethylene glycol ether (EO 4 moles).
Tergitol NP-14 [878]

Nonylphenol polyethylene glycol ether (EO 7 moles).
Tergitol NP-27 [878]

Nonylphenol polyethylene glycol ether (EO 40 moles).
Tergitol NP-44 [878]

Nonyl phenol polyglycolether.
Chimipal WN [514]
Elfapur N 50, N 70, N 90, N 120, N 150 [16]
Etophen 103, 105, 106, 107, 108, 109, 110, 112, 114, 120 [948]
Lorapal HV 8, HV 9, HV 10, HV 14, HV 25 [274]
Marlophen 83, 84, 85, 86, 86S, 87, 88, 89, 810, 811, 812, 814, 820, 825, 830, 850 [178]
Rewopal HV 5, HV 8 [274], [725]
Rewopal HV 9, HV 10 [274]
Rewopal HV 14, HV 25 [274], [725]
Rewopal NOS 5 [725]
Steinapal HV 5, HV 8, HV 9, HV 10, HV 14, HV 25 [274]

Nonylphenol-1.5-polyglycolether.
Merpoxen NO 15 [285]

Nonylphenol 4 polyglycolether.
Merpoxen NO 40 [285]

Nonylphenol 6 polyglycolether.
Merpoxen NO 60 [285]

Nonylphenol 8 polyglycolether.
Merpoxen NO 80 [285]

Nonylphenol 9.5 polyglycolether.
Merpoxen NO 95 [285]

Nonylphenol 15 polyglycolether.
Merpoxen NO 150 [285]

Nonylphenol 20 polyglycolether.
Merpoxen NO 200 [285]

Nonylphenol 30 polyglycolether.
Merpoxen NO 300 [285]

Nonylphenol 60 polyglycolether.
Merpoxen NO 600 [285]

Nonylphenol polyglycol ether (EO 4 to 30 mol).
Arkopal N Brands [416]

Nonyl phenol polyglycol ether (EO 4).
Rewopal HV 4 [725a]
Serdox NNP 4 [788]

Nonylphenol polyglycol ether (EO 5).
Serdox NNP 5 [788]

Nonylphenol polyglycol ether (EO 6).
Rewopal HV 6 [725a]
Serdox NNP 6 [788]

Nonylphenol polyglycol ether (EO 7).
Serdox NNP 7 [788]

Nonylphenol polyglycol ether (EO 8.5).
Serdox NNP 8.5 [788]

Nonylphenol polyglycol ether (EO 9).
Serdox NNP 9 [788]

Nonylphenol polyglycol ether (EO 9.5 moles).
Neutronyx 600 [663]

Nonylphenol polyglycol ether (EO 10).
Serdox NNP 10 [788]

Nonylphenol polyglycol ether (EO 11.0 moles).
Neutronyx 656 [663]
Serdox NNP 11 [788]

Nonylphenol polyglycol ether (EO 12).
Serdox NNP 12 [788]

Nonylphenol polyglycol ether (EO 13).
Serdox NNP 13 [788]

Nonylphenol polyglycol ether (EO 14).
Serdox NNP 14 [788]

Nonylphenol polyglycol ether (EO 15).
Serdox NNP 15 [788]

Nonylphenol polyglycol ether (EO 20).
Serdox NNP 20 [788]
Sermul EN 20 [788]

Nonylphenol polyglycol ether (EO 25).
Serdox NNP 25 [788]

Nonylphenol polyglycol ether (EO 30).
Serdox NNP 30, NNP 30/70 [788]

Nonylphenol polyglycol ether (EO 30) *(cont'd).*
Sermul EN 30, EN 145 [788]

Nonylphenol polyglycol ether (EO 50).
Serdox NNP 50 [788]

Nonylphenol 4-polyglycolether hydrogen phosphate.
Merpiphos EP [285]

Nonylphenol 6-polyglycolether hydrogen phosphate.
Merpiphos LP [285]

Nonylphenol 9.5-polyglycolether hydrogen phosphate.
Merpiphos GP [285]

Nonyl phenol polyglycol ether phosphate.
Rewophat NP 90 [725]

Nonylphenolpolyglycolether sulfate.
Rewopol NOS 10, NOS 25 [725]

Nonyl phenol, polyoxyethylated.
Synthrapol N [438]

Nonylphenoxyacetic acid.
Corrosion Inhibitor NPA [190]

Nonyl phenoxy ethanol, phosphated.
ESI-Terge 320 [294]
ESI-Terge P-320 [892]

Nonylphenoxy polyether alcohol.
Poly-tergent B-150 [659]

Nonyl phenoxy polyethoxyethanol.
Arnox 912 [76]
Carsonon N-4, N-6, N-9, N-10, N-11, N-12, N-30, N-40, N-50, N-100 [529]
Gradonic N-95 [365]
Poly-Tergent® B-150, B-200, B-300, B-350, B-500 [659]
Surfonic N-10, N-31.5, N-40, N-60, N-95, N-100, N-102, N-120, N-150, N-200, N-300, N-400, NB-5, NB-14 [854]
Triton® N-42, N-57 [725]
Triton® N-87, N-101, N-111, N-128, N-401 [742]

Nonylphenoxypolyethoxyethanol (EO 1.5).
Triton N-17 [742]

Nonylphenoxypolyethoxyethanol (EO 4).
Triton N-42 [742]

Nonylphenoxy polyethoxy ethanol (EO 5).
Triton N-57 [742]

Nonylphenoxy polyethoxy ethanol (EO 6).
Triton N-60 [742]

Nonylphenoxypolyethoxyethanol (EO 8.5).
Triton N-87 [742]

Nonylphenoxy polyethoxy ethanol (EO 9-10).
Triton N-101 [742]

Nonylphenoxy polyethoxy ethanol (EO 11).
Triton N-111 [742]

Nonylphenoxy polyethoxy ethanol (EO 15).
Triton N-150 [742]

Nonylphenoxypolyethoxyethanol (EO 30).
Triton N-302 [742]

Nonylphenoxy polyethoxy ethanol (EO 40).
Triton N-401 [742]

Nonylphenoxypolyethoxyethanol (EO 100).
Triton N-998 [742]

Nonyl phenoxypolyethoxy ethanol-iodine complex.
Bardyne-20 [529]

Nonylphenoxy polyethoxy ethanol, phosphated.
ESI-Terge 320 [294]

Nonyl phenoxy poly (ethyleneoxy) ethanol.
Arkopal N040, N060, N080, N090, N100, N110, N130 [55]
Conco NI Series [209]
Igepal CO-210, CO-430, CO-520, CO-530, CO-610, CO-620, CO-630, CO-660, CO-710, CO-720, CO-730, CO-850, CO-880. CO-887, CO-890, CO-897, CO-970, CO-977, CO-980, CO-985, CO-987, CO-990, CO-997 [338]
Mars NI-100 [554]
Siponic NP4, NP6, NP9, NP10.5, NP15, NP40, NP75 [29]

Nonylphenoxy poly (ethyleneoxy) ethanol-iodine complex.
Biopal NR-20, VRO-20 [338]

Nonylphenoxypoly (ethyleneoxy) ethanol-iodine complex, iodophor concentrate.
Bio-Surf I-20 [120]

Nonylphenoxypoly (ethyleneoxy) ethanol, sulfated, ammonium salt.
Alipal CO-436, EP-110, EP-115, EP-120 [338]

Nonylphenoxypoly-(ethyleneoxy) ethanol, sulfated, sodium salt.
Alipal CO-433 [338]

p-Nonylphenoxy-polyglycidol.
Glycidol Surfactant 10G [659]

Nonyl phenoxy polyoxyethylene ethanol.
Solar NP [840]

Norbornene-2-methanol.
Dihydrocyclol [452]

Norea.
Herban [406]

Norepinephrine bitartrate.
Levophed Bitartrate [133]

Norethindrone acetate.
Aygestin [97]
Norlutate [674]

Norethindrone preparations.
Brevicon 21-Day Tablets, 28-Day Tablets [844]
Loestrin 21 1/20, Fe 1/20, 21 1.5/30, Fe 1.5/30 [674]
Micronor Tablets [666]
Modicon 21 Tablets, 28 Tablets [666]
Norinyl 1 + 35 Tablets 21-Day, 1 + 35 Tablets 28-Day, 1 + 50 21-Day, 1 + 50 28-Day, 1 + 80 21-Day, 1 + 80 28-Day, 2 mg. [844]
Norlestrin 21 1/50, 21 2.5/50, 28 1/50, Fe 1/50, Fe 2.5/50 [674]
Norlutin [674]
Nor-Q.D. [844]
Ortho-Novum 1/35□21, 1/35□28, 1/50□21, 1/50□28, 1/80□21, 1/80□28, 10/11□..21 Tablets, 10/11□..28 Tablets, Tablets 2 mg□21 [666]
Ovcon-35, -50 [572]

Norethynodrel.
Enovid 5 mg, 10 mg, -E 21 [782]

Norgestrel.
Lo/Ovral Tablets, -28 Tablets [942]
Ovral Tablets, -28 Tablets [942]
Ovrette Tablets [942]

Nortriptyline hydrochloride.
Aventyl HCI [523]
Pamelor [762]

Noscapine = guaifenesin; 30 mg:200 mg.
Actol Expectorant® Tablets [114]

Novobiocin sodium.
Biodry® [884]
Drygard® [884]

NTA.
SEQ NT-15 [903]

NTA, trisodium salt.
Cheelox® NTA-14, NTA-Na_3 [338]

Nylidrin.
Arlidin Tablets [891]

Nylidrin hydrochloride.
Nylidrin Tablets [344]
Nylidrin HCI Tablets [240]
Nylidrin HCI Tablets [771]

Nylon.
Milvex 1000 [400]
Standamid Resin BC-1283 [400]
Versamid 930 [400]

Nylon 6.
Capran Films [36]
Ertalon 6SA, 6SAU [180]
Firestone Nylon 200-001, 210-001L, 213-001, 228-001 [321]
Fosta Nylon 438, 471, 523, 589 [56]

Nylon 6, heat-stabilized.
Fosta Nylon 870, 1047 [56]

Nylon 6, nucleated.
Fosta Nylon 446 [56]

Nylon 6, plasticized.
Fosta Nylon 567 [56]

Nylon 6, plasticized, heat stabilized.
Fosta Nylon 1379, 1525 [56]

Nylon 6/6.
Cerex® Spunbonded Nylon Fabrics [602]

Nylon 6/10.
Migralube Q-1000 [526]

Nylon 11.
Ertalon 11SA [180]

Nylon 66.
Ertalon 66SA [180]

Nylon-phenolic laminate.
Insurok T-819 [812]

Nystatin.
Korostatin Vaginal Tablets [945]
Mycolog Cream and Ointment [817]
Mycostatin Cream & Ointment, Oral Suspension, Oral Tablets, Topical Powder, Vaginal Tablets [817]
Mytrex Cream & Ointment [767]
Nilstat Oral Suspension, Oral Tablets, Topical Cream & Ointment, Vaginal Tablets [516]
Nystaform Ointment [585]
Nystatin Cream, Oral & Vaginal Tablets [769]
Nyst-olone Cream & Ointment [769]
O-V Statin [817]

Octabromodiphenyl oxide.
Great Lakes DE-79 [369]
Saytex 111 [302]

(E,Z) & (Z,Z)-3,13-Octadecadien-1-ol acetate.
No Mate Boren Gard [21]

Octadecanamide.
Armid 18 [79]

n-Octadecanol.
Cyclogol Stearyl Alcohol [234]

(2,E,E,) 9, 11, 13-Octadecatrienoic acid, methyl α-eleostearate, methyl ester.
Bollex [122]

n-Octadecylamine.
Amine 18D [487]
Lilamin 18D [487]

Octadecyl amine acetate.
Kemamine A 990 [429a]
Nissan Cation SA [636]

Octadecylamine acetate, distilled.
Armac 18D [79]

Octadecylamine + EO 2 moles.
Teric 18M2 [439]

Octadecylamine + EO 5 moles.
Teric 18M5 [439]

Octadecylamine + EO 10 moles.
Teric 18M10 [439]

Octadecylamine + EO 20 moles.
Teric 18M20 [439]

Octadecylamine + EO 30 moles.
Teric 18M30 [439]

Octadecylamine primary.
Armeen 18, 18D [79]
Kemamine P-990 [429a]
Nissan Amine AB [636]

Octadecylamine primary, distilled.
Kemamine P-990 D [429a]

Octadecyl B (3,5-t-butyl 4-hydroxy-phenyl) propionate.
Irganox/076 [190]

Octadecyl 3,5-di-t-butyl-4-hydroxyhydrocinnamate.
Irganox 1076 [190]

Octadecyldiethanol methyl ammonium chloride.
M-Quat 32 [564]

Octadecyldihydroxyethyl methyl ammonium chloride.
Tomah Q-18-2 [866]

Octadecyldimethylamine.
Adma™ 8 [302]
Armeen DM18D [16]
Nissan Tertiary Amine AB [636]

Octadecyldimethylamine oxide.
Aromox DM18D-W [16]

Octadecyldimethylbenzylammonium chloride.
Arquad DM18B-90 [16]
Nissan Cation S_2-100 [636]

Octadecyl meth-acrylate.
SR-324 [766]

Octadecyl poly (15) oxyethylene methyl ammonium chloride.
Tomah Q-18-15 [866]

N-Octadecyl sulfo-succinamate, sodium salt.
Astromid 18, 18 LV [28]

Octadecyltrimethyl ammonium chloride.
Nissan Cation AB [636]

2-Octadodecanol.
Eutanol G [399], [400]

Octadodecanol stearate.
Standamul 7063 [399], [400]

Octaethylene glycol dodecyl ether.
Nikkol BL-8SY [631]

Octane 99+%.
Ultralog 68-4660-00 [172]

Octasodium diethylene triamine penta (methylene phosphonate).
Dequest 2066 [602]

Octocrylene.
Uvinul N-539 [109]

Octomethylcyclotetrasiloxane.
Abil K 4 [856]

Octoxyglyceryl behenate.
Mexanyl GQ [185]

Octoxynol-1.
Alkasurf OP-1 [33]
Triton X-15 [742]

Octoxynol-3.
Igepal CA-420 [338]
Triton X-35 [742]

Octoxynol-5.
Alkasurf OP-5 [33]
Igepal CA-520 [338]
Triton X-45 [742]

Octoxynol-7.
Igepal CA-620 [338]

Octoxynol-9.
Conco NIX-100 [209]
Hyonic PE-250 [252]
Igepal CA-630 [338]
Neutronyx 605 [663]
Triton X-100 [742]

Octoxynol-10.
Alkasurf OP-10 [33]

Octoxynol-13.
Igepal CA-720, CA-730 [338]
Triton X-102 [742]

Octoxynol-40.
Alkasurf OP-40 [33]
T-DET O-407 [859]

Octrizole.
Spectra-Sorb UV-5411 [50]

n-Octyl alcohol.
Kalcohl 08H [478]

Octyl alcohol with EO 4 mol.
Dehydol O4 [401]

n-Octylamine.
Amine 8D [487]
Lilamin 8D [487]

Ocytlamine primary.
Armeen 8D [79]

Octyl chloride.
Barchlor 8S [529]

n-Octyl, n-decyl adipate.
Monoplex NODA [379]
PlastHall NODA [379]
Staflex NODA [722]

Octyl decyl dimethyl ammonium chloride.
Bardac 20 W (Water Solvent System), 2050, 2080 [529]

Octyl-decyl methacrylate.
Empicryl OM [23], [25]

N-Octyl-n-decyl phthalate.
Staflex NODP, ODP, NONDTM [722]

Octyldimethylamine.
Adma™ C8 [302]

Octyl dimethyl PABA.
Escalol 507 [895]

Octyl diphenyl phosphate.
Disflamoll DPO [597]
Santicizer 141 [602]

Octyldodecanol.
Cyclal G20 [939]
Eutanol G [399], [400], [401]
Michel XO-143 [581]
Primarol 1208 [400]
Standamul G [399], [400]

Octyldodeceth-20.
Seodol E-2020 [628]

Octyldodeceth-25.
Seodol E-2025 [628]

N-Octyldodecyl dimethyl ammonium chloride.
BTC 812 [663]

Octyldodecyl myristate.
MOD [798]

Octyldodecyl neodecanoate.
Nikkol Neodecanoate-20 [631]

Octyl dodecyl stearate.
Standamul 7063 [400]

Octyldodecyl stearoyl stearate.
Ceraphyl 847 [895]

Octyl epoxy stearate.
Drapex 3.2 [74]

Octyl epoxy tallate.
Drapex 4.4 [74]
Flexol Plasticizer EP-8 [878]

Octyl fatty acid ester.
Liponol O [178]

Octyl-hydroxystearate.
Crodamol OHS [223]
Naturechem OHS [158]
Wickenol 171 [926]

Octyl isononanoate.
Emerest 2300 [291]
Isolanoate [507]
Kessco Octyl Isononanoate [79]

2-n-Octyl-4-isothiazolin-3-one.
Kathon® 4200, LM, LP [742]
Micro-Cheek 11 [315]

Octyl methoxycinnamate.
Neo Heliopan AV [378]
Parsol MCX [353]

Octyl palmitate.
Cegesoft C24 [375]
Ceraphyl 368 [895]
Schercemol OP [771]
Wickenol 155 [926]

Octyl pelargonate.
Emerest 2307 [291]

Octyl phenol + EO 5 moles.
Teric X5 [439]

Octyl phenol + EO 7.5 moles.
Teric X7 [439]

Octyl phenol + EO 8.5 moles.
Teric X8 [439]

Octyl phenol + EO 10 moles.
Teric X10 [439]

Octyl phenol + EO 11 moles.
Teric X11 [439]

Octyl phenol + EO 13 moles.
Teric X13 [439]

Octyl phenol + EO 16 moles.
Teric X16 [439]

Octyl phenol + EO 40 moles.
Teric X40 [439]

Octyl phenol ethoxylate.
Alkasurf OP-1, OP-5, OP-8, OP-10, OP-12, OP-16, OP-30, OP-40, OP-70, OP-70-50% [33]
Chemax OP-5, OP-9, OP-40 [170]
Nonal 310 [864]
Rexol 45/1, 45/3, 45/5, 45/7, 45/10, 45/12, 45/16, 45/307, 45/407 [390]
Soprofor BC/Series [550]
Synperonic OP10, OP11 [443]
Syntopon 8 A, 8 B [937a], [938]
Syntopon 8 C [934], [937a], [938]
Syntopon 8 D [937a], [938]
Triton X-45, X-100, X-102, X-114 [743]

Octyl phenol ethoxylate (EO 3-4 moles).
T-Det O-4 [859]

Octyl phenol ethoxylated, (EO 5).
Iconol OP-5 [109]

Octyl phenol ethoxylate (EO 6-7 moles).
T-Det O-6 [859]

Octyl phenol, ethoxylated, (EO 7).
Iconol OP-7 [109]

Octyl phenol ethoxylate (EO 7-8 moles).
T-Det O-8 [859]

Octyl phenol ethoxylate (EO 9 moles).
T-Det O-9 [859]

Octylphenol, ethoxylated, (EO 10).
Iconol OP-10 [109]

Octylphenol, ethoxylated, (EO 40).
Iconol OP-40, OP-40-70% [109]
T-Det O-40 [859]

Octylphenol ethylene oxide condensate.
Synperonic OP10, OP11, OP16, OP40 [441]

Octylphenol with ethylene oxide, 40 moles.
T-Det O-407 [859]

Octylphenol polyethoxyethanol.
Triton X-165 [743]

Octylphenol polyethoxylated, (EO 4).
Hyonic OP-40 [252]

Octylphenol polyethoxylated, (EO 7).
Hyonic OP-70 [252]

Octylphenol polyethoxylated, (EO 9).
Hyonic OP-100 [252]

Octylphenol polyethoxylated, (EO 40).
Hyonic OP-407 [252]

Octylphenol polyethoxylated, (EO 70).
Hyonic OP-705 [252]

Octylphenol polyglycol ether (EO 9).
Serdox NOP 9 [788]

Octylphenol polyglycol ether (EO 30).
Serdox NOP 30/70 [788]

Octylphenol polyglycol ether (EO 5) EO 40.
Serdox NOP 40/70 [788]

Octylphenoxy polyethoxy ethanol.
Triton® AG-98, X-165, X-305, X-405, X-705 [725]

Octylphenoxy polyethoxy ethanol (EO 1).
Triton X-15 [742]

Octylphenoxy polyethoxy ethanol (EO 3).
Triton X-35 [742]

Octylphenoxy polyethoxy ethanol, ethylene oxide 5 moles.
Alkasurf OP-5 [33]
Triton X-45 [742]

Octylphenoxy polyethoxy ethanol, ethylene oxide 7-8 moles.
Alkasurf OP-8 [33]
Triton X-114 [742]

Octylphenoxy polyethoxy ethanol (EO 9-10).
Triton X-100, X-120 [742]

Octylphenoxy polyethoxy ethanol, ethylene oxide 10 moles.
Alkasurf OP-10 [33]

Octylphenoxy polyethoxy ethanol, ethylene oxide 12-13 moles.
Alkasurf OP-12 [33]
Triton X-102 [742]

Octyl phenoxy polyethoxy ethanol, ethylene oxide 30 moles.
Alkasurf OP-30 70% [33]

Octylphenoxy polyethoxy ethanol, ethylene oxide 40 moles.
Alkasurf OP-40 [33]

Octylphenoxypoly(ethyleneoxy)ethanol.
Igepal CA-210, CA-420, CA-520, CA-620, CA-630, CA-720, CA-880, CA-887, CA-890, CA-897, CA-950 [338]
Siponic OP 1.5, OP7, OP10, OP30, OP40 [29]

Octyl salicylate.
Sunarome WMO [312]

Octyl stearate.
Cetiol 868 [400]
Radia 7131 [658]
Wickenol 156 [926]

Octyl tallate.
PlastHall R-9 [379]

Octyl tallate, epoxidized.
Epoxol 5-2E [840]

Oleamide.
Armid O [79]
Armoslip CPM [16], [653]

Oleamide *(cont'd)*.
Crodamide O, OR [223]
Kemamide O, U [429]
Petrac Slip-Eze [688]

Oleamide DEA.
Active #18 [123]
Airosol O [190]
Aminol OF, OL, OT [320]
Calamide O [697]
Carsamide O [529]
Chimipal OLD [514]
Clindrol 100-O, 200-O [197]
Comperlan OD [400]
Condensate OFO [174]
Cyclomide DO 280, DO 280/S [234]
Cyclomide OD [939]
Hetamide DO [407]
Jordamide 201 [474]
Mackamide O [567]
Mazamide O-20 [564]
Ninol 201 [825]
Product LT 10-8-1, WRS 1-66 [197]
Protamide OFO, T [713]
Rewo-Amid DO 280 [725]
Richamide 5085 [732]
Sandrol 200-O [197]
Schercomid ODA, SO-A [771]
Unamide O [529]
Witcamide 511C [935]

Oleamide MEA.
Aminol OM [320]
Cyclomide OM [939]
Hetamide MO [407]
Rewo-Amid O 280 [725]

Oleamide MIPA.
Cyclomide OP [939]
Mackamide OP [567]
Rewo-Amid IPE 280 [725]
Schercomid OMI [771]
Witcamide 61 [934]

Oleamido betaine.
Mackam HV [567]

Oleamidopropyl betaine.
Cycloteric BET O-30 [234]
Incronam OP-30 [223b]

Oleamidopropyl dimethylamine.
Incromine OPB [223b]
Lexamine O13 [453]
Mazeen OA [564]
Schercodine O [771]

Oleamidopropyl dimethylamine glycolate.
Naetex 118, 135 [507]

Oleamidopropyl dimethylamine hydrolyzed animal protein.
Lexein CP 125 [453]

Oleamidopropyl dimethylamine lactate.
Incromate ODL [223b]

Oleamidopropyl dimethylamine oxide.
Incromine Oxide O [223b]

Oleamidopropyl ethyldimonium ethosulfate.
Foamquat ODES [41]

Oleamidopropyl hydroxysultaine.
Lonzaine OS [529]

Oleamine.
Armeen O, OD [79]
Kemamine P-989 [429a]

Oleamine oxide.
Conco XAO [209]
Standamox 01 [399], [400]

α-Olefin oligomer.
Nikkol Synceiane 30 [631]

α-Olefin, SO_3 sulfonated, sodium salt.
Conco AOS-40 [209]

α-Olefin sulfonate.
Allfoam [179]
Bio Terge [825]
Lipolan 327 F, 440, 1400, AO, AOL, G [524]
Soft Detergent 95 [524]
Surco AOS [663]
Witconate AOS [935], [937a], [938]

α-Olefin sulfonate, sodium salt.
Hostapur OS Brands [416]

α-Olefin sulfonate, sodium salt, C_{14}-C_{16}.
DI-AOS-46-40 [248]

Oleic acid.
Acme 105, 110, 120 [6]
Century CD Fatty Acid [876]
Emersol 210 [288], [289]
Emersol 213, 221 [288]
Emersol 233 [289]
Emersol 233LL [288]
Emersol 6321 [288]
Groco 2, 4, 5L, 6 [374]
Hy-Phi 1055, 1088, 2066, 2088, 2102 [241]
Industrene 105 [429]
Industrene 205 [429a]
Industrene 206 FG [429]
Neo-Fat 90-04, 92-04 [79]
Nilox OE [722]
WGS-Oleic 100 [916]
Wochem #320 [941]

Oleic acid amide ethoxylate.
Lutensol FSA 10 [106]

Oleic acid amide polyglycolether.
Lorapal O 8 [274]
Rewopal O 8 [274]

Oleic acid amide polyglycolether *(cont'd).*
Steinapal O 8 [274]

Oleic acid chloride.
S-Oil [285]

Oleic acid condensate, sodium salt.
Merpinol S [285]

Oleic acid decylester.
Cetiol V [401]

Oleic acid dibutylamide.
Rewocor RA 280 [725a]

Oleic acid diethanolamide.
Aminol OF [320]
Clindrol 200-O [197]
Comperlan OD [401]
Lauridit OD [16], [17]
Loramine DO 280/SE [274]
Loropan OD [858]
Marlamid D 1885 [178]
Merpinamid OD [285]
Rewomid DO 280/SE [274]
Schercomid SO-A [771]
Steinamid DO 280/SE [274]
Varamide A-7 [796]

Oleic acid + EO 6 moles.
Teric OF6 [439]

Oleic acid + EO 8 moles.
Teric OF8 [439]

Oleic acid, ethoxylated.
Alkasurf O-2, O-5, O-7, O-9, O-14 [33]

Oleic acid, ethoxylated-t amine.
Ethomeen O/12, O/15, O/25 [79]

Oleic acid, ethylene oxide condensate.
Ethofat O/20 [79]
Lubrol 90 [442]

Oleic acid imidazoline.
Nopcogen 22-O [252]

Oleic acid isopropanolamide sulfosuccinate.
Sole-Terge 8 [415]

Oleic acid oleyl ester.
Cetiol [401]

Oleic acid, oxidized, ammonium salt.
Avirol® AOO-1080 [399], [400]

Oleic acid oxyethylate.
Emulan A [106]

Oleic acid polydiethanolamide.
Loramine DC 280 [274]
Rewomid DO 280 [274]
Serdolamide POF 61, POF 61C [788]
Steinamid DO 280 [274]

Oleic acid, polyethylene glycol (400) ester.
Nonisol 210 [190]

Oleic acid polyglycerol ester.
Hostacerin DGO [55]

Oleic acid polyglycol ester.
Rewopal EO 70 [725a]

Oleic acid 6 polyglycol ester.
Merpoxen OFS 60 [285]

Oleic acid 8 polyglycol ester.
Merpoxen OFS 80 [285]

Oleic acid, polyoxyethylated.
Emulphor VN-430 [338]

Oleic acid, primary amine.
Radiasurf 6172 [658]

Oleic acid, sodium sulfosuccinate, amido-ester.
Schercopol OMIS-Na [771]

Oleic acid, substituted imidazoline.
Mazoline OA [564]
Miramine® OC [592]
Monazoline O [599]
Schercozoline O [771]

Oleic acid, sulfated.
Actrasol ISRD, SR75, SR-606, SRK 75, SRS-85 [811]

Oleic acid, sulfonated.
Sulfonic 800 [194], [852]

Oleic acid, sulfonated sodium salt.
Sul-fon-ate OA5 [194]
Sul-fon-ate OA5-R [852]

Oleic acid, sulfonated, sodium salt, amyl ester.
Sul-fon-ate OE 500 [194]

Oleic acid, white.
Industrene 205 [429]

Oleic alkanolamide.
Mazamide O-20, T-20 [564]

Oleic diethanolamide.
Alkamide SDO [33]
Calamide O [697]
Crillon ODE [223]
Cyclomide DO280, DO280/S [234]
Emid 6545 [289]
Felsamid-OPD [823]
Hartamide 9137 [390]
Incromide OPD [223b]
Rewomid DO 280/SE [725]
Serdolamide POF 61 [788]
Varamide A-7 [796]

Oleic ester, sulfated.
Densol P-82 [365]

Oleic fatty acid.
Industrene 105 [430]

Oleic fatty acid diethanolamide.
Merpinamid OD [285]

Oleic fatty acid ethoxylate.
Alkasurf O-2, O-5, O-9, O-14 [33]
Chemax E200MO, E400MO, E600MO, E1000MO [170]

Oleic fatty acid poly-diethanolamide.
Ninol AC-201 [762], [825a]

Oleic fatty acid, white.
Industrene 205 [430]

Oleic hydroxyethyl imidazoline.
Alkazine O [33], [34]
Varine O [796]

Oleic imidazoline.
Crodazoline O [223]
Schercozoline O [771]

Oleic imidazoline quaternary.
Alkaquat O [33]

Oleic imidazolinium.
Finazoline OAQ [320]

Oleic isopropanolamide.
Felsamid-OI [823]
Witcamide 61 [935], [937a], [938]

Oleic (kernel) glycerides, ethoxylated.
Labrafil M 1944 CS [339]

Oleic monoethanolamide.
Felsamid-OM [823]
Incromide OM, OPM [223b]

Oleic monoisopropanolamide.
Loramine IPE 280 [274]
Rewomid IPE 280 [274]
Steinamid IPE 280 [274]

Oleic (olive) glycerides, ethoxylated.
Labrafil M 1969 CS, M 1980 CS [339]

Oleic polyamine condensate.
Nopcogen 16-O [253]

Oleic sulfosuccinate.
Emcol® 4161 L [935]
Mackanate OD, OM [567]

Oleic super amide.
Calamide O [697]

Oleic (tri-oleine) glycerides, ethoxylated.
Labrafil M 2735 CS [339]

Oleoamphocarboxypropionate.
Miranol O2M-SF [592]

Oleoamphoglycinate.
Miranol OM [592]
Rewoteric AM-20 [725]

Oleoamphopropionate.
Miranol OM-SF Conc. [592]
Schercoteric OS-SF [771]

Oleoamphopropylsulfonate.
Miranol OS [592]
Sandopan TFL Conc. [762]

Oleocetylsulfate.
Montopol CST, OC Paste [711]

Oleo-fatty imidazoline.
Textamine O-5 [399], [400]

Oleolinoleic (corn) glycerides, ethoxylated.
Labrafil M 2125 CS [339]

Oleostearine.
Acme TOS [6]

Oleoyl diethanolamide.
Nitrene NO [399], [400]

Oleoyl methyl tauride, sodium salt.
Hostapon T Powder h.c. [55]

Oleoyl sarcoside.
Arkomon SO [55]

Oleoyl sarcosine.
Hamposyl O [363], [364b]
Maprosyl O [663]
Nikkol Sarcosinate OH [631]
Sarkosyl O [190]

Oleth-2.
Alkasurf OA-2 [33]
Ameroxol OE-2 [44]
Brij 92, 93 [438]
Hodag Nonionic 1802 F [415]
Lanycol-92 [507]
Lipocol O-2 [525]
Macol OA-2 [564]
Procol OA-2 [713]
Simulsol 92 [785]
Siponic Y-050 [29]

Oleth-3.
Ethoxol 3 [507]
Hetoxol OL-3 [407]
Hostacerin O-3 [55]
Lipocol O-3 [525]
Procol OA-3 [713]
Volpo 3 [223]

Oleth-3 phosphate.
Briphos 03D [548]
Crodafos N.3 Acid [223]
Hodag PE-1803 [415]

Oleth-4.
Hetoxol OL-4 [407]
Macol OA-4 [564]
Procol OA-4 [713]

Oleth-4 phosphate.
Protaphos 400-A [713]

Oleth-5.
Eumulgin 05 [399], [400]
Hetoxol OL-5 [407]
Hostacerin O-5 [55]
Lipocol O-5 [525]
Procol OA-5 [713]
Standamul 05 [399], [400]

Oleth-5 *(cont'd).*
Volpo 5 [223]

Oleth-6.
Empilan KL6 [548]
Emulgator GO 9 [375]

Oleth-7.
Nikkol BO-7 [631]

Oleth-8.
Emalex 508 [628]

Oleth-9.
Emulgator GO 12 [375]

Oleth-10.
Alkasurf OA-10 [33]
Ameroxol OE-10 [44]
Brij 96, 97 [438]
Empilan KL10 [548]
Ethoxol 10 [507]
Eumulgin 010 [399], [400]
Lipocol O-10 [525]
Macol OA-10 [564]
Nikkol BO-10TX [631]
Procol OA-10 [713]
Simulsol 96 [785]
Standamul 010 [399], [400]
Volpo 10 [223]

Oleth-10 phosphate.
Crodatos N.10 Acid [223]
Hodag PE-1810 [415]

Oleth-12.
Ethoxol 12 [507]

Oleth-15.
Emalex 515, 515P [628]
Procol OA-15 [713]

Oleth-20.
Ameroxol OE-20 [44]
Brij 98, 99 [438]
Empilan KL20 [548]
Emulphor ON-870 [338]
Ethoxol 20 [507]
Hostacerin O-20 [55]
Lanycol-98 [507]
Lipal 20OA [715]
Lipocol O-20 [525]
Macol OA-20 [564]
Nikkol BO-20TX [631]
Procol OA-20 [713]
Simulsol 98 [785]
Standamul 020 [399], [400]
Volpo 20 [223]

Oleth-20 phosphate.
Hodag PE-1820 [415]

Oleth-23.
Emalex 523 [628]
Hetoxol OL-23 [407]
Procol OA-23 [713]

Oleth-25.
Lipocol O-25 [525]
Siponic Y-500, Y-500-70 [29]

Oleth-44.
Ethoxol 44 [507]

Oleyl alcohol.
Adol 34, 80, 85, 90, 320, 330, 340 [796]
Cachalot O-3, O-8, O-15 [581]
Croacol O, A.10 [223]
Fancol OA, OA 50, OA 70, OA 90, OA 95 [308]
HD Eutanol, Oleyl Alcohol CG, Oleyl Alcohol 70/75, Oleyl Alcohol 80/85, Oleyl Alcohol 90/95 [400]
Lancol [507]
Lipocol O-80, O-90 [525]
Novol [223]

Oleyl alcohol ethoxylate.
Alkasurf OA-2, OA-10, OA-20 [33]
Chemal OA-5, OA-9, OA-20, OA-20/70 [170]
Ethosperse OA-2 [358]
Ethoxol (3-10-20-44 moles) [507]
Genapol O-200 [55]
Trycol OAL-23 [290]

Oleyl alcohol, ethoxylated (2 mole).
Ameroxol® OE-2 [44], [45]

Oleyl alcohol, ethoxylated (10 mole).
Ameroxol® OE-10 [44], [45]

Oleyl alcohol, ethoxylated (20 mole).
Ameroxol® OE-20 [44], [45]
Industrol LG-100, OAL-20 [109]
Siponic Y501 [29]

Oleyl alcohol ethoxylate (EO 25 moles).
Siponic Y500-70 [29]

Oleyl alcohol, polyethoxylated (2).
Simulsol 92 [711]

Oleyl alcohol, polyethoxylated (10).
Simulsol 96 [711]

Oleyl alcohol, polyethoxylated (20).
Simulsol 98 [711]

Oleyl alcohol polyglycol ether.
Genapol O-050, O-080, O-100, O-120 [55]

Oleyl alcohol, polyoxyethylated.
Emulphor ON-877 [338]

Oleyl alcohol, polyoxyethylated (20).
Emulphor ON-870 [338]

Oleyl alcohol polyoxyethylene ether; EO, 10 moles.
Ameroxol OE-10 [44]

Oleyl alcohol polyoxyethylene ether; EO, 20 moles.
Ameroxol OE-20 [44]

Oleyl alcohol, sulfonated.
Hipochem LCA [410]

Oleyl alcohol, super-refined.
Lancol [507]

Oleyl alkanolamide.
Alkamide SDO [33]

Oleyl amide (EO 5), ethoxylated.
Icomid O-5 [109]

Oleyl amido betaine.
Schercotaine OAB [771]

N(Oleyl amido ethyl) N(ethanol amine).
Servamine KEO 260 [788]

Oleyl amido ethyl oleyl imidazoline.
Servamine KOO 330 B [788]

Oleyl amidopropyl betaine.
Lexaine O [453]

Oleylamine.
Amine OLD [487]
Armeen OD [16]
Crodamine 1.OD [223]
Lilamin OLD [487]
Nissan Amine OB [636]
Noram O [162]
Radiamine 6172 [658]

Oleylamine acetate.
Armac OD [16], [79]
Noramac O [162]

Oleyl amine, distilled.
Radiamine 6173 [658]

Oleyl amine, ethoxylated.
Chemeen O-30, O-30/80 [170]
Lutensol FA 12 [106]

Oleyl amine ethoxylate (EO 30).
Icomeen O-30, O-30-80% [109]

Oleylamine oxethylate.
Genamine O-020, O-080, O-100, O-150, O-250 [55]

Oleyl amine, polyoxyethylated (30).
Katapol OA-860 [338]

Oleyl amine, primary.
Armeen O [79]
Kemamine P-989 [429], [429a]
Nissan Amine OB [636]

Oleyl amine primary, distilled.
Kemamine P-989 D [429a], [430]

Oleyl arachidate.
Carson Wax OA [529]

Oleyl betaine.
Mackam OB [567]
Mirataine ODMB-35 [592]
Standapol OLB-30, OLB-50 [400]
Velvetex OLB-30 [400]
Velvetex OLB-50 [399], [400]

Oleyl-cetyl alcohol, sulfated, sodium salt.
Elfan 680 [16], [764]

Oleylcetyl polyglycol ether.
Merpemul 3034, 3050, 3200 [285]

Oleylcetyl sulfate.
Cortapol OC [711]

Oleyl diamine.
Dicrodamine 1.O [223]
Lilamin 572 [522]
Radiamine 6572 [658]

N-Oleyl-1,3-diamino-propane.
Duomeen O [79]
Duomeen OX [16]

Oleyl dimethyl t-amine (distilled).
Lilamin 372 D [522]

Oleyl dimethylamine oxide.
Conco XA-O [209]
Incromine Oxide OD-50 [223b]

Oleyl dimethyl benzyl ammonium chloride.
Ammonyx KP [663]
Empigen BCJ-50 [24]
Noramium O 85 [162]

Oleyl (distilled) ether, phosphated (EO 3 moles).
Crodafos N3 Acid [223]

Oleyl (distilled) ether, phosphated (EO 5 moles).
Crodafos N5 Acid [223]

Oleyl (distilled) ether, phosphated (EO 10 moles).
Crodafos N10 Acid [223]

Oleyl ether, phosphated.
Crodafos N-3 Acid, N-3 Neutral, N-10 Acid, N-10 Neutral [223]

Oleyl ether, phosphated (EO 2 moles).
Crodafos 02 Acid [223]

Oleyl ether, phosphated (EO 5 moles).
Crodafos 05 Acid [223]

Oleyl ether, phosphated (EO 10 moles).
Crodafos 010 Acid [223]

Oleyl ethoxylate.
Merpemul 3020 [285]

Oleyl hydroxyethyl imidazoline.
Unamine O [529]

Oleyl imidazoline.
Amine O [190]
Miramine OC [592]
Schercozoline O [771]

Oleyl imidazoline *(cont'd)*.
Textamine O-I [399], [400]

Oleyl imidazoline methyl sulfate.
Adogen 473 [796]

Oleyl imidazoline, quaternized.
Quaternary O [190]

N-Oleyl imidazolinium hydrochloride.
Norfox IM-38 [650]

Oleyl methyl tauride, sodium salt.
Hostapon T Paste 33 [55]

Oleyl nitrile.
Arneel OD [79]
Plasticizer OLN [392]

Oleyl oleate.
Cetiol [400]
Schercemol OLO [771]
Wickenol 143 [926]

Oleyl palmitamide.
Kemamide P-181 [429]

Oleyl phosphate, acid form.
Servoxyl VPFZ 100 [788]

Oleyl phosphate esters, ethoxylated.
Hodag PE-1803, PE-1810, PE-1820 [415]

Oleyl polyether, phosphated.
Crodofos N Series [223]

Oleyl polyglycol ether.
Emulsogen MS-12 [55]

Oleyl polyglycol ether (EO 2).
Serdox NOL 2 [788]

Oleyl polyglycol ether (EO 8).
Serdox NOL 8 [788]

Oleyl polyglycol ester (EO 10).
Serdox NOG 440 [788]

Oleyl polyglycol ether (EO 15).
Serdox NOL 15 [788]

N-Oleyl-1,3-propanediamine.
Duomeen O [79]

N-Oleyl propylene diamine.
Dinoram O [162]
Kemamine D-989 [430]

N-Oleyl propylene diamine acetate.
Dinoramac O [162]

N-Oleyl propylene diamine dioleate.
Inipol 002 [162]

Oleyl sarcosine.
Sarkosyl O [190]

Oleyl stearate.
Carsowax OS [529]

Oleylsulfo succinamate sodium.
Cosmopon BN [514]

Oleyltrimethyl ammonium chloride.
Noramium MO 50 [162]

Oleyltrimethyl ammonium chloride/dicocodimethyl ammonium chloride, 1:1.
Arquad S-2C-50 [16]

Organophosphonic acid.
Fostex U [400]

Organopolysiloxane.
Dow Corning FF-412, FF-414 [262]
Silicone Rubber Compound SWS-7801U, SWS-7802 [841]

Organosiloxane.
Dow Corning FF-400 [262]

Organotin carboxylate.
Fomrez UL-2 [935]

Organotin catalyst.
Fomrez UL-6, UL-8, UL-22, UL-28, UL-29, UL-32 [935]

Organotin comp.
Advastab LS-202 [192]

Organotin, liquid non-sulfur.
Mark 1044 [74]

Orphenadrine citrate.
Myotrol [518]
Norflex Tablets [510], [734]
Norgesic & Norgesic Forte [734]
X-Otag S.R. Tablets [723]

Orphenadrine hydrochloride.
Disipal Tablets [734]

Ouricury wax.
Ouricury Wax-STRALPITZ [830]

Oxabicyclo (2,2,1) heptane 2,3-dicarboxylic acid.
Accelerate [682]
Aquathol [682]

Oxacillin.
Oxacillin Capsules, Solution [119]

Oxacillin sodium.
Bactocill® Capsules: 250 mg., 500 mg.; for Injection—Buffered: 1 Gm., 2 Gm., 4 Gm., 10 Gm., 500 mg. [114]
Oxacillin Sodium Capsules & Powder [769]
Prostaphlin Capsules, Oral Solution, for Injection [134]

Oxalyl bis(benzylidenehydrazide).
Eastman Inhibitor OABH [278]

2,2′-Oxamido bis-[ethyl 3-(3,5-di-t-butyl-4-hydroxyphenyl)propionate].
Naugard XL-1 [881]

Oxandrolone.
Anavar [782]

Oxazepam.
Serax Capsules, Tablets [942]

8- α-12-Oxido-13,14,15,16 tetra-norlabdane.
Ambroxan [399], [400]

Oxime, powdered.
Troykyd Anti-Skin Odorless Powder [869]

Oxoalcohol polyglycol ethers, C_{13}.
Marlipal 013/30, 013/50, 013/60, 013/80, 013/100, 013/120, 013/170 [178]

Oxoalcohol polyglycol ethers, C_{13}/C_{14}.
Marlipal 34/30, 34/50, 34/60, 34/70, 34/100, 34/110, 34/120, 34/140, 34/200, 34/300 [178]

Oxoalcohol polyglycol ether, sulfated, sodium salt.
Lutensit AS 3330 [106]

Oxtriphyline.
Brondecon [674]
Choledyl, Choledyl Pediatric Syrup, Choledyl SA Tablets [674]

Oxybenzone.
PreSun 8 Lotion, Creamy & Gel, 15 Creamy Sunscreen, 15 Sunscreen Lotion [920]
Solbar Plus 15 Sun Protectant Cream, Sun Protective Cream [686]

p,p-Oxybis-(benzenesulfonyll-hydrazide).
Celogen OT [881]

Oxybutynin chloride.
Ditropan Syrup, Tablets [551]

Oxycodone hydrochloride.
Oxycodone Hydrochloride & Acetaminophen Tablets [751]
Oxycodone Hydrochloride Oral Solution & Tablets [751]
Oxycodone Hydrochloride, Oxycodone Terephthalate & Aspirin Tablets (Half & Full Strengths) [751]
Percocet [269]
Percodan & Percodan-Demi Tablets [269]
SK-Oxycodone with Acetaminophen Tablets, with Aspirin Tablets [805]
Tylox Capsules [570]

Oxycodone Terephthalate.
Oxycodone Hydrochloride, Oxycodone Terephthalate & Aspirin Tablets (Half & Full Strengths) [751]
SK-Oxycodone with Aspirin Tablets [805]

N-Oxydiethylene-2-benzothiazole sulfenamide.
Akrochem OBTS [15]
Amax [362], [894]
OBTS [15]
Royal OBTS [752]

Oxy-ethyl-alkl ammonium phosphate.
Dehyquart SP [401]

Oxyethylene dodecylamine.
Nissan Nymeen L-201 [636]

Oxyfatty acid monoglyceride.
Softigen 701 [275]

Oxymetazoline hydrochloride.
Dristan Long Lasting Nasal Spray, Regular & Menthol [922]
Oxymeta-12 Nasal Spray [769]

Oxymetholone.
Anadrol-50 [844]

Oxymolybdenum organophosphorodithioate, sulfurized.
Molyvan L [894]

Oxymorphone hydrochloride.
Numorphan [269]

Oxyphenbutazone.
Oxalid Tablets [891]
Tandearil [340]

Oxyphencyclimine hydrochloride.
Daricon® Tablets, 10 mg. [114]

Oxyphencyclimine HCI and hydroxyzine HCI, 5 mg:25 mg.
Enarax® 5 Tablets [114]

Oxyphencyclimine HCI and hydroxyzine HCI, 10 mg:25 mg.
Enarax® 10 Tablets [114]

Oxyphencyclimine hydrochloride/phenobarbitol, 5 mg:15 mg.
Daricon® PB Tablets [114]

Oxyphenonium bromide.
Antrenyl Bromide Tablets [189]

Oxyquinoline sulfate.
Aci-Jel Therapeutic Vaginal Jelly [666]
Otipyrin Otic Solution (sterile) [498]
Triva Combination, Douche Powder, Jel [132]

Oxystearin.
Chill-Ox 100 [272]

Oxytetracycline.
Oxymycin Injectable [661]
Terramycin Capsules [510]
Terramycin Film-coated Tablets, Intramuscular Solution, Ointment [690]
Urobiotic-250 [739]

Oxytetracycline hydrochloride.
E.P. Mycin Capsules [281]
Oxytetracycline HCI Capsules [769]
Terra-Cortril Ophthalmic Suspension, Topical Ointment [690]
Terramycin Capsules, Intramuscular Solution, Ointment, with Polymyxin B Sulfate Ophthalmic Ointment [690]

Oxytocin (injection).
Oxytocin Injection, in Tubex [942]
Pitocin Injection [674]
Syntocinon Injection [762]

Oxytocin (nasal spray).
Syntocinon Nasal Spray [762]

Ozokerite.
Ozokerite Wax-STRALPITZ, 170 D-STRALPITZ, 170 M.F.-STRALPITZ [830]
White Ozokerite Wax-STRALPITZ [830]
Yellow Ozokerite Wax-STRALPITZ, S Special-STRALPITZ [830]

Padimate O (Octyl dimethyl PABA).
Herpecin-L Cold Sore Lip Balm [144]
PreSun 4 Creamy Sunscreen, 8 Lotion, Creamy & Gel, 15 Creamy Sunscreen, 15 Sunscreen Lotion [920]

Palladium catalyst.
Pd-0803 E 1/16″ [389]

Palladium(II) oxide.
Ultralog 69-4920-00 [172]

Palm kernel alcohol.
Laurex PKH [548]

Palm kernel diethanolamide/1:1.
Cyclomide DP240/S [234]
Mackamide PK [567]

Palm kernel fatty acid polydiethanolamide.
Lauridit PPD [16]

Palm kernelamide DEA.
Mackamide PK [567]
Norfox KD [650]

Palm oil glyceride.
Monomuls 90-30 [375]
Myverol 18-35K Distilled Monoglycerides [278]

Palm oil glyceride, hydrogenated.
Monomuls 90-35 [375]
Myverol 18-04 Distilled Monoglycerides, 18-04K Distilled Monoglycerides [278]

Palm oil, partially hydrogenated.
Durkee 27 [272]

Palmamide DEA.
Cyclomide PD [939]

Palmamide MEA.
Cyclomide PM [939]

Palmamide MIPA.
Cyclomide PP [939]

Palmitamide MEA.
Nikkol PMEA [631]

Palmitamidopropyl betaine.
Incronam P-30 [223b]
Schercotaine PAB [771]

Palmitamidopropyl dimethylamine.
Lexamine P-13 [453]
Schercodine P [771]

Palmitamine.
Armeen 16D [79]
Kemamine P-890 [429]

Palmitamine oxide.
Ammonyx CO (Onyx) [662]
Conco XAC [209]

Palmitic acid.
Acme 370, 380, 390 [6]
Emersol 142, 144 [288]
Hydrofol Acid 1690 [796]
Hystrene 7016, 9016 [429]
Neo-Fat 16, 16-S, 16-54, 16-56 [79]
PRIFAC 2960 [875]

Palmitic acid, 45%.
Industrene 4516 [429]

Palmitic acid, 80%.
Hystrene 8016 [429]

Palmitic acid, 90%.
Hydrofol Acid 1690 [796]

Palmitic amido alkyl dimethyl amine.
Schercodine P [771]

Palmitic fatty acid, 90%.
Hystrene 9016 [430]

Palmitoyl milk protein, hydrolyzed.
Lipacide PCA [730]

Palmitoylcaseinique acid.
Lipacide PCO [730]

Palmityl amido betaine.
Schercotaine PAB [771]

N (Palmityl amido propyl) N,N dimethyl amine.
Servamine KEP 350 [788]

N (Palmityl amido propyl) N,N,N trimethyl ammonium chloride.
Servamine KEP 4527 [788]

Palmityl amine primary.
Kemamine P-880D [430]

Palmityl dimethyl tertiary amine (distilled).
Lilamin 316 D [522]

Palmityl glyceryl ether.
Nikkol Chimyl Alcohol 100 [631]

Palmityl-trimethyl ammonium chloride.
Radiaquat 6444 [658]
Variquat E228 [796]

Pancreatic preparations.
Arco Lase, Arco-Lase Plus [72]
Cotazym, Cotazym-S [664]
Donnazyme Tablets [737]
Entozyme Tablets [737]
Enzobile Improved Formula [541]
Enzypan [649]
Glucagon for Injection Ampoules [523]
Kanulase [257]
Karbokoff Tablets [78]
Ku-Zyme HP [499]
Pancrease [570]
Pancreatin Tablets 2400 mg. N.F. (High Lipase) [907]
Phazyme Tablets, Phazyme-95 Tablets, Phazyme-PB Tablets [719]
Viokase [905]

Pancreatic preparations *(cont'd).*
Zypan Tablets [819]

Pancuronium bromide injection.
Pavulon [664]

Panthenol.
Albafort Injectable [105]
Eldertonic [563]
Ilopan Injection [7]
D-Panthenol Ampul Type No. 63912, DL-Panthenol Cosmetic Grade No. 63920, D-Panthenol Regular Type No. 63909, DL-Panthenol No. 63915 [738]
DL-Panthenol TK [867]

Pantothenate, calcium.
Mega-B [72]
Natalins Rx, Tablets [572]

Pantothenic acid.
Dexol T.D. Tablets [518]

Papain.
Cardioguard Natural Liptropic Dietary Supplement [589]
Panafil Ointment, -White Ointment [759]

Papain, purified.
Panol [297a]

Papaverine.
Papaverine T.D. Capsules [344]

Papaverine hydrochloride.
Cerebid-150 TD Capsules, -200 TD Capsules [765]
Cerespan Capsules [891]
Papverine HCl Tablets [240]
Papaverine HCl T.D. Capsules [240]
Papaverine HCl Timed Capsules [769]
Papaverine Hydrochloride Capsules [751]
Pavabid Capsules, HP Capsulets [551]
Pavatym Capsules [304]

Papaverine hydrochloride 200 mg.
Pavatab [535]

Papaverine HCL T.D.
Paverolan Lanacaps [510]

Paper-phenolic laminate.
Insurok T-308 [812]

Paraben esters.
Protasens [713]

Parachlorometaxylenol.
Otic-HC Ear Drops [394]

Paraffin.
Paraffin Wax-STRALPITZ [830]
Parvan 2730, 3150, 3830, 4450, 4550, 5250, 6550 [306]

Paraffin, chlorinated.
Cereclor 42, 42P, 50LV, 51L, 52P, 70L, AP45, AP52, LP4446, LP4985, S45, S-52 [445]
Clorafin®, 40, 50 [406]
CPF-0001, CPF-0003, CPF-0008, CPF-0019, CPF-0022 [678]
Flexchlor 0001, 0002, 0008, 0018, 0023 [678]

Paraffin emulsion.
Shearlon AC, N [630]

Paraffin oil, chlorinated.
Chloroflo 42 [259]

Paraffin wax.
Attafin Series [89]

Paraffin wax emulsion.
Paracol 404A, 404D, 404G, 404N, 447K, 505A, 505G, 800N, 802A, 802N, 802NW, 810N, 810NP, 815N [406]

Paraffin wax, fully refined.
CS-2043 [231]

Paraffin wax, refined.
Akrowax 130, 145 [15]

Paraffinic oil.
Savo [881]
Shellflex 210, 310, 790 [791]
Sunflex 250, 3340, 110, 115 [835]
Sunpar 110, 150, 2280 [835]
Sunpreme 100, 200 [835]

Paraldehyde.
Paraldehyde [287]

Paramethadione.
Paradione [3]

Paregoric.
Paregoric [751]
Parepectolin [746]

Pareth-15-3.
Tergitol 15-S-3 [878]

Pareth-15-3 oleate.
Carsemol T-530 [529]

Pareth-15-3 stearate.
Carsemol T-538 [529]

Pareth-15-5.
Tergitol 15-S-5 [878]

Pareth-15-7.
Tergitol 15-S-7 [878]

Pareth-15-9.
Tergitol 15-S-9 [878]

Pareth-15-12.
Tergitol 15-S-12 [878]

Pareth-15-12 stearate.
Carsemol T-528 [529]

Pareth-15-20.
Tergitol 15-S-20 [878]

Pareth-15-30.
Tergitol 15-S-30 [878]

Pareth-15-40.
Tergitol 15-S-40 [878]

Pareth-23-3.
Neodol 23-3 [791]

Pareth-23-7.
Neodol 23-6.5 [791]

Pareth-25-2 phosphate.
Nikkol TDP-2 [631]

Pareth-25-3.
Alkasurf LA-3 [33]
Neodol 25-3 [791]
Sterling Emulsifier #3 [145]
Tergitol 25-L-3 [878]

Pareth 25-5.
Tergitol 25-L-5 [878]

Pareth 25-7.
Alkasurf LA-7 [33]
Neodol 25-7 [791]
Tergitol 25-L-7 [878]

Pareth-25-9.
Neodol 25-9 [791]
Tergitol 25-L-9 [878]

Pareth-25-9 hydrogenated tallowate.
Carsemol T-900 [529]

Pareth-25-10.
Sterling XE [145]

Pareth 25-12.
Alkasurf LA-12 [33]
Neodol 25-12 [791]
Tergitol 25-L-12 [878]

Pareth-25-12 oleate.
Carsemol N-520 [529]

Pareth-25-15.
Sterling Emulsifier #15 [145]

Pareth-45-7.
Neodol 45-7 [791]

Pareth-45-11.
Neodol 45-11 [791]

Pareth-45-13.
Neodol 45-13 [791]

Pareth-91-3.
Neodol 91-2.5 [791]

Pareth-91-6.
Neodol 91-6 [791]

Pareth-91-8.
Neodol 91-8 [791]

Pargyline hydrochloride.
Eutonyl Filmtab Tablets [3]

Peanutamide MEA.
Rewo-Amid OM 101/G [725]

Peanutamide MIPA.
Rewo-Amid OM 101/IG, OM 101/IG/ER [725]

Pectin.
Donnagel, -PG [737]
Kaolin, Pectin, Belladonna Mixture [769]
Kaolin-Pectin-Concentrated [751]
Kaolin-Pectin Mixture, PG Mixture [769]
Kaolin-Pectin Suspension [751]
Parepectolin [746]

Pectinase.
Extractase™ L5X, P20X [313]

Pectinase (fungal).
Extractase L5X, P15X [313]

Pectolytic enzyme.
Pectinex 1XL, 2XL, 3XL [654]
Ultrazym 40 L, 100 G [654]

PEG.
Alkapol PEG-200, PEG-300, PEG-600, PEG-1000, PEG-1500, PEG-3350, PEG-6000, PEG-8000 [33]
Empilan BO 100 [25]
Lamigen ET-180 [237]
Merpoxen PEG 6000 S [285]
Nissan PEG [636]
Pluracol E-200, E-300, E-400, E-600, E-1500, E-4000, E-6000 [109]
Pluriol E 200, E 300, E 400, E 600, E 1500, E 4000, E 6000, E 9000, P 600, P 900, P 2000, P 4000 [106]
Pogol 200, 300, 400, 400 USP, 600, 1540, 1570 [390]
Soprofor PE/Series [550]

PEG alkyl amine ether.
Amiladin, C-1802 [237]

PEG alkyl ether.
Noigen ET-65, ET-77, ET-95, ET-97, ET-107, ET-115, ET-127, ET-135, ET-147, ET-157, ET-165, ET-167, ET-187, ET-207 [237]

PEG, di-ester.
Actrol 4DP [811]

PEG dilauric fatty acid ester.
Alkamuls 400-DL [33]

PEG dimethacrylate.
SR-210 [766]

PEG dimethacrylate/hydroquinone inhibitor, 75±25.
SR-210 [766]

PEG dioctoate.
Flexol Plasticizer 4GO [878]

PEG dioleic fatty acid ester.
Alkamuls 400-DO, 600-DO [33]

PEG distearate.
Nissan Nonion DS-60HN [636]
Rewopal PEG 6000 DS [725a]

PEG distearic acid ester.
Noigen DS601 [237]

PEG-distearic ester.
Alkamuls 400-DS [33]

PEG distearic fatty acid ester.
Alkamuls 200-DS, 600-DS [33]

PEG dodecyl phenol ether.
Noigen EA33, EA73, EA83, EA143 [237]

PEG ester.
Actrol 4MP, 628 [811]
Kessco PEG Series [79]
Hercules AR 150 [406]
Marvanol RE-1824 [552]
Rycofax® 614, "O" [756]

PEG fatty acid ester.
Emanon 1112, 3115, 3199, 3299 R, 4110 [478]

PEG fatty acid ester, unsaturated.
Ethylan C40AH, C 160, VPK [509]

PEG fatty ester.
Dymsol L [253]
Modicol L [252]

PEG lanolin alcohol ether.
Lamigen ES-180, ET-20, ET-70, ET-90 [237]

PEG lanolin fatty acid ester.
Lamigen ES-30, ES-60, ES-100 [237]

PEG lauryl ether.
Noigen ET83, ET102, ET143, ET160, ET170, ET190, YX400, YX500 [237]

PEG (M = 0.3 kg/mol).
Serdox NDO 5.4 [788]

PEG (M = 0.4 kg/mol).
Serdox NDO 7.7 [788]

PEG (M = 1 kg/mol).
Serdox NDO 22 [788]

PEG methylether, fatty alcohol.
Rewopal MT 65 [725]

PEG monolaurate.
Empilan AQ 100 [23]
Nissan Nonion L-2, L-4 [636]

PEG monolauric fatty acid ester.
Alkamuls 400-ML, 600-ML [33]

PEG monooleate.
Nissan Nonion O-2, O-3, O-4, O-6 [636]

PEG monooleate ester.
Empilan BQ 100 [23]

PEG monopalmitate.
Nissan Nonion P-6 [636]

PEG monostearate.
Alkamuls EGMS [33]
Nissan Nonion S-2, S-4, S-6, S-10, S-15, S-15.4, S-40 [636]

PEG monotallow acid ester.
Nippon Nonion T-15 [636]

PEG nonylphenol ether.
Noigen EA50, EA70, EA80, EA80E [237]

PEG octyl phenol ether.
Noigen EA92, EA102, EA110, EA112, EA120, EA120B, EA130T, EA140, EA140L, EA/142, EA150, EA152, EA160, EA160P, EA170, EA190D [237]

PEG oleic acid ester.
Noigen ES90, ES120, ES140, ES160 [237]

PEG oleyl ether.
Noigen ET60, ET80, ET100, ET120, ET140, ET150, ET180, ET190S, O 100 [237]

PEG palmito-stearate.
Tefose 63, 1500 [339]

PEG-2 cocamine.
Alkaminox C-2 [33]
Chemox C-2 [174]
Ethomeen C/12 [79]
Hetoxamine C-2 [407]
Mazeen C-2 [564]
Protox C-2 [713]

PEG-2 laurate.
Diglycol Laurate ASE [79]
Diglycol Laurate S [358]
Kessco Diglycol Laurate A-Neutral, ASE, N, N-Syn [79]
Lipo-PEG 1-L [525]
Mapeg DGL [564]
Pegosperse 100 L, 100 ML [358]

PEG-2 laurate SE.
Kessco Diglycol Laurate SE [79]
Lipo DGLS [525]

PEG-2M.
Polyox WSR N-10 [878]

PEG-2 monooleate.
Hetoxamate MO-2 [407]

PEG-2 oleamine.
Ethomeen O/12 [79]
Hetoxamine O-2 [407]
Protox O-2 [713]

PEG-2 oleate.
Emalex 200 [628]
Lipo-Peg 1-O [525]
Pegosperse 100 O [358]

PEG-2 oleate SE.
Kessco Diglycol Oleate L-SE [79]

PEG-2 ricinoleate.
Pegosperse 100 MR [358]

PEG-2 soyamine.
Mazeen S-2 [564]
Protox S-2 [713]

PEG-2 stearamine.
Chemox HTA-2 [174]
Ethomeen 18/12 [79]
Hetoxamine ST-2 [407]
Protox HTA-2 [713]

PEG-2 stearate.
Clindrol SDG [197]
Cyclochem DGS [234]
Diglycol Stearate [360]
Glicosterina DPG [901]
Hodag DGS-N [415]
Hydrine [339]
Kessco Diethylene Glycol Monostearate, Stearate Neutral [79]
Lipo-PEG 1-S [525]
Nikkol DEGS, MYS-2 [631]
Pegosperse 100 S [358]
Product DG19 [197]
Schercemol DEGMS [771]

PEG-2 stearate SE.
CPH-104-DC-SE, CPH-104-DG-SE [379]
Hodag DGS-C [415]
Lipo DGS-SE [525]

PEG-2 tallow amine.
Mazeen T-2 [564]
Protox T-2 [713]

PEG-3 castor oil.
Nikkol CO-3 [631]

PEG-3 cocamide.
Alkamidox CME-2 [33]
Amidox C-2 [825]
Hetoxamide CD-4 [407]
Mazamide C-2 [564]
Oramide MLM 02 [785]
Unamide C-2 [529]

PEG-3 cocamine.
Lowenol C-243 [531]

PEG-3 dipalmitate.
Nikkol PEARL-3216 [631]

PEG-3 distearate.
Genapol TS Powder [55]
Nikkol PEARL-3222 [631]

PEG-3 lauramide.
Amidox L-2 [825]
Mazamide L-2 [564]
Unamide L-2 [529]

PEG-3 lauramine oxide.
Empigen OY [548]

PEG-3 sorbitan oleate.
Emalex 103 [628]
Emalex EG-2854 (OL) [628]

PEG-3 (tallow) aminopropylamine.
Ethoduomeen T/13 [79]

PEG-4.
Carbowax 200 [878]
Hetoxide PEG-200 [407]
Mapeg 200X [564]
Pluracol E-200 [109]
Polyglycol E-200 [261]

PEG-4 dilaurate.
Cyclochem PEG 200 DL [234]
Emerest 2704 [290]
Kessco PEG 200 dilaurate [79]
Lexemul 200 DL [453]
Lipo-PEG 2 Di Laurate [525]
Lipopeg 2-DL [525]
Lonzest PEG-4DL [529]
Mapeg 200 DL [564]
Pegosperse 200-DL [358]
Protamate 400-DL [713]

PEG-4 dioleate.
Kessco PEG 200 Dioleate [79]
Mapeg 200 DO [564]

PEG-4 distearate.
Cyclochem PEG 200 DS [234]
Kessco PEG 200 Distearate [79]
Mapeg 200 DS [564]

PEG-4 laurate.
Chemester 200-ML [174]
CPH-27-N [379]
Cyclochem PEG 200 ML [234]
Emerest 2703 [290]
Hetoxamate LA-4 [407]
Hodag 20-L [415]
Kessco PEG 200 Monolaurate [79]
Lipo-PEG 2-L [525]
Mapeg 200 ML [564]
Pegosperse 200 ML [358]
Protamate 200 ML [713]

PEG-4 oleate.
Chemester 200-OC [174]
Kessco PEG 200 Monooleate [79]
Mapeg 200 MO [564]
Pegosperse 200 MO [358]
Protamate 200 OC [713]

PEG 4 sorbitan monolaurate.
Hetsorb L-4 [407]

PEG 4 sorbitan monostearate.
Hetsorb S-4 [407]

PEG-4 stearate.
Chemester 200-DPS [174]
Kessco PEG 200 Monostearate [79]
Lipo-PEG 2-S [525]
Mapeg 200 MS [564]
Nikkol MYS-4 [631]
Pegosperse 200 MS [358]
Protamate 200 DPS [713]

PEG-4 tallate.
Protamate 200 T [713]

PEG-5 castor oil.
Surfactol 318 [640a]

PEG-5 castor oil, hydrogenated.
Nikkol HCO-5 [631]

PEG-5 cocamide.
Eumulgin C4 [400]

PEG-5 cocamine.
Chemox C-5 [174]
Ethomeen C/15 [79]
Hetoxamine C-5 [407]
Mazeen C-5 [564]
Protox C-5 [713]

PEG-5 cocoate.
Ethofat C/15 [79]

PEG-5 coconut amide.
Hetoxamide C-4 [407]

PEG-5 lanolate.
Ritalafa 5 [735]
Sklirate 5 [223]

PEG-5M.
Polyox Resin WSR 35, Polyox WSR N-80 [878]

PEG-5 oleamine.
Chemox O-5 [174]
Ethomeen O/15 [79]
Hetoxamine O-5 [407]
Protox O-5 [713]

PEG-5 oleate.
Emulphor VN-430 [338]
Ethofat O/15 [79]
Hetoxamate MO-5 [407]

PEG-5 pentaerythritol ether.
PME [901]

PEG-5 sorbitan isostearate.
Montanox 71 [785]

PEG 5 sorbitan monooleate.
Hetsorb O-5 [407]

PEG-5 soyamine.
Mazeen S-5 [564]
Protox S-5 [713]
Varionic L-205 [796]

PEG-5 soya sterol.
Generol 122 E 5 [399], [400]

PEG-5 stearamine.
Ethomeen 18/15 [79]
Hetoxamine ST-5 [407]

PEG-5 stearate.
Ethofat 60/15 [79]
Hetoxamate SA-5 [407]
Industrol MS-5 [923]

PEG-5 stearyl ammonium chloride.
Genamin KS5 [56]

PEG-5 tallow amine.
Katapol PN-430 [338]
Mazeen T-5 [564]
Protox T-5 [713]

PEG-6.
Carbowax 300 [878]
Hetoxide PEG-300 [407]
Polyglycol E-300 [261]

PEG-6-caprylic/capric glyceride.
Softigen 767 [482]

PEG-6-32.
Carbowax 1450 [878]
Mapeg 1500X [564]

PEG-6-32 dilaurate.
Pegosperse 1500 DL [358]

PEG-6-32 dioleate.
Pegosperse 1500 DO [358]

PEG-6-32 stearate.
Hodag 150-S [415]
Lipo-PEG 15-S [525]
Pegosperse 1500 MS [358]
Protamate 1500 DPS [713]
Tefose 1500 [339]

PEG-6 cocamide.
Alkamidox CME-5 [33]
Amidox C-5 [825]
Empilan MAA [548]
Hetoxamide CD-6 [407]
Mazamide C-5 [564]
Rewopal C6 [725]
Unamide C-5 [529]

PEG-6 distearate.
Cyclochem PEG 300 DS [234]
Kessco PEG 300 Distearate [79]

PEG-6 isolauryl thioether.
Siponic 260 [29]

PEG-6 isopalmitate.
Isopalm K [494]

PEG-6 isostearate.
Polysolve 6 [494]

PEG-6 lauramide.
Amidox L-5 [825]
Lanidox-5 [502]
Mazamid L-5 [564]
Mazamide L-5 [564]
Unamide L-5 [529]

PEG-6 laurate.
Kessco PEG 300 Monolaurate [79]

PEG-6 methyl ether.
Carbowax 350 [878]

PEG-6 oleate.
Chemester 300-OC [174]
Grocor 300 [374]
Kessco PEG 300 Monooleate [79]
Lipo-Peg 3-O [525]
Pegosperse 300 MO [358]
Protamate 300 OC [713]

PEG-6 palmitate.
Polymulse 6 [494]

PEG-6 soya sterol undecylenate.
Undelene [901]

PEG-6 stearate.
Kessco PEG 300 Monostearate [79]
Lipo-PEG 3-S [525]
Polystate [339]
Protamate 300 DPS [713]

PEG-7 cocamide.
Oramide MLM 06 [785]

PEG-7-glyceryl cocoate.
Cetiol HE [399], [400]
Mazol 159 [564]
Standamul HE [400]

PEG-7M.
Polyox WSR N-750 [878]

PEG-7 monostearate.
Hetoxamate SA-7 [407]

PEG-7 oleamide.
Ethomid O/17 [79]

PEG-7 oleate.
Industrol MO-7 [923]

PEG-7 stearate.
Pegosperse 350 MS [358]

PEG-8.
Carbowax 400 [878]
Hodag PEG 400 [415]
Mapeg 400X [564]
Pluracol E-400 [109]
Polyglycol E-400 [261]

PEG-8 castor oil.
Mapeg CO-8 [564]
Protachem CA-8 [713]

PEG-8 cocoate.
Pegosperse 400 MC [358]

PEG-8 dilaurate.
CPH-79-N, CPH-361-N [379]
Cyclochem PEG 400 DL [234]
Emerest 2706 [291]
Kessco PEG 300 Dilaurate [79]
Kessco PEG 400 Dilaurate [79]
Lexemul 300 DL [453]
Lexemul 400 DL [453]
Lipo-PEG 3 Di Laurate [525]
Lipo-PEG 4 Di Laurate [525]
Lipopeg 4-DL [525]
Mapeg 400 DL [564]
Pegosperse 400 DL [358]

PEG-8 dioleate.
CPH-211-N [379]
Cyclochem PEG 400 DO [234]
Emerest 2708 [291]
Hodag 32-O [415]
Kessco PEG 300 Dioleate [79]
Kessco PEG 400 Dioleate [79]
Lipo-PEG 4 Di Oleate [525]
Lipopeg 4-DO [525]
Lonzest PEG-4DO [529]
Mapeg 400 DO [564]
Nonisol 210 [190]
Pegosperse 400 DO [358]
Protamate 400 DO [713]

PEG-8 distearate.
Cyclochem PEG 400 DS [234]
Emerest 2712 [291]
Kessco PEG 400 DS-356, PEG 400 Distearate [79]
Lipopeg 4-DS [525]
Mapeg 400 DS [564]
Pegosperse 400 DS [358]
Protamate 400 DS [713]

PEG-8 ditallate.
Pegosperse 400 DOT [358]

PEG-8 ditriricinoleate.
Pegosperse 400 DTR [358]

PEG-8 isolauryl thioether.
Siponic SK [29]

PEG-8 laurate.
Chemester 400-ML [174]
CPH-30-N [379]
Cyclochem PEG 400 ML [234]
Emerest 2705 [291]
Kessco PEG 400 Monolaurate [79]
Lipopeg 4-L [525]

PEG-8 laurate *(cont'd).*
Lonzest PEG-4L [529]
Mapeg 400 ML [564]
Pegosperse 400 ML [358]
Protamate 400 ML [713]

PEG-8 oleate.
Chemester 400-OC [174]
CPH-40-N, CPH-46-N, CPH-233-N [379]
Cyclochem PEG 400 MO [234]
Emerest 2707 [291]
Empilan BQ100 [548]
Kessco E112, PEG 400 Monooleate [79]
Lipo-PEG 4-O [525]
Lonzest PEG-40 [529]
Mapeg 400 MO [564]
Pegosperse 400 MO [358]
Protamate 400 OC [713]

PEG-8 palmitoyl methyl.
Stepanrinse 700 [825]

PEG-8 propylene glycol cocoate.
Emulsynt 900 [895]

PEG-8 sesquioleate.
Mapeg 400 SO [564]

PEG-8 sorbitan beeswax.
Nikkol GBW-8 [631]

PEG-8 stearate.
Alkasurf ST-8 [33]
Chemester 400-DPS [174]
CPH-50-N [379]
Emerest 2711 [290]
Eumulgin ST-8 [399], [400]
Grocor 400 [374]
Hodag 40-S [415]
Kessco PEG 400 Monostearate [79]
Lexemul 400 MS [453]
Lipo-PEG 4-S [525]
Lonzest PEG-4S [529]
Mapeg 400 MS [564]
Myrj 45 [438]
Pegosperse 400 MS [358]
Protamate 400 DPS [713]
Simulsol M 45 [785]
Witconol H35A [935]

PEG-8 tallate.
Mapeg 400 MOT [564]
Pegosperse 400 MOT [358]
Protamate 400 T [713]

PEG-9.
Alkapol PEG-400 [33]

PEG-9 castor oil.
Protachem CA-9 [713]

PEG-9 laurate.
Hetoxamate LA-9 [407]
Industrol ML-9 [923]

PEG-9 oleamide.
Rewopal O8 [725]

PEG-9 oleate.
Alkamuls 400-MO [33]
Hetoxamate MO-9 [407]
Industrol MO-9 [923]
Pegosperse 450 MO [358]

PEG-9 stearate.
Industrol MS-9 [923]

PEG 10 castor oil.
Incrocas 10 [223]
Nikkol CO-10 [631]

PEG-10 cocamine.
Chemox C-10 [174]
Ethomeen C/20 [79]
Mazeen C-10 [564]
Protox C-10 [713]

PEG-10 isolauryl thioether.
Siponic 218 [29]

PEG-10 lanolate.
Ritalafa 10 [735]
Sklirate 10 [223]

PEG-10 oleate.
Ethofat O/20 [79]
Protamate 600 OC [713]

PEG-10 sorbitan laurate.
Atlas G-7596J [438]
G-7596J [438]
Hetsorb L-10 [407]
Liposorb L-10 [525]

PEG-10 soya sterol.
Generol 122 E 10 [400]

PEG-10 soyamine.
Mazeen S-10 [564]
Protox S-10 [713]

PEG-10 stearamine.
Ethomeen 18/20 [79]

PEG-10 stearate.
Ethofat 60/20 [79]
Nikkol MYS-10 [631]

PEG-10 (tallow) aminopropylamine.
Ethoduomeen T/20 [79]

PEG-11 cocamide.
Oramide MLM 10 [785]

PEG-12.
Carbowax 600 [878]
Mapeg 600X [564]
Pluracol E-600 [109]
Polyglycol E-600 [261]

PEG-12 dilaurate.
Kessco PEG 600 Dilaurate [79]

PEG-12 dioleate.
CPH-213-N [379]

PEG-12 dioleate *(cont'd).*
Kessco PEG 600 Dioleate [79]
Mapeg 600 DO [564]

PEG-12 distearate.
Cyclochem PEG 600 DS [234]
Kessco PEG 600 Distearate [79]
Mapeg 600 DS [564]
Protamate 600 DS [713]

PEG-12 ditallate.
Pegosperse 600 DOT [358]

PEG-12 glyceryl laurate.
Lamacit GML-12 [375]

PEG-12 laurate.
CPH-43-N [379]
Kessco PEG 600 Monolaurate [79]
Lipo-PEG 6-L [525]
Mapeg 600 ML [564]
Pegosperse 600 ML [358]

PEG-12 oleate.
Chemester 600-OC [174]
CPH-41-N [174]
Kessco PEG 600 Monooleate [79]
Lipo-Peg 6-O [525]
Mapeg 600 MO [564]
Pegosperse 600 MO [358]

PEG-12 stearate.
Kessco PEG 600 Monostearate [79]
Lipo-PEG 6-S [525]
Mapeg 600 MS [564]
Pegosperse 600 MS [358]
Protamate 600 DPS [713]

PEG-12 tallate.
Protamate 600 T [713]

PEG-13 monostearate.
Hetoxamate SA-13 [407]

PEG-14.
Alkapol PEG-600 [33]

PEG-14 laurate.
Industrol ML-14 [923]

PEG-14M.
Carbowax Polyethylene Glycol 14M [878]
Polyox WSR-205, WSR N-3000 [878]

PEG-14 oleate.
Hetoxamate MO-14 [407]
Industrol MO-14 [923]
Pegosperse 700 TO [358]

PEG-14 stearate.
Industrol MS-14 [923]

PEG-15 cocamine.
Chemox C-15 [174]
Ethomeen C/25 [79]
Hetoxamine C-15 [407]
Icomeen C-15 [923]
Mazeen C-15 [564]
Protox C-15 [713]

PEG-15 cocoate.
Ethofat C/25 [79]

PEG-15 cocomonium chloride.
Ethoquad C/25 [79]

PEG-15 glyceryl ricinoleate.
Tagat R1 [360]

PEG-15 oleamine.
Chemox O-15 [174]
Ethomeen O/25 [79]
Hetoxamine O-15 [407]
Protox O-15 [713]

PEG-15 oleamonium chloride.
Ethoquad O/25 [79]

PEG-15 soyamine.
Mazeen S-15 [564]
Protox S-15 [713]

PEG-15 stearamine.
Chemox HTA-15 [174]
Ethomeen 18/25 [79]
Hetoxamine ST-15 [407]
Protox HTA-15 [713]

PEG-15 stearmonium chloride.
Ethoquad 18/25 [79]

PEG-15 tallow amine.
Katapol PN-730 [338]
Mazeen T-15 [564]
Protox T-15 [713]

PEG-15 (tallow) aminopropylamine.
Ethoduomeen T/25 [79]

PEG-15 tallow polyamine.
Polyquart H [399], [400]

PEG-16 soya sterol.
Generol 122 E 16 [400]

PEG-16 tallate.
Renex 20 [438]

PEG-18 palmitate.
Pegosperse 900 MP [358]

PEG-18 stearate.
Pegosperse 900 MS [358]

PEG-20.
Carbowax 1000 [878]
Mapeg 1000X [564]

PEG-20 castor oil.
Nikkol CO-20TX [631]

PEG-20 dilaurate.
Kessco PEG 1000 Dilaurate [79]

PEG-20 dioleate.
Kessco PEG 1000 Dioleate [79]

PEG-20 distearate.
Kessco PEG 1000 Distearate [79]

PEG-20 distearate *(cont'd)*.
Lipo-PEG 10 Di Stearate [525]
Mapeg 1000 DS [564]

PEG-20 glyceryl laurate.
Lamacit GML-20 [375]
Tagat L2 [360]

PEG-20 glyceryl oleate.
Tagat O2 [360]

PEG-20 glyceryl stearate.
Aldosperse MS-20 [358]
Arosurf GMS-E20 [796]
Cutina E-24 [400]
Mazol 80 MG [564]
Tagat S2 [360]
Varonic LI 42 [796]

PEG-20 isostearate.
Emerest 2701 [290]

PEG-20 lanolate.
Sklirate 1000 [223]

PEG-20 lanolin.
Laneto 20 [735]

PEG-20 laurate.
Kessco PEG 1000 Monolaurate [79]

PEG-20M.
Polyox WSR-1105 [878]

PEG-20 oleate.
Chemester 1000-OC [174]
Kessco PEG 1000 Monooleate [79]
Protamate 1000 OC [713]

PEG-20 sorbitan isostearate.
Montanox 70 [785]
T-Maz 67 [564]

PEG (20) sorbitan monoisostearate.
Nikkol TI-10 [631]

PEG 20 sorbitan monopalmitate.
Hetsorb P-20 [407]

PEG 20 sorbitan trioleate.
Hetsorb TO-20 [407]

PEG 20 sorbitan tristearate.
Hetsorb TS-20 [407]

PEG-20 stearate.
Cerasynt 840 [895]
Chemester 1000-DPS [174]
CPH-90-N [379]
Emerest 2713 [290]
Kessco PEG 1000 Monostearate [79]
Lipo-PEG 10-S [525]
Mapeg 1000 MS [525]
Pegosperse 1000 MS [358]
Protamate 1000 DPS [713]
Simulsol M 49 [785]

PEG-20 tallate.
Protamate 1000 T [713]

PEG-20 tallow amine.
Katapol PN-810 [338]

PEG-23 glyceryl laurate.
Aldosperse ML-23 [358]

PEG-23 monostearate.
Hetoxamate SA-23 [407]

PEG-24 glyceryl stearate.
Cutina E-24 [399]

PEG-24 lanolin, hydrogenated.
Lipolan 31 [525]

PEG-25 castor oil.
Mapeg CO-25 [564]
Pegosperse 1100 CO [358]
Protachem CA-25 [713]
Ricino Viscoil [901]

PEG-25 castor oil, hydrogenated.
Mapeg CO-25-H [564]
Protachem CAH-25 [713]

PEG-25 glyceryl oleate.
Lamacit GMO-25 [375]

PEG-25 glyceryl trioleate.
Tagat TO [360]

PEG-25 propylene glycol stearate.
Atlas G-2162 [438]
G-2162 [438]
Simulsol PS 20 [785]

PEG-25 soya sterol.
Generol 122 E 25 [400]

PEG-25 stearate.
Nikkol MYS 25 [631]

PEG-27 lanolin.
Lanete 27 [735]
Lanogel 21 [44]

PEG-28 glyceryl tallowate.
Varonic LI 2 [746]

PEG 30 castor oil.
Hetoxide C-30 [407]
Incrocas 30 [223b]
Protachem CA-30 [713]

PEG-30 glyceryl cocoate.
Varonic LI 63 [796]

PEG-30 glyceryl laurate.
Tagat L [360]

PEG-30 glyceryl oleate.
Tagat O [360]

PEG-30 glyceryl stearate.
Tagat S [360]

PEG-30 lanolin.
Lanalox L 30 [562]
Laneto 30 [735]
Nikkol TW-30 [631]

PEG-30 oleamine.
Katapol OA-860 [338]

PEG-30 sorbitan tetraoleate.
Nikkol GO-430 [631]

PEG-32.
Mapeg 1540X [564]
Pluracol E-1540 [109]

PEG-32 dilaurate.
Kessco PEG 1540 Dilaurate [79]

PEG-32 dioleate.
Kessco PEG 1540 Dioleate [79]

PEG-32 distearate.
Kessco PEG 1540 Distearate [79]
Mapeg 1540 DS [564]

PEG-32 laurate.
Kessco PEG 1540 Monolaurate [79]

PEG-32 oleate.
Kessco PEG 1540 Monooleate [79]

PEG-32 stearate.
Kessco PEG 1540 Monostearate [79]
Mapeg 1540 MS [564]
Protamate 1540 DPS [713]

PEG-35 stearate.
Hetoxamate SA-35 [407]
Lipo-PEG 16-S [525]

PEG-36 castor oil.
Mapeg CO-36 [564]

PEG 40 castor oil.
Incrocas 40 [223]
Mapeg CO-40 [564]
Nikkol CO-40TX [631]
Pegosperse 2000 CO [358]
Protachem CA-40 [713]
Simulsol ELL, OL 50 [785]
Surfactol 365 [640a]
T-DET C-40 [859]

PEG-40 castor oil, hydrogenated.
Eumulgin RO-40 [399]
Hetoxide HC-40 [407]
Mapeg CO-40-H [564]
Nikkol HCO-40 [631]
Protachem CAH-40 [713]
Tagat R40 [360]

PEG-40 lanolin.
Ethoxygel 50, 100 [290]
Lan-Aqua-Sol Hydrophilic, Hydrophilic 50, 100 [308]
Laneto 40 [735]
Lanogel 31 [44]

PEG 40 monostearate.
Hetoxamate SA-40 (DF) [407]

PEG-40 sorbitan peroleate.
Arlatone T [438]

PEG-40 sorbitan tetraoleate.
Nikkol GO-440 [631]
Rheodol 440 [480]

PEG-40 stearate.
Alkasurf ST-40 [33]
Emerest 2715 [290]
Industrol MS-40 [923]
Lanoxide-52 [507]
Lipal 39S [715]
Lip PEG-39-S [525]
Mapeg S-40 [564]
Myrj 52, 52S [438]
Nikkol MYS-40 [631]
Pegosperse 1750-MS [358]
Protamate 2000 DPS [713]
Simulsol M 52 [785]

PEG-40 tallow amine.
Protox T-40 [713]

PEG-44 sorbitan laurate.
Atlas G-7596DJ [438]
G-7596DJ [438]
Hetsorb L-44 [407]

PEG-45 stearate.
Lipo-PEG 20—S [525]
Nikkol MYS-45 [631]
Protamate 4000 DPS [713]

PEG-50 castor oil.
Nikkol CO-50TX [631]

PEG-50 castor oil, hydrogenated.
Nikkol HCO-50 [631]

PEG-50 lanolin.
Lan-Aqua-Sol Hydrophilic-Plus 50, 100 [308]
Lan-Aqua-Solxtra Hydrophilic [308]
Laneto 49 [735]

PEG (50) monostearate.
Pegosperse 50-MS [358]

PEG-50 stearamine.
Chemox HTA-50 [174]
Ethomeen 18/60 [79]
Icomeen 18-50 [923]
Protox HTA-50 [713]

PEG-50 stearate.
Atlas G-2153 [438]
G-2153 [438]
Myrj 53 [438]

PEG-50 tallow amide.
Methomid 60 [771]
Protox T-50 [713]
Schercomid HT-60 [771]
Varonic U-250 [796]

PEG-50 tallow amide (monoalkanolamide).
Schercomid HT-60 [771]

PEG 60 castor oil.
Incrocas 60 [223b]
Nikkol CO-60TX [631]

PEG 60 castor oil, hydrog..
Hetoxide HC-60 [407]
Nikkol HCO-60 [631]
Tagat R60 [360]

PEG-60 lanolin.
Laneto 60 [735]
Solan, Solan 50 [223]

PEG-60 sorbitan tetraoleate.
Nikkol GO-460 [631]
Rheodol 460 [480]

PEG-60 sorbitan tetrastearate.
Nikkol GS-460 [631]

PEG-66 trihydroxy stearin.
Surfactol 575 [158], [640a], [641a]

PEG-75.
Carbowax 3350 [878]
Mapeg 4000X [564]
Pluracol E-4000 [109]

PEG-75 dilaurate.
Kessco PEG 4000 Dilaurate [79]

PEG-75 dioleate.
Kessco PEG 4000 Dioleate [79]
Pegosperse 4000 DO [358]

PEG-75 distearate.
Kessco PEG 4000 Distearate [79]
Mapeg 4000 DS [564]

PEG-75 lanolin.
Ethoxylan® 50, 100 [290]
Lanalox L 75, L75/50 [562]
Lan-Aqua-Sol xtra-Hydrophilic 50, 100 [308]
Laneto 50, 100 [735]
Lanogel 41 [44]
Lantox 55, 110 [507]
Solangel 401 [223]
Solulan 75, L-575 [44]

PEG-75 lanolin oil.
Lanalox AWS [562]
Lanoil Water/Alcohol Soluble Lanolin [507]
Lanotein AWS 30 [308]
Ritalin AWS [735]
Vigilan AWS [308]

PEG-75 lanolin wax.
Lanfrax WS55 [290]

PEG-75 laurate.
Kessco PEG 4000 Monolaurate [79]

PEG-75 oleate.
Kessco PEG 4000 Monooleate [79]

PEG-75 stearate.
Chemester 4000-DPS [174]
Kessco PEG 4000 Monostearate [79]
Mapeg 400 MS [564]
Pegosperse 4000 MS [358]

PEG-80 castor oil, hydrogenated.
Nikkol HCO-80 [631]

PEG-80 sorbitan laurate.
G-4280 [438]
Hetsorb L80-72% [407]

PEG-80 sorbitan palmitate.
G-7426 HJ [438]

PEG-85 lanolin.
Lan-Aqua-Sol Super-Hydrophilic 50, 100, Super-Hydrophilic [308]
Laneto 85 [735]
Lanogel 61 [44]

PEG-90M.
Polyox WSR-301 [878]

PEG-90 stearate.
Hetoxamate SA-90 [407]

PEG 100 castor oil.
Incrocas 100 [223b]
Protachem CA-100 [713]

PEG-100 castor oil, hydrogenated.
Nikkol HCO-100 [631]
Protachem CAH-100 [713]

PEG-100 lanolin.
Laneto 99 [735]

PEG 100 monolaurate.
Nopalcol 1-L [252]
Pegosperse 100-L, 100-ML [358]

PEG (100) monooleate.
Pegosperse 100-O [358]

PEG (100) monostearate.
Pegosperse 100-S [358]

PEG 100 monotallowate.
Nopalcol 1-TW [252]

PEG-100 stearate.
Emerest 2717 [290]
Lipo PEG-100-S [525]
Myrj 59 [438]
Simulsol M 59 [785]

PEG-150.
Carbowax 8000 [878]
Mapeg 6000X [564]
Pluracol E-6000 [109]

PEG-150 dilaurate.
Kessco PEG 6000 Dilaurate [79]

PEG-150 dioleate.
Kessco PEG 6000 Dioleate [79]

PEG-150 distearate.
Emulvis [379]
Kessco PEG 6000 Distearate, Kessco X-211 [79]
Lipopeg 6000-DS [525]
Mapeg 6000 DS [564]
Pegosperse 6000-DS [358]
Protamate 6000 DS [713]
Rewopal PEG-6000 DS [725]
Witconol L32-45 [935]

PEG-150 laurate.
Kessco PEG 6000 Monolaurate [79]
Mapeg 6000 ML [564]

PEG-150 oleate.
Kessco PEG 6000 Monooleate [79]

PEG-150 stearate.
Kessco PEG 6000 Monostearate [79]
Mapeg 6000 MS [564]

PEG (mol. wt. 200).
Merpoxen PEG 200 [285]
Nopalcol 200 [252]

PEG 200 castor oil.
Hetoxides C-200-50% [407]
Mapeg CO-200 [564]
Protachem CA-200 [713]

PEG-200 castor oil, hydrogenated.
Protachem CAH-200 [713]

PEG 200 diacrylate/hydroquinone inhibitor/100-150 ppm.
SR-259 [766]

PEG 200 dibenzoate.
Benzoflex P-200 [898]

PEG-200-diisostearate.
Scher Esters [771]

PEG 200 dilaurate.
Alkamuls 200 DL [33]
Cithrol 2 DL [223]
CPH-71-N [379]
Emerest 2622 [289]
Emerest® 2704 [290]
Hodag 22-L [415]
Lipopeg 2 DL [525]
Mapeg 200DL [564]
Pegosperse® 200 DL [358]
Scher Esters [771]

PEG 200 dioleate.
Alkamuls 200 DO [33]
Chemax PEG 200 DO [170]
Cithrol 2 DO [223]
CPH-249-N, CPH-306-N [379]
Mapeg 200DO [564]

PEG 200 distearate.
Alkamuls 200 DS [33]
Cithrol 2 DS [223]
CPH-55-N [379]
Mapeg 200DS [564]

PEG 200 laurate.
Tegin 200 L [360]

PEG-200-monoisostearate.
Emerest 2625 [289]
Scher Esters [771]

PEG 200 monolaurate.
Alkamuls 200 ML [33]
Cithrol 2 ML, 2 MO [223]
CPH-27-N [379]
Emerest 2620 [289]
Emerest® 2703 [290]
Hodag 20-L [415]
Mapeg 200ML [564]
Nopalcol 2-L [252]
Pegosperse® 200 ML [358]
Radiasurf 7422 [658]

PEG 200 monolaurate, ethoxylated (EO 5).
Industrol ML-5 [109]

PEG 200 monooleate.
Alkamuls 200 MO [33]
Cithrol 2MO [223]
Collemul H2 [37]
CPH-39-N, CPH-125-N, CPH-217-N [379]
Emerest 2624 [289]
Mapeg 200MO [564]
Pegosperse® 200 MO [358]
Radiasurf 7402 [658]

PEG 200 monooleate, ethoxylated (EO 5).
Industrol MO-5 [109]

PEG 200 monostearate.
Alkamuls 200 MS [33]
Cithrol 2MS [223]
CPH-54-N [379]
Lexemul 200 MS [453]
Mapeg 200 MS [564]
Radiasurf 7412 [658]

PEG 200 monostearate, ethoxylated (EO 5).
Industrol MS-5 [109]

PEG (200) monotallate.
Mapeg 200 MOT [564]

PEG 200 oleate, nonionic.
Ethylan A2 [509]

PEG 200 ricinoleate.
CPH-100-N [379]

PEG-200 tall oil dioleate.
Polyfac TDO-5 [917a]

PEG 200 trihydroxystearin.
Naturechem THS-200 [158]
Surfactol 590 [640a], [641a]

PEG (220) monotallate.
Mapeg Tao-5 [564]

PEG 250 monooleate, ethoxylated (EO 6).
Industrol MO-6 [109]

PEG 300, nonionic.
Merpoxen PEG 300 [285]
PGE-300 [397]

PEG 300 di laurate.
CPH-75-N [379]

PEG 300 di oleate.
CPH-49-N, CPH-49-TL-N [379]

PEG 300 distearate.
Cithrol 3 DS [223]
CPH-174-N [379]

PEG 300 mono laurate.
CPH-76-N [379]
Emerest 2630 [289]

PEG 300 monooleate.
Acconon 300 MO [81], [829]
CPH-80-N, CPH-80-TL-N [379]
Emerest 2632 [289]
Pegosperse® 300 MO [358]
Radiasurf 7431 [658]

PEG 300 monooleate, ethoxylated (EO 7).
Industrol MO-7 [109]

PEG-300-monopelargonate.
Emerest 2634 [289]

PEG 300 monostearate.
Cithrol 3MS [223]
CPH-106-N [379]
Emerest 2636 [289]
Radiasurf 7432 [658]

PEG 300 monostearate, ethoxylated (EO 7).
Industrol MS-7 [109]

PEG 300 oleate.
Ethylan A3 [509]

PEG 300 ricinoleate.
CPH-101-N [379]

PEG-350.
Carbowax 20M [878]

PEG (350) monostearate.
Mapeg 350 MS [564]

PEG 350 monostearate, ethoxylated (EO 8).
Industrol MS-8 [109]

PEG 400.
Merpoxen PEG 400 [285]
Nopalcol 400 [252]
PGE-400 [397]

PEG 400 cocoate.
Nopalcol 4-C, 4-CH [252]

PEG 400 di-2-ethylhexoate.
TegMeR™ 809 [379]

PEG-400-diisostearate.
Scher Esters [771]

PEG 400 dilaurate.
Alkamuls 400-DL [33]
Cithrol 4 DL [223]
CPH-79-N [379]
Emerest 2652 [289]
Emerest® 2706 [290]
Industrol DL-9 [109]
Lipal 400 DL [715]
Lipopeg 4 DL [525]
Mapeg 400DL [564]
Pegosperse® 400 DL [358]
Scher Esters [771]
Varonic 400DL [796]

PEG 400 dioleate.
Alkamuls 400-DO [33], [34]
Chemax PEG 400 DO [170]
Cithrol 4 DO [223]
CPH-170-N, CPH-210-N, CPH-211-N [379]
Emerest 2648 [289]
Emerest® 2708 [290]
Industrol DO-9 [109]
Lexemul 400 DO [453]
Lonzest PEG-4DO [529]
Mapeg 400 DO [564]
Nonisol 210 [190]
Pegosperse® 400 DO [358]
PGE-400-DO [397]
Radiasurf 7443 [658]
Scher Esters [771]

PEG 400 distearate.
Alkamuls 400-DS [33], [34]
Cithrol 4 DS [223]
CPH-58-N [379]
Cyclochem PEG 400 DS [234]
Emerest 2642 [289]
Emerest® 2712 [290]
Kessco-PEG 400 Distearat [16]
Lipal 400 DS [266]
Mapeg 400 DS [564]
Pegosperse® 400 DS [358]
PGE-400-DS [397]
Radiasurf 7453 [658]
Scher Esters [771]

PEG 400 distearate, polyethylene glycol ester.
Lipal 400 DS [715]

PEG (400) ditallate.
Mapeg 400 DOT [564]
Pegosperse® 400 DOT [358]
Witconol H-33 [934], [935]

PEG 400 ditallowate.
Nopalcol 4-DTW [252]

PEG 400 di tri ricinoleate.
Pegosperse 400 DTR [358]

PEG 400 ester.
Eccoterge 200 [277]

PEG 400 monococoate.
Alkamuls 400-MC [33]
Nopalcol 4-C, 4-CH [252]
Pegosperse® 400 MC [358]

PEG-400-monoisostearate.
Emerest 2644 [289]
Industrol MIS-9 [109]
Scher Esters [771]

PEG 400 monolaurate.
Alkamuls 400-ML [33], [34]
Cithrol 4ML [223]
Collemul L4 [37]
CPH-30-N [379]
Cyclochem PEG 400 ML [234]
Emerest 2650 [289]
Emerest® 2705 [290]
Gradonic 400-ML [365]
Industrol ML-9 [109]
Lipopeg 4-L [525]
Lonzest PEG-4L [529]
Mapeg 400ML [564]
Nonisol 100 [190]
Nopalcol 4-L [252]
Pegosperse® 400 ML [358]
PGE-400-ML [397]
Radiasurf 7423 [658]
Scher Esters [771]

PEG 400 monooleate.
Acconon 400 MO [81], [829]
Alkamuls 400-MO [33], [34]
Cithrol 4MO [223]
Collemul H4 [37]
CPH-30-N, CPH-40-N, CPH-46-N, CPH-233-N [379]
Cyclochem PEG 400 MO [234]
Dur PEG 400MO [272]
Emerest 2646 [289]
Emerest® 2707 [290]
Emulsan O [724]
Kessco-PEG 400 Monooleat [16]
Lipopeg 4-O [525]
Lonzest PEG-4O [529]
Mapeg 400MO [564]
Nopalcol 4-O [252]
Pegosperse® 400 MO, 400 MOT [358]
PGE-400-MO [397]
Radiasurf 7403 [658]
Scher Esters [771]
Witconol H-31A [934], [935]

PEG 400 monooleate, ethoxylated (EO 9).
Industrol MO-9 [109]

PEG 400 monooleate (technical).
Cithrol A [223]

PEG-400-monopelargonate.
Emerest 2654 [289]

PEG 400 monostearate.
Alkamuls 400-MS [33], [34]
Cithrol 4MS [223]
CPH-50-N [379]
Emerest 2640 [289]
Emerest® 2711 [290]
Lexemul 400 MS [453]
Lipal 400 S [266]
Lipopeg 4-S [525]
Mapeg 400MS [564]
Nonisol 300 [190]
Nopalcol 4-S [252]
Pegosperse® 400 MS [358]
PGE-400-MS [397]
Radiasurf 7413 [658]
Scher Esters [771]
Varonic 400 MS [796]
Witconol H-35A [934], [935]

PEG 400 monostearate, ethoxylated (EO 9).
Industrol MS-9 [109]

PEG 400 monostearate, polyethylene glycol ester.
Lipal 400 S [715]

PEG® 400 monostearate USP XIV.
Pegosperse® 400 MS USP [358]

PEG 400 monotallate.
Industrol 400-MOT [109]
Mapeg 400 MOT [564]
Pegosperse® 400 MOT [358]
Witconol H-31 [934], [935]

PEG 400 oleate; nonionic.
Ethylan A4 [509]
Witconol H 31 A [937a], [938]

PEG 400 ricinoleate.
CPH-47-N [379]

PEG 400 sesquioleate.
Emerest® 26647 [290]

PEG-400-stearate.
Cremophor S 9 [106]
Witconol H 35 A [937a], [938]
WL 817 [339]

PEG (440) monotallate.
Mapeg Tao-10 [564]

PEG 550.
PGE-550 [397]

PEG 600.
Merpoxen PEG 600 [285]

PEG 600 *(cont'd).*
Nopalcol 600 [252]
PGE-600 [397]

PEG 600 dilaurate.
Alkamuls 600 DL [33]
Cithrol 6 DL [223]
CPH-95-N [379]
Mapeg 600 DL [564]

PEG 600 dioleate.
Alkamuls 600-DO [33]
Chemax PEG 600 DO [170]
Cithrol 6 DO [223]
CPH-96-N, CPH-213-N [379]
Emerest 2665 [289]
Industrol DO-13 [109]
Mapeg 600DO [564]
Nopalcol 6-DO [252]

PEG 600 distearate.
Alkamuls 600 DS [33]
Cithrol 6 DS [223]
CPH-56-N [379]
Mapeg 600 DS [564]
PGE-600-DS [397]
Radiasurf 7454 [658]
Scher Eters [771]

PEG 600 ditallate.
Industrol DT-13 [109]
Mapeg 600 DOT [564]
Nopalcol 6-DTW [252]
Pegosperse® 600 DOT [358]

PEG-600-monoisostearate.
Emerest 2664 [289]
Emerest® 2701 [290]

PEG 600 monolaurate.
Alkamuls 600-ML [33]
Cithrol 6ML [223]
CPH-43-N [379]
Emerest 2661 [289]
Mapeg 600 ML [564]
Nopalcol 6-L [252]
Pegosperse 600 ML [358]
PGE-600-ML [397]

PEG 600 monolaurate, ethoxylated (EO 14).
Industrol ML-14 [109]

PEG 600 monooleate.
Alkamuls 600-MO [33]
Cithrol 6MO [223]
Collemul H6 [37]
CPH-41-N, CPH-127-N, CPH-225-N [379]
Emerest 2660 [289]
Industrol MO-13 [109]
Mapeg 600MO [564]
Nopalcol 6-O [252]
PGE-600-MO [397]
Radiasurf 7404 [658]

PEG 600 monoricinoleate.
Nopalcol 6-R [252]

PEG 600 monostearate.
Alkamuls 600-MS [33]
Cithrol 6MS [223]
CPH-57-N [379]
Emerest 2662 [289]
Lipal 600 S [266]
Mapeg 600MS [564]
Nopalcol 6-S [252]
Pegosperse® 600 MS [258], [358]
Radiasurf 7414 [658]

PEG 600 monotallate.
Mapeg 600 MOT [564]

PEG 600 oleate.
Ethylan A6 [509]

PEG 600 ricinoleate.
CPH-94-N [379]

PEG 600 sesquioleate.
Industrol SO-13 [109]

PEG-600 tall oil dioleate.
Polyfac TDO-14 [917a]

PEG 660 monotallate.
Mapeg TAO-15 [564]

PEG® 700 monotallate.
Pegosperse® 700 MOT, 700 TO [358]

PEG 800.
Merpoxen PEG 800 [285]

PEG 1000.
Merpoxen PEG 1000 [285]
PGE-1000 [397]

PEG 1000 castor oil.
Nopalcol 10-CO, 12-R [252]

PEG 1000 castor oil (hydrogenated).
Nopalcol 10-COH [252]

PEG 1000 dilaurate.
Cithrol 10 DL [223]

PEG 1000 dioleate.
Cithrol 10 DO [223]
CPH-136-N [379]

PEG 1000 distearate.
Cithrol 10 DS [223]
CPH-135-N [379]

PEG 1000 laurate.
Ethylan L10 [509]

PEG 1000 monolaurate.
Cithrol 10ML [223]

PEG 1000 monooleate.
Cithrol 10MO [223]
CPH-81-N [379]

PEG 1000 monostearate.
Cerasynt 840 [895]
Cithrol 10MS [223]
CPH-90-N [379]
Emerest® 2713 [290]
Lipopeg 10-S [525]
Mapeg 1000 MS [564]
Pegosperse 1000 MS [358]
PGE-1000-MS [397]
Varonic 1000MS [796]

PEG 1000 monostearate, ethoxylated (EO 23).
Industrol MS-23 [109]

PEG 1200 castor oil.
Nopalcol 12-CO [252]

PEG 1200 castor oil (hydrogenated).
Nopalcol 12-COH [252]

PEG 1200 monoricinoleate.
Dacospin [252]

PEG 1500.
Merpoxen PEG 1500 [285]
PGE-1500 [397]

PEG 1500 dioleate.
Pegosperse 1500 DO [358]

PEG 1500 monostearate.
Cithrol 15MS [223]
Lipopeg 15-S [525]
Mapeg 1500 MS [564]
Pegosperse 1500 MS [358]
Radiasurf 7417 [658]

PEG 1540 distearate.
Mapeg 1540DS [564]

PEG® 1750 monostearate.
Pegosperse® 1750 MS [358]

PEG 1750 monostearate, ethoxylated (EO 40).
Industrol MS-40 [109]

PEG 1760 monostearate, nonionic.
Mapeg S-40 [564]

PEG 1800 monostearate.
Varonic 1800 MS [796]

PEG 1900 castor oil.
Nopalcol 19-CO [252]

PEG 2000.
PGE-2000 [397]

PEG 3000 monostearate.
Nopalcol 30-S [252]

PEG 3000 monotallate.
Nopalcol 30-TWH [252]

PEG 4000 dioleate.
Kessco Polyethylene Glycol Esters [825]
Pegosperse® 4000 DO [358]

PEG 4000 distearate.
CPH-146-N [379]
Kessco Polyethylene Glycol Esters [825]

PEG 4000 monooleate.
Cithrol 40MO [223]
Kessco Polyethylene Glycol Esters [825]

PEG 4000 monostearate.
Cithrol 40MS [223]
CPH-121-N [379]
Kessco Polyethylene Glycol Esters [825]
Mapeg 4000 MS [564]
Pegosperse 4000 MS [358]

PEG 4400 monostearate, ethoxylated (EO 100).
Industrol MS-100 [109]

PEG 6000.
Merpoxen PEG 6000 [285]

PEG 6000 dilaurate.
Kessco Polyethylene Glycol Esters [825]

PEG 6000 distearate.
Alkamuls 6000 DS [33]
Chimipal DS 6000 [514]
CPH-393-N [379]
Kessco-PEG 6000 Distearat [16]
Kessco Polyethylene Glycol Esters [825]
Lipopeg 6000 DS [525]
Mapeg 6000DS [564]
Plurest DS 150 [109]
Witconol L32-45 [935], [937a], [938]

PEG 6000 monolaurate.
Cithrol 60ML [223]

PEG 6000 monoleate.
Cithrol 60MO [223]
Kessco Polyethylene Glycol Esters [825]

PEG 6000 monostearate.
Kessco Polyethylene Glycol Esters [825]
Mapeg 6000 MS [564]

Pemoline.
Cylert Tablets [3]

Penicillamine.
Cuprimine Capsules [578]
Depen Titratable Tablets [911a]

Penicillin G, benzathine.
Bicillin C-R 900/300, C-R 900/300 in Tubex, C-R Injection, C-R in Tubex, L-A Injection [942]
Permapen Isoject [690]

Penicillin G potassium.
Penicillin G Potassium for Injection USP [817]
Penicillin G Potassium Solution, Tablets [119]
Pentids for Syrup, Tablets, Pentids 400 & 800 Tablets [817]

Penicillin G potassium *(cont'd)*.
Pfizerpen for Injection [690]
SK-Penicillin G Tablets [805]

Penicillin G procaine.
Bicillin 900/300, 900/300 in Tubex, C-R Injection, in Tubex [942]
Crysticillin 300 A.S. & Crysticillin 600 A.S. [817]
Penicillin G Procaine Suspension, Sterile, Vials [523]
Pfizerpen-AS Aqueous Suspension [690]
Wycillin, Wycillin in Tubex, Wycillin & Probenecid Tablets & Injection [942]

Penicillin G sodium.
Penicillin G Sodium for Injection USP [817]

Penicillin (oral).
Cyclapen-W [942]
Omnipen Capsules, Oral Suspension, Pediatric Drops [942]
Pathocil Capsules, Oral Suspension [942]
Pentids for Syrup, Tablets, Pentids 400 & 800 Tablets [817]
Pen-Vee K, for Oral Solution & Tablets [942]
Principen Capsules, for Oral Suspension, with Probenecid Capsules [817]
Unipen Injection, Capsules, Powder for Oral Solution, & Tablets [942]
Veetids for Oral Solution, Tablets [817]
Wymox Capsules & Oral Suspension [942]

Penicillin (repository).
Bicillin C-R 900/300, C-R 900/300 in Tubex, C-R Injection, C-R in Tubex, L-A Injection [942]

Penicillin V potassium.
Beepen-VK Oral Solution, 125 mg/5 ml, 250 mg/5 ml; -VK Tablets, 250 mg, 500 mg [114]
Betapen-VK [134]
Lanacillin VK [510]
Penicillin V Potassium Solution, Tablets [119]
Penicillin VK Powder for Oral Solution & Tablets [769]
Pen-Vee K, for Oral Solution & Tablets [942]
Robicillin VK [737]
SK-Penicillin VK for Oral Solution & Tablets [805]
Uticillin VK® [884]
V-Cillin K for Oral Solution & Tablets [523]
Veetids for Oral Solution, Tablets [817]

Pentabromochlorocyclohexane.
FR-651-A [261]

Pentabromodiphenyl oxide.
DE-71 [369]
Great Lakes DE-71™ [369]

Pentabromo diphenyl oxide/nyacol AGO-40, 2:1.
Nyacol HA-9 [655]

Pentabromoethylbenzene.
Saytex 105 [767a]

Pentachloronitrobenzene.
Botrilex [112]
Brassicol [416]
Terraclor [659]
Tilcarex [112]
Tritisan [416]

Pentachlorophenol.
Dowicide FC-7 [261]
Santoborite [602]
Pentachlorophenol DP-2 [261]

Pentadecalactone.
Exaltex [323]
Exaltolide [323]

Pentadecylbenzene sulfonic acid, high purity branched.
Petrostep A-70 [825]

Pentaerythritol ester.
Camphor Tech (Syn), USP (Syn) [358]
Glycox PETC [358]
Hercoflex 707, 707A [406]
Ichthammol NF [358]
Precipitated Sulfur USP [358]

Pentaerythritol monooleate.
Alkamuls PEMO [33]
Radiasurf 7156 [658]

Pentaerythritol monostearate.
Nikkol PEMS [631]
Radiasurf 7175 [658]

Pentaerythritol rosinate.
Pentalyn A [406]

Penta-erythritol stearate-partial ester.
Rilanit PES [401]

Pentaerythritol tetraabietate.
Foral 105 [406]

Pentaerythritol tetraacrylate.
SR-295 [766]

Pentaerythritol tetra-acrylate/MEHQ inhibitor, 300-400 ppm.
SR-295 [766]

Pentaerythritol tetrabehenate.
Liponate PB-4 [525]
Radia 7514 [658]

Pentaerythritol tetra-C_7.
Radiasyn 7177 [658]

Pentaerythritol tetra-C_8/C_{10}.
Radiasyn 7178 [658]

Pentaerythritol tetra caprylate/caprate.
Crodamol PTC [223]

Pentaerythritol tetraester.
Pentalan [223]

Penta-erythritol tetra-fatty acid ester.
Rilanit PEC 4 [401]

Pentaerythritol tetra isostearate.
Crodamol PTIS [223]

Pentaerythritol tetrakis (B-laurylthiopropionate).
Seenox 412S [934]

Pentaerythritol tetrakis (mercaptopropionate).
Mercaptate Q-43 Ester [192]

Pentaerythritol tetrakis (thioglycolate).
Mercaptate Q-42 Ester [192]

Pentaerythritol tetramethacrylate.
SR-367 [766]

Pentaerythritol tetranitrate.
Cartrax Tablets [739]
Duotrate Plateau Caps, 45 Plateau Caps [551]
Pentaerythritol Tetranitrate Tablets (PETN) Timed Capsules & Tablets [769]
Pentritol [891]
Pentylan Tablets [510]
Peritrate SA, Tablets 10 mg., 20 mg., and 40 mg. [674]
P.E.T.N. S.R. Tablets [344]

Pentaerythritol tetraoleate.
Cyclochem PETO [234]
Liponate PO-4 [525]
Radia 7171 [658]

Pentaerythritol tetrapelargonate.
Emerest 2485 [289]

Pentaerythritol tetrastearate.
Cyclochem PETS [234]
Liponate PS-4 [525]
Radia 7176 [658]

Pentaerythritol triacrylate/MEHQ inhibitor, 300-400 ppm.
SR-444 [766]

Pentaerythritol-tris-(B- (N-azidridinyl) propionate).
XAMA® -7 [215]

Pentaerythrityl hexylthiopropionate.
Mark 2140 [75]

Pentaethylene glycol dodecyl ether.
Nikkol BL-5SY [631]

Pentagastrin.
Peptavlon [97]

Pentahydrate borax.
Three Elephant® V-BOR® [490]

Pentahydroxyethyl stearyl ammonium chloride; cationic.
Genamin K S5 [55]

3,6,6,9,9-Pentamethyl-3-n-butyl-1,2,4,5,tetraoxy cyclononane.
USP-130 [890]

1,1,3,3,5-Pentamethyl-4,6-dinitroindane (q.v.).
Moskene [353]

3,6,6,9,9-Pentamethyl-3-(ethyl acetate)-1,2,4,5-tetraoxy cyclononane.
USP-138 [890]

N,N,N′,N′,N′-Pentamethyl-n-tallow-1,3-propanadiammonium dichloride.
Duoquad T-50 [16]

N,N,N′,N′,N′-Penta methyl-N-tallow-1,3-propanediammonium dichloride; cationic.
Duoquad T-50 [79]

2,4-Pentanedione peroxide.
Lupersol 224 [536]
Trigonox® 40, 44P [16], [653]

Pentaoxyethyl stearyl ammonium chloride.
Genamin KS 5 [416]

Pentasodium amino tris (methylene phosphonate).
Dequest 2006 [602]

Pentasodium diethylenetriamine pentaacetate; anionic.
Kalex Penta [390]

Pentasodium diethylenetriaminepentaacetate aq. sol'n., tech. grade.
Chelon 80 [821]

Pentasodium diethylenetriaminepenta-acetic acid.
Chel DTPA-41 [189]

Pentasodium pentetate.
Chel DTPA-4 [189]
Hamp-ex 80 [364b]
Perma Kleer 140 [663]
Versenex 80 [261]

Pentasodium salt D.T.P.A..
Tetralon B [37]

Pentasodium triphosphate.
Empiphos STP, STP Gran./M, STP/D, STP/L [548]

Pentazocine hydrochloride.
Talacen [932]
Talwin Compound, Talwin Nx [932]

Pentazocine lactate.
Talwin Injection [932]

Pentetic acid.
Chel DTPA [190]
Hamp-ex Acid [364b]

PEO (30) POP (6) decylteradecylether.
Nikkol PEN-4630 [631]

Pepsin.
Donnazyme Tablets [737]
Entozyme Tablets [737]
Enzobile Improved Formula [541]
Enzypan [649]
Kanulase [649]
Muripsin [649]
Zypan Tablets [819]

Perc-biphenyl-benzoate; nonionic.
Amacarrier CE [48]
Perkare [705]
PPG Perchlor [705]

Perchlorethylene, emulsified.
Hipochem AS [410]
Rexoclean PCL [292]

Perchloroethylene.
Perk [821]

Perchloroethylene, emulsified.
Marvanol Carrier PK [552]

Perchloropentacyclodecane.
Dechlorane [421]

Peroxide preparations.
Debrox Drops [551]
Gly-Oxide Liquid [551]

Perphenazine.
Etrafon Tablets [772]
Triavil [578]
Trilafon Tablets, Repetabs Tablets, Concentrate & Injection [772]

Pertussis immune globulin (human).
Pertussis Immune Globulin (Human) Hypertussis [232]

Petrolatum.
Fonoline [934]
Fybrene [934]
Markpet [934]
Markpet I [75]
Mineral Jelly No. 5, No. 10, No. 15, No. 20, No. 25 [684]
Pennsoline Soft White, Soft Yellow [684]
Penreco Amber, Blond, Cream, Frost, Lilly, Regent, Royal, Snow, Super, Ultima, White [684]
Perfecta [936]
Protopet [934]
Protopet, Alba, White 3C, White 2L, White 1S, Yellow 2A, Yellow 1E [936]
Snow White Petrolatum [826]
Sonojell #4, #9 [936]

Petroleum distillate.
Penreco No. 2251 Oil, No. 2263 Oil [684]

Petroleum naphthas.
Lacolene [136]

Petroleum oil, ASTM type 102, aromatic.
Flexon 340, 391 [306]

Petroleum oil, ASTM type 104B, paraffinic.
Flexon 815, 845 [306]

Petroleum sulfonate.
Aristonate 430, 460, 500 [697]
Petrostep 420, 465, HMW, MMW [825]

Petroleum sulfonate of molecular weight 340/360.
Pyronate 40 [936]

Petroleum wax, fully refined.
Sun Wax, Anti-Chek [835]

PGMS.
PMS-33 [397]

Phenacemide.
Phenurone [3]

Phenacetin.
A.P.C. with Codeine, Tabloid brand [139]
Propoxyphene Compound 65 [769]
Soma Compound, Compound w/ Codeine [911a]

Phenalopyridine HCL.
Urogesic [281]

Phenazopyridine.
Azotrex Capsules [134]
Urobiotic-250 [739]

Phenazopyridine hydrochloride.
Azo Gantanol Tablets, Gantrisin Tablets [738]
Papd [19a]
Pyridium, Pyridium Plus [674]
Sul-Azo Tablets [512]
Thiosulfil Duo-Pak, Thiosulfil-A Forte [97]
Urogesic Tablets [281]

Phendimetrazine.
Phenazine Tablets & Capsules [518]
PT 105 Capsules [518]

Phendimetrazine tartrate.
Bacarate Tablets [723]

Phendimetrazine tartrate *(cont'd)*.
Bontril PDM, Slow Release [153]
Dyrexan-OD Capsules [868]
Hyrex-105 [436]
Melfiat Tablets, 105 Unicelles [723]
Plegine [97]
Prelu-2 Timed Release Capsules [125]
Slyn-LL Capsules, Capsules CIII [281]
SPRX-105 Capsules [723]
Trimcaps [563]
Trimstat Tablets [513]
Trimtabs [563]
Wehless-105 Timecelles [394]
X-Trozine Capsules & Tablets, LA-105 Capsules [726]

Phenelzine sulfate.
Nardil [674]

Phenindamine tartrate.
Nolahist [153]
Nolamine Tablets [153]
P-V-Tussin Syrup, Tablets [723a]

Pheniramine maleate.
Citra Forte Capsules, Syrup [132]
Dristan Nasal Spray, Regular & Menthol [922]
Fiogesic Tablets [258]
Poly-Histine-D Capsules, Elixir [124]
Ru-Tuss Expectorant, Plain, with Hydrocodone [127]
S-T Forte Syrup & Sugar-Free [779]
Triaminic Juvelets, Oral Infant Drops, TR Tablets (Timed Release) [257]
Verstat Capsules [768]

Phenmetrazine hydrochloride.
Preludin Endurets, Tablets [125]

Phenobarbital.
Antispasmodic Capsules, Elixir & Tablets [769]
Antrocol Tablets, Capsules & Elixir [703]
Arco-Lase Plus [72]
Bronkolixir, Bronkotabs [133]
Chardonna-2 [746]
Isordil (10 mg.) w/Phenobarbital (15 mg.) [463]
Isuprel Hydrochloride Compound Elixir [133]
Levsin/Phenobarbital Tablets, Elixir & Drops [499]
Levsinex/Phenobarbital Timecaps [499]
Luminal [932]
Mudrane GG Elixir, GG Tablets, Tablets [703]
Phazyme-PB Tablets [719]
Phenobarbital Elixir, Tablets [751], [769]
Phenobarbital Tablets [240]
Primatene Tablets-P Formula [922]
Pro-Banthine w/Phenobarbital [782]
Quadrinal Tablets & Suspension [495]
SK-Phenobarbital Tablets [805]
Solfoton Tablets & Capsules [703]
T-E-P Tablets [769]
Theofedral Tablets [240]
Valpin 50-PB [269]

Phenobarbital sodium.
Phenobarbital Sodium Injection [287]
Phenobarbital Sodium in Tubex [942]

Phenobarbital; theophyllin, hydrous, + ephedrin HCl; 8 mg:.12 Gm:25 mg.
Thalfed® Tablets [114]

Phenol.
Anbesol Gel Antiseptic Anesthetic, Liquid Antiseptic Anesthetic [922]
Castellani Paint [679]
Chloraseptic Liquid, Lozenges [709]
Derma Cas Gel [412]
Osti-Derm Lotion [679]

Phenol, alkoxylated aromatic.
Poly-Tergent® LF-405 [659]

Phenol, high molecular weight.
Topanol CA [438]

Phenol, hindered.
Mastermix Stabilizer B 1017 MBB [392]
Stabilite 49-466-Antioxidant, 49-467-Antioxidant, 49-470-Antioxidant [722]

Phenol, alkylated.
Nevastain 21, A, B [623]

Phenol, alkylated.
Stabilite 49-455-White liquid Antioxidant [722]

Phenol alkyl sulfonic ester.
Mesamoll [112]

Phenol, butylated, octylated.
Wingstay T [362]

Phenol ethoxylate.
Carsonon P-30M [529]

Phenol, ethoxylated.
Phenoxy-ethanol [290]

Phenol, ethoxylated (EO 5).
Iconol P-5 [109]

Phenol, ethoxylated (EO 6).
Iconol P-6 [109]

Phenol ethoxylate, phosphate ester.
Emphos TS-230 [935]

Phenol=formaldehyde condensate.
Corephen 10 [309]

Phenol-formaldehyde condensate, carboxylic and phenolic.
Duolite CS-100 [252]

Phenol hindered phosphited.
Wytox 604 [659]

Phenol, oxyethylated.
Aktaflo-S [641]

Phenol polyalkylated.
Stabiwhite Powder (49-454) [379]

Phenol polyglycol ether.
Rewopal MPG 12, 40 [725a]

Phenol, styrenated.
Naugard SP [881]
Nevastain 21 [623]
Prodox® 120 [315]
Stabilite 49-464-SP Antioxidant [722]
Wingstay S [362]

Phenol sulfonic acid.
Eltesol PSA [25]
Eltesol PSA65 [23]

Phenolphthalein.
Agoral, Raspberry & Marshmallow Flavors [674]
Evac-Q-Kit, Evac-Q-Kwik [7]
Evac-U-Gen [911]
Prulet [593]
Sarolax [765]
Trilax [267]

Phenothiazine derivatives.
Compazine [805]
Largon in Tubex [942]
Phenergan Expectorant Plain (without Codeine), Expectorant w/Codeine, Expectorant, w/Dextromethorphan, Pediatric, Injection, in Tubex, Tablets, Syrup & Rectal Suppositories, VC Expectorant Plain (without Codeine), VC Expectorant w/Codeine [942]
Sparine Injection in Tubex [942]
Stelazine [805]
Temaril [805]
Thorazine [805]

Phenoxy compounds.
Iconol™ WA-1, WA-2, WA-4 [109]

Phenoxybenzamine hydrochloride.
Dibenzyline Capsules [805]

3-Phenoxybenzyl-d-cis, trans chrysanthemate.
Multicide Sumithrin [832]

3-Phenoxybenzyl-2-(2,2-dichloro-vinyl)-3,3-dimethyl cyclopropane-1-carboxylate.
Outflank [794]

3-Phenoxybenzyl 2-(2,2-dichlorovinyl)-3,3-dimethyl cyclopropane-1-carboxylate permethrin.
Outflank [794]

3-Phenoxybenzyl-2,2-dimethyl-3-(2,2-dichlorovinyl) cyclopropane carboxylate.
Outflank [794]
Permethrin [794]
Talcord [794]

Phenoxyethanol.
Dowanol EPH [261]
Emery 6705 [289]
Emthox 6705 [290]
Phenoxetol [632]
Rewopal MPG [725]

Phenoxyisopropanol.
Dowanol PPH [261]
Propylene Phenoxetol [632]

3-(Phenoxyphenyl) methyl ± cis, trans-3-(2,2-dichloroethenyl)-2,2-dimethyl cycloprane-carboxylate, trans: cis = 60:40.
Ambush Ectiban [444]
Eksmin [832]
Kafil Perthrine [444]
Pounce [331a]

Phenoxypoly (ethyleneoxy) ethanol.
Igepal OD-410, RC-520, RC-620, RC-630 [338]

Phensuximide.
Milontin Kapseals [674]

Phentermine hydrochloride.
Adipex-P Tablets [519]
Fastin Capsules, 30 mg. [114]
Oby-Trim 30 Capsules [726]
Phentermine HCl Capsules [769]
Teramine Capsules [518]

D-Phenylalanine.
Endorphenyl [872]

Phenylazo-diamino-pyridine HCl.
Phenazodine Tablets [510]

Phenylazo-2-naphthol.
Oil Orange [270]

Phenylbis [1- (2-methyl) aziridinyl] phosphine oxide (q.v.).
Phenyl MAPO [452]

Phenylbutazone.
Azolid Capsules & Tablets [891]
Butazolidin Capsules & Tablets [340]
Phenylbutazone Capsules [344]
Phenylbutazone Capsules & Tablets [769]
Phenylbutazone Tablets [240], [769]

N-Phenyl-N′ cyclohexyl-p-phenylenediamine.
Flexzone 6H [881]

Phenyl didecyl phosphite.
Mark PDDP [934]

Phenyl diisodecyl phosphite.
Weston PDDP [129]

Phenyl dimethicone.
Dow Corning 556 Fluid [262]
Silicone 556 Fluid [262]

N-Phenyl-N′ (l,3 dimethyl butyl) ppheny-lene diamine.
Flexzone 7L or 7F [881]
UOP 588 [883]

3-Phenyl-1,1-dimethylurea.
Beet Kleen [792]
Dozer [423]
Fikure 62-U [318]

m-Phenylenediamine.
Rodol MPD [531]

p-Phenylene diamine.
Rodol D [531]

m-Phenylenediamine sulfate.
Rodol MPDS [531]

p-Phenylenediamine sulfate.
Rodol DS [531]

N,N′-m-Phenylenedimaleimide.
HVA-2 [269]

Phenylephrine.
Donatussin Drops [513]
Quadrahist Pediatric Syrup, Syrup & Timed Release Tablets [769]
Sinovan Timed [267]

Phenylephrine bitartrate.
Duo-Medihaler [734]

Phenylephrine hydrobromide.
Albatussin [105]
Anatuss Tablets & Syrup [563]

Phenylephrine hydrochloride.
Brocon C.R. Tablets [334]
Bromphen Compound Elixir - Sugar Free, Compound Tablets, DC Expectorant, Expectorant [769]
Codimal DH, DM, PH [166]
Colrex Compound Capsules, Elixir [750]
Comhist LA Capsules, Liquid, Tablets [652]
Congespirin [134]
Coryban-D Cough Syrup [690]
Dallergy Syrup, Tablets and Capsules [513]
Dimetapp Elixir, Extentabs [737]
Donatussin DC Syrup [513]
Dristan, Advanced Formula Decongestant/Antihistamine/Analgesic Tablets, Nasal Spray, Regular & Menthol [922]
Dura Tap-PD, Dura-Vent/DA [271]
E.N.T. Syrup, Tablets [814]
E-Tapp Elixir [281]
Entex Capsules, Liquid [652]
Extendryl Chewable Tablets, Sr. & Jr. T.D. Capsules, Syrup [326]
4-Way Nasal Spray [134]
Histalet Forte Tablets [723a]
Histaspan-D Capsules, -Plus Capsules [891]
Histor-D Timecelles [394]
Hycomine Compound [269]
Korigesic Tablets [868]
Naldecon [134]
Neo-Synephrine Hydrochloride 1% Injection, Hydrochloride Ophthalmic [932]
Neotep Granucaps [723]
Ocu-Phrin [655a]
P-V-Tussin Syrup [723a]
Pediacof [133]
Phenergan VC Expectorant Plain (without Codeine), VC Expectorant w/ Codeine [942]
Protid [512]
Quelidrine Syrup [3]
Ru-Tuss Expectorant, Plain, Tablets, with Hydrocodone [127]
Rymed Capsules, -JR Capsules, Liquid, -TR Capsules [281]
S-T Forte Syrup & Sugar-Free [779]
Sine-Aid Sinus Headache Tablets [569]
Singlet [580]
T-Dry Capsules, Jr. Capsules [930]
Tamine S.R. Tablets [344]
Tussar DM [891]
Tympagesic Otic Solution [7]

Phenylephrine hydrochloride (0.12%).
Ocu-Phrin [655a]

Phenylephrine tannate.
Ryantuss Tablets & Pediatric Suspension [911a]
Rynatan Tablets & Pediatric Suspension [911a]

Phenylglyoxylonitrile oxime o,o-diethyl phosphorothioate phoxim.
Bathion [112]
Baythion Volaton [112]

N-Phenyl-N′ isopropyl-p-phenylene diamine.
Vulkanox 4010NA [597a]

Phenyl mercury acetate.
Troysan PMA-10-SEP, PMA-100 [869]

Phenyl mercury acetate solubilized.
Troysan PMA-30 [869]

Phenyl mercury ammonium acetate, 3.5%.
Setrete [869]

Phenyl mercury ammonium propionate.
Merkyl MAP [903]

Phenyl mercury oleate.
PMD-10 [851]
Troysan PMO-30 [869]

Phenyl mercury triethanol ammonium lactate.
Merkyl PM-TL [903]

(ξ5-(Phenyl methyl)-3-furanyl ξ methyl 2,2-dimethyl-3-(2-methyl-1-propenyl) cyclopropanecarboxylate.
Chrysron [832]

N-Phenyl, N,-1 methylheptyl-p-phenylenediamine.
UOP 688 [883]

Phenyl methyl silicone oils, linear.
Abil AV 20, 200, 300, 350, 1000 [856]

Phenyl- α-naphthyl amine.
Additin 30 [597]
Akrochem Antioxidant PANA [15]
Naugard PAN [881]
Vulkanox PAN [597]

Phenyl neopentylene glycol phosphite.
Weston PNPG [129]

2-Phenyl phenol.
Dowicide 1 [261]

o-Phenyl phenol.
Anthrapole 73 [77]
Dowicide 1 [261]

o-Phenyl phenol, aqueous emulsion.
Matexil CA-OPE [442]

o-Phenyl phenol, sodium salt.
Dowicide A [261]

N-Phenyl-p-phenylenediamine.
Rodol Gray B Base, Gray BS [531]

Phenylpropanolamine.
Bromphen Compound Eixir - Sugar Free, DC Expectorant, Expectorant [769]
Cafamine T.D. 2X Capsules [518]
Decongestant Elixir, Expectorant, -AT (Antitussive) Liquid [769]
Dexatrim Capsules, Extra Strength, Plus Vitamins, 18 Hour, 18 Hour Caffeine-Free [860]
Quadrahist Pediatric Syrup, Syrup & Timed Release Tablets [769]
Symptrol TD Capsules [765]
Tuss-Ade Timed Capsules [769]

Phenylpropanolamine hydrochloride.
Anatuss Tablets & Syrup, with Codeine Syrup [563]
Appedrine, Maximum Strength [860]
Bayer Children's Cold Tablets, Cough Syrup for Children [355]
Brocon C.R. Tablets [334]
Bromphen Compound Tablets [769]
Chlorpheniramine (PPA) [19a]
Codimal Expectorant [166]
Comtrex [134]
Conex, Conex with Codeine [661]
Congesprin Liquid Cold Medicine [134]
Control Capsules [860]
Coryban-D Capsules [690]
CoTylenol Children's Liquid Cold Formula [569]
Cremacoat 3, 4 [902]
Dehist [661]
Dexatrim Capsules, Extra Strength, Extra Strength Caffeine-Free [860]
Dieutrim Capsules [518]
Dimetapp Elixir, Extentabs [737]
Dorcol Pediatric Cough Syrup [257]
Dristan, Advanced Formula Decongestant/Antihistamine/Analgesic Capsules [922]
Drize Capsules [86]
Dura Tap-PD, -Vent, -Vent/A [271]
E.N.T. Syrup, Tablets [814]
E-Tapp Elixir [281]
Entex Capsules, LA Tablets, Liquid [652]
4-Way Cold Tablets [134]
Fiogesic Tablets [258]
Head & Chest [709]
Help [900]
Histalet Forte Tablets [723a]
Hycomine Pediatric Syrup, Syrup [269]
Korigesic Tablets [868]
Kronohist Kronocaps [314]
Naldecon, Naldecon-CX Suspension, -DX Pediatric Syrup, -EX Pediatric Drops [134]
Nolamine Tablets [153]
Norel Plus Capsules [888]
Ornade Spansule Capsules [805]
Phenate [541]
Poly-Histine Expectorant Plain, Expectorant with Codeine, -D Capsules, -D Elixir [124]
Prolamine Capsules, Maximum Strength [860]
Propagest & Propagest Syrup [153]
Protid [512]
Resaid T.D. Capsules [344]
Rescaps-D T.D. Capsules [344]
Rhindecon Capsules, -G Capsules [566]
Rhinolar Capsules, -EX Capsules, -EX 12 Capsules [566]
Ru-Tuss Expectorant, Plain, Tablets, II Capsules, with Hydrocodone [127]

Phenylpropanolamine hydrochloride *(cont'd).*
Rymed Capsules, -JR Capsules, Liquid, -TR Capsules [281]
S-T Forte Syrup & Sugar-Free [779]
Sinubid [674]
Sinulin Tablets [153]
T-Dry Capsules, Jr. Capsules [930]
Tamine S.R. Tablets [344]
Tavist-D Tablets [258]
Triaminic Cold Syrup, Cold Tablets, Expectorant, Expectorant w/Codeine, Juvelets, Oral Infant Drops, TR Tablets (Timed Release, -DM Cough Formula, -12 Tablets [257]
Triaminicol Multi-Symptom Cold Syrup, Multi-Symptom Cold Tablets [257]
Tuss-Ornade Liquid, Spansule Capsules [805]

Phenyl salicylate.
Trac Tabs, Tabs 2X [436]

n-Phenyl-N′-1,2,3-thiadiazol-5-yl urea.
Thidiozuron [645]

Phenyltoloxamine.
Quadrahist Pediatric Syrup, Syrup & Timed Release Tablets [769]

Phenyltoloxamine citrate.
Comhist LA Capsules, Tablets [652]
Magsai Tablets [888]
Naldecon [134]
Percogesic Analgesic Tablets [902]
Poly-Histine-D Capsules, Elixir [124]
Sinubid [674]

Phenyltoloxamine dihydrogen citrate.
Kutrase Capsules [499]
Norel Plus Capsules [888]

1-Phenyl -1.2.4-triazoyl-3-(o.o-diethylthiorophosphate).
Hostathion [416]

Phenyl trimethyl polysiloxane.
Abil B 8853 [856]

Phenytoin.
Dilantin Infatabs, -30 Pediatric/Dilantin-1 Suspension [674]

Phenytoin sodium.
Dilantin Kapseals, Parenteral, with Phenobarbital [674]
Phenytoin Sodium Capsules-Prompt Action [769]
Phenytoin Sodium Injection [287]

Phosphate ester, organic.
Marlophor ND-Acid, T 10-Acid [178]

Phosphate rock.
KM® [490]

Phosphate surfactant, potassium salt.
Triton® H-55, H-66 [742]

Phosphatide.
Driltreat [641]

Phosphite.
Mark 217, 1178, 1178B, 1500 [75]
Vanstay SC [894]

Phosphite, phenolic.
Mark 1409, 1409S [75]
Naugard P, PHR [882]

Phosphite stabilizer.
Mark 366, C [74]

Phosphonate.
Arquest 674-S, 675-S [76]

Phosphone alkyl amide.
Pyrovatex CP [190]

Phosphonic acid, organic.
Turpinal SL [401]

Phosphonitrilic fluoroelastomer.
PNF [322]

2-Phosphono-butane-tricarboxylic acid-1,2,4.
Bayhibit-AM [597]

N-(Phosphonomethyl) glycine.
Roundup [601]

Phosphoric acid-(7-chloro-bicyclo- ξ3.2.0ξ hepta-2,6-dien-6-yl)-dimethyl ester.
Hostaquick Ragadan [416]

Phosphoric acid-(7-chloro-bicyclo- ξ3.2.0ξ hepta-2,6-dien-6-yl)-dimethyl ester-(7-chloro-bicyclo ξ3.2.0ξ hepta-2.6-dien-6-yl) dimethyl phosphate (IUPAC).
Hostaquick [416]

Phosphoric acid, complex organic ester.
Chimin P50 [514]
Hostaphat F Brands, KL340 N, KO300 N, KO380, KW340 N [416]

Phosphoric acid ester.
Divalin SA 4 [948]
Phosfetal 201, 204, 600, 601, 602, 603 [948]

Phosphoric acid ester, organic.
Hostaphat L Brands [416]

Phosphoric acid ester, potassium salt.
Dapral AS [16]
G-2203 [92]

Phosphoric acid ester salt.
Primasol NB-NF [109]

Phosphoric acid; free acid, organic ester.
Concofac 972 [209]

Phosphoric acid monoester.
Sokalan® SMK, SNC [106]

Phosphoric acid, partially esterified.
Marlowet 5300, 5301, 5305, 5306 [178]

Phosphorus oxychloride.
Extrema Grade IC Grade [774]

Phosphorus preparations,
K-Phos M.F. (Modified Formula) Tablets, Newutral Tablets, No. 2 Tablets, Original Formula 'Sodium Free' Tablets [113]
Neutra-Phos Powder & Capsules, -K Powder & Capsules [929]
Phos-Flur Oral Rinse Supplement [427]
Thera-Flur Gel-Drops [427]
Thiacide Tablets [113]
Uro-KP-Neutral [820]
Uroqid-Acid Tablets, No. 2 Tablets [113]

Phosphorus tribromide.
Extrema Grade [774]

Phosphotungstomolybdic acid salt.
Atlas A8906 Brilliant Green Toner [496]

Phthalate ester.
Staflex BDP, DIDP [722]

Phthalate esters, emulsified.
Hipochem Jet Dye R-50 [410]

Phthalate, linear.
Palatinol 91P [99]

Phthalic anhydride.
Vulkalent B/C [597]

Phthalic glycerol alkkyl resin.
Triton® B-1956 [742]

Phthalocyanine dye-stuffs.
Heliogen [338]

Physostigmine salicylate.
Antilirium Injectable [661]

Phytonadione.
AquaMEPHYTON [578]
Konakion Injectable [738]

Phytosterols.
Generol [400]

Pilocarpine hydrochloride.
Ocu-Carpine [655a]

Pindolol.
Visken [761]

Pine lignin, purified free acid.
Indulin AT [918]

Piperacetazine.
Quide [580]

Piperacillin sodium.
Pipracil [516]

N,N'- ξ1,4-Piperazinediyl-bis (2,2,2-trichloro ethyl-idene)ξ -bis ξformermideξ.
Funginex [332]
Saprol [162a]

Piperazine preparations.
Antepar [139]

Piperidinium pentamethylene dithiocarbamate.
Accelerator P [392]
Vanax 552 [894]

Piperocaine hydrochloride, U.S.P. (q.v.).
Metycaine [523]

Piperonal.
Heliotropin [12]

Piperonyl butoxide.
A-200 Pyrinate Liquid Gel [646]
Pyrinyl Liquid [769]
Rid Liquid Pediculicide [690]

Piroxicam.
Feldene Capsules [691]

Pitressin tannate.
Pitressin Tannate in Oil [674]

2-Pivalyl-1,3-inandione.
Pindone Pival [609]

Plasma fractions, human.
Antihemophilic Factor (Human) (Factor VIII, AHF, AHG) Koate [232]
Autoplex, Anti-Inhibitor Coagulant Complex Dried [434]
Buminate 5%, Normal Serum Albumin (Human), U.S.P., 5% Solution; 25%, Normal Serum Albumin (Human), U.S.P., 25% Solution [434]
Factor IX Complex (Human) (Factors II, VII, IX, and X) Konyne [232]
Hemofil, Antihemophilic Factor (Human), Method Four, Dried [434]
Hepatitis B Immune Globulin (Human) HyperHep [232]
Hu-Tet, Tetanus Immune Globulin (Human) U.S.P. [434]
Immuglobin [767]
Immune Globulin Intravenous, 5% (In 10% Maltose) Gamimune [232]
Immune Serum Globulin (Human) Gamastan [232]
Immune Serum Globulin (Human), U.S.P. Gamma Globulin [434]
Imogam Rabies Immune Globulin (Human) [579]
Pertussis Immune Globulin (Human) Hypertussis [232]
Plasma-Plex [80]
Proplex, Factor IX Complex (Human) (Factors II, VII, IX & X), Dried [434]

Plasma fractions, human *(cont'd)*.
Proplex SX [434]
Protenate 5%, Plasma Protein Fraction (Human), U.S.P., 5% Solution [434]
Rabies Immune Globulin (Human) Hyperab [232]
Rabies Immune Globulin (Human), Imogam Rabies [579]
Rh_0-D Immune Globulin (Human) HypRho-D, Immune Globulin (Human) HypRho-D Mini-Dose [232]
Tetanus Immune Globulin (Human) Hyper-Tet [232]

Plastic molding powders.
Hi-fax [406]

Plicamycin.
Mithracin [585]

Pneumoccoccal vaccine, polyvalent.
Pneumovax 23 [578]
Pnu-Imune [516]

Podophyllin.
Cantharone Plus [786]
Pod-Ben-25 [199]
Verrex [199]
Verrusol [199]

POE alkyl amide.
Nissan Nymide MT-215 [636]

POE alkyl amine.
Amiet 105, 308, 320 [478]
Atlas G-3780A, G-4961, G-4962 [92]
Consostat DKM [207]
G-3780A, G-4961, G-4962 [92]
Levenol A Conc. [478]
Nissan Nymeen [636]
Pegnol HA-120 [864]

POE alkyl amine ether.
Pegnol OA-400 [864]

POE alkyl aryl ether.
Emulgen 810, 905, 910, 920, 935, 950 [478]
Newcol 607, 610, 614, 623, 704, 707, 710, 714, 723 [635]
Nissan Dispanol N-100, Nonion HS, Nonion NS [636]
Nonionic E-4, E-5, E-6, E-7, E-10, E-12, E-20, E-30 [415]

POE alkyl aryl ether sulfate.
Nissan Trax H-45, N-300 [636]

POE alkyl ester.
SM-151 TH [864]

POE alkyl ether.
Atlox 4873, 4883, 4991, 4995 [92]
Emulgen 106, 120, 147, 210, 220, 306, 320, 408, 420 [478]
Honol GA [849]
Levelene [190]
Levenol PW [478]
Liponox LCF, LCR, N-105, OCS [524]
Neoscoa 203C [864]
Newcol 1515, 1525, 1545 [635]
Nissan Dispanol 16 A, Dispanol LS-100, Dispanol TOC [636]
Renex 702, 703, 704, 707, 711, 714, 720 [92]
Scourol 700 [478]

POE alkyl ether sulfate.
Nissan Trax K-40, Trax K-300 [636]
Sunnol 605 S, 710 H [524]

POE alkyl ether sulfate, sodium salt.
Nikkol NES-203 [631]

POE alkyl ether sulfate, triethanolamine salt.
Nikkol NES-303 [631]

POE alkyl ether, sulfated oil.
Honol 405 [849]

POE alkyl phenol ether.
Emulsit 16, 25, 49, 100 [237]
Liponox NC 6E, NCG, NCI, NCT [524]

POE alkyl phenol phosphate.
Chemfac PC-099 [170]
Geronol PRH/4-A, PRH/4-B, PRH/4-C [550]

POE alkylphenyl ether.
Nonipol BX [764]

POE alkylphenyl ether sulfate.
Sunnol DOS, NES [524]

POE alkylphenyl sulfoacetate.
Newkalgen NX 405 H [849]

POE alkyl phosphate.
Pelex RP [478]

POE amine.
Crodamet [223]
Synthrapol LN-CW [438]

POE aryl ether.
Emulgen A-60, A-500, S-90 [478]
Pycal 94 [445]

POE aryl phenol phosphate.
Geronol TZ/13 [550]

POE aryl phenol phosphate amine salt.
Geronol TZ/14 [550]

POE castor oil.
Arlatone 285, 650 [92]

POE castor oil, hydrogenated.
Arlatone 289 [92]

POE cetyl ether.
Lamacit CA [176]
Newcol 1610, 1620 [635]

POE cetyl ether *(cont'd).*
Nissan Nonion E-205, Nonion E-206, Nonion E-215, Nonion E-220, Nonion E-230, Nonion P-208, Nonion P-210, Nonion P-213, Nonion S-206, Nonion S-207, Nonion S-215, Nonion S-220 [636]
Pegnol C-14, C-18, C-20 [864]

POE cholesterol, ethoxylated, 24 moles.
Solulan C-24 [45]

POE coconut fatty acid.
Atlas G-2109 [445]

POE coconut fatty acid amide.
Profan ME-20 [764]

POE dialkyl ester.
Nissan Nonion DS-60HN [636]

POE dilaurate.
Ionet DL-200 [764]

POE dioleate.
Ionet DO-200, DO-400, DO-600, DO-1000 [764]

POE distearate.
Ionet DS-300, DS-400 [764]

POE dodecyl amine.
Nissan Nymeen L-202, Nymeen L-207 [636]

POE dodecyl monomethyl ammonium chloride.
Nissan Cation L-207 [636]

POE ester.
Atlox 4875 [92]

POE ether.
Ethosperse [358]
Sponto 200 [935]
Swanic D-16, D-24, D-52, D-60, D-80, D-120 [837]

POE ether alcohol.
Chemcol JL-60, JL-120, JL-180 [171]
Renex 30, 36 [438]

POE ether phosphate.
Chemfac PB-082, -106, -109, -135, -184, -804 [170]

POE fatty acid.
Synthrapol SN-SPE [438]

POE fatty acid ester.
Arlatone 975, 980, 983S [92]
Mazon 1045A, 1086, 1096 [564]

POE fatty acid ester, saturated.
Arlacel 989 [92]

POE fatty alcohol.
Arosurf 66-PE-12 [796]
Geropon K/202 [550]

POE fatty alcohol ether.
Lexemul CS-20 [453]
Promulgen D, G [44]
Ritapro 100, 200, 300 [735]

POE fatty amine.
Atlas G-3684 [445]
Milstat N-20 [438]

POE fatty ester.
Crestex Emulsifier S-11 [222]
Emkatex-11, -21 [242]

POE fatty glyceride.
Arlacel 989 [445]
Arlatone 285, 289 [445]
Arlatone G [438], [445]
Atlas G-1292, G-1295 [445]

POE glyceride ester.
Atlas G-1300 [445]

POE glyceride monostearate, EO 20 mol.
Cutina E 24 [401]

POE glycerine sorbitan fatty acid ester.
Arlacel 988 [445]

POE glycerol monoisostearate.
Tagat I, I 2 [360], [856]

POE glycerol monolaurate.
Tagat L, L 2 [360], [856]

POE glycerol monooleate.
Tagat O, O 2 [360], [856]

POE glycerol monoricinoleate.
Tagat R 1 [360], [856]

POE glycerol monostearate.
Tagat S, S 2 [360], [856]

POE glycerol sorbitan fatty acid ester.
Arlacel 988 [92]

POE glycerol trioleate.
Tagat TO [360], [856]

POE glyceryl ester.
Glycerox [223]

POE glycol.
Witbreak DPG-15, DPG-25, DPG-40, DPG-482 [934]

POE glycol, diester.
Geronol 40/BB [550]

POE glycol mono-dioleate.
Lactomul 843 [176]

POE glycol mono-distearate.
Lactomul 461 [176]

POE glycol monolaurate.
Lamacit GML-12, GML 20 [176]
Nissan Nonion L-2, Nonion L-4 [636]

POE glycol monooleate.
Lamacit GMO 25 [176]
Nissan Nonion O-2 [636]

POE glycol oleate.
Cithrol A [223]

POE glycol, propoxylated.
Alkatronic EGE 17-1, EGE 17-2, EGE 17-4, EGE 17-8, EGE 25-1, EGE 25-2, EGE 25-4, EGE 25-5, EGE 25-8 [33], [34]

POE glycol (300), stearate.
Polystate C [339]

POE glycol (400).
Alkapol PEG 400 [33]

POE glycol (400) dilaurate.
Hodag 42-L [415]

POE glycol (400) dioleate.
Hodag 42-O [415]

POE glycol (400) distearate.
Hodag 42-S [415]

POE glycol (400) fatty acid monoester.
PMSA-33 [397]

POE glycol (400) monolaurate.
Hodag 40-L [415]
PML-33 [397]

POE glycol (400) monooleate.
Hodag 40-O [415]
PMO-33 [397]

POE glycol (400) monoricinoleate.
Hodag 40-R [415]

POE glycol (400) monostearate.
Hodag 40-S [415]

POE glycol (600).
Alkapol PEG 600 [33]

POE glycol (600) dioleate.
Hodag 62-O [415]

POE glycol (600) distearate.
Hodag 62-S [415]

POE glycol (600) monolaurate.
Hodag 60-L [415]

POE glycol (600) monostearate.
Hodag 60-S [415]

POE glycol (1000) monostearate.
Hodag 100-S [415]

POE glycol (1500) dioleate.
Hodag 152-O [415]

POE glycol (1500) monostearate.
Hodag 150-S [415]

POE higher alcohol ether.
Emulmin L-380, 40, 50, 70, 140, 240, 862 [764]
Nonipol Soft SS-50, SS-70, SS-90 [764]

POE isohexadecyl ether.
Arlasolve 200 [438]

POE lanolin.
Lanpol 5 [223]
Solan, Solan E, E 50 [223]

POE lanolin alcohol.
Polychol 5, 20-40 [223]

POE laurate.
Atlas G-2109, G-2127 [92]
Atlas G-2127 [445]
CPH-376-N [379]
G-2109, G-2127 [92]
Hallco C-7065 [379]
Newcol 150 [635]

POE lauryl alcohol.
Atlas G-3707 [92]
G-3707 [92]

POE lauryl amine.
Newcol 405, 410, 420 [635]

POE lauryl ether.
Newcol 1100, 1110, 1120, 1203, 1204 [635]
Nissan Nonion DN-202, DN-203, DN-209, K-202, K-203, K-204, K-207, K-211, K-215, K-220, K-230 [636]
Pegnol L-6, L-8, L-10, L-12, L-15, L-20 [864]

POE lauryl ether sulfate.
Sunnol LBN [524]

POE lauryl ether sulphated amino salt.
Soprofor RGS/21 [550]

POE linear alcohol.
Tex-Wet 1048 [460]

POE mono coco-fatty acid ester.
Value-1209C [556]

POE monolaurate.
Nissan Nonion L-2 [636]

POE monooleate.
Ionet MO-200, MO-400 [764]
Nissan Nonion O-2, O-4, O-6 [636]
Value 1407, 1414 [556]

POE monostearate.
Ionet MS-400, MS-600, MS-1000 [764]
Nissan Nonion S-2, S-15 [636]

POE monotallate.
Aconol X6 [390]

POE naphthyl ether.
Newcol B4, B10, B18 [635]

POE nonyl phenol ether.
Renex 647, 649, 679 [92]

POE nonyl phenol ether sulfated, sodium salt.
Octaron PS 80 [711]

POE nonylphenyl ether.
Eleminol HA-100, HA-161 [764]

POE nonylphenyl ether *(cont'd).*
Newcol 506, 508, 560, 561H, 562, 564, 565, 566 [635]
Nissan Nonion NS-202, NS-204.5, NS-206, NS-208.5, NS-209, NS-210, NS-212, NS-215, NS-220, NS-230, NS-240, NS-250, NS-270 [636]
Nonipol 20, 40, 55, 60, 70, 85, 95, 100, 110, 120, 130, 160, 200, 400, D-160 [764]
Value-3706, -3710 [556]

POE octadecyl amine.
Nissan Nymeen S-202, S-204, S-210, S-215, S-220 [636]

POE octyl ether.
Newcol 1010, 1020, 1100, 1105, 1110 [635]

POE octyl phenol ether.
Renex 751 [92]

POE octylphenyl ether.
Newcol 804, 808, 860, 862, 864, 865 [635]
Nissan Nonion HS-204.5, HS-206, HS-208, HS-210, HS-215, HS-220, HS-240, HS-270 [636]
Octapol 60, 100, 300, 400 [764]
Value-3608 [556]

POE oleate.
Newcol 170 [635]

POE oleate/laurate.
Emulsynt 2400 [895]

POE oleyl ether.
Newcol 1120, 1200, 1203, 1204, 1208, 1210 [635]
Nissan Nonion E-205, E-208, E-215, E-230 [636]
Pegnol O-6, O-16 [864]

POE palmitate.
Polymulse 6 [494]

POE phenol phosphate.
Chemfac PC-006 [170]

POE polyamine.
Jeffamine ED-600, ED-900, ED-2001 [854]

POE polyaryl ether.
Soprofor PIN/56 [550]

POE polyol fatty acid ester.
Arlatone T [438]
Rheodol 430, 440, 450 [478]

POE-POP cetyl ether.
Nikkol PBC-31, PBC-32, PBC-33, PBC-34, PBC-41, PBC-42, PBC-43, PBC-44, PBC-44(FF) [631]

POE-POP ether.
Nissan Plonon 102, 104, 108, 171, 172, 201, 204, 208 [636]

POE propylene glycol stearate.
Atlas G-2162 [438]

POE sorbitan (4).
Alkamuls PSML-4 [33]

POE sorbitan ester.
Crillet [223]

POE sorbitan fatty acid ester.
Arlatone 970 [445]
Laxan [507]

POE sorbitan fatty acid ester, saturated.
Arlatone 970 [92]

POE sorbitan laurate.
Atlas G-7596H [445]
Newcol 25 [635]

POE sorbitan monoisostearate.
Crillet 6 [223]

POE sorbitan monolaurate.
Alkamuls PSML-4 [33]
Crillet 1, 41 [223]
Emasol L-106 [478]
Hodag PSML-20 [415]
Ionet T-20 C [764]
Mulsifan RT 141 [948]
Nissan Nonion LT-221 [636]
Sorbon T-20 [864]
Sorgen TW20 [237]

POE sorbitan monolaurate (specially deodorized).
Tween 20 SD [92]

POE sorbitan (20) monolaurate.
Alkamuls PSML-20 [33], [34]

POE sorbitan monooleate.
Alkamuls PSMO-30 [33]
Emasol O-105 R [478]
Hodag PSMO-20 [415]
Ionet T-80 C [764]
Mulsifan RT 146 [948]
Nissan Nonion OT-221 [636]
Sorbon T-80 [864]
Sorgen TW80 [237]

POE sorbitan (5) monooleate.
Alkamuls PSMO-5 [33], [34]

POE sorbitan (20) monostearate.
Alkamuls PSMS-20 [33], [34]
Hodag SVS-18 [415]

POE sorbitan monopalmitate.
Hodag PSMP-20 [415]
Nissan Nonion PT-221 [636]
Sorbon T-40 [864]

POE sorbitan monostearate.
Hodag PSMS-20 [415]
Ionet T-60 C [764]
Nikkol TS-10 [631]
Nissan Nonion ST-221 [636]
Sorbon T-60 [864]
Sorgen TW60 [237]

POE sorbitan (4) monostearate.
Alkamuls PSMS-4 [33], [34]

POE sorbitan monotallate.
Witcomul 1557 [935]

POE sorbitan oleate.
Newcol 85 [635]

POE sorbitan palmitate.
Newcol 45 [635]

POE sorbitan stearate.
Newcol 65 [635]

POE sorbitan trioleate.
Hodag PSTO-20 [415]

POE sorbitan (20) trioleate.
Alkamuls PSTO-20 [33], [34]

POE sorbitan tristearate.
Hodag PSTS-20 [415]

POE sorbitan (20) monooleate.
Alkamuls PSMO-20 [33], [34]
Hodag SVO-9 [415]

POE sorbitan (20) tristearate.
Alkamuls PSTS-20 [33]

POE sorbitan (30) monooleate.
Alkamuls PSMO-30 [33]

POE sorbitol distearate.
Atlas G-1052 [445]

POE sorbitol ester.
Atlox 775, 1045A [438]
Sorbax HO-50, MO-50 [170]

POE sorbitol hexaoleate.
Atlox 1096 [92]
Atlas G-1086 [92], [445]
Atlas G-1096 [92]
G-1086, G-1096 [92]

POE sorbitol laurate.
Atlas G-1045 [445]

POE sorbitol monotallate.
Armul 21 [76]

POE sorbitol oleate.
Atlox 1196 [445]
Atlas G-1186 [445]
Sorbon TR 814, TR 843 [864]

POE sorbitol oleate laurate.
Atlox 1045A [92], [445]

POE sorbitol polyoleate.
Ahco EO-102, EO-114 [438]
Atlas G-1087 [438]

POE sorbitol polyoleate laurate.
Atlas G-1045A [438]

POE sorbitol septaoleate.
Atlox 1087 [438]

POE stearate.
Ahco DHS-111 [438]
Emulvis [379]
Newcol 180, 180T [635]

POE stearoyl ether.
Lamecreme SA 7 [176]

POE stearyl ether.
Newcol 1807, 1820 [635]
Nissan Nonion S-207, S-220 [636]

POE styril phenol.
Geronol M30 P [550]

POE synthetic alcohol.
Dovanox 23H, 23M, 25I, 25N [524]

POE tallow alkyl amine.
Nissan Nymeen T_2-206, T_2-210, T_2-260 [636]

POE tallow alkyl propylene diamine.
Nissan Nymeen DT-203, DT-208 [636]

POE TAM tallow amine.
Trymeen SAM-50, TAM-15, TAM-20, TAM-40 [290]

POE thio ethers.
Siponic-218, -260, SK [29]

POE tridecyl ether.
Newcol 1305, 1310, 1515, 1525 [635]

POE triglyceride.
Atlas 1281, G-1284, G-1285, G-1288, G-1289, G-1292, G-1295, G-1300, G-1304 [92]
Atlox 1285 [92]
G-1281, G-1284, G-1285, G-1288, G-1289, G-1292, G-1295, G-1300, G-1304 [92]

POE (1) lauryl alcohol.
Siponic L 1 [29]

POE (1) monostearate.
Nikkol MYS-1EX [631]

POE (1) nonyl phenol.
Trycol NP-1 [289]

POE (1.5) nonyl phenol.
Chemax NP-1.5 [170]

POE-(2)-cetyl alcohol.
Brij 52 [92]
Ethosperse CA2 [358]

POE (2) cetyl ether.
Brij 52 [438]
CE-55-2 [397]

POE (2) cetyl ether *(cont'd).*
Lipocol C-2 [525]
Nikkol BC-2 [631]
Siponic C 20 [29]

POE (2) cetyl stearyl ether.
Macol CSA-2 [564]

POE (2) coco amine.
Chemeen C-2 [170]
Crodamet 1.C2 [223]
Mazeen C 2 [564]

POE (2) lauryl ether.
Nikkol BL-2 [631]

POE (2) monooleate.
Nikkol MYO-2 [631]

POE (2) monostearate.
Nikkol MYS-2 [631]

POE-(2)-oleyl alcohol.
Brij 92 [92]
Ethosperse® OA 2 [358]
Macol OA-2 [564]

POE (2) oleyl amine.
Crodamet 1.02 [223]

POE (2) oleyl ether.
Brij 92, 93 [438]
Lipocol O-2 [525]
Macol OA-2 [564]
Nikkol BO-2 [631]
OL-55-F-2 [397]

POE (2) oleyl ether with BHA 0.01% and citric acid 0.005%.
Brij 92, 93 [445]

POE (2) sorbitan trioleate.
TO-55-A [397]

POE (2) soya amine.
Mazeen S-2 [564]

POE-(2)-stearyl alcohol.
Brij 72 [92]
Macol SA-2 [564]

POE (2) stearyl amine.
Chemeen 18-2 [170]

POE (2) stearyl ether.
Brij 72 [438]
Lipocol S-2 [525]
Macol SA-2 [564]
ST-55-2 [397]

POE (2) stearyl ether with BHA 0.01% and citric acid 0.005%.
Brij 72 [445]

POE-(2)-synthetic primary alcohol C_{13}/C_{15}.
Renex 702 [92]

POE (2) tallow amine.
Chemeen T-2 [170]
Crodamet 1.T2 [223]

POE (2) tallow amine, hydrogenated.
Crodamet 1.HT2 [223]

POE (2) tridecyl phenol.
Trycol TP-2 [289]

POE (3) C_{12}-C_{15} ether.
Volpo 25 D3 [223]

POE (3) castor oil.
Nikkol CO-3 [631]

POE (3) ceto stearyl ether.
Volpo CS3 [223]

POE (3) coco amine.
Mazeen C-3 [564]

POE (3) lauryl ether.
LA-55-3 [397]
Volpo L3 Special [223]

POE (3) myristyl alcohol.
Emthox 5969 [290]

POE (3) myristyl ether.
Hetoxol M-3 [407]

POE (3) octyl phenol.
Chemax OP-3 [170]

POE (3) octyl-phenyl ether.
Nikkol OP-3 [631]

POE-(3) oleyl ether.
Hetoxyol OA-3 Special [407]
Volpo O3 [223]

POE (3) oleyl ether, distilled.
Volpo N3 [223]

POE (3) sorbitan monoostearate.
Emsorb 6906 [289]

POE-(3)-synthetic primary alcohol C_{13}/C_{15}.
Renex 703 [92]

POE (3) tallow diamine.
Chemeen DT-3 [170]

POE (3) tallow diamine, hydrogenated.
Dicrodamet 1.HT3 [223]

POE (3) tridecyl alcohol.
Macol TD-3 [564]
Siponic TD3 [29]
Trycol TDA-3 [289]

POE (3) tridecyl ether.
Chemal TDA-3 [170]
Lipocol TD-3 [525]
Volpo T3 [223]

POE (4) alkyl ether.
Syntens KMA 40 [397]

POE (4) cetearyl ether.
Hetoxol CS-4 [407]

POE (4) cetyl ether.
Lipocol C-4 [525]

POE (4) cetyl/stearyl alcohol.
Siponic E 2 [29]

POE (4) cetyl stearyl ether.
Macol CSA-4 [564]

POE (4) decyl alcohol.
Trycol DA-4 [289]

POE (4) decyl ether.
Chemal DA-4 [170]

POE (4) 2-ethyl hexyl ether.
Hetoxol CD-4 [407]

POE (4) isostearic acid.
Trydet ISA-4 [289]

POE (4) lauric acid.
Crodet L4 [223]

POE-(4)-lauryl alcohol.
Brij 30 [92]
Emthox 5882 [290]
Ethosperse® LA 4 [358]
Macol LA-4 [564]
Siponic L 4 [29]

POE (4) lauryl ether.
Brij 30 [438]
Brij 30, 30 SP [445]
Chemal LA-4 [170]
LA-55-4 [397]
Lipocol L-4 [525]
Macol LA-4 [564]
Volpo L4 [223]

POE (4) monostearate.
Nikkol MYS-4 [631]

POE (4) myristyl ether.
Lipocol M-4 [525]

POE (4) nonyl phenol.
Chemax NP-4 [170]
Renex 647 [92]
Trycol NP-4 [289]

POE (4) nonyl phenol ether.
Nutrol 622 [198]

POE (4) nonyl phenyl ether.
Macol NP-4 [564]
NP-55-40 [397]

POE (4) oleic acid.
Crodet O4 [223]

POE (4) oleyl alcohol.
Macol OA-4 [564]

POE (4) oleyl ether.
Chemal OA-4 [170]
Macol OA-4 [569]

POE (4) sorbitan monolaurate.
Ahco 7596D [438]
Crillet 11 [223]
Emasol L-106 [478]
Emsorb 6916 [289]
Glycosperse® L4 [358]
Tween 21 [92], [438], [445]

POE (4) sorbitan monostearate.
Ahco DFS-96 [438]
Crillet 31 [223]
Emasol S-106 [478]
Emsorb 6906 [290]
T-Maz 61 [564]
Tween 61 [92], [438], [445]

POE (4) sorbitol hexaoleate.
Crodasorb 40 HO [223]

POE (4) stearic acid.
Crodet S4 [223]

POE (4) stearyl cetyl ether.
Lipocol SC-4 [525]

POE (4) tridecyl alcohol.
Macol TD-4 [564]

POE (4.2) lauryl ether.
Nikkol BL-4.2 [631]

POE (5) behenyl ether.
Nikkol BB-5 [631]

POE (5) C_{12}-C_{15} ether.
Volpo 25 D5 [223]

POE (5) castor oil.
Chemax CO-5 [170]

POE (5) castor oil, hydrogenated.
Chemax HCO-5 [170]
Nikkol HCO-5 [631]
Trylox HCO-5 [289]

POE (5) cetearyl ether.
Hetoxol CS-5 [407]

POE (5) ceto stearyl ether.
Volpo CS5 [223]

POE (5) coco amine.
Chemeen C-5 [170]
Crodamet 1.C5 [223]
Ethomeen CC/15 [79]
Mazeen C-5 [564]

POE (5) coco fatty acid.
Chemax E-200 ML [170]

POE (5) coconut ester, ethoxylated polyol.
Lipal CE 38 [715]

POE (5) glycol cocoate.
Ethofat C/15 [79]

POE (5) glycol oleate.
Ethofat O/15 [79]

POE (5) glycol stearate.
Ethofat 60/15 [79]

POE (5) lanolin acids, distilled.
Lanpol 5 [223]

POE (5) lanolin alcohol.
Nikkol BWA-5 [631]

POE (5) lanolin alcohols, distilled.
Polychol 5 [223]

POE (5) laurate.
Trydet LA-5 [289]

POE-(5)-nonyl phenol.
Renex 648 [92]

POE (5) nonyl phenol ether.
Renex 648 [445]

POE (5) nonyl phenyl ether.
NP-55-50 [397]

POE (5) octadecyl amine.
Ethomeen 18/15 [79]

POE (5) octyl phenol.
Chemax OP-5 [170]

POE (5) oleamide.
Ethomid O/15 [79]

POE (5) oleic acid.
Trydet OA-5 [289]

POE (5) oleic fatty acid.
Chemax E-200 MO [170]

POE (5) oleyl amine.
Crodamet 1.05 [223]
Ethomeen O/15 [79]

POE-(5) oleyl ether.
Hetoxol OA-5 Special [407]
Macol OA-5 [564]
Volpo O5 [223]

POE (5) oleyl ether, (distilled).
Volpo N5 [223]

POE (5)-phytosterol.
Nikkol BPS-5 [631]

POE (5) sorbitan monooleate.
Ahco DFO-100 [438]
Crillet 41 [223]
Emasol O-106 [478]
Emsorb 6901 [289]
Glycosperse O-5 [358]
Liposorb O-5 [525]
Lonzest SMO-5 [529]
Sorbax PMO-5 [170]
T-Maz 81 [564]
Tween 81 [92], [438], [445]

POE (5) soya amine.
Ethomeen S/15 [79]
Mazeen S-5 [564]

POE (5) stearic acid.
Trydet SA-5 [289]

POE (5) stearic fatty acid.
Chemax E-200 MS [170]

POE (5) stearyl alcohol.
Macol SA-5 [564]

POE (5) stearyl amine.
Chemeen 18-5 [170]

POE (5) stearyl ether.
Macol SA-5 [564]

POE (5) tall oil monooleate.
Polyfac TMO-5 [917a]

POE (5) tallow amine.
Chemeen T-5 [170]
Crodamet 1.T5 [223]
Ethomeen T/15 [79]

POE (5) tallow amine, hydrogenated.
Chemeen HT-5 [170]
Crodamet 1.HT5 [223]

POE (5) tallow diamine, hydrogenated.
Dicrodamet 1.HT5 [223]

POE (5) tridecyl ether.
Volpo T5 [223]

POE (5.5) alkyl ether.
Syntens KMA 55 [397]

POE (5.5) cetyl ether.
Nikkol BC-5.5, BC-5.5(FF) [631]

POE (6) cetyl/stearyl alcohol.
Siponic E 3 [29]

POE (6) coco monoethanolamide.
Mazamide C-5 [564]

POE (6) decyl alcohol.
Trycol DA-69 [289]

POE (6) decyl ether.
Chemal DA-6 [170]

POE (6) lauric monoethanolamide.
Mazamide L-5 [564]

POE (6) monooleate.
Nikkol MYO-6 [631]

POE (6) nonyl phenol.
Chemax NP-6 [170]
Renex 697 [92]
Trycol NP-6 [289]

POE (6) nonyl phenol ether.
Renex 697 [445]

POE (6) nonyl phenyl ether.
Macol NP-6 [564]
NP-55-60 [397]

POE-(6)-octyl phenol.
Renex 756 [92]

POE (6) PPG (2.5) C_9-C_{11} alcohols ether.
Hetoxol 916 P [407]

POE (6) PPG (3) tridecyl ether.
Hetoxol TDEP-63 [407]

POE (6) sorbitan monooleate.
Nikkol TO-106 [631]

POE (6) sorbitol.
Atlas G-2240 [445]

POE (6) tridecyl alcohol.
Macol TD-6 [564]
Renex 36 [42]
Siponic TD6 [29]
Trycol TDA-6 [289]

POE (6) tridecyl ether.
Ahcowet DQ-114 [438]
Chemal TDA-6 [170]
Lipocol TD-6 [525]
Renex 36 [445]

POE (6) tridecyl phenol.
Trycol TP-6 [289]

POE (7) cetyl ether.
Nikkol BC-7 [631]

POE (7) lauric acid.
Trydet LA-7 [289]

POE (7) lauryl alcohol.
Siponic L7-90 [29]

POE (7) lauryl ether.
Macol LA-790 [564]

POE (7) nonyl phenol.
Tycol NP-7 [289]

POE (7) octyl phenol.
Chemax OP-7 [170]

POE (7) oleic acid.
Trydet OA-7 [289]

POE (7) oleyl ether.
Nikkol BO-7 [631]

POE (7) stearic acid.
Trydet SA-7 [289]

POE-(7)-synthetic primary alcohol C_{13}/ C_{15}.
Renex 707 [92]

POE (7) tall oil monooleate.
Polyfac TMO-7 [917a]

POE (7.5) castor oil, hydrogenated.
Nikkol HCO-7.5 [631]

POE (7.5) nonylphenyl ether.
Nikkol NP-7.5 [631]

POE (8) alkyl ether.
Syntens KMA 80 [397]

POE (8) dinonyl phenol.
Chemax DNP-8 [170]
Trycol DNP-8 [289]

POE (8) glyceryl mono laurate.
Glycerox L8 [223]

POE (8) lauric acid.
Crodet L8 [223]

POE (8) monostearate.
Pegosperse 350-MS [358]

POE-(8)-nonyl phenol.
Renex 688 [92]

POE (8) nonyl phenol ether.
Nutrol 611 [198]
Renex 688 [445]

POE (8) nonyl phenyl ether.
NP-55-80 [397]

POE (8) oleic acid.
Crodet O8 [223]

POE (8) stearate.
Alkasurf S-8 [33]
Myrj 45 [92], [438]
Pegosperse® 350 MS [358]

POE (8) stearic acid.
Crodet S8 [223]
Trydet SA-8 [289]

POE (8) tallow amine.
Trymeen TAM-8 [289]

POE (8) tallow ester, ethoxylated polyol.
Lipal TE 43 [715]

POE (8) tridecyl alcohol.
Macol TD-8, TD-610 [564]
Trycol TDA-8 [289]

POE (8) triethanol amine.
Syn Fac TDA-92 [588]

POE (8.5) nonyl phenyl ether.
NP-55-85 [397]

POE (9) coco fatty acid.
Chemax E-400 ML [170]

POE (9) decyl ether.
Chemal DA-9 [170]

POE (9) dioleate.
Trydet DO-9 [289]

POE (9) isostearic acid.
Trydet ISA-9 [289]

POE (9) lauryl ether.
Chemal LA-9 [170]
Macol LA-9 [564]
Nikkol BL-9EX, BL-9EX(FF) [631]

POE (9) monolaurate.
Pegosperse 400-ML [358]

POE (9) monooleate.
Pegosperse 400-MO [358]

POE (9) monostearate.
Pegosperse 400-MS [358]

POE (9) nonyl phenol.
Chemax NP-9 [170]
Renex 698 [92]
Trycol NP-9 [289]

POE (9) nonyl phenol ether.
Nutrol 600 [198]

POE (9-9.5) nonyl phenol ether.
Renex 698 [445]

POE (9) nonyl phenyl ether.
NP-55-90 [397]

POE (9) octyl phenol.
Chemax OP-9 [170]
Renex 759 [92]

POE (9) octyl phenol ether.
Nutrol 100 [198]

POE (9) oleic fatty acid.
Chemax E-400 MO [170]

POE (9) oleyl alcohol.
Ethosperse OA-9 [358]

POE (9) oleyl ether.
Chemal OA-9 [170]

POE (9) pelargonic acid.
Trydet MP-9 [289]

POE (9) sesquioleate.
Trydet SO-9 [289]

POE (9) stearate.
RS-55-9 [397]

POE (9) stearic fatty acid.
Chemax E-400 MS [170]

POE (9) tall oil monooleate.
Polyfac TMO-9 [917a]

POE (9) tridecyl alcohol.
Siponic TD9 [29]

POE (9) tridecyl ether.
Chemal TDA-9 [170]
Hetoxol TD-9 [407]

POE (9.5) nonyl phenol.
Syn Fac 905 [588]

POE (9.5) nonyl phenyl ether.
Macol NP-9.5 [564]
NP-55-95 [397]

POE (10) alkyl ether.
Syntens KMA 100 [397]

POE (10) behenyl ether.
Nikkol BB-10 [631]

POE (10) C_{12}-C_{15} ether.
Volpo 25 D10 [223]

POE (10) castor oil.
Etocas 10 [223]
Nikkol CO-10 [631]
Trylox CO-10 [289]

POE (10) castor oil, hydrogenated.
Croduret 10 [223]
Nikkol HCO-10 [631]

POE (10) ceto stearyl ether.
Volpo CS10 [223]

POE-(10)-cetyl alcohol.
Brij 56 [92]

POE (10) cetyl ether.
Brij 56 [438]
Hetoxol CA-10 [407]
Lipocol C-10 [525]
Nikkol BC-10TX [631]

POE (10) cetyl/stearyl alcohol.
Siponic E 5 [29]

POE (10) cetyl/stearyl ether.
Macol CSA-10 [564]

POE (10) coco amine.
Chemeen C-10 [170]
Crodamet 1.C10 [223]
Ethomeen [79]
Mazeen C 10 [564]
Trymeen CAM-10 [289]

POE (10) coconut fatty acids.
Crodet C10 [223]

POE (10) glycol oleate.
Ethofat O/20 [79]

POE (10) glycol stearate.
Ethofat 60/20 [79]

POE (10) glycol tallate.
Ethofat 142/20 [79]

POE (10) lanolin.
Nikkol TW-10 [631]

POE (10) lanolin acids, distilled.
Lanpol 10 [223]

POE (10) lanolin alcohol.
Nikkol BW A-10 [631]

POE (10) lanolin alcohol, acetylated.
Ethoxyol® AC [290]

POE (10) lanolin alcohols, distilled.
Polychol 10 [223]

POE (10) monooleate.
Atlas G2143 [445]
Nikkol MYO-10 [631]

POE (10) monostearate.
Nikkol MYS-10 [631]

POE (10) monotallate.
Polyfac MT-610 [917a]

POE (10) nonyl phenol.
Chemax NP-10 [170]
Renex 690 [92]
Trycol NP-9 [289]

POE (10) nonyl phenol ether.
Nikkol NP-10 [631]
Renex 690 [445]

POE (10) octadecyl amine.
Ethomeen 18/20 [79]

POE-(10)-octyl phenol.
Renex 750 [92]

POE (10) octyl phenyl ether.
Nikkol OP-10 [631]

POE (10) oleic acid.
Trydet OA-10 [289]

POE-(10)-oleyl alcohol.
Brij 96 [92]
Macol OA-10 [564]

POE (10) oleyl amine.
Crodamet 1.010 [223]

POE (10) oleyl ether.
Brij 96, 97 [438]
Hetoxol OA-10 Special, OL-10, OL-10H [407]
Lipocol O-10 [525]
Macol OA-10, OA-20 [564]
Nikkol BO-10TX [631]
OL-55-F-10 [397]
Volpo O10 [223]

POE (10) oleyl ether (distilled).
Volpo N10 [223]

POE (10) oleyl ether with BHA 0.01% and citric acid 0.005%.
Brij 97 [445]

POE (10)-phytosterol.
Nikkol BPS-10 [631]

POE (10) PPG (15) tridecyl ether.
Hetoxol TDEP-15 [407]

POE (10) sorbitan monolaurate.
Atlas G-7596-J [445]
Liposorb L-10 [525]

POE (10) soya amine.
Ethomeen S/20 [79]
Mazeen S 10 [564]

POE-(10)-stearyl alcohol.
Brij 76 [92]
Ethosperse® ST 10 [358]
Macol SA-10 [564]

POE (10) stearyl/cetyl ether.
Lipocol SC-10 [525]

POE (10) stearyl ether.
Brij 76 [438]
Hetoxol STA-10 [407]
Lipocol S-10 [525]
Macol SA-10 [564]

POE (10) tall oil fatty acid.
Chemax TO-10 [170]

POE (10) tallow amine.
Chemeen T-10 [170]
Crodamet 1.T10 [223]

POE (10) tallow amine, hydrogenated.
Crodamet 1.HT10 [223]

POE (10) tallow diamine, hydrogenated.
Dicrodamet 1.HT10 [223]

N,N′,N′-POE (10)-N-tallow-1,3-diaminopropane.
Ethoduomeen T/20 [16]

POE (10) tridecyl alcohol.
Macol TD-10 [564]

POE (10) tridecyl ether.
Volpo T10 [223]

POE (11) alkyl ether.
Syntens QSA 110 [397]

POE (11) coconut ester, ethoxylated polyol.
Lipal CE 55 [715]

POE (11) nonyl phenol.
Trycol NP-11 [289]

POE (11) nonyl phenol ether.
Nutrol 656 [198]

POE-(11)-octyl phenol.
Renex 751 [92]

POE-(11)-synthetic primary alcohol C_{13}/C_{15}.
Renex 711 [92]

POE (12) lauric acid.
Crodet L12 [223]

POE (12) lauryl alcohol.
Crodamet LA-12 [223]
Emthox 5967 [290]
Ethosperse® LA 12 [358]
Macol LA-12 [564]
Siponic L 12 [29]

POE (12) lauryl ether.
Chemal LA-12 [170]
Hetoxol L-12 [407]
Lipocol L-12 [525]
Macol LA-12 [564]

POE-(12)-nonyl phenol.
Renex 682 [92]

POE (12) nonyl phenyl ether.
NP-55-120 [397]

POE (12) oleic acid.
Crodet O12 [223]

POE (12) POP (6) decyltetradecyl ether.
Nikkol PEN-4612 [631]

POE (12) safflower ester, ethoxylated polyol.
Lipal OE 55 [715]

POE (12) stearic acid.
Crodet S12 [223]

POE (12) tallow ester, ethoxylated polyol.
Lipal TE 55 [715]

POE-(12)-tridecyl alcohol.
Renex 30 [92]
Siponic TD12 [29]

POE (12) tridecyl ether.
Ahcowet DQ-145 [438]
Chemal TDA-12 [170]
Lipocol TD-12 [525]
Renex 30 [445]

POE (12.5) nonyl phenyl ether.
Macol NP-12.5 [564]

POE (12.5) tallow amide, hydrogenated.
Ethomid HT/23 [79]

POE-(13)-nonyl phenol.
Renex 679 [92]

POE (14) coco fatty acid.
Chemax E-600 ML [170]

POE (14) isostearic acid.
Trydet ISA-14 [289]

POE (14) oleic fatty acid.
Chemax E-600 MO [170]

POE (14) stearic fatty acid.
Chemax E-600 MS [170]

POE (14) tall oil monooleate.
Polyfac TMO-14 [917a]

POE (15) C_{12}-C_{15} ether.
Volpo 25 D15, T15 [223]

POE (15) ceteary1 ether.
Hetoxol 15 CSA, CS-15 [407]

POE (15) ceto stearyl ether.
Volpo CS15 [223]

POE (15) cetyl ether.
Nikkol BC-15TX, BC-15TX(FF) [631]

POE (15) cetyl/stearyl alcohol.
Siponic E 7 [29]

POE (15) cetyl/stearyl ether.
Macol CSA-15 [564]

POE (15) coco amine.
Chemeen C-15 [170]
Crodamet 1.C15 [223]
Ethomeen C/25 [79]
Mazeen C-15 [564]
Trymeen CAM-15 [289]

POE (15) dinonyl phenol.
Chemax DNP-15 [170]

POE (15) glyceryl monolaurate.
Glycerox L15 [223]

POE (15) glycol cocoate.
Ethofat C/25 [79]

POE (15) glycol stearate.
Ethofat 60/25 [79]

POE (15) glycol tallate.
Ethofat 242/25 [79]

POE (15) lanolin alcohols, distilled.
Polychol 15 [223]

POE (15) monotallate.
Polyfac MT-615 [917a]

POE (15) nonyl phenol.
Chemax NP-15 [170]
Renex 678 [92]

POE (15) nonyl phenol ether.
Nikkol NP-15 [631]
Nutrol 640 [198]
Renex 678 [445]

POE (15) nonyl phenyl ether.
NP-55-150 [397]

POE (15) octadecyl amine.
Ethomeen 18/25 [79]

POE (15) oleyl amine.
Crodamet 1.015 [223]
Ethomeen O/25 [79]

POE (15) oleyl ether.
Nikkol BO-15TX [631]
Volpo O15 [223]

POE (15) oleyl ether (distilled).
Volpo N15 [223]

POE (15) soya amine.
Ethomeen S/25 [79]
Hetoxamine S-15 [407]

POE (15) stearyl alcohol.
Macol SA-15 [564]

POE (15) stearyl amine.
Ethomeen HT/25 [16]
Hetoxamine ST-15 [407]

POE (15) stearyl/cetyl ether.
Lipocol SC-15 [525]

POE (15) stearyl ether.
Macol SA-15 [564]

POE (15) tallow amine.
Chemeen T-15 [170]
Crodamet 1.T15 [223]
Ethomeen T/25 [79]
Trymeen TAM-15 [289]

POE (15) tallow amine, hydrogenated.
Chemeen HT-15 [170]
Crodamet 1.HT15 [223]

POE (15) tallow diamine.
Chemeen DT-15 [170]

N,N',N'-POE (15)-N-tallow-1,3-diaminopropane.
Ethoduomeen T/25 [16]

POE (15) N-tallow 1,3,propylene diamine.
Trymeen 6640 [289]

POE (15) tridecyl ether.
Renex 31 [445]
Volpo T15 [223]

POE (16) castor oil.
Chemax CO-16 [170]
Trylox CO-16 [289]

POE (16) castor oil, hydrogenated.
Chemax HCO-16 [170]
Trylox HCO-16 [289]

POE-(16)-cetyl alcohol.
Atlas G-3816 [92]
G-3816 [92]

POE (16) coconut ester, ethoxylated polyol.
Lipal CE 64 [715]

POE (16) lanolin alcohol.
Ethoxyol® 16 [290]

POE (16) octyl phenol alcohol.
Siponic F160 [29]

POE (16) sorbitan tristearate.
Atlas G-7166P [445]

POE (16) tall oil fatty acid.
Chemax TO-16 [170]

POE (17) nonyl phenyl ether.
NP-55-300 [397]

POE (17) safflower ester, ethoxylated polyol.
Lipal OE 64 [715]

POE (17) sorbitan trioleate.
TO-55-EL [397]

POE (18) nonyl phenol ether.
Nikkol NP-18TX [631]

POE (18) sorbitan trioleate.
TO-55-E [397]

POE (18) tridecyl alcohol.
Trycol TDA-12, TDA-18 [289]

POE (20) alkyl ether.
Syntens KMA 200 [397]

POE (20) behenyl ether.
Nikkol BB-20 [631]

POE (20) C_{12}-C_{15} ether.
Volpo 25 D20, T20 [223]

POE (20) castor oil.
Etocas 20 [223]
Nikkol CO-20TX [631]

POE (20) castor oil, hydrogenated.
Nikkol HCO-20 [631]

POE (20) ceto stearyl ether.
Cetomacrogol 1000 BP [223]
Volpo CS20 [223]

POE-(20)-cetyl alcohol.
Brij 58 [92]

POE (20) cetyl ether.
Brij 58 [438]
CE-55-20 [397]
Lipocol C-20 [525]
Nikkol BC-20TX, BC-20TX(FF) [631]

POE (20) cetyl/stearyl ether.
Macol CSA-20 [564]

POE (20) glycerol monostearate.
Aldosperse MS-20, MS-20 FG [358]
Radiasurf 7000 [658]

POE (20) GMS/GMS hi mono, 40:60.
Aldosperse 40/60 FG [358]

POE-(20) isohexadecyl alcohol.
Arlasolve 200 [92]

POE (20) isohexadecyl ether.
Arlasolve 200 [445]

POE (20) lanolin.
Nikkol TW-20 [631]

POE (20) lanolin acids, distilled.
Lanpol 20 [223]

POE (20) lanolin alcohol.
Nikkol BWA-20 [631]

POE (20) lanolin alcohols, distilled.
Polychol 20 [223]

POE (20) mannitan monolaurate.
Atlas G-9046T [445]

POE (20) methyl glucoside sesquistearate.
Glucamate SSE-20 [44]

POE (20) nonyl phenol.
Chemax NP-20 [170]
Renex 649 [92]
Trycol NP-20 [289]

POE (20) nonyl phenyl ether.
Macol NP-20 [564]

POE-(20)-oleyl alcohol.
Brij 98 [92]
Ethosperse® OA 20 [358]
Macol OA-20 [564]

POE (20) oleyl ether.
Ahco 3998 [438]
Brij 98, 99 [438], [445]
Chemax OA-20, OA-20/70 [170]
Hetoxol OA-20 Special, OL-20 [407]
Lipocol O-20 [525]
Macol OA-20 [564]

POE (20) oleyl ether *(cont'd)*.
Nikkol BO-20 [631]
OL-55-F-20 [397]
Volpo O20 [223]

POE (20) oleyl ether (distilled).
Volpo N20 [223]

POE (20) palmitate.
Atlas G-2079 [445]

POE (20)-phytosterol.
Nikkol BPS-20 [631]

POE (20) POP (6) decyltetradecyl ether.
Nikkol PEN-4620 [631]

POE (20) sorbital monolaurate.
Accosperse 20 [829]

POE (20) sorbitan monoisostearate.
Crillet 6 [223]
Emsorb 6912 [290]
T-Maz 67 [564]

POE-(20) sorbitan monolaurate.
Accosperse 20 [81]
Ahco 7596T [438]
Armotan PML 20 [16]
Crillet 1 [223]
Drewmulse POE-SML [715]
Emasol L-120 [478]
Emsorb 6915 [289]
Glycosperse L-20 [358]
Industrol L-20-S [109]
Lonzest SML-20 [529]
Nikkol TL-10, TL-10(FF) [631]
Sorbax PML-20 [170]
T-Maz 20 [564]
Tween 20 [438]

POE (20) sorbitan monolaurate, anhydrous.
Glycosperse L-20X [358]

POE (20) sorbitan monolaurate (polysorbate 20).
Capmul POE-L [829]
Tween 20 [92], [438], [445]

POE-(20) sorbitan monooleate.
Accosperse 80 [81]
Ahco DFO-150 [438]
Amerchol Polysorbate 80 [44]
Armotan PMO 20 [16]
Crillet 4 [223]
Drewmulse POE-SMO [715]
Drewpone 80 [715]
Durfax 80, 80K [272]
Emasol O-120 [478]
Emsorb 6900 [289]
Flo Mo SMO-20 [247]
Glycosperse O-20 [358]
Industrol O-20-S [109]
Lonzest SMO-20 [529]
Nikkol TO-10, TO-10(FF) [631]
Sorbax PMO-20 [170]
T-Maz 80 [564]
Tween 80 [438]

POE (20) sorbitan monooleate, anhydrous.
Glycosperse O-20X [358]

POE (20) sorbitan monooleate (polysorbate 80).
Capmul POE—O [829]
T-Maz 80 [564]
Tween 80 [92], [438], [445]

POE-(20)-sorbitan monooleate, tech.
Atlas G-4905 [92]
G-4905 [92]

POE (20) sorbitan monooleate, vegetable grade.
Glycosperse O-20 VEG [358]

POE (20) sorbitan monopalmitate.
Ahco DFP-156 [438]
Crillet 2 [223]
Emsorb 6910 [289]
Emasol P-120 [478]
Glycosperse P-20 [358]
Liposorb P-20 [525]
Lonzest SMP-20 [529]
Sorbax PMP-20 [170]
T-Maz-40 [564]
Tween 40 [445]

POE (20) sorbitan monopalmitate (polysorbate 40).
Tween 40 [92], [438]

POE-(20) sorbitan monostearate.
Accosperse 60 [81]
Ahco DFS-149 [438]
Alkamuls PSMS-20 [33]
Armotan PMS20 [16]
Crillet 3 [223]
Drewmulse POE-SMS [715]
Drewpone 60 [715]
Durfax 60K [272]
Emasol S-120 [478]
Emsorb 6905 [289]
Glycosperse S-20 [358]
Incrosorb S-60 [223]
Industrol S-20-S [109]
Lonzest SMS-20 [529]
Nikkol TS-10, TS-10(FF) [631]
Sorbax PMS-20 [170]
T-Maz-60 [564]
Tween 60 [438], [445]

POE (20) sorbitan monostearate (polysorbate 60).
Capmul POE-S [829]
Durfax 60 [272]
T-Maz 60 [564]
Tween 60 [92], [438]

POE (20) sorbitan monotallate.
Flo Mo SMT-20 [247]
T-Maz 90 [564]

POE (20) sorbitan trioleate.
Ahco DFO-110 [438]
Crillet 45 [223]
Emasol O-320 [478]
Emsorb 6903 [289]
Glycosperse TO-20 [358]
Lonzest STO-20 [529]
Nikkol TO-30 [631]
Sorbax PTO-20 [170]
T-Maz 85 [564]
Tween 85 [92]

POE (20) sorbitan trioleate, (polysorbate 85).
TO-55-F [397]
Tween 85 [438], [445]

POE (20) sorbitan tristearate.
Ahco 7166T [438]
Alkamuls PSTS-20 [33]
Crillet 35 [223]
Drewmulse POE-STS [715]
Drewpone 65 [715]
Durfax 65K [272]
Emasol S-320 [478]
Emsorb 6907 [289]
Glycosperse TS-20 [358]
Industrol STS-20-S [109]
Lonzest STS-20 [529]
Nikkol TS-30 [631]
Sorbax PTS-20 [170]
T-Maz 65 [564]

POE (20) sorbitan tristearate (polysorbate 65).
Durfax 65 [272]
T-Maz 65 [564]
TS-55-F [397]
Tween 65 [92], [438]

POE (20) sorbitan tritallate.
Polyfac SEE-340 [917a]

POE (20) sorbitan tritallato.
T-Maz 95 [564]

POE (20) sorbitol.
Atlas G-2320 [445]
Ethosperse SL-20 [358]
Trylox SS-20 [289]

POE-(20)-stearate.
Myrj 49 [92]

POE-(20)-stearyl alcohol.
Brij 78 [92]
Macol SA-20 [564]
Trycol SAL-20 [289]

POE (20) stearyl/cetyl ether.
Lipocol SC-20 [525]

POE (20) stearyl ether.
Brij 78 [438]
Hetoxol STA-20 [407]
Lipocol S-20 [525]
Macol SA-20 [564]
ST-55-20 [397]

POE (20) stearyl ether with BHA 0.01% and citric acid 0.005%.
Brij 78 [445]

POE-(20)-synthetic primary alcohol C_{13}/C_{15}.
Renex 720 [92]

POE (20) tallow amine.
Chemeen T-20 [170]

POE (20) tridecyl ether.
Volpo T20 [223]

POE (21) coconut ester, ethoxylated polyol.
Lipal CE 71 [715]

POE (21) lauryl ether.
Nikkol BL-21 [631]

POE-(21)-stearyl alcohol.
Brij 721 [92]

POE (21) stearyl ether.
Brij 721 [438]

POE (21) stearyl ether with (BHA and citric acid, .01:.005%).
Brij 721 [438]

POE (22) tallow ester, ethoxylated polyol.
Lipal TE 70 [715]

POE (23) cetyl ether.
Nikkol BC-23, BC-23(FF) [631]

POE (23) glyceryl monolaurate.
Aldosperse™ ML 23 [358]

POE-(23)-lauryl alcohol.
Brij 35 [92]
Ethosperse® LA 23 [358]
Macol LA-23 [564]
Trycol LAL-23 [289]

POE (23) lauryl ether.
Brij 35 [438], [445]
Chemal LA-23 [170]
LA-55-23 [397]
Lipocol L-23 [525]
Macol LA-23 [564]
Volpo L23 [223]

POE (23) lauryl ether with BHA 0.01% and citric acid 0.005%.
Brij 35 SP [445]

POE (23) oleyl alcohol.
Trycol OAL-23 [289]

POE (23) safflower ester, ethoxylated polyol.
Lipal OE 70 [715]

POE (23) stearic acid.
Trydet SA-23 [289]

POE (23) stearic fatty acid.
Chemax E-1000MS [170]

POE (24) lanolin, hydrogenated.
Ethoyxol® 24 [290]

POE (24) lauric acid.
Crodet L24 [223]

POE (24) oleic acid.
Crodet O24 [223]

POE (24) stearic acid.
Crodet S24 [223]

POE (25) alkyl ether.
Syntens QSA 250 [397]

POE (25) C_{12}-C_{15} ether.
Volpo 25D20 [223]

POE (25) castor oil.
Chemax CO-25 [170]
Trylox CO-25 [289]

POE-(25)-castor oil, hydrogenated.
Arlatone G [92]
Chemax HCO-25 [170]
Mapeg CO-25H [564]

POE (25) cetyl ether.
Nikkol BC-25TX [631]

POE (25) glyceride ester.
Syn Lube 107 [588]

POE (25) lauryl ether.
Nikkol BL-25 [631]

POE (25) monostearate.
Nikkol MYS-25, MYS-25(FF) [631]

POE (25) oleyl ether.
Siponic Y500-70 [29]

POE (25)-phytosterol.
Nikkol BPS-25 [631]

POE (25) propylene glycol monostearate.
Atlas G-2160 [445]

POE (25) propylene glycol stearate.
Atlas G-2162 [445]

POE (25) tridecyl ether.
Hetoxol TD-25 [407]

POE (26) glycerol.
Ethosperse G-26 [359]

POE (28) castor oil.
Chemax CO-28 [170]

POE (30) behenyl ether.
Nikkol BB-30 [631]

POE (30) castor oil.
Chemax CO-30 [170]
Etocas 30 [223]

POE (30) castor oil, hydrogenated.
Croduret 30 [223]
Nikkol HCO-30, HCO-30(FF) [631]

POE (30) cetyl ether.
Nikkol BC-30TX [631]

POE (30) cetyl/stearyl alcohol.
Siponic E 15 [29]

POE (30) decyl tetradecyl ether.
Nikkol BEG-2430 [631]

POE (30) 2-hexyldecyl ether.
Nikkol BEG-1630 [631]

POE (30) lanolin.
Nikkol TW-30 [631]

POE (30) nonyl phenol.
Chemax NP-30, NP-30/70 [170]
Renex 650 [92]
Trycol NP-30, NP-307 [289]

POE (30) nonyl phenol ether.
Renex 650 [445]

POE (30) nonyl phenyl ether.
Macol NP-30 [564]

POE (30) octylphenyl alcohol.
Siponic F300 [29]

POE (30) octylphenyl ether.
Nikkol OP-30 [631]

POE (30) oleyl amine.
Chemeen O-30 [170]
Trymeen OAM-30/80 [289]

POE (30) oleyl amine, 80%.
Chemeen O-30/80 [170]

POE (30)-phytosterol.
Nikkol BPS-30 [631]

POE (30) POP (6) decyltetradecyl ether.
Nikkol PEN-4630 [631]

POE (30) sorbitol tetraoleate.
Nikkol GO--430 [631]

POE (30) stearate.
Atlas G-2151 [445]
Myrj 51 [92]

POE (30) stearyl ether.
Hetoxol STA-30 [407]

POE (30) tallow diamine.
Chemeen DT-30 [170]

POE (30) tallow ester, ethoxylated polyol.
Lipal TE 76 [715]

POE (31) nonyl phenyl ether.
Macol NP-307 [564]

POE (35) castor oil.
Etocas 35 [223]

POE (35) castor oil, hydrogenated.
Croduret 35 [223]

POE (36) castor oil.
Chemax CO-36 [170]
Trylox CO-36 [289]

POE (40) castor oil.
Chemax CO-40 [170]
Etocas 40 [223], [631]
Nikkol CO-40TX [631]
Trylox CO-40 [289]

POE (40) castor oil, hydrogenated.
Croduret 40 [223]
Nikkol HCO-40, HCO-40(FF) [631]

POE (40) cetyl ether.
Nikkol BC-40TX [631]

POE (40) cetyl/stearyl alcohol.
Macol SA-40 [564]

POE (40) cetyl/stearyl ether.
Incropol CS-40 [223b]

POE (40) glyceryl monolaurate.
Glycerox L40 [223]

POE (40) lanolin alcohol.
Nikkol BWA-40 [631]

POE (40) lanolin alcohols, distilled.
Polychol 40 [223]

POE (40) lauric acid.
Crodet L40 [223]

POE (40) monostearate.
Emerest® 2715 [290]
Mapeg S-40 [564]
Nikkol MYS-40, MYS-40(FF) [631]
Pegosperse 1750-MS [358]

POE (40) nonyl phenol.
Chemax NP-40, NP-40/70 [170]
Trycol NP-40, NP-407 [289]

POE (40) octyl phenol.
Chemax OP-40/70 [170]
Trycol OP-407 [289]

POE (40) octyl phenyl alcohol.
Siponic F400 [29]

POE (40) oleic acid.
Crodet O40 [223]

POE (40) sorbitan hexaoleate.
Glycosperse® HO 40 [358]

POE (40) sorbitan hexatallate.
Glycosperse® HTO 40 [358]

POE (40) sorbitan monolaurate.
Emasol 21 [478]

POE (40) sorbitol hexaoleate.
Atlas G-1086 [438]

POE-(40)-sorbitol septaoleate.
Arlatone T [92], [445]

POE (40) sorbitol tetraoleate.
Nikkol GO-440 [631]

POE (40) stearate.
Atlas G-2198 [445]
Lipopeg 39-S [525]
Myrj 52, 52C, 52S [92]
RS-55-40 [397]

POE (40) stearic acid.
Crodet S40 [223]
Trydet SA-40 [289]

POE (40) stearyl ether.
Macol SA-40 [564]

POE (45) monostearate.
Nikkol MYS-45 [631]

POE (50) castor oil.
Etocas 50 [223]

POE (50) castor oil, hydrogenated.
Nikkol HCO-50, HCO-50(FF) [631]

POE (50) cetearyl alcohol.
Hetoxol CS-50 [407]

POE (50) cetearyl ether.
Hetoxol CS-50 Special [407]

POE (50) nonyl phenol.
Chemax NP-50, NP 50/70 [170]
Trycol NP-50, NP-507 [289]

POE (50) octadecyl amine.
Ethomeen 18/60 [79]

POE (50) oleyl alcohol.
Ethosperse® OA 50 [358]

POE (50) oleyl ether.
Nikkol BO-50 [631]

POE (50) sorbitol hexaoleate.
Atlas G-1096 [438], [445]

POE-(50)-stearate.
Myrj 53 [92], [438]

POE (50) stearic acid.
Trydet SA-50/30 [289]

POE (50) stearyl amine.
Chemeen 18-50 [170]
Hetoxamine ST-50 [407]
Trymeen SAM-50 [289]

POE (50) tallow amine.
Ethomeen T/60 [79]

POE (50) tallow amine, hydrogenated.
Chemeen HT-50 [170]

POE (55) monostearate.
Nikkol MYS-55 [631]

POE (55) sorbitol.
Crodasorb 55 HO [223]

POE (60) castor oil.
Etocas 60 [223]
Nikkol CO-60TX [631]

POE (60) castor oil, hydrogenated.
Croduret 60 [223]
Nikkol HCO-60, HCO-60(FF) [631]

POE (60) cetyl/stearyl ether.
Incropol CS-60 [223]

POE (60) sorbitol tetraoleate.
Nikkol GO-460 [631]

POE (60) sorbitol tetrastearate.
Nikkol GS-460 [631]

POE (70) nonyl phenyl ether.
Macol NP-70 [564]

POE (70) octyl phenyl alcohol.
Siponic F700 [29]

POE (80) castor oil.
Chemax CO-80 [170]
Trylox CO-80 [289]

POE (80) castor oil, hydrogenated.
Nikkol HCO-80 [631]

POE (80) glyceride ester.
Syn Lube 109 [588]

POE (80) sorbitan monolaurate.
T-Maz 28 [564]

POE (100) alkyl ether.
Syntens QSA 1000 [397]

POE (100) castor oil.
Etocas 100 [170]

POE (100) castor oil, hydrogenated.
Croduret 100 [223]
Nikkol HCO-100, HCO-100(FF) [631]

POE (100) lauric acid.
Crodet L100 [223]

POE (100) monostearate.
Emerest® 2717 [290]
Mapeg S-100 [564]

POE (100) nonyl phenol.
Chemax NP-100, NP-100/70 [170]
Macol NP-100 [564]
Trycol NP-1007 [289]

POE (100) oleic acid.
Crodet O100 [223]

POE (100) stearate.
Lipopeg 100-S [525]
Myrj 59 [92]
RS-55-100 [397]

POE (100) stearic acid.
Crodet S100 [223]

POE-(100)-stearyl alcohol.
Brij 700 [92]

POE (100) stearyl ether.
Brij 700 [438]

POE (100) stearyl ether/0.01% BHA/ 0.005% citric acid.
Brij 700 [438]

POE (100) tridecyl ether.
Macol TD-100 [564]

POE (150) dinonyl phenol.
Chemax DNP-150 [170]
Trycol DNP-150 [289]

POE (200) castor oil.
Chemax CO-200/50 [170]
Etocas 200 [223]
Trylox CO-00 [289]

POE (200) castor oil, hydrogenated.
Chemax HCO-200/50 [170]
Croduret 200 [223]

POE (200) castor oil sol'n.
Trylox CO-200/50 [289]

POE (200) castor oil sol'n, hydrogenated.
Trylox HCO-200/50 [289]

POE (200) dilaurate.
Nonex DL-2 [390]

POE (200) glyceride ester.
Syn Lube 106 (60%) [588]

POE (300)-monostearate.
Nonex S3E [390]

POE (400)-dioleate.
Nonex DO-4 [390]

POE (400)-monooleate.
Nonex O4E [390]

POE (600) monolaurate.
Pegosperse 600ML [358]

Poison ivy extract.
Rhus Tox Antigen [519]

Poison ivy, oak, sumac extract combined.
Rhus All Antigen - Poison Ivy, Oak, Sumac Combined [104]

Poliomyelitis vaccine purified, trivalent types 1,2,3 (Salk).
Poliomyelitis Vaccine (Purified) [815]

Poliovirus vaccine, live, oral, trivalent, types 1,2,3, (Sabin).
Orimune Poliovirus Vaccine, Live, Oral, Trivalent [516]

Poloxamer 101.
Pluronic L-31 [109]

Poloxamer 105.
Pluronic L-35 [109]

Poloxamer 108.
Pluronic F-38 [109]

Poloxamer 122.
Pluronic L-42 [109]

Poloxamer 123.
Pluronic L-43 [109]

Poloxamer 124.
Pluronic L-44 [109]

Poloxamer 181.
Pluronic L-61 [109]

Poloxamer 182.
Pluronic L-62 [109]

Poloxamer 183.
Pluronic L-63 [109]

Poloxamer 184.
Pluronic l-64 [109]

Poloxamer 185.
Pluronic P-65 [109]

Poloxamer 188.
Pluronic F-68 [109]

Poloxamer 212.
Pluronic L-72 [109]

Poloxamer 215.
Pluronic P-75 [109]

Poloxamer 217.
Pluronic F-77 [109]

Poloxamer 231.
Pluronic L-81 [109]

Poloxamer 234.
Pluronic P-84 [109]

Poloxamer 235.
Pluronic P-85 [109]

Poloxamer 237.
Pluronic F-87 [109]

Poloxamer 238.
Pluronic F-88 [109]

Poloxamer 282.
Pluronic L-92 [109]

Poloxamer 284.
Pluronic P-94 [109]

Poloxamer 288.
Pluronic F-98 [109]

Poloxamer 331.
Pluronic L-101 [109]

Poloxamer 333.
Pluronic P-103 [109]

Poloxamer 334.
Pluronic P-104 [109]

Poloxamer 335.
Pluronic P-105 [109]

Poloxamer 338.
Pluronic F-108 [109]

Poloxamer 401.
Pluronic L-121 [109]

Poloxamer 402.
Pluronic L-122 [109]

Poloxamer 403.
Pluronic P-123 [109]

Poloxamer 407.
Pluronic F-127 [109]

Poloxamine 304.
Tetronic 304 [109]

Poloxamine 504.
Tetronic 504 [109]

Poloxamine 701.
Tetronic 701 [109]

Poloxamine 702.
Tetronic 702 [109]

Poloxamine 704.
Tetronic 704 [109]

Poloxamine 707.
Tetronic 707 [109]

Poloxamine 901.
Tetronic 901 [109]

Poloxamine 904.
Tetronic 904 [109]

Poloxamine 908.
Tetronic 908 [109]

Poloxamine 1101.
Tetronic 1101 [109]

Poloxamine 1102.
Tetronic 1102 [109]

Poloxamine 1104.
Tetronic 1104 [109]

Poloxamine 1301.
Tetronic 1301 [109]

Poloxamine 1302.
Tetronic 1302 [109]

Poloxamine 1304.
Tetronic 1304 [109]

Poloxamine 1307.
Tetronic 1307 [109]

Poloxamine 1501.
Tetronic 1501 [109]

Poloxamine 1502.
Tetronic 1502 [109]

Poloxamine 1504.
Tetronic 1504 [109]

Poloxamine 1508.
Tetronic 1508 [109]

Polyacrylamide.
Cyanamer A-370, P-26, P-250 [50a]

Polyacrylamide *(cont'd)*.
Gelamide 250F [50]
Reten 420 [406]
Separan AP273P, AP30 [261]

Polyacrylamide, hydrolyzed.
Bonaril [261]

Polyacrylate.
Cyanacryl C, L, R [50]

Polyacrylate ammonium salt.
Serpol QPA 150 [788]

Polyacrylate sodium salt.
Serpol QPA 160 [788]

Polyacrylic acid.
Acrysol A-3 [742]
Alcosperse 404 and 409 [27]
Antiprex 461 [37]
Carbopol 907, 910, 934, 940, 941 [361]
Nopcosize N-30 [252]
Sokalan® PAS [106]

Polyacrylic acid, sodium salt.
Sokalan CP 5 [106]

Polyacrylic rubber.
Hycar Acrylic [361]

Polyadipate.
Ultramoll II, III [112]

Polyalkanolamide unsaturated fatty.
Lauramide S [946]

Polyalkyl ammonium sulfate.
Hipochem Finish 57 [410]

Polyalkyl methacrylate.
Empicryl 6051, 6052, 6058, 6059, DH122, DH135, DH145, PPT37, PPT38, PPT144, PPT145, PT1334, PT1345, PT1397, PT1544, PT1764/D [23]

Polyalkylene glycol.
Dehydran 240, 241 [401]

Polyalkylene glycol ester.
Pronal 502 [864]

Polyalkylene glycol ether.
Polylube WS [390]
Tergitol XD, XH, XJ [878]
Witconol 171, 172 [934]

Polyalkylene glycol ether, fatty alcohol.
Aethoxal B [401]
Propetal 99, 103, 241, 254, 281 [948]

Polyalkylene oxide ether, fatty alcohol.
Merpoxen T-2-3, T-4-8, T-5-5, T-5-7 [285]

Polyamine.
Softal 300 [865]

Polyamine condensate.
Fastgene PNG-708 [865]
Marvanol RE-1277 [552]

Polyamine condensate, fatty acid.
Avistin FD, PN [178]
Marlamid A 18, A 18E, AS 18, O 18, OS 18 [178]

Polyamine condensate, quarternized.
Marvanol RE-1511 [552]

Polyamino carboxylate.
Conco Seq. Conc. [209]

1-Polyaminoethyl-2-n-alkyl-2-imidazoline.
Witcamine 235 [935]

Polyamino glycol stearate.
Polytex 10 [494]

Polyarylate.
Arylef U 100 [809]

1,2 Polybutadiene.
Ricon 150, 151, 152, 153, 154 [201]

Polybutadiene rubber.
Cisdene [59a]
Diene, 35 NFA/AC, 55 NFA/AC [322]

cis-1,4-Polybutadiene rubber.
Budene [362]

(cis)-Polybutadiene rubber.
Ameripol CB [62], [361]

Polybutene.
Indopol, H-100 [64]

Polybutylene terphthalate.
Gafite LWX-4424S, LWX-4612R [338]

Polycaprolactone, branched (f=2.4).
CAPA 304 [809]

Polycaprolactone, branched (f=3).
CAPA 305 [809]

Polycaprolactone, linear.
CAPA 200, 205, 210, 212, 215, 220, 223, 231, 240, 520, 600, 600M, 601, 601M [809]

Polycarbonate (q.v.)
Polycarbafil [316]
Makrolon 2400, 2809, 3100, 3103, 3105, 3108, 3119, 3200, 3203, 6553, 6555, 6557, 6655, 6870 [112]
Makrofol E, N, SN [112]

Polycarbonate with glass staple fibers 20%.
Makrolon 8020 [112]

Polycarbonate with glass staple fibers 35%.
Makrolon 8344 [112]

Polycarboxylic acid, ammonium salt.
Sperse Polymer I, II, III, IV [95]

Polycarboxylic acid, sodium salt.
Geropon TA/72/S [550]

Polycarboxylic acid, sodium salt ***(cont'd).***
Peroxide Stabilizer H [416]

Polychlorobenzoic acid, dimethylamine salt.
Benzac 354 [878]
Zobar [270]

Polychloro-copper-phthalocyanine.
Atlas A5776 Monofast Green Toner [496]

Polychlorodicyclopentadiene isomers.
Bandane [898]

Polychloroprene.
Skyprene CCB30/31 Y30/31, CCR10/22, B5/B10 G Series [386a]

Polychloroprene (2-chlorobutadiene 1,3).
Neoprene AH, FB, GN, GNA, GRT, WD, WHV, WHV-A, WK, WRT [269]

Polychloroprene (CR).
Baypren 210, 211 [597]

Polychloroprene latex.
Baypren Latex B, GK, MKB, SK, T [112]

Polychloroprene latex with methacrylic acid 4%.
Baypren Latex 4 R [112]

Polychloroprene rubber.
Baypren 110 [597]

Polychloroprene rubber (CR).
Baypren 230 [597]

Poly-dibromophenylene oxide.
Great Lakes PO-64P [369]
PO-64P [369]

Polydimethyldiphenyl siloxane.
SF 1153, 1154 [341]

Polydimethyl siloxane.
Abil K O 3 [360]
AF 66, 72, 75 [341]
SF 18, 81, 96, 97, 99, 1093, 1173 [341]
Silicone AF-30 FG [859]

Poly(dipropylene glycol) phenyl phosphite.
Weston DHOP [129]

Polyester adipate.
Paraplex G-51 [379]
Plasthall P-630, HA7A, P-633, P-640, P-643, P-644, P-650, P-670, P-686 [379]

Polyester glutarate.
Plasthall P-530, P-550P, P-7035M, P-7035P, P-7046P, P-7092D, P-7092P [379]

Polyester poly ol.
Baycoll 12, 14, 22 [112]

Polyestradiol phosphate.
Estradurin [97]

Polyether acid phosphate.
Korantin LUB [106]

Polyether alcohol.
Deceresol Surfactant NI Conc. [52]

Polyether alcohol, phosphated, free acid form.
Mars PA-40 [554]

Polyether carboxylic acid.
Emery 5340 [290]

Polyether polyol.
Colorin 102, 104, 202, 301, 302 [764]

Polyether polythioether.
Emulvin S [597]

Polyether solvent, aliphatic.
Marsolve I [554]

Polyethoxy adduct.
Arnox LF-11, LF-12, LF-13 [76]
Triton CF-54, CF-76 [742]

Polyethoxy alkylphenol sulfonate triethanolamine salt.
Cellopal 100 [842]

Polyethoxy amine surfactant.
Triton RW-10, RW-50 [742]

Polyethoxy amine surfactant (EO 2).
Triton RW-20 [742]

Polyethoxy amine surfactant (EO 7.5).
Triton RW-75 [742]

Polyethoxy amine surfactant (EO 10).
Triton RW-100 [742]

Polyethoxy amine surfactant (EO 15).
Triton RW-150 [742]

Polyethoxy polypropoxy polyethoxy ethanol-iodine complex iodophor concentrate.
Bio-Surf I-21LF [120]

Polyethoxy tallate.
Teox 120 [659]

Polyethylene.
A-C Polyethylene 6, 6A, 8, 8A, 9, 9A, 617, 617A, 629A [36]
Astroturf® Doormats, H-D heavy-duty matting [602]
Fortiflex [163]
Interflo Porous Plastic [187]
Irrathene [342]
Microthene ML-733, MN-714, MN-722 [889]
Polywax 500, 655, 1000, 2000 [101]

Polyethylene, chlorosulfonated.
Hypalon [270]

Polyethylene, chlorosulfonated *(cont'd)*.
Hypalon 20 [269]
Hypalon 30, 40, 40S, 4085, LD-999 [269]

Polyethylene (500) crystalline, aliphatic saturated.
Polywax® 500 [101]

Polyethylene (700) crystalline, aliphatic saturated.
Polywax® 665 [101]

Polyethylene (1000) crystalline, aliphatic, saturated.
Polywax® 1000 [101]

Polyethylene (2000) crystalline, aliphatic, saturated.
Polywax® 2000 [101]

Polyethylene dispersion.
Ceranine L [762]

Polyethylene, emulsion.
Hipochem PE-25 [410]
Marvanol RE-1287, RE-1298 [552]
Melatex PE-820 [865]

Polyethylene films.
Zendel [878]

Polyethylene foam.
Ethafoam [261]

Polyethylene, glass fiber-reinforced.
Ethofil [316]

Polyethylene glycol *(See* **PEG***)*.

Polyethylene, oxidized high density.
Polywax® E-2020 [101]

Polyethylene sorbitan monolaurate.
Sorgen TW-20 [237]

Polyethylene sorbitan monostearate.
Sorgen TW-60 [237]

Polyethylene terephthalate.
Dacron [270]

Polyethylene wax.
BASF Wax FB [106]

Polyethylene wax emulsion.
NT-42 [630]

Polyethyleneimine.
Corcat [215]

Polyethyleneimine, mw. 1200.
Corcat P-12 [215]

Polyethyleneimine, mw. 1800.
Corcat P-18 [215]

Polyethyleneimine, mw. 10,000.
Corcat P-150 [215]

Polyethyleneimine, mw >60,000.
Corcat P-600 [215]

Polyglycerol-10 hexaoleate.
Polyaldo DGHO [358]

Polyglycerol ester.
Admul WOL, 1405, 1411 [704]
Caprol [829]
Drewmulse M [266]
Drewpol 4-1S [266]

Polyglycerol ester, fatty acid.
Hostacerin DGO and DGS [416]
Witconol 14, 14F, 18F, 18L [935]

Polyglycerol ester, saturated.
Admul 1405 [704]

Polyglycerol ester, unsaturated.
Admul 1411 [704]

Polyglycerol oleate.
Emcol 14 [937a], [938]
Plurol Oleique [339]

Polyglycerol polyricinoleate.
Admul WOL 1403 [704]
Cithrol PR [223]
Crester PR [223]
Radiamuls POLY [658]

Polyglycerol stearate.
GMS-333, -333-SES [397]
Hodag PGS [415]
Plurol Stearique [339]
Tefose [339]

Polyglycerol stearate self-emulsifying.
GMS-333-SES [397]

Polyglyceryl esters.
Polyaldo [358]

Polyglyceryl sorbitol.
EmCon G [308]

Polyglyceryl-2-PEG-4 stearate.
Hoe S 2720 [55]

Polyglyceryl-2 sesquiisostearate.
Hoe S 2721 [55]

Polyglyceryl-2 sesquioleate.
Hostacerin DGO [55]

Polyglyceryl-3 diisostearate.
Emerest 2452 [290]
Lameform TGI [375]

Polyglyceryl-3 oleate.
Caprol 3GO [147]
Grindtek PGE 25 [373]
Santone 3-1-SH [272]

Polyglyceryl-3 stearate.
Caprol 3GS [147]
Grindtek PGE 55 [373]
Polyaldo TGMS [358]
Santone 3-1-S [272]

Polyglyceryl-3 stearate SE.
Grindtek PGE 55-6 [373]

Polyglyceryl-4 isostearate.
Witconol 18L [935]

Polyglyceryl-4 oleate.
Protachem 100 [713]
Witconol 14 [935]

Polyglyceryl-4 stearate.
Witconol 18F [935]

Polyglyceryl-6 dioleate.
Caprol 6G2O [147]

Polyglyceryl-6 distearate.
Caprol 6G2S [147]
Polyaldo HGDS [358]

Polyglyceryl-8 oleate.
Santone 8-1-O [272]

Polyglyceryl-8 stearate.
Santone 8-1-S [272]

Polyglyceryl-10 decalinoleate.
Drewpol 10-10-LN [715]

Polyglyceryl-10 decaoleate.
Drewpol 10-10-O [715]
Mazol PGO-1010 [564]
Polyaldo DGDO [358]
Santone 10-10-O [272]

Polyglyceryl-10 decastearate.
Drewpol 10-10-S [715]

Polyglyceryl-10 tetraoleate.
Caprol 10G40 [147]
Drewpol 10-4-O [715]

Polyglycol.
Surflo-S41 [641], [641a]

Polyglycol ester.
Cithrol L, O Range, S Range [223]
Rychem® 17 [756]
Rychem® 21 [249], [756]
Rychem® 33, 70 [756]

Polyglycol ester, fatty.
Syntofor A03, A04, AB03, AB04, AL3, AL4, B03, B04 [937a], [938]

Polyglycol ester, fatty acid.
Alphoxat O 102, O 110, O 115, S 110, S 120 [948]
Emulgator 100, 109, 209, 309, 600 [150]
Emulgin TL 55, Ti 60 [401]
Emulsifier DMR [416]
Emulsogen A [55]
Genapol TS Powder [416]
Leomin OR [416]
Letocil LP 80, SP 1000 [182]
Marlipal BS, FS [178]
Marlosol 183, 189, 1820, 1825, 2414, OL 7, OL 10, OL 15, OL 20 [178]
Mulsifan RT 1, RT 2, RT 113 [948]
Oxypon 2145 [948]
Rewopal ER 100 [725]
Sponto® H-44C [935]
Zoharex A10, B10 [946]

Polyglycol esters, fatty series.
Secosov DO/600, MA/300, ML/300, MO/400 [825a]

Polyglycol ether.
Antistatic Plasticizer KA [112]
Arbyl R Conc. [176]
Chupol C [849]
Emulvin W [597]
Imerol XN [762]
Newlon K-1 [849]
Sandopan 2N [762]

Polyglycol ether acetate, fatty alcohol.
Rewopol CT [725]

Polyglycol ether, aliphatic.
Ekaline F [762]
Nilo O [762]

Polyglycol ether, aromatic.
Emulvin W [597]

Polyglycol ether, C_9-C_{11} alcohol, 2.2.
Merpoxen UN 22 [285]

Polyglycol ether, C_9-C_{11} alcohol, 4.0.
Merpoxen UN 40 [285]

Polyglycol ether, C_9-C_{11} alcohol, 5.
Merpoxen UN 50 [285]

Polyglycol ether, C_9-C_{11} alcohol, 6.5.
Merpoxen UN 65 [285]

Polyglycol ether, C_{10}-C_{12} alcohol, 5.2.
Merpoxen DL 52 [285]

Polyglycol ether, C_{12}-C_{14} alcohol, 2.
Merpoxen LM 20 [285]

Polyglycol ether, C_{12}-C_{14} alcohol, 3.
Merpoxen LM 30 [285]

Polyglycol ether, C_{12}-C_{15} alcohol, 6.
Merpoxen LP 60 [285]

Polyglycol ether, C_{12}-C_{15} alcohol, 7.
Merpoxen LP 70 [285]

Polyglycol ether, C_{13}-C_{15} alcohol, 10.
Merpoxen TP 185 [285]

Polyglycol ether, C_{16}-C_{18} alcohol, 11.
Merpoxen T 110 [285]

Polyglycol ether, C_{16}-C_{18} alcohol, 18.
Merpoxen T 180 [285]

Polyglycol ether, C_{16}-C_{18} alcohol, 25.
Merpoxen T 250 [285]

Polyglycol ether, C_{16}-C_{18} alcohol, 50.
Merpoxen T 500 [285]

Polyglycol ether, C_{16}-C_{18} alcohol, 80.
Merpoxen T 800 [285]

Polyglycol ether, C_{18}-C_{22} alcohol.
Merpemul 9060 [285]

Polyglycol ether carboxylic acid, fatty alcohol.
Akypo 23 Q 38, RLM 100, RLM 130, RLM 160, RLM Q33, RLM Q38, RO 40, RO 50, RO 90, RT 20, RT 60, RLM 25, RLM 45 [182]

Polyglycol ether carboxylic acid, fatty alcohol, monoethanolamine salt.
Akypo RLM 45 M [182]

Polyglycol ether carboxylic acid, fatty alcohol, sodium salt.
Akypo RLM 45 N [182]

Polyglycol ether, fatty acid.
Mulsifan RT 248 [948]
Osimol Grunau 560 [176]

Polyglycol ether, fatty alcohol.
Akyporox RTO 70, RC 80, RLM 22, RLM 40, RLM 80, RLM 100, RLM 160 [182]
Arbyl 18/50, N [176]
Dehydol 737, LS 2, LS 3, LS 4, LT 2, LT 3, LT 4, LT 5, LT 7, TA 5, TA 14, WM, WM 90 [401]
Diazital O Extra Conc. [253]
Eumulgin 05, 010, 535, M8 [401]
Hostacerin T-3 [416]
Marlipal MG, ML SU, KE, KF [178]
Mulsifan CPA, RT 3, RT 11, RT 19, RT 24, RT 27, RT 160, RT 203/80, RT 258 [948]
Oxetal C 110, C 114 Spez., D 104, O 108/112/120, S 125, T 106/110, T 107/110, TG 111/118/125 [948]
Product 100 [401]

Polyglycol ether, fatty amine.
Marlazin L 10, OL 20, S 10, S 15, S 20, S 25, S 40, T 10 [178]

Polyglycol ether hydrogen phosphate, C_{8-10} alcohol-5.
Merpiphos B 59 [285]

Polyglycol ether hydrogen phosphate, C_{12}-C_{14} alcohol 4.
Merpiphos RP [285]

Polyglycol ether hydrogen phosphate, C_{16}-C_{18} alcohol-8.
Merpiphos MP [285]

Polyglycol ether hydrogen phosphate (higher unsaturated), C_{16}-C_{18} alcohol-8.
Merpiphos VP [285]

Polyglycol ether hydrogen phosphate with hydrotropic addition, C_{16}-C_{18} alcohol-4.
Merpiphos TP [285]

Polyglycol ether hydrogen phosphate (unsaturated), C_{16}-C_{18} alcohol-8.
Merpiphos MP [285]

Polyglycol ether phosphate, fatty alcohol.
Rewophat OP 80, TD 40, TD 70 [725]

Polyglycol ether phosphate, sodium salt with hydrotropic component, C_{10}-C_{12} alcohol-5.
Merpiphos BP 80 [285]

Polyglycol ether, quaternary fatty amine.
Marlazin L 410, T 410 [178]

Polyglycol ether, sulfated natural alcohol, sodium salt.
Lutensit AS 2230, AS 2270 [106]

Polyglycol ether, (synth.) EO 10 mol., C_{12}-C_{15} alcohol.
Merpoxen LP 109 [285]

Polyglycol ether, (synth.) EO 12 mol., C_{12}-C_{15} alcohol.
Merpoxen LP 120 [285]

Polyglycol ether tridecylalcohol.
Produkt RT 38 [948]

10 Polyglycol ether, unsaturated C_{16-18} alcohol.
Merpoxen SP 100 [285]

20 Polyglycol ether, unsaturated C_{16-18} alcohol.
Merpoxen SP 200 [285]

23 Polyglycol ether, unsaturated C_{16-18} alcohol.
Merpoxen SP 230 [285]

Polyglycol ether, unsaturated, fatty acid.
Merpemul 6060, 6080 [285]

Polyglycol laurate.
Chimipal APG 400 [514]

Polyglycol-400-monolaurate.
Rilanit GL 401 [401]

Polyglycol-400-monooleate.
Rilanit GO 401 [401]

Polyglycol oleate.
Cithrol A [223]

Polyglycol 200 oleate.
Ethylan A2 [253], [938]

Polyglycol 300 oleate.
Ethylan A3 [253], [938]

Polyglycol 400 oleate.
Ethylan A4 [253], [938]

Polyglycol 600 oleate.
Ethylan A6 [253], [938]

Polyglycol, polyalkoxylated.
Pluracol W-170, W-660, W-2000, W-3520N, W-3520N-RL, W-5100 [109]

Polyglycol-polyamine condensation product.
Polyquart H81 [401]

Polyglycolic ether, ammonium salt sulfated.
Geropon LIV/30, LIV/SS [550]

Polyimide.
Envex 1000, 1115, 1228, 1315, 1330, 1620, 3540 [740]

Polyimide film.
Kapton Type H [269]

Polyimide molding compound.
Gemon 3010 [342]

Polyisobutene hydrogenated.
Polysynlane (Parleam) [636]

Polyisobutylene.
Vistanex MM LSO, MM L-100, MM L-120, MM L-140 [305]

Polyisocyanate, aromatic.
Bonding Agent 2001, 2005 [112]
Desmodur RC, RU [112]

Polyisoprene rubber.
Amcycle Isopreme [61]
Isolene [385]

(cis)-Polyisoprene rubber.
Ameripol SN [361]

cis-1, 4-Polyisoprene, rubber.
Coral [322]
Isolene D-40, D-75, D-400, DPR-40, DPR-75, DPR-400 [385]

Poly(methyl vinyl ether).
Gantrez M-154, M-555, M-556, M-574 [338]

Poly(methyl vinyl ether/maleic acid).
Gantrez® S-95, S-97 [338]

Poly (methyl vinyl ether/maleic anhydride).
Gantrez AN, Gantrez® AN-119, Gantrez AN-139, AN-149, AN-169, AN-179 [338]

Polymethylene polyaniline.
Curithane 103 [885]

Polymethylene polyphenylisocyanate.
Papi [884]
Papi 135, 580 [885]

Polymyxin B sulfate.
Aerosporin Powder [139]
BaySporin Otic Suspension [111]
Chloromyxin [674]
Cortisporin Cream, Ointment, Ophthalmic Ointment, Ophthalmic Suspension, Otic Solution, Otic Suspension [139]
Cortixin Otic Suspension [498]
Neo-Polycin [580]
Neosporin Aerosol, G.U. Irrigant, Ointment, Ophthalmic Ointment Sterile, Opthalmic Solution Sterile, Powder, -G Cream [139]
Ophthocort [674]
Otobiotic Otic Solution [772]
Polymyxin B Sulfate [139]
Polysporin Ointment, Ophthalmic Ointment [139]
Pyocidin-Otic Solution [115]
Terramycin Ointment, with Polymyxin B Sulfate Ophthalmic Ointment [690]
Tri-Thalmic Ophthalmic Solution [769]

Polymyxin B sulfate neomycin sulfate dexamethasone 10,000 u/g (equiv. 3.5 mg as base/g) 0.1%.
Ocu-Trol [655a]

Polymyxin B sulfate neomycin sulfate gramicidin, 10,000 u/cc (equiv. 1.75 mg as base/cc) 0.025%.
Ocu-Spor G [655a]

Polymyxin B sulfate neomycin sulfate hydrocortisone 10,000 u/cc (equiv. to 3.5 mg as base/cc) 1%.
Oti-Sore [655a]

Polymyxin B sulphate neomycin sulphate zinc bacitracin 5,000 u/g (equivalent 3.5 mg as base/g), 400 u/g.
Topimycin [655a]

Polymyxin B sulfate neomycin sulfate zinc bacitracin 10,000 u/cc (equiv. 3.5 mg as base/g) 400 u/g.
Ocu-Spor B [655a]

Polymyxin B sulfate neomycin sulfate zinc bacitracin hydrocortisone 10,000 u/g (equiv. 3.5 mg as base/g) 1%.
Ocu-Cort [655a]

Polymyxin preparations.
Neopsorin G.U. Irrigant [139]
Polysporin Ointment [139]
Pyocidin-Otic Solution [115]

Polynorbornene.
Norsorex Powder [160]

Poly (octadecyl vinyl ether/maleic anhydride).
Gantrez AN-8194 [338]

Polyol.
Mazon 90 [564]

Polyol borate.
Emulbon GB-90 [864]

Polyol ester ethoxylate.
Cirrasol ALN-GM [849]

Polyol ester, fatty acid.
Cetiol HE [401]

Polyol oleate.
Atsurf 2801 [445]

Polyol oleate, ethoxylated.
Atsurf 2821 [445]

Polyol polyborate.
Emulbon LB-75 [864]

Polyol, polypropoxylated.
Witconol CC-43, CD-17 [935]
Witconol CD-18 [937a], [938]

Polyol stearate.
Atsurf 2803 [445]

Polyol stearate, ethoxylated.
Atsurf 2823 [445]

Polyol trioleate.
Atsurf 2802 [445]

Polyol trioleate, ethoxylated.
Atsurf 2822 [445]

Polyorgano-siloxane.
Coagulant WS [597]

Polyoxyalkylene alcohol.
Rexonic P-4, P-5, P-6, P-7 [390]

Polyoxyalkylene condensate.
Fibertex WS [390]
Polylube GK [390]

Polyoxyalkylene DEA amide ester.
Nitrene OE [400]

Polyoxyalkylene ether.
Armul 03, 16 [76]

Polyoxyalkylene ether ester.
Mazon 86LF [564]

Polyoxyalkylene glycol.
Carsonon LF-46 [155]
Elfacos E200, ST 9 [764]
Elfacos ST 9, ST 37 [16]
Hartopol 25R2, 31R1, L42, L44, L62, L62LF, L64, L81, LF-5, P65, P85 [390]
Leocon 114Y, 703X, 710X, 1705W [524]
Pluradot HA 410, 420, 430, 440, 450, 510, 520, 530, 540, 550 [109]
Witbreak DTG-62 [934]

Polyoxyalkylene glycol polyol.
Pluracol V-10 [109]

Polyoxyalkylene lanolins.
Lanogel® 21, 31, 41, 61 [44]

Polyoxyalkylene laurate.
Emulsynt 900 [895]

Polyoxyalkylene oleate/laurate.
Emulsynt 1055 [895]

Polyoxyalkylene propylene glycol condensate.
Hartopol LF-1, LF-2 [390]

Polyoxyalkylone amide ester.
Nitrene OE [399], [400]

Polyoxyethylene *(See* **POE***).*

Polyoxyl (40) stearate.
Myrj 52, 52S [438]

Poly-p-oxybenzoate.
Ekonol [383]

Polyphenolic sodium lignosulfonate.
Maratan SNV [47]

Polyphenyl ether.
OS-124 [602]
Santovac 5 vacuum diffusion pump fluid [602]

Polyphenylene oxide.
PPO [342]

Polyphenylene sulfide.
Liquinite PPS [527]

Polyphosphoric ester, acid anhydride, complex organic.
Strodex MO-100, MR-100, P-100, SE-100 [250]

Polyphosphoric ester, acid anhydride, complex organic, potassium salt.
Strodex MOK-70, MRK-98, PK-90, SEK-50 [250]

Polyphthalate.
Ultramoll PP, TGN [112]

Polypropoxy quaternary ammonium acetate.
Emcol CC-55 [935]

Polypropoxy quaternary ammonium chloride.
Emcol CC-9, CC-36, CC-42, CC-55, CC-422 [935]

Polypropoxy quaternary ammonium phosphate.
Emcol CC-57 [935]

Polypropoxylate.
Degressal SD 20 [106]

Polypropylene.
Hostalen PP 920, PP 927, PP 933, PP 934, PP 936, PP 941, PP 998 [56]
PRO-FAX special series products, 6000 Series, 7000 Series, 8000 Series, mineral-filled [406]

Polypropylene (2000) oleate.
Witconol F 26-46 [937a], [938]

Polypropylene, amorphous.
A-Fax, A-Fax 500, 600, 800, 940 [406]

Polypropylene, amorphous *(cont'd)*.
Stan Wax AP [392]

Polypropylene film.
BX [406]
BXT [406]

Polypropylene, glass fiber-reinforced.
Profil [316]

Polypropylene glycol *(See* **PPG***)*.

Polypropylene, mica-filled.
Micalite PP-H2MFS-6 [914]

Polypropylene olefin fibers.
Herculon [406]

Polypropylene thermoplastic.
Escon [296]

Polyquaternary ammonium chloride.
Mirapol A-15 [592]

Polyquaternium-1.
Onamer M [663]

Polyquaternium-2.
Mirapol A-15 [592]

Polyquaternium-4.
Celquat H60, L200 [619]

Polyquaternium-5.
Catamer Q [732]
Emcol Q [935]
Reten 210, 220, 230, 240, 1104, 1105, 1106 [406]

Polyquaternium-6.
Merquat 100 [577]

Polyquaternium-7.
Merquat 550 [577]

Polyquaternium-10.
Leogard G [524]
UCARE Polymer JR-30M, JR-125, JR-400, LR 30M LR 400 [878]

Polyquaternium-11.
Gafquat 734, 755 [338]

Polyquaternium-14.
Reten 300 [406]

Polyquaternium-18.
Mirapol AZ-1 [592]

Polysiloxane.
Perenol S4, S5 [400]

Polysiloxane polyglycol ether.
Skinotan S 10 [948]

Polysorbate 20.
Armotan PML-20 [79]
Drewmulse POE-SML [715]
Durfax 20 [272]
Emsorb 2720 [290]
Hetsorb L-20 [407]
Hodag PSML-20 [415]
Industrol L20S [923]
Laxan ESL [507]
Liposorb L-20 [525]
Lonzest SML-20 [529]
Montanox 20 [785]
Nikkol TL-10, TL-10EX [631]
Protasorb L-20 [713]
T-Maz 20 [564]
Tween 20 [413]

Polysorbate 20/POE (20) sorbitan monolaurate.
ML-55-F [397]

Polysorbate 21.
Hetsorb L-4 [407]
Industrol L4S [923]
Protasorb L-5 [713]
T-Maz 21 [564]
Tween 21 [438]

Polysorbate 21/POE (4) sorbitan monolaurate.
ML-55-F-4 [397]

Polysorbate 40.
Hetsorb P-20 [407]
Hodag PSMP-20 [415]
Industrol P20S [923]
Laxan ESP [507]
Liposorb P-20 [525]
Lonzest SMP-20 [529]
Montanox 40 [785]
Protasorb P-20 [713]
T-Maz 40 [564]
Tween 40 [438]

Polysorbate 60.
Alkamuls PSMS-20 [33]
Armotan PMS-20 [79]
Capmul POE-S [147]
Drewmulse POE-SMS [715]
Durfax 60 [272]
Emrite 6125 [290]
Emsorb 2728 [290]
Hetsorb S-20 [407]
Industrol S20S [923]
Laxan ESS [507]
Liposorb S-20 [525]
Lonzest SMS-20 [529]
Montanox 60 [785]
Nikkol TS-10 [631]
Protasorb S-20 [713]
T-Maz 60 [564]
Tween 60 [438]

Polysorbate 60K.
Emrite 6126 [290]

Polysorbate 61.
Alkamuls PSMS-4 [33]
Industrol S4S [923]
Liposorb S-4 [525]

Polysorbate 61 *(cont'd).*
T-Maz 61 [564]
Tween 61 [438]

Polysorbate 65.
Alkamuls PSTS-20 [33]
Drewmulse POE-STS [715]
Durfax 65 [272]
Emsorb 2729 [290]
Hetsorb TS-20 [407]
Industrol STS-20S [923]
Liposorb TS-20 [525]
Montanox 65 [785]
T-Maz 65 [564]
Tween 65 [438], [445]

Polysorbate 65/glycerol monostearate, 20:80.
Aldosperse TS-20 FG [358]

Polysorbate 80.
Alkamuls PSMO-20 [33]
Armotan PMO-20 [79]
Capmul POE-O [147]
Drewmulse POE-SMO [715]
Durfax 80 [272]
Emrite 6120 [290]
Emsorb 2722 [290]
Hetsorb O-20 [407]
Industrol O20S [923]
Laxan ESO [507]
Liposorb 0-20 [525]
Lonzest SMO-20 [529]
Montanox 80 [785]
Nikkol TO-10 [631]
Protasorb O-20 [713]
T-Maz 80 [564]
Tween 80 [438]

Polysorbate 80/glycerol monostearate, 20:80.
Aldosperse O-20 FG [358]

Polysorbate 80K.
Emrite 6121 [290]

Polysorbate 80/POE (20) sorbitan monooleate.
MO-55-F [397]

Polysorbate 81.
Alkamuls PSMO-5 [33]
Industrol O5S [923]
Liposorb O-5 [525]
Lonzest SMO-5 [529]
Montanox 81 [785]
Protasorb O-5 [713]
T-Maz 81 [564]
Tween 81 [438]

Polysorbate 81/POE (5) sorbitan monooleate.
MO-55-F-5 [397]

Polysorbate 85.
Alkamuls PSTO-20 [33]
Emsorb 2723 [290]
Hetsorb TO-20 [407]
Industrol STO-20S [923]
Laxan EST [507]
Liposorb TO-20 [525]
Lonzest STO-20, STS-20 [529]
Montanox 85 [785]
Protasorb TO-20 [713]
T-Maz 85 [564]
Tween 85 [438]

Polysorbate 85 POE (20) sorbitan trioleate.
TO-55-F [397]

Polystyrene.
Cosden Polysytrene 500, 525, 550, 625, 710, 825, 825 E [217]
Dorvon FR [261]
Dylene [497]
Evenglo [497]
Fosta Tuf-Flex 240, 271, 283, 326, 329, 474, 717, 721-M, 742, 782, 840, 880, 929 [56]
Lustrex 3350, 4220 [603]
Luxtrex 2220, 4300 [603]

Polystyrene beads, expandable.
Fostafoam Type 3775, 4775, 5775 [56]

Polystyrene film.
Kardel [878]

Polystyrene, high flow.
Fostarene 817 [56]

Polystyrene, high heat.
Fostarene 50, 57, 58 [56]

Polystyrene latex.
Darex 670L [363]
Lytron 614, 621 [602]

Polystyrene latex, modified.
Lytron 614 Latex [602]
Lytron Latex 621 [603]

Polystyrene, light stable.
Fostalite [56]

Polystyrene, low monomer content.
Fosta Tuf-Flex 730 [56]

Polysulfide, proprietary.
Union Carbide Organofunctional Silane Y-9194 [878]

Polytetrafluoroethylene.
Hostaflon TF Series [416]
Klingerflon PTFE (60% bronze filled), PTFE (25% glass filled), PTFE (Virgin) [493]
Tetran [682]
TL-115, TL-126 [392]

Polytetrafluoroethylene/carbon 25%, electrographitized.
Hostaflon TF 4215 [416]

Polytetrafluoroethylene natural graphite 25%.
Hostaflon TF 4303 [416]

Polythiazide.
Minizide Capsules [691]
Renese, Renese-R [691]

Polythioether.
Emulvin S [597]

Polytrifluorochloroethylene.
TL-340 [527]

Polyurethane.
Ultramoll PU [112]
Vedoc VP-180 [315]

Polyurethane elastomer.
Genthane-S [343]

Polyurethane emulsion.
Elaslene 600 [865]

Polyurethane, hydroxyl.
Desmocoll 110, 130, 176, 400, 406, 420, 500, 510, 530 [112]

Polyurethane system.
Stepanfoam AX-64, BX-105 Series, BX-250 (A-D) Series, BX-289, BX-341A, BX-341B, BX-345, BX-350-7, BX-352, C-600 Series [825]

Polyurethane, thermoplastic.
Liquinite LPU [527]

Polyvinyl acetate.
UCAR Latex Resin WC-130 [878]
Vinac 880 [13]

Polyvinyl acetate emulsion.
Griffco [643]

Polyvinyl alcohol.
Elvanol, Elvanol 85-50, 85-60, 85-80, T-25, T-66 [269]
Gelvatol [602]
Lemoflex [128]
Lemol [128]
Vinol 523, 540 [13]

Polyvinyl alcohol, hydrolyzed.
Elvanol 71-30, 75-15, 85-82, 90-50, HV [269]

Polyvinyl chloride, chlorinated.
Geon CPVC [361]

Polyvinylidene chloride emulsion.
Daran 220, 229, 820 [363]

Polyvinyl fluoride.
Dalvor [251]

Polyvinyl fluoride film.
Tedlar [270]

Polyvinyl imidazolinium acetate.
Resin QR-686 [742]

Poly(vinyl isobutyl ether).
Gantrez B-773 [338]

Polyvinyl laurate.
Mexomere PP [185]

Polyvinyl methyl ether.
Gantrez M [338]

Polyvinyl platisol.
Enthonite [297]

Polyvinylidene fluoride.
TL-450 [527]

Polyvinylpyrrolidone.
Luviskol K12, K17, K30, K60, K80, K90 [106]
Peregal ST [338]
Plasdone [338]
Plasdone C-30, K-26/28, K-29/32 [338]
Polyclar AT [338]
Polyplasdone XL [338]
PVP [338]
PVP K-15, K-30, K-60, K-90 [338]

POP ether.
Pronal 5500, 3300, ST-1 [864]

POP, ethoxylated + 5.7 moles.
Teric PE61 [439]

POP, ethoxylated + 17 moles.
Teric PE62 [439]

POP, ethoxylated + 25.5 moles.
Teric PE64 [439]

POP, ethoxylated + 150 moles.
Teric PE68 [439]

POP glycol, ethoxylated.
Alkatronic PGP 10-1, PGP 12-2, PGP 18-1, PGP 18-2, PGP 18-2D, PGP 18-2LF, PGP 18-4, PGP 18-8, PGP 18-8LF, PGP 23-5, PGP 23-7, PGP 23-8, PGP 33-1, PGP 33-8, PGP 40-7 [33]
Alkatronic PGP 10-1, PGP 12-2, PGP 18-1, PGP 18-2, PGP 18-2D, PGP 18-2LF, PGP 18-4, PGP 18-8, PGP 23-5, PGP 23-7, PGP 23-8, PGP 33-1, PGP 33-8, PGP 40-7 [34]
Poly-Tergent P-17A [659]

POP glycol monoether.
Nissan Unilube MB [636]

POP polyamine.
Jeffamine D-230, D-400, D-2000, T-403 [854]

POP polyamine, urea condensate.
Jeffamine BuD-2000, DU-700, DU-1700, DU-3000 [854]

POP (10) cetyl ether.
Procetyl 10 [223]

POP (15) stearyl alcohol.
Arlamol E [92]

POP (15) stearyl ether.
Arlamol E [445]
Prostearyl 15 [223]

POP (20) cetyl ether.
Procetyl 20 [223]

POP (30) cetyl ether.
Procetyl 30 [223]

POP (50) cetyl ether.
Procetyl 50 [223]

POP (1750), ethoxylated + 6 moles.
Teric PE Series [439]

POP (1750), ethoxylated + 17 moles.
Teric PE Series [439]

POP (1750), ethoxylated + 25 moles.
Teric PE Series [439]

POP (1750), ethoxylated + 150 moles.
Teric PE Series [439]

POP (2250), ethoxylated + 52 moles.
Teric PE75 [439]

POP (2250), ethoxylated + 124 moles.
Teric PE87 [439]

Potash muriate.
Trona KM [490]

Potash sulfate.
Trona [490]

Potassium acid phosphate.
K-Phos M.F. (Modified Formula) Tablets, Neutral Tablets, No. 2 Tablets, Original Formula 'Sodium Free' Tablets [113]
Thiacide Tablets [113]

Potassium alginate.
Kelmar, Improved Kelmar [484], [485]

Potassium alkylaryl sulfonate, high molecular weight.
Ninate 415 [762]
Ninate 415 [825a]

Potassium alkyl (C_{15}) benzene sulfonate, branched.
Polystep A-12 [825]

Potassium alkyl carboxylate fluorinated.
Fluorad FC-128 [822]
Fluorad FC-129 [862]

Potassium aluminum polyacrylate.
Permasorb 7710, 7730, 7731, 7733 [619]

Potassium aluminum silicate, anhydrous.
Frianite [143]

Potassium animal protein coco hydrolyzed.
Lamepon S, S-2 [375]
Lexein S-620 [453]
Maypon 4C [825]
May-Tein C [562]

Potassium animal protein undecylenoyl hydrolyzed.
Lamepon UD [375]
Maypon UD [825]

Potassium bicarbonate.
Klorvess Effervescent Granules, Effervescent Tablets [762]
K-Lyte & K-Lyte DS, K-Lyte/Cl & K-Lyte Cl 50 [572]
Vitaplex C [872]

Potassium bitartrate.
Ceo-Two Suppositories [117]

Potassium chloride.
Infalyte [682]
Kaochlor 10% Liquid, S-F 10% Liquid (Sugar-free) [7]
Kaon Cl, Cl-10, -Cl Tabs, -Cl 20% [7]
Kato [518]
Kay Ciel Elixir, Ciel Powder [115]
Klor-Con Powder, -Con/25 Powder, -Con 20%, -10% [886]
K-Lor Powder [3]
Klorvess Effervescent Granules, Effervescent Tablets, 10% Liquid [762]
Klotrix [572]
Kolyum Liquid, Powder [682]
K-Lyte/Cl & K-Lyte/Cl 50 [572]
KM [490]
K-Tab [3]
Micro-K Extencaps [737]
Potage [519]
Potassium Chloride Concentrate, Powder & Liquid [769]
Potassium Chloride Elixir [344]
Potassium Chloride for Oral Solution (Flavored), Oral Solution, Powder (Unflavored) [751]
Potassium Chloride Injection [287]
Potassium Chloride Oral Solution, Powder & for Oral Solution [751]
Rum-K [326]
SK-Potassium Chloride Oral Solution [805]
Slow-K [189]
Trona [490]
Ultralog 75-1520-00 [172]

Potassium chloride elixir, 10%.
Potasalan Elixir [510]

Potassium chloride liquid, 10%.
Kaochlor [7]

Potassium citrate.
Bi-K [891]
K-Lyte & K-Lyte DS, K-Lyte/Cl & K-Lyte/Cl 50 [572]
Polycitra Syrup, -K Syrup, -LC—Sugar-Free [929]
Quik-Prep [468]
Twin-K, -K-Cl [127]

Potassium cocoate.
Protachem LP-40 [713]

Potassium coconut oil soap.
Norfox 1101 [650]

Potassium cocoyl collagen peptide.
Nikkol CCK-40 [631]

Potassium dichloroisocyanurate.
ACL 59 [602]

Potassium dichromate.
Acculog 75-2300-00 [172]

Potassium dimethyldithiocarbamate.
Vulnopol KM [27]

Potassium gluconate.
Bi-K [891]
Kaon Elixir, Grape Flavor, Tablets [7]
Kolyum Liquid, Powder [682]
Potassium Gluconate Elixir [751], [769]
Twin-K, -K-Cl [127]

Potassium gluconate elixir.
Kaylixir [510]

Potassium guaiacolsulfonate.
Albatussin [105]
Ambenyl Expectorant [551]
Bromanyl Expectorant [769]
Codiclear DH Syrup [166]
Entuss Expectorant Tablets & Liquid [394]
Phenergan Expectorant Plain (without Codeine), Expectorant w/Codeine, Expectorant w/Dextromethorphan, Pediatric, VC Expectorant Plain (without Codeine), VC Expectorant w/Codeine [942]

Potassium hetacillin.
Versapen Oral Suspension, Pediatric Drops, -K Capsules [134]

Potassium hydroxide.
Acculog 75-4200-00, 75-4200-10 [172]

Potassium N-hydroxymethyl-N-methydithio carbamate.
Bunema [137]

Potassium iodate.
Acculog 75-4300-00 [172]

Potassium iodide.
Iodo-Niacin Tablets [661]
Iomag [542]
Isuprel Hydrochloride Compound Elixir [133]
KIE Syrup [513]
Mudrane Tablets, -2 Tablets [703]
Pediacof [133]
Pedi-Bath Salts [679]
Pima Syrup [326]
Potassium Iodide Liquid [751]
Quadrinal Tablets & Suspension [495]
SSKI [886]

Potassium lauryl sulfate.
Conco Sulfate P [209]

Potassium oleate.
Emkapol PO-18 [292]
Norfox KO [650]

Potassium perfluoroalkyl sulfonate.
Fluorad FC-95, FC-98 [862]

Potassium permanganate.
Acculog 75-5700-00, 75-5700-10 [172]
Cairox [157]

Potassium phosphate, dibasic.
Neutra-Phos Powder & Capsules, -K Powder & Capsules [929]
Potassium Phosphates Oral Solution [751]

Potassium phosphate, monobasic.
K-Phos M.F. (Modified Formula) Tablets, Neutral Tablets, Original Formula Sodium Free Tablets [113]
Neutra-Phos Powder & Capsules, -K Powder & Capsules [929]
Potassium Phosphates Oral Solution [751]
Thiacid Tablets [113]

o-Potassium phosphite.
Ultralog 75-6060-00 [172]

Potassium preparations.
Effervescent Potassium Tablets [769]
Kaochlor 10% Liquid, S-F 10% Liquid (Sugar-free) [7]
Kaon Cl-10, Elixir, Grape Flavor, Tablets, -Cl Tabs, -Cl 20% [7]
K-Lor Powder [3]
Klorvess Effervescent Granules, Effervescent Tablets, 10% Liquid [762]
Kolyum Liquid, Powder [682]
Pima Syrup [326]

Potassium propionate.
Guard Sodium Propionate [65]

Potassium ricinoleate.
Solricin 135 [158], [641a], [642]

Potassium salicylate.
Pabalate-SF Tablets [737]

Potassium salt, phosphoric acid ester.
Atlas G-2203 [92]

Potassium silicates.
Kasil [706]

Potassium soap.
Dresinate 92, 93, 94 [406]

Potassium soap, aq. sol'n.
Dresinate 90 Liquid Soap, 91 Liquid Soap, 94 Liquid Soap, 95 Liquid Soap [406]

Potassium sorbate.
Alflaban KSG [602]
Guard Potassium Sorbate [65]
Monitor [602]
Preservastat K [867]
Sorbistat-K [691]
Summer's Eve Medicated Douche, Medicated Vaginal Suppositories [325]

Potassium stearate.
Witco Potassium Stearate [935]

Potassium sulfate.
Trona [490]

Potassium tallow ether sulfate.
Elfan NS 682 KS [764]

Potassium thiocyanate.
Acculog 75-7600-00 [172]

Potassium thiosulfate.
Ultralog 75-7840-00 [172]

Potassium titanate.
Tipersul [270]

Potassium toluene sulphonate.
Eltesol PT45 [23]
Eltesol PT63, PT93 [25]
Eltesol PT 93 [26]
Eltesol PX 93 [25]
Naxonate KT, 4KT, 5KT [754]

Potassium xylene sulfonate.
Conco PXS [209]
Eltesol PX 40 [23]
Eltesol PX 93 [25], [26]
Reworyl KXS 40 [775]

Potassium xylene sulfonate, 40% aq. sol'n.
Manro KXS 40 [545]

Povidone-iodine.
Betadine Aerosol Spray, Disposable Medicated Douche, Douche, Helafoam Solution, Ointment, Skin Cleanser, Solution, Surgical Scrub, Viscous Formula Antiseptic Gauze Pad [714]
Neotec [260]

PPG.
Alkapol PPG-425, PPG-1200, PPG-2000, PPG-4000 [33]
Monolan PPG 440, PPG 1100, PPG 2200 [253], [509]
Multranols [597]
Pluracol P-410, P-710, P-1010, P-2010, P-3010, P-4010 [109]
Pluriol P 600, P 900, P 2000, P 4000 [106]

PPG ceteth (3) acetate.
Hetester PCA [407]

PPG, ethoxylated.
Chimipal PE 300, PE 302 [514]
Macol 1, 2, 4 [564]

PPG fatty acid ester.
Witconol F26-46 [934], [935]

PPG isoceteth (3) acetate.
Hetester PHA [407]

PPG methyl ether.
Newpol 5PM [764]

PPG myristyl ether.
Promyristyl PM [223]

PPG-PEG ether.
Epan 485, 710, 720, 740, 750, 785, U-102, U-103, U-104, U-105, U-108 [237]

PPG (2) buteth (3).
UCON 50-HB-55 [878]

PPG (2) dibenzoate.
Benzoflex 9-88 [898]

PPG (2) lanolin ether.
Solulan PB-2 [44]

PPG (2) methyl ether.
Dowanol DPM [261]

PPG (2) myristyl ether propionate.
Crodamol PMP [223]

PPG (3) buteth (5).
UCON 50-HB-100 [878]

PPG (3) isosteareth (9).
Arosurf CLA1 [796]

PPG (3) methyl ether.
Dowanol TPM [261]

PPG (3) myreth (3).
Witconol APEM [935]

PPG (3) myristyl ether.
Promyristyl PM3 [223]
Witconol APM [934]

PPG (4) ceteareth (12).
Arosurf 63-PE16, CL-744 [796]

PPG (4) ceteth (1).
Nikkol PBC-31 [631]

PPG (4) ceteth (5).
Nikkol PBC-32 [631]

PPG (4) ceteth (10).
Nikkol PBC-33 [631]

PPG (5) buteth (7).
UCON 50-HB-170 [878]

PPG (5) butyl ether.
UCON LB-65 [878]

PPG (5) ceteth (10) phosphate.
Crodafos SG [223]

PPG (5) ceteth (20).
Procetyl AWS [223]

PPG (5) lanolin ether.
Solulan PB-5 [44]

PPG (5) lanolin wax.
Propoxyol 5 [290]

PPG (6) pareth (28-11).
Plurafac D-25 [109]

PPG (7) buteth (10).
UCON 50-HB-260 [878]

PPG (8) ceteth (1).
Nikkol PBC-41 [631]

PPG (8) ceteth (2).
Procetyl AWS Modified [223]

PPG (8) ceteth (5).
Nikkol PBC-42 [631]

PPG (8) ceteth (10).
Nikkol PBC-43 [631]

PPG (8) ceteth (20).
Nikkol PBC-44 [631]

PPG (8) oleate.
Witconol H-31A [934]

PPG (9).
Macol P-400 [564]
Pluracol P-410 [109]
Witconol PPG 400 [935]

PPG (9) buteth (12).
UCON 50-HB-400 [878]

PPG (9) butyl ether.
UCON LB-135 [878]

PPG (9) diethylmonium chloride.
Emcol CC-9 [934]

PPG (9) steareth (3).
Witconol APES [935]

PPG (10) ceteareth (20).
Standamul OXL [400]

PPG (10) cetyl ether.
Procetyl 10 [223]

PPG (10) cetyl ether phosphate.
Crodafos CAP [223]

PPG (10) lanolin ether.
Solulan PB-10 [44]

PPG (10) methyl glucose ether.
Glucam P-10 [44]

PPG (10) oleyl ether.
Hodag CSA-91 [415]
Provol 10 [223]

PPG (11) stearyl ether.
Witconol APS [934], [935]

PPG (12).
Pluracol P-710 [109]

PPG (12) buteth (16).
Pluracol W-660 [109]
UCON 50-HB-660 [878]

PPG (12) PEG (50) lanolin.
Laneto AWS [735]
Lanexol AWS [223]
Lanoil AWS [507]

PPG (14) butyl ether.
Fluid AP [878]
UCON LB-165, LB-200, LB-250 [878]
Witconol APB [935]

PPG (15) buteth (20).
UCON 50-HB-1000 [878]

PPG (15) butyl ether.
UCON LB-285 [878]

PPG (15) stearyl ether.
Arlamol E [438]
Hetoxol SP-15 [407]

PPG (16) butyl ether.
UCON LB-300 [878]

PPG (17).
Pluracol P-1010 [109]

PPG (17) dioleate.
CPH-327-N [379]

PPG (18) butyl ether.
UCON LB-385 [878]

PPG (20).
Macol P-1200 [564]
Polyglycol P-1200 [261]

PPG (20) buteth (30).
UCON 50-HB-2000 [878]

PPG (20) decyltetradeceth (10).
S-Safe 2010 [636]

PPG (20) lanolin ether.
Solulan PB-20 [44]

PPG (20) methyl glucose ether.
Glucam P-20 [44]

PPG (20) methyl glucose ether acetate.
Glucacet AP-20 [44]

PPG (22) butyl ether.
UCON LB-525 [878]

PPG (23) oleyl ether.
UCON LO-500 [878]

PPG (24) buteth (27).
Tergitol XD [878]

PPG (24) butyl ether.
UCON LB-625 [878]

PPG (25) diethylmonium chloride.
Emcol CC-36 [934], [935]

PPG (25) laureth (25).
ADF Oleile [901]

PPG (26).
Jeffox PPG-2000 [854]
Macol P-2000 [564]
Pluracol P-2010 [109]
Polyglycol P-2000 [261]

PPG (26) buteth (26).
Witconol APEB [935]

PPG (26) oleate.
Hodag CSA-80 [415]
OP-2000 [109]

PPG (27) glyceryl ether.
Witconol CD-18 [934]

PPG (28) buteth (35).
Pluracol W-3520 [109]
UCON 50-HB—3520 [878]

PPG (30).
Macol P-4000 [564]
Pluracol P-4010 [109]

PPG (30) butyl ether.
Nissan Unilube MB-38 [636]

PPG (30) cetyl ether.
Wickenol 707 [926]

PPG (33) buteth (45).
Pluracol W-5100 [109]
UCON 50-HB-5100 [878]

PPG (33) butyl ether.
UCON LB-1145 [878]

PPG (34).
Witconol CD-17 [934]

PPG (36) oleate.
Witconol F26-46 [934], [935]

PPG (37) oleyl ether.
UCON LO-1000 [878]

PPG (40) butyl ether.
Ambiflo L-317 [261]
UCON LB-1715 [878]
Unilube MB-370 [636]

PPG (40) diethylmonium chloride.
Emcol CC-42 [934], [935]

PPG (50) cetyl ether.
Procetyl 50 [223]

PPG (50) oleyl ether.
Provol 50 [223]

PPG (53) butyl ether.
UCON LB-3000 [878]

PPG (55) glyceryl ether.
Witconol CC-43 [935]

PPG (400) dilaurate.
CPH-296-N [379]

PPG (400) monooleate.
CPH-309-N [379]

PPG (1000) ditallate.
Industrol 1025-DT [109]

PPG (1025) dilaurate.
CPH-270-N [379]

PPG (1025) dioleate.
CPH-327-N, CPH-387-N [379]

PPG (2025) dilaurate.
CPH-328-N [379]

PPG (2025) dioleate.
CPH-239-N [379]

PPG (2025) monolaurate.
CPH-116-N [379]

PPG (2025) monooleate.
CPH-238-N [379]

PPG (2025) monostearate.
CPH-346-N [379]

Pralidoxime chloride.
Protopam Chloride [97]

Pramoxine hydrochloride.
Anusol Ointment, Suppositories [674]
Derma-Smoothe/FS, -Sone Cream [412]
F-E-P Creme [127]
Fleet Relief [325]
Otic-HC Ear Drops [394]
Pramosone Cream, Lotion & Ointment [314]
Prax Cream & Lotion [314]
Pricort Cream [78]
ProctoFoam/non-steroid [719]
Tronolane Anesthetic Hemorrhoidal Cream, Anesthetic Hemorrhoidal Suppositories, Hydrochloride [3]

Prazepam.
Centrax [674]

Praziquantel.
Biltricide [585]

Prazosin hydrochloride.
Minipress [691]
Minizide Capsules [691]

Prednisolone.
Delta-Cortef Tablets [884]
Prednisolone Tablets [240], [344], [751], [769]

Prednisolone acetate.
Metimyd Ophthalmic Ointment - Sterile, Ophthalmic Suspension [772]
Ocu-Pred A [655a]
Predate L.A.S.A., -100 [518]

Prednisolone sodium phosphate.
Metreton Ophthalmic/Otic Solution-Sterile [772]
Ocu-Pred, Ocu-Pred Forte [655a]
Predate L.A.S.A. [518]

Prednisolone sodium succinate.
Solu-Delta-Cortef [884]

Prednisolone tebutate.
Hydeltra-T.B.A. Suspension [578]

Prednisone.
Deltasone [884]
Deltasone Tablets [884]
Liquid Pred Syrup [612]
Orasone [750]
Prednisone Tablets [240], [344], [751], [769]
SK-PRednisone Tablets [805]
Sterapred Uni-Pak [563]

Prilocaine hydrochloride.
Citanest Solutions [90]

Primaquine phosphate.
Aralen Phosphate with Primaquine Phosphate [932]

Primidone.
Mysoline [97]
Mysoline Tablets [510]
Primidone Tablets [240], [344], [769]

Pristane.
Robuoy [736]

Probenecid.
Ampicillin-Probenecid Suspension [119]
Benemid Tablets [578]
ColBENEMID [578]
Col-Probenecid Tablets [240]
Polycillin-PRB [134]
Principen with Probenecid Capsules [815]
Probenecid Tablets [240], [344], [769]
Probenecid with Colchicine Tablets [344], [769]
SK-Probenecid Tablets [805]
Wycillin & Probenecid Tablets & Injection [942]

Probucol.
Lorelco [580]

Procainamide.
Procainamide Capsules [344]

Procainamide hydrochloride.
Procainamide HCl Capsules [240], [769]
Procan SR [674]
Pronestyl Capsules and Tablets, Injection, -SR Tablets [815]

Procaine hydrochloride.
Novocain [932]
Novocain Hydrochloride, Novocain Hydrochloride for Spinal Anesthesia [133]
Procaine HCl Injection [287]

Procarbazine hydrochloride.
Matulane Capsules [738]

Prochlorperazine.
Combid Spansule Capsules [805]
Compazine [805]
Prochlor-Iso Timed Release Capsules [769]
Prochlorperazine Tablets [344]

Prochlorperazine edisylate.
Prochlorperazine Edisylate Injection [287]
Prochlorperazine Edisylate in Tubex [942]

Prochlorperazine maleate.
Pro-Iso Capsules [344]

Procyclidine hydrochloride.
Kemadrin [139]

Progesterone.
BayProgest Injection [111]
Progestasert Intrauterine Contraceptive System [40]

Promazine hydrochloride.
Sparine Injection in Tubex [942]

Promethazine.
Dihydrocodeine Compound Tablets [769]
Maxigesic Capsules [560]
Promethazine DM (Ped) Expectorant, Expectorant Plain, w/Codeine Expectorant, VC Expectorant, VC w/ Codeine Expectorant [344]

Promethazine hydrochloride.
BayMeth [111]
Compal Capsules [723a]
Mepergan Injection, in Tubex [942]
Phenergan Expectorant Plain (without Codeine), Expectorant w/Codeine, Expectorant w/Dextromethorphan, Pediatric, Injection, in Tubex, Tablets, Syrup & Rectal Suppositories, VC Expectorant Plain (without Codeine), VC Expectorant w/Codeine [942]

Promethazine hydrochloride *(cont'd).*
Promethazine HCl Injection [287]
Remsed Tablets [269]
Stopayne Capsules, Syrup [814]
ZiPan-25 & ZiPan-50 [769]

Propamocarb hydrochloride.
Banol [884]

Propane.
A-108 [9], [695]
Hydrocarbon Propellant A-108 [695]

Propantheline bromide.
Pro-Banthine Tablets, with Phenobarbital [782]
Propantheline Bromide Tablets [240], [344], [751], [769]
SK-Propantheline Bromide Tablets [805]

Proparacaine hydrochloride.
Ocu-Caine [655a]

Propiomazine hydrochloride injection.
Largon in Tubex [942]

Propionic acid.
Tenox P Grain Preservative, Tenox P [278]

Propoxyphene hydrochloride.
Darvon, Compound, Compound-65 [523]
Darvon with A.S.A. [344]
Propox 65 w/APAP Tablets [344]
Propoxyphene & Apap Tablets 65/650, Compound 65 [769]
Propoxyphene Compound 65 Capsules, HCl Capsules [344]
Propoxyphene HCl Capsules [751], [769]
SK-65 APAP Tablets, SK-65 Capsules, Compound Capsules [805]
Wygesic Tablets [942]

Propoxyphene napsylate.
Darvocet-N 50, 100 [523]
Darvon-N, with A.S.A. [523]

Propranolol hydrochloride.
Inderal Tablets & Injectable, LA Long Acting Capsules [97]
Inderide [97]

S-Propyl butylethyl thiocarbamate.
Tillam [821]

Propyl ξ3-(dimethylamino) propyl carbamate monohydrochloride.
Banol [870]
Prevex [645]
Previcur N [645], [772]

S-Propyl dipropyl thiocarbamate.
Surpass [821]
Vernam [821]

Propyl gallate.
Lexgard PG [453]
Progallin P [632]
Tenox PG [278]

Propyl gallate (n-propyl 3,4,5 trihydroxybenzoate).
Sustane PG [883]

Propyl gallate (3,4,5-trihydroxybenzoic acid n-propyl ester).
Tenox PG [278]

Propyl oleate.
Emerest 2302 [289]
Grocor 3000 [374]

Propyl oleate, sulfated.
Cinwet BOS [193]
Hipochem Dispersol SP [410]
Marvanol SPO 60% [552]

Propyl paraben.
Cosept P [218a]
Lexgard P [453]
Nipasol M [632]
Preservaben P [867]
Propyl Aseptoform [370]
Protaben P [713]

Propyl sorbitol, hydroxylated.
Atlas G-2402 [445]

n-Propylene alkyl, fatty diamines.
Diamin AB, C, HT, O, S, T [803]

1,3-Propylene diamine, N-hydrogenated tallow.
Kemamine D-970 [430]

1,3-Propylene diamine (N-vegetable).
Kemamine D-999 [429]

Propylene glycol.
Otic-HC Ear Drops [394]
Water-Soluble Vegetol Broad Leaved Lime, Hop, Marshmallow, Cornflower, Hawthorn, BirchTree, Marigold, Common Nettle, Camomile [339]

Propylene glycol alginate.
Kelcoloid [484]
Kelcoloid HVF, D, LVF, O, S [485]

Propylene glycol alginate-carrageenan.
Dricoid K [682]

Propylene glycol/BHA/propyl gallate/citric acid: 70%:20%:60%:4%.
Sustane 3 [883]

Propylene glycol dibenzoate.
Benzoflex 284 [898]

Propylene glycol dicaprylate.
Crodamol PC [223]

Propylene glycol dicaprylate/dicaprate.
Cyclochem PGDC [234]
Hodag CC-22, CC-22-S [415]
Liponate PC [525]
Neobee M-20 [715]
Standamul 302 [400]

Propylene glycol dicocoate.
Dodecalene [901]

Propylene glycol diisononanoate.
Liponate L [525]

Propylene glycol dilaurate.
CPH-110-N [379]

Propylene glycol dioleate.
CPH-73-N [379]

Propylene glycol dipelargonate.
DPPG [339]
Emerest 2388 [290]
Schercemol PGDP [771]

Propylene glycol distearate.
CPH-155-N [379]
Kessco Alcohol Esters [825]
Mapeg PGDS [564]

Propylene glycol esters.
Radiamuls PG 2201, 2206 [658]

Propylene glycol fatty acid ester.
Witconol F26-46 [934]
Witconol RHP [935]

Propylene glycol hydroxystearate.
Naturechem PGHS [158]
Paricin 9 [640a]

Propylene glycol laurate.
Drewmulse PGML [715]
Kessco PGML-X533 [79]
Mapeg PGML, PGMS [564]
Schercemol PGML [771]
Tegin PL [360]

1,2-Propylene glycol mono-distearate.
Propylenglycolmonostearat Protegin [856]
Tegin P, P 412 [856]
Tegomuls P 411 [856]
Witconol RHP [937a], [938]

Propylene glycol, monoester.
Myverol P-06 [278]

Propylene glycol monofatty acid esters.
Canamulse 55 [145]

Propylene glycol monolaurate.
CPH-5-SE, CPH-32-J-N, CPH-32-N [379]
Drewmulse PGML [715]
Hodag PGML [415]

Propylene glycol monolaurate-E.
Kessco Alcohol Esters [825]

Propylene glycol monolaurate, non-emulsifying.
Cithrol PGML N/E [223]

Propylene glycol monolaurate, self-emulsifying.
Cithrol PGML S/E [223]

Propylene glycol monomyristate.
Radiasurf 7196 [658]

Propylene glycol monooleate.
Atlas G-2185 [445]
CPH-3-SE, CPH-3M-SE, CPH-150-N, CPH-243-N [379]
Radiasurf 7206 [658]
Rilanit PMO [401]

Propylene glycol monooleate, non-emulsifying.
Cithrol PGMO N/E, PGMO S/E [223]

Propylene glycol monooleate, self-emusifying.
Cithrol PGMO S/E [223]

Propylene glycol monopalmitate.
Atlas G-2183 [445]

Propylene glycol monoricinoleate.
Cithrol PGMR N/E, PGMR S/E [223]
Flexricin 9 [158], [642]

Propylene glycol monoricinoleate, non-emulsifying.
Cithrol PGMR N/E [223]

Propylene glycol monoricinoleate, self-emulsifying.
Cithrol PGMR S/E [223]

Propylene glycol monostearate.
Aldo PGHMS, PMS CG [358]
Cerasynt PA [895]
CPH-51-N, CPH-52-SE, CPH-301-N [379]
Cyclochem PGMS [234]
Drewlene 10 [715]
Drewmulse PGMS [266]
Emerest 2300 [290]
Emerest 2381 [289]
Hodag PGMS [415]
Homotex PS-90 [478]
Kessco Alcohol Esters [825]
Kessco-Propylen-glycol-monostearate [16]
Lexemul P [453]
Mapeg PGMS [564]
Mazol PGMS [564]
Nikkol PMS-1C [631]
PMS-33 [397]
Radiasurf 7201 [658]
Schercemol PGMS [771]
Tegin P [360]
Tegomuls P 610 [856]

Propylene glycol monostearate, non-emulsifying.
Cithrol PGMS N/E [223]

Propylene glycol monostearate, self-emulsifying.
Cithrol PGMS N/E, PGMS S/E [223]
Emerest 2381 [290]
Nikkol PMS-1CSE [631]

Propylene glycol oleate.
CPH-3-SE [379]

Propylene glycol ricinoleate.
Flexricin 9 [640a]
Naturechem PGR [158]

Propylene glycol soyate.
Kessco 3283 [79]

Propylene glycol stearate.
Aldo PMS [358]
Cerasynt PA [895]
Cyclochem PGMS [234]
Durpro 107-55 [272]
Emerest 2380 [290]
Grindtek PGMS 90 [373]
Hodag PGMS [415]
Kessco PGMS-R, Propylene Glycol, Monostearate Pure [719]
Lipo PGMS [525]
Monosteol [339]
Schercemol GMS, PGMS [771]
Tegin P61 [360]
Witconol PS-50 [935]

Propylene glycol stearate (hi mono).
Aldo PGHMS [358]
PGHMS [358]

Propylene glycol stearate, self-emulsifying.
CPH-52-SE [379]
Emerest 2381 [290]
Kessco PGMS-8615, PGMS-X174, PGMS-X534F [79]
Lexemul P [453]
Tegin P [360]
Tesal [339]
Witconol RHP [934]

Propylene oxide-ethylene oxide alkoxylate.
Sandoxylate SX-208 [761]

S-Propylidpropylthiocarbamate.
Vernam [821]
Vernolate [821]

2-Propynyl (E,E) 3,7,11-trimethyl-2,4-dodecadienoate.
Enstar [947]

Protamine sulfate.
Protamine Sulfate [523]
Protamine Sulfate for Injection, USP, Sterile Solution [884]

Protease.
Papain [313]
Protease-110C [54]
Protoferm A (acid), N (neutral) [313]

Protease amylase enzyme supplement, diastatic.
Fermex [913]

Protease enzyme.
Amerzyme A400-LC [54]

Protease, fungal.
Protease-75B [54]
Protoferm [313]

Protease (papain).
Papain P-100 [313]

Proteinase, bacteria.
Neutrase 0.5 L, 1.5 G [654]

Proteolytic enzyme.
Fungozyme AN3, AN6 [54]
Papain 3000 [54]
Rennilase 11 L, 14 L, 46 L, 150 L type T [654]
Texzyme [707]

Proteolytic preparations.
Ananase [746]
Arco-Lase, Plus [72]
Biozyme-C Ointment [80]
Celluzyme Chewable Tablets [239]
Cotazym, -S [664]
Festal II [418]
Festalan [418]
Gustase [350]
Kutrase Capsules [499]
Ku-Zyme Capsules, HP [499]
Panafil Ointment, -White Ointment [759]
Pancreatin Tablets 2400 mg. N.F. (High Lipase) [907]
Travase Ointment [327]
Tri-Cone Capsules, Plus Capsules [354]

Protirelin.
Thypinone [2]

Protriptyline hydrochloride.
Vivactil [578]

Pseudoephedrine hydrochloride.
Actifed-C Expectorant [139]
Ambenyl-D Decongestant Cough Formula [551]
Anafed Capsules & Syrup [304]
Anamine Syrup, T.D. Caps [563]
Brexin Capsules & Liquid, L.A. Capsules [767]
Bromfed Capsules (Timed Release), -PD Capsules (Timed Release), Tablets [612]
Cardec DM Drops & Syrup [769]

Pseudoephedrine hydrochloride *(cont'd).*
Chlorafed H.S. Timecelles, Liquid, Timecelles [394]
Codimal-L.A. Capsules [166]
Congress Jr. & Sr. T.D. Capsules [326]
CoTylenol Cold Medication Tablets & Capsules, Liquid Cold Medication [569]
Deconanime Tablets, Elixir, SR Capsules, Syrup [115]
Dristan Ultra Colds Formula Aspirin-Free Analgesic/Decongestant/Antihistamine/Cough Suppressant Nighttime Liquid, Tablets and Capsules [922]
Ex-Span Tablets [748]
Fedahist Expectorant, Gyrocaps, Syrup & Tablets [746]
Guaifed Capsules (Timed Release) [612]
Histalet DM Syrup, Syrup, X Syrup & Tablets [723a]
Isoclor Timesule Capsules [49]
Kronofed-A Jr. Kronocaps, -A Kronocaps [314]
Novafed A Capsules, A Liquid, Capsules, Liquid [580]
Novahistine DH, Expectorant [580]
Nucofed Capsules, Expectorant, Pediatric Expectorant, Syrup [114]
Poly-Histine-DX Capsules, -DX Syrup [124]
Pseudo-Bid Capsules [420]
Pseudoephedrine HCl Tablets [240], [769]
Pseudoephedrine Hydrochloride Tablets [751]
Pseudoephedrine Hydrochloride & Triprolidine Hydrochloride Syrup, Tablets [751]
Pseudo-Hist Capsules, Expectorant, Liquid [420]
Respaire-SR Capsules [513]
Rienafed Capsules, -EX Capsules [566]
Robitussin-DAC [737]
Ryna Liquid, C Liquid, CX Liquid [911a]
T-Moist Tablets [930]
Triafed-C Expectorant [769]
Trifed Tablets & Syrup [344]
Trinex Tablets [560]
Tripodrine Tablets [240]
Tussend Expectorant, Liquid & Tablets [580]
Tylenol, Maximum-Strength, Sinus Medication Tablets & Capsules [569]
Zephrex Tablets, -LA Tablets [124]

Pseudoephedrine hydrochloride + chlorpheniramine maleate; 60 mg:4 mg.
Cotrol-D Tablets [114]

Pseudoephedrine preparations.
Chlorafed H.S. Timecelles, Liquid, Timecelles [394]
Co-Pyronil 2 [254]
Deconamine Tablets, Elixir, SR Capsules, Syrup [115]
Deproist Expectorant w/ Codeine [344]
Detussin Expectorant, Liquid [769]
Disobrom Tablets [344]
Fedahist Expectorant, Gyrocaps, Syrup & Tablets [746]
Poly-Histine-DX Capsules, Syrup [124]
Probahist Capsules [518]
Pseudoephedrine S.R. Capsules, Tablets [344]
Rondec Drops, Syrup, Tablet, -DM Drops, -TR Tablet [747]
Triafed Syrup & Tablets [769]

Pseudoephedrine sulfate.
Trinalin Repetabs Tablets [772]

Psyllium.
Hydrocil [750]

Psyllium preparations.
Effersyllium [831]
Fiberall, Natural Flavor, Orange Flavor [757]
Hydrocil Instant [750]
Konsyl, -D (formerly L.A. Formula) [503]
Metamucil, Instant Mix, Orange Flavor, Instant Mix, Regular Flavor, Powder, Orange Flavor, Powder, Regular Flavor, Powder Strawberry Flavor [782]
Modane Bulk [7]
Nuggets [468]
Perdiem, Plain Granules [746]

PVP.
Luviskol K30, K90 [109]

Pyrantel pamoate.
Antiminth Oral Suspension [690]

Pyrazine 99+%.
Ultralog 77-5300-00 [172]

Pyrazolone red pigments, light, medium, and deep shades.
Plasticone Red [797]

Pyrethrin-rotenone plant spray concentrates.
Foliafume [680]

Pyridostigmine bromide.
Mestinon Injectable, Syrup, Tablets, Timespan Tablets [738]
Regonol [664]

Pyridoxine.
Aminoxin [872]
Herpecin-L Cold Sore Lip Balm [144]

Pyridoxine *(cont'd)*.
Rodex T.D. Capsules [518]

Pyridoxine dilaurate.
Nikkol Pyridoxine Dilaurate [631]

Pyridoxine dipalmitate.
Nikkol Pyridoxine Dipalmitate [631]

Pyridoxine hydrochloride.
Alba-Lybe [105]
Al-Vite [267]
Beelith Tablets [113]
Besta Capsules [394]
Eldertonic [563]
Glutofac Tablets [489]
Hemo-Vite, Liquid [267]
Mega-B [72]
Neuro B-12 Forte Injectable [505]
Nu-Iron-V Tablets [563]
Rodex [518]
Vicon-C Capsules, -Plus Capsules [354]

Pyrilamine maleate.
Albatussin [105]
Citra Forte Capsules, Syrup [132]
Codimal DH, DM, PH [166]
Excedrin P.M. [134]
4-Way Nasal Spray [134]
Fiogesic Tablets [258]
Histalet Forte Tablets [723]
Kronohist Kronocaps [314]
Poly-Histine-D Capsules, Elixir [124]
Primatene Tablets-M Formula [922]
P-V-Tussin Syrup [723]
Ru-tuss Expectorant, Plain, with Hydrocodone [127]
Triaminic Juvelets, Oral Infant Drops, TR Tablets (Timed Release) [257]
WANS (Webcon Anti-Nausea Supprettes) [915]

Pyrilamine tannate.
Rynatan Tablets & Pediatric Suspension [911a]

Pyrimethamine.
Daraprim [139]
Fansidar Tablets [738]

Pyrocatechin.
Rodol C [531]

Pyrocatechol.
Rodol C [531]

Pyrogallol.
Rodol PG [531]

Pyrophyllite-aluminum silicate.
Pyrax [894]

Pyrophyllite aluminum silicate, hydrated.
Pyrax A,B, WA [894]

2-Pyrrolidone.
2-Pyrol [338]

Pyrrolidone, substituted.
Solvofen HM [338]

Pyrvinium pamoate.
Povan Filmseals [674]

Rabies antiserum.
Antirabies Serum (equine), Purified [778]

Rabies immune globulin (human).
Imogam Rabies Immune Globulin (Human) [579]
Rabies Immune Globulin (Human) Hyperab [232]
Rabies Immune Globulin (Human), Imogam Rabies [579]

Rabies vaccine.
Imovax Rabies Vaccine, Rabies Vaccine I.D. [579]
Rabies Vaccine Human Diploid Cell, Imovax Rabies, Imovax Rabies I.D. [579]
Wyvac Rabies Vaccine [942]

Racemethionine.
Pedameth Capsules, Liquid [661]

Ranitidine hydrochloride.
Zantac Tablets [354]

N-Rapeseed-(3-amido-propyl)-N-N dimethyl-N-(2,3 epoxypropyl)-ammonium chloride.
Schercoquat ROEP [771]

Rapeseedamidopropyl benzyldimonium chloride.
Schercoquat ROAB [771]

Rapeseedamidopropyl ethyldimonium ethosulfate.
Schercoquat ROAS [771]

Rauwolfia preparations.
Harmonyl [3]
Raudixin Tablets [817]
Rauwiloid Tablets [734]

Rauwolfia serpentina.
Raudixin Tablets [817]
Rauzide Tablets [817]

Rauwolfia serpentina, whole root.
Raudixin [817]

Red petrolatum.
RVP Ointment, RVPaba Lip Stick, RVPaque Ointment [284]

Rescinnamine.
Moderil [691]

Reserpine.
Chloroserpine 250 & 500 Tablets [769]
Chlorothiazide w/Reserpine Tablets [344]
Demi-Regroton Tablets [891]
Diupres [578]
Diutensen-R Tablets [911a]
H-H-R Tablets [769]
Hydrochlorothiazide, Hydralazine HCl, Reserpine Tablets [240]
Hydrochlorothiazide/Reserpine Tablets [240]
Hydrochlorothiazide w/Reserpine Tablets [344]
Hydro-Fluserpine Tablets #1 & #2 [769]
Hydromox R Tablets [516]
Hydropres [578]
Hydroserpine Tablets #1 & #2 [769]
Metatensin [580]
Naquival Tablets [772]
Regroton Tablets [891]
Renese-R [691]
Reserpine Tablets [769]
Rezide tablets [281]
SK-Reserpine Tablets [805]
Salutensin/Salutensin-Demi [134]
Ser-Ap-Es [189]
Serpasil Parenteral Solution, Tablets, -Apresoline, -Esidrix [189]
Unipres Tablets [723a]

Reserpine .1 mg; hydralazine HCL 25 mg; hydrochlorothiazide 15 mg.
Rezide Tablets [281]

Resorcinol.
Castellani Paint [679]
Derma Cas Gel [412]
Hedal H-C Suppositories [78]
Rodol RS [531]

Resorcinol-isocyanate.
PA-53-086 [701a]

Resorcinol monobenzoate.
Eastman Inhibitor RMB [278]

Retinyl palmitate.
Aquapalm No. 63841 [738]

Rh (D) immune globulin (human).
Gamulin Rh [80]
MICRhoGAM [665]
Mini-Gamulin Rh [80]
Rh_0-D Immune Globulin (Human) HypRho-D, Immune Globulin (Human) HypRho-D Mini-Dose [232]
RhoGAM [665]

Riboflavin.
Riboflavin, No. 602995200 [738]

Ricinole amide alkyl betaine.
Rewoteric AM R 40

Ricinoleamide DEA.
Alkamide RDO [33]
Aminol CA-2 [320]
Condensate CA [174]
Cyclomide RD [939]
Mackamide R [567]
Protamide CA [713]
Rewo-Amid DR 280 [725]

Ricinoleamide DEA *(cont'd).*
Rodea [197]

Ricinoleamide MEA.
Cyclomide RM [939]
Emid 6573 [291]
Rewo-Amid R 280 [725]

Ricinoleamide MIPA.
Cyclomide RP [939]

Ricinoleamidopropyl betaine.
Mackam RA [567]
Rewoteric AM-R40 [725]
Varion AM-R40 [796]

Ricinoleamidopropyl dimethylamine.
Mackine 201 [567]

Ricinoleamidopropyl dimethylamine lactate.
Mackalene 216 [567]
Richamate 3555 [732]

Ricinoleamidopropyl ethyldimonium ethosulfate.
Lipoquat R [525]

Ricinoleic acid.
Aci-Jel Therapeutic Vaginal Jelly [666]
P-10 Acid [158]
P-10 Acid [640a]
Wochem #110 [941]

Ricinoleic acid alkanolamide.
Aminol CA-2 [320]

Ricinoleic acid alkanolamide sulfosuccinate.
Rewoderm S 1333/P [725a]

Ricinoleic acid monoethanolamide.
Emid 6573 [290]
Rewomid R 280 [725a]

Ricinoleic acid propylamido trimethyl ammonium methosulfate.
Rewoquat RTM 50 [725a]

Ricinoleic acid, technical.
P-10 Acids [640a], [641a]

Ricinoleic alkanolamide.
Alkamide RDO [33]

Ricinoleic diethanolamide.
Alkamide RDO [33]
Mackamide R [567]
Rodea [197]

Ricinoleic glycerides, ethoxylated.
Labrafil WL 1958 CS [339]

Ricinoleic sulfosuccinate (40% conc.).
Mackanate RM [567]

Ricinoleoamphoglycinate.
Rewoteric AM-2R [725]

Ricinoleth-40.
Poliglycoleum [901]

Riconoleamido amine.
Mackine 201 [567]

Rifampin.
Rifadin [580]
Rifamate [580]
Rimactane Capsules [189]

Ritodrine hydrochloride.
Yutopar Intravenous Injection, Tablets [90]

Rubella & mumps virus vaccine, live.
Biavax$_{II}$ [578]

Rubella virus vaccine, live.
Meruvax$_{II}$ [578]

Rubidium carbonate 99.9%.
Ultralog Grade 79-1900-00 [172]

Rubidium chloride 99.9%.
Ultralog Grade 79-1940-00 [172]

Rubidium iodide 99.9%.
Ultralog Grade 79-1960-00 [172]

Rubidium sulfate 99.9%.
Ultralog 79-1980-00 [172]

Rum attapulgite.
Refinex [330]

Rutile, synthetic.
KM [490]

Saccharin (powder and granular).
Saccharin SL, US [797]
Syncal [797]

Saccharose distearate.
Sucro Ester W.E. 7 [339]

Saccharose mono-distearate.
Sucro Ester W.E. 11 [339]

Saccharose monopalmitate.
Sucro Ester W.E. 15 [339]

Safflower glyceride.
Myverol 18-98 Distilled Monoglycerides, 18-98K Distilled Monoglycerides [278]

Safflower oil.
Lipovol SAF [525]

Safflower oil, hybrid.
Neobee 18 [715]

Salicylamide.
Codalan [510]
Espasmotex Tablets [78]
Korigesic Tablets [868]
Os-Cal-Gesic Tablets [551]
Sinulin Tablets [153]
Uromide [281]

Salicylic acid.
Acne-Dome Medicated Cleanser [585]
Aveenobar Medicated [214]
Barseb HC Scalp Lotion, Thera=Spray [103]
Cantharone Plus [786]
Duofilm [827]
Fostex Medicated Cleansing Bar, Medicated Cleansing Cream [920]
Hydrisalic Gel [679]
Keralyt Gel [920]
Komed Acne Lotion, HC Lotion [103]
Pernox Lotion, Medicated Lathering Scrub [920]
Salactic Film [679]
Sebucare [920]
Sebulex & Sebulex Cream Shampoo, Shampoo with Conditioners [920]
Sebutone & Sebutone Cream Shampoo [920]
Tineasol Lotion [78]
Tinver Lotion [103]
Verrex [199]
Verrusol [199]
Viranol [53]
Wart-Off [690]

Salicylic acid, treated.
Vulcatard SA [212]

Salicylsalicylic acid.
Disalcid [734]
Mono-Gesic Tablets [166]

Sarcoside, fatty acid.
Medialan Brands [416]
Sarkosine Grades [55]

Sarcoside, sodium salt, fatty acid (60% conc.).
Arkomon A Conc. [416]

Sarcosinate, fatty acid (40% conc.).
Sarkosine KF [55]

Sarcosine surfactants.
Sarkosyl [190]

Scopolamine hydrobromide.
Ru-Tuss Tablets [127]
Scopolamine Hydrobromide Injection [287]
Urogesic Tablets [281]

Scopolamine preparations.
Dallergy Syrup, Tablets and Capsules [513]
Transderm-Scop Transdermal Therapeutic System [189]

Sebacic acid, inorganic ester.
Harchem 2-SL [912]

Secobarbital sodium.
Secobarbital Sodium in Tubex [942]
Seconal Sodium Pulvules & Vials [523]
Tuinal [523]

Secretin.
Secretin-Kabi [693]

Selenious acid, 40 mcgm./ml.
Selenitrace [80]

Selenium.
Added Protection III Multi-Vitamin & Multi-Mineral Supplement [589]
Enviro-Stress with Zinc & Selenium [907]
Selenium Tablets (200 mcg) [907]
Total Formula [907]
Vandex [894]

Selenium, elementary.
Vandex [894]

Selenium diethyldithiocarbamate.
Ethyl Selenac [894]

Selenium dimethyldithiocarbamate.
Methyl Selenac [894]

Selenium sulfide.
Exsel Lotion [403]
Selsun Blue Lotion, Selsun Lotion [3]

Senna.
Perdiem [746]
Senokot Syrup [714]

Senna concentrates.
Senokot Tablets/Granules, -S Tablets [714]

Senna concentrates *(cont'd).*
X-Prep Liquid [368]

Sesame oil.
Lipovol SES [525]

Shea butter.
Karite Butter [784]

Shea butter, unsaponifiable.
Karite Nonsaponifiable [784]

Silica.
Aerosil 130, 200, 255, 300, 380, 400, R812, R972 [244]
CAB-O-Sil Fumed Silica, CAB-O-Sil HS-5 [141]
Earth, Diatomaceous [804]
FK 300 DS, FK 320, FK 320 DS [244]
Flo-Gard [705]
Hi-Sil 404, Hi-Sil EP [705]
Min-U-Sil [681]
Nyacol 830, 1430 [706]
Silica, Anhydrous 31, Crystalline 216 [924]
Siltex [479]
Sipernat 22, 22S, D17 [244]
Supersil [681]

Silica, amorphous.
Flo-Gard AG 110, AG 130, AG 150, CC 120, CC 140, CC 160, FF 310, FF 320, FF 330, FF 350, FF 370, FF 390 [705]
Imsil A-10, A-15, A-25, A-108 [448]

Silica, diatomaceous.
Celite 21-a, 110, 209, 219, 263, 266, 270, 281, 289, 292, 305, 315, 321, 321A, 350, 388, 400, 408, 410, 499, 1200, CAS-30KR, CS-22R, F.C., HSC, Snow Floss, Super Fine Super Floss, White Mist, Super Floss [472]

Silica, diatomaceous diatomite.
Dicalite 103 PS, SSF-5, WB-5 [371]

Silica, dispersion.
Ludox HS-30, SM [269]
Nyacol 215, 1440, 2030 EC, 2040 NH_4, 2046 EC, 9950 [706]

Silica, fibrous.
Sil-Temp [395]

Silica, fumed.
HDK [402]

Silica, fumed, colloidal.
Cabosil [141]

Silica, gel.
Britesorb [706]

Silica gel, bulk.
Natrosorb T [611a]

Silica, hydrated.
Amorphous Silica 0, 54, 200, 250, 1160, 1240 [448]
QUSO G30, G32, WR50, WR82 [706]
Syloid 63, 63 FP, 72, 72 FP, 74, 74 FP, 244, 244 FP, 266, 266 FP [242]
Zeo 49 [428]
Zeodent, Zeodent 113 [428]
Zeofree, Zeofree 80, 153 [428]
Zeosyl, Zeosyl 200 [428]
Zeothix, Zeothix 95, 265 [428]

Silica, hydrated, amorphous.
Hi-Sil, Hi-Sil 210, 215, 233, 250, 260, 262, 422 [705]
Silene D [705]
Quso [694]

Silica, hydrous.
Zeo [428]

Silica, powder.
Levilite [730]

Silica, precipitated.
32 Silica [188]
Sipernat D 17 [244]

Silica, precipitated, amorphous.
Quso [706]

Silica, precipitated, hydrated, amorphous.
Hi-Sil EP, 210, 215, 233 [392], [705]
Silene D [392], [705]

Silica, precipitated, spray dried.
Sipernat 22, 22 S [244]

Silica, pyrogenic, aq. dispersion.
Cab-O-Sperse [141]

Silica, synthetic amorphous.
Hi-Sil T-600 [705]

Silica alumina, microspherical.
M-S [364a]

Silicon carbide.
Ferrocarbo [364]
Globar [364]
SiC [8]

Silicon carbide brick.
Harbide [364]

Silicon complex.
Chromosol SS [637]

Silicon dioxide.
Aerosil Colloidal Silicas, Aerosil R-972 [244]
Sipernat 22, 22S, D10, D17 [244]
Zeothix 265 [428]

Silicon dioxide, amorphous.
Aerosil [244]
Cab-O-Sil [141]
Imsil [448]

Sodium alkylate sulfonate, linear (50% conc.).
Bio Sof D-60 [825]

Sodium alkylate sulfonate, linear (60% conc.).
Bio Sof D-62 [825]
Ultrawet 60K [72]

Sodium alkylate sulfonate, linear (90% conc.).
Ultrawet DS, K, KX [72]

Sodium alkyl benzene sulfate, linear.
Polyfac LAS-50S [918]

Sodium alkyl benzene sulfate slurry, linear.
Polyfac LAS-40S, LAS-50S [918]

Sodium alkyl benzene sulfonate.
Gardilene S25-L [23]
Marlon ARL [178]
Nansa HS40-AU [24], [178]
Nansa HS40 Soft [26]
Nansa HS80 [24]
Nansa HS80-AU [23], [24]
Nansa HS80P [23], [24], [26]
Nansa HS80/S, HS80SK [23]
Nansa HS80SK [24]
Nansa HS80 Soft, HS 85/S [26]
Nansa SL30, SS30, SS50, SS60, 1042/P, 1106/P, 1169/P, 1292, 1293 [23]
Zohar 60 SD [946]

Sodium alkyl benzene sulfonate (40% conc.).
Nansa HS 40-AU [24]

Sodium alkyl benzene sulfonate (78% conc.).
Nansa HS 80-AU [24]

Sodium alkyl benzene sulfonate (80% conc.).
Nansa HS 80 P [24]

Sodium alkyl benzene sulfonate, (aq. sol'n.).
Agent 768-82 [825]

Sodium alkyl benzene sulfonate, branched.
Nansa SB62 [23]

Sodium alkyl benzene sulfonate, branched (22% conc.).
Polystep A-16-22 [825]

Sodium alkyl benzene sulfonate, branched (30% conc.).
Polystep A-16 [825], [825a]

Sodium alkyl (C_{12}) benzene sulfonate, linear.
Polystep A-4, A-6, A-15, A-16 [825]

Sodium alkyl benzene sulfonate, linear (22% conc.).
Polystep A-15 [825], [825a]

Sodium alkyl benzene sulfonate, linear (soft) (34-36% conc.).
Ufasan 35 [874]

Sodium alkyl benzene sulfonate, linear (39% conc.).
Polystep A-7 [825]

Sodium alkyl benzene sulfonate, linear (50% conc.).
Polystep A-4 [825], [825a]

Sodium alkyl benzene sulfonate, linear (55% conc.).
Cedepon SS-55 [256]

Sodium alkyl benzene sulfonate, linear (soft) (64-66% conc.).
Ufasan 65 [874]

Sodium alkyl benzene sulfonate, linear (soft) (83-87% conc.).
Ufaryl DL85 [874]

Sodium alkyl benzene sulfonate, solubilized.
Gardilene S25/L [24]

Sodium alkyl benzene sulfonate, solubilized (25% conc.).
Gardilene S25/L [24]

Sodium alkyl diaminoethyl glycine.
Lebon 15 [764]

Sodium alkyl diphenyl ether disulfonate.
Eleminol MON-2 [764]
Newcol 261 A, 271 A [635]

Sodium alkyl diphenyl ether disulfonate (45% conc.).
Newcol 261 A, 271 A [635]

Sodium alkylether sulfate.
Empimin LSM 30 [23]
Manro BES 27, BES 70 [545]
Polyfac PCW-92 [918]
Sulfotex PAI-S [400]

Sodium alkylether sulfate (26.5% conc.).
Sandet END, ENM [764]

Sodium alkylether sulfate (27% conc.).
Witcolate AE 35 [937a], [938]

Sodium alkylether sulfate (C_{12}-C_{14}) (3 EO) (27% conc.).
Serdet DCK 30 [788]

Sodium alkylether sulfate (C_{12}-C_{15}) (3 EO) (27% conc.).
Serdet DPK 30 [788]

Sodium alkylether sulfate (28% conc.).
Steol 7N [825]

Sodium alkylether sulfate (30% conc.).
Empimin LSM 30 [25]

Sodium alkylether sulfate (31.5% conc.).
Sandet EN [764]

Sodium alkylether sulfate (40% conc.).
Berol 475 [116]

Sodium alkylether sulfate (58% conc.).
Steol CS760 [825a]

Sodium alkylether sulfate (60% conc.).
Cedepal FS-406 [256]
Polystep B-23 [825], [825a]
Sterling Super A [145]

Sodium alkylether sulfate (65% conc.).
Avirol SE-3012 [399], [400]

Sodium alkylether sulfate (C_{12}-C_{14}) (2 EO) (67% conc.).
Serdet DCK 2/65 [788]

Sodium alkylether sulfate (C_{12}-C_{14}) (2 EO) (70% conc.).
Serdet DCK 3/70 [788]

Sodium alkylether sulfate (C_{12}-C_{15}) (3 EO) (70% conc.).
Serdet DPK 3/70 [788]

Sodium alkylether sulfate, industrial grade, aq. sol'n.
Empimin LSM30 [25]

Sodium alkylethoxy sulfate.
Empimin LSM30, LSM30-AU [23]
Empimin LSM30-AU [24]

Sodium alkylethoxy sulfate, industrial grade.
Empimin LSM30-AU [24]

Sodium alkyl naphthalene sulfonate.
Alkanol XC [269]
Eccowet LF Conc. [338]
Nansa BXS [26]
Nekal BA-77, BX-78, NF [338]
NSAE Powder [663]
Rexowet MS [292]
Sellogen HR, W, WL [252]
Soitem B [807]
Titazole SA [863]

Sodium alkyl naphthalene sulfonate (35.5% conc.).
Sellogen HR-90 Liquid [252]

Sodium alkyl naphthalene sulfonate (36% conc.).
Sellogen WL Liquid [252]

Sodium alkyl naphthalene sulfonate (50% conc.).
Pelex NBL [478]

Sodium alkyl naphthalene sulfonate (65% conc.).
Wettol NT 1 [106]

Sodium alkyl naphthalene sulfonate (70% conc.).
Sellogen W [252]

Sodium alkyl naphthalene sulfonate (75% conc.).
Nekal BA-77, BX-78 [338]

Sodium alkyl naphthalene sulfonate (79% conc.).
Sellogen K [252]

Sodium alkyl naphthalene sulfonate, formaldehyde condensate.
Atlox 4862 [92]

Sodium alkyl phenol sulfate (28% conc.).
Polystep B-27 [825]

Sodium alkyl phenol sulfate (4 EO).
Polystep B-27 [825]

Sodium alkyl phenoxy polyoxyethylene sulfate.
Gardisperse AC [23]
Gardisperse AC [24]

Sodium alkyl phenoxy polyoxyethylene sulfate (30% conc.).
Gardisperse AC [24]

Sodium alkyl polyether sulfate (40% conc.).
Polystep B-28 [825]

Sodium alkyl polyether sulfate, ethoxylated.
Polystep B-14, B-28 [825]

Sodium alkyl polyglycol ether phosphate.
Chimin P40, P45 [514]
Marlophor FC-Na-Salt [178]

Sodium alkyl polyglycol ether phosphate (30% conc.).
Chimin P45 [514]

Sodium alkyl polyglycol ether phosphate (75% conc.).
Chimin P40 [514]

Sodium alkyl polyglycol ether phosphate (85% conc.).
Marlophor FC-Na-Salt [178]

Sodium alkyl sulfate.
Cosmopon 35 [514]
Drewfax 220 [266]
Empimin LR 28 [23]
IML-1 [270]
Manro DS 35 [545]
Perlankrol O [509]
Polyfac SLS-30, SLS-30 Special [918]
Richonol 7031 [801]
Sandopan KD, KD Conc. [762]
Vanfre IL-1 [894]

Sodium alkyl sulfate (5% conc.).
Emersal 6462 [289]

Sodium alkyl sulfate (28% conc.).
Empimin LR 28 [25]

Sodium alkyl sulfate (28-30% conc.).
Drewfax 220 [266]

Sodium alkyl sulfate (30% conc.).
Emal 40 [478]
Sulfopan WA 1 Special [399]

Sodium alkyl sulfate (C_{12}-C_{14}) (31% conc.).
Serdet DFK 30 [788]

Sodium alkyl sulfate (38% conc.).
Emersal 6462 [290]

Sodium alkyl sulfate (39% conc.).
Sipex CAV [29]

Sodium alkyl sulfate (40% conc.).
Perlankrol O [253]
Witcolate D5-10 [938]

Sodium alkyl sulfate (C_{12}-C_{14}) (41% conc.).
Serdet DFK, DFK 40 [788]

Sodium alkyl sulfate (90% conc.).
Drewfax 250 [266]

Sodium alkyl sulfate, aq. sol'n.
Agent 613-95 [825]
Empimin LR28 [25]
Polyfac SLS-30, SLS-30 Special [918]

Sodium alkyl sulfate, modified.
Sandopan KD, KD Conc. [762]

Sodium alkyl sulfate, modified (50% conc.).
Alphenate TFC-76 [253]

Sodium alkyl sulfosuccinamate.
Empimin MHH, MKK, MKK 98, MKK/L, MTT [23]

Sodium alkyl sulfosuccinamate (35% conc.).
Empimin MKK, MKK/L [25]

Sodium alkyl sulfosuccinamate (40% conc.).
Empimin MHH, MTT [25]

Sodium alkyl sulfosuccinamate (98% conc.).
Empimin MKK 98 [25]

Sodium alkyl sulfosuccinate.
Newcol 290K, 290M, 290P [635]

Sodium alkyl triethoxysulfate (25% conc.).
Empimin SQ25 [24]

Sodium alkyl triethoxysulfate (68% conc.).
Empimin SQ70 [24]

Sodium alumino silicate zeolite.
Linde ZB-100, ZB-300, ZB-400 [880]

Sodium aluminum chlorohydroxy lactate.
Choracel [721]

Sodium aluminum phosphate.
Levn-Lite [602]
Pan-O-Lite [602]
Stabil-9 [602]

Sodium arsenite.
Pennite [682]

Sodium ascorbate.
Cee-500 (Sodium Ascorbate) [518]
Sodium Ascorbate Fine Granular, No. 6047709, Fine Powder, No. 6047708, Type AG, No. 6047710 [738]

Sodium bicarbonate.
Arm & Hammer [78a]
Ceo-Two Suppositories [117]
Infalyte [682]
Pedi-Bath Salts [679]
Quik-Prep [468]
Sodium Bicarbonate [134]

Sodium bifluoride.
Na-0101 T 1/8″ [389]

Sodium biphosphate.
Sodium Phosphates Oral Solution [751]
Uro-Phosphate Tablets [703]

Sodium bis-naphthalene sulfonate.
Disrol SH [635]

Sodium bis-tridecyl sulfosuccinate.
Aerosol TR 70 [50], [233]
Emcol 4600 [935]

Sodium bis-tridecyl sulfosuccinate (70% conc.).
Aerosol TR 70 [233]

Sodium borate.
Trichotine Liquid, Vaginal Douche [719]

Sodium butabarbital.
Butisol Sodium Elixir & Tablets [911a]

Sodium butoxyethoxy acetate.
Mirawet B [592]

Sodium butyl biphenylsulfonate.
Aresket [602a]

Sodium n-butyl naphthalene sulfonate (75% conc.).
Morwet B [688]

Sodium butyl-ortho-phenylphenol sulfonate.
Areskap [602a]

Sodium cacodylate/cacodylic acid 29:5, with surfactant.
Rad-E Cate 35 [904]

Sodium cacodylate dimethyl arsenic acid (cacodylic acid).
Clean-Boll [904]
Dutch-Treat [904]
Rad-E-Cate 16, 25, 35 [904]

Sodium caparyl lactylate.
Pationic 122A [674a], [675], [735]
Pationic R-122A [676]

Sodium carbonate.
Acculog 80-6710-00 [172]

Sodium carboxymethylcellulose.
Aqualon C, R [406]
Blanose Cellulose Gum, Refined CMC [405]
Carbose [109]
Carmethose [191]
Cellex [616]
CMC-6CTL, CMC-T, CMC Warp Size, CMTC-T [406]
Hercules Cellulose Gum, Cellulose Gums, CMC [406]
Qualex [270]
Sarcell [764a]
Tylose C, CB Series, CBR Series, CR [416]

Sodium carboxymethylcellulose, creped sheet form.
Aquasorb Absorbent Sheet [406]

Sodium carboxymethylcellulose, crude.
Hercules CMC-6CTL, CMC-6-DG-L [406]

Sodium carrageenan.
Viscarin [549]

Sodium caseinate.
Vi-Mate [908]

Sodium castor oil (66% conc.).
Servo Brillant Olie B AZ 75 [788]

Sodium castor oil sulfonate (45% conc.).
Servo Brillant Olie B AN 50 [788]

Sodium cetearyl sulfate.
Dehydag Wax E [400]
Lanette E [399]

Sodium cetyl oleyl sulfate.
Empicol CHC 30 [26]

Sodium cetyl stearyl sulfate.
Empicol TAS 30 [23]
Polystep B-26 [825]

Sodium cetyl stearyl sulfate (25% conc.).
Sipex EC 111 [29]

Sodium cetyl stearyl sulfate (30% conc.).
Empicol TAS 30 [25]

Sodium cetyl stearyl sulfosuccinate (45% conc.).
Secosol AS [825a]

Sodium cetyl sulfate.
Conco Sulfate C [209]
Nikkol SCS [631]
Sipex EC-111 [29]

Sodium cetyl sulfate (30% conc.).
Conco Sulfate C [209]

Sodium chlorate.
Altacide [444]
Harvest-Aid [490]
KM [490]
Tumbleaf [490]

Sodium chloride.
Acculog 80-7000-00 [172]
BaySaline [111]
Infalyte [682]
Pedi-Bath Salts [679]
Quik-Prep [468]
Sodium Chloride, Bacteriostatic in Tubex [942]
Sodium Chloride Inhalation [751]
Sodium Chloride Injection (Unpreserved) [287]
Sodium Chloride Injection, Bacteriostatic [287]
Trichotine Powder, Vaginal Douche [719]

Sodium chloride, micron.
Microsized Salt [251a]

Sodium chloride tablets 250 mg.
Ocu-Disal [655a]

Sodium 5-ξ2 chloro-4 (trifluoromethyl) phenoxyζ-2-nitrobenzoate.
Blazer, 2L, 2S [730a]

Sodium citrate.
Bicitra—Sugar-Free [929]
Polycitra Syrup, -LC—Sugar-Free [929]
Quik-Prep [468]
Tussar SF, -2 [891]

Sodium cocaminopropionate.
Deriphat 151 [399]

Sodium-N-cocoaminopropionate.
Deriphat 151 [400]

Sodium coco-hydrolyzed animal protein.
Maypon SK [825]

Sodium coco isethionate (83% conc.).
Tauranol I-78 [320]

Sodium coco monoglyceride sulfate.
POEM-LS-90 [733]

Sodium N-coconut-acid-N-methyl taurate (24% conc.).
Igepon TC-42 [338]

Sodium coconut monoglyceride sulfonate.
SCMS [413]

Sodium coconut oil soap (92% conc.).
Norfox Coco Powder [650]

Sodium coconut sulfosuccinamate (35% conc.).
Alkasurf SS-CA [33]

Sodium cocoyl collagen peptide (30% conc.).
Nikkol CCN-40 [631]

Sodium cocoyl glutamate.
Acylglutamate CS-11 [14]

Sodium cocoyl isethionate.
Igepon AC-78 [338]

Sodium N-cocoyl methyl taurate (30% conc.).
Nikkol CMT-30 [631]

Sodium cocoyl sarcosinate.
Hamposyl C-30 [364a]
Sarkosine KA Conc [55]

Sodium cocoyl sarcosinate (30% conc.).
Hamposyl C-30 [363]

Sodium-N-cocoyl sarcosinate (30-31% conc.).
Closyl 30 [198]

Sodium cumene sulfonate.
Eltesol SCS40 [23]
Na-Cumene Sulfonate 40, Sulfonate Powder [178]
Reworyl NCS 40 [725], [725a]
Stepanate C-S [825]
Ultra SCS Liquid [935]
Witconate SCS [934], [935]

Sodium cumene sulfonate (40% conc.).
Na-Cumene Sulfonate 40 [178]

Sodium cumene sulfonate (45% conc.).
Naxonate 45SC [754]
Witconate SCS Liquid [935]

Sodium cumene sulfonate (93% conc.).
Na-Cumene Sulfonate Powder [178]
Naxonate SC [754]
Witconate SCS Powder [935]

Sodium cyanide.
Cymag [444]

Sodium cyanide (98% conc.).
Cyanobrik [270]
Cyanogran [270]

Sodium N-cyclohexyl-N-palmitoyl taurate.
Igepon CN-42 [338]

Sodium N-cyclohexyl-N-palmitoyl taurate (23% conc.).
Igepon CN-42 [338]

Sodium decanol sulfate.
Empimin SDS [23], [25]

Sodium deceth sulfate.
Richonol 7093 [732]

Sodium decyl benzene sulfonate.
Ultrawet DS [72]

Sodium n-decyl diphenyl ether disulfonate.
Dowfax 3B2 [262]

Sodium n-decyl diphenyl oxide disulfonate (45% conc.).
Conco Sulfate 3B2 Acid [209]
Dowfax 3B2 [261]

Sodium decyl ether sulfate.
Carsonol SDES [529]

Sodium decyl sulfate.
Polystep B-25 [825]
Sulfotex 110 [400]

Sodium n-decyl sulfate (30% conc.).
Avirol SA-4110 [399], [400]

Sodium n-decyl sulfate (31-33% conc.).
Sulfotex 110 [399]

Sodium decyl sulfate (38% conc.).
Polystep B-25 [825]

Sodium decyl sulfate (40% conc.).
Merpinal D 40 [285]

Sodium dehexyl sulfosuccinate.
Alcopol OS [37]

Sodium deisobutyl sulfosuccinate.
Alcopol OB [37]

Sodium dextrothyroxine.
Choloxin [327]

Sodium diacetate.
Dykon [824]

Sodium dialkyl naphthalene sulfonate (70% conc.).
Sellogen W [253]

Sodium dialkyl naphthalene sulfonate (75% conc.).
Sellogen HR [252], [253]

Sodium dialkyl naphthalene sulfonate (90% conc.).
Sellogen HR [253]

Sodium dialkyl sulfosuccinate.
Empimin OT [25]

Sodium dialkyl sulfosuccinate (55% conc.).
Stantex T-15 [399], [400]

Sodium dialkyl sulfosuccinate (63% conc.).
Empimin MA [25]

Sodium diamyl sulfosuccinate.
Aerosol AY, AY-65 [50]
Cyanasol AY Surface Active Agent [50]

Sodium diamyl sulfosuccinate *(cont'd)*.
Geropon AY [550]

Sodium diamyl sulfosuccinate (70% conc.).
Geropon AY [550]

Sodium di-n-butyl dithio carbamate.
Butyl Namate [894]
Tepidone [270]

Sodium dibutyl dithio carbamate, in water.
Butyl Namate [894]

Sodium dibutyl naphthalene sulfonate.
Geropon NK [550]

Sodium dibutyl naphthalene sulfonate (65% conc.).
Geropon NK [550]

Sodium dichloro isocyanurate.
Diamin C [112]

Sodium dichloro isocyanurate dihydrate.
ACL 56 [602]

Sodium dichromate, dry, granular, orange.
Diesel-Treat [142]

Sodium dicyclohexyl sulfosuccinate.
Aerosol A-196 [50], [233]

Sodium dicyclohexyl sulfosuccinate (40% conc.).
Protowet 4337 [710]

Sodium dicyclohexyl sulfosuccinate (85% conc.).
Aerosol A 196 [233]

Sodium (diester) sulfosuccinate.
Emcol 4580PG [935]

Sodium (diester, asymmetrical) sulfosuccinate.
Emcol 4930 [935]

Sodium di-2-ethyl hexyl sulfosuccinate.
Serwet WH 170 [788]

Sodium di-2 ethyl-hexyl sulfosuccinate (65% conc.).
Serwet WH 170 [788]

Sodium di-2-ethyl hexyl sulfosuccinate (70% conc.).
Servoxyl VLA 2170 [788]

Sodium di 2-ethyl hexyl sulfosuccinate (75% conc.).
Protowet D-75 [710]

Sodium di-2-ethyl-hexyl sulfosuccinate (98% conc.).
Nikkol OTP-100S [631]

Sodium dihexyl sulfosuccinate.
Aerosol MA-80 [50], [233]
Alcopol OS [37]
Empimin MA [23]
Monawet MM-80 [599]

Sodium dihexyl sulfosuccinate (60% conc.).
Alcopol OS [37]

Sodium dihexyl sulfosuccinate (79% conc.).
Protowet 4604 [710]

Sodium dihexyl sulfosuccinate (80% conc.).
Aerosol M-80 [233]

Sodium dihexyl sulfosuccinate (contains ethanol).
Lankropol KMA [509]

Sodium dihexyl sulfosuccinate sol'n.
Empimin MA [23]

Sodium 2,2′-dihydroxy-4,4′-dimethoxy-5-sulfobenzophenone.
Uvinul DS-49 [338]

Sodium dihydroxyethyl glycinate.
Hampshire DEG [363]

Sodium dihyroxyethyl glycine.
Hampshire DEG [363]

Sodium diisobutyl sulfosuccinate.
Aerosol IB-45 [50]
Alcopol OB [37]
Condanol SB-DO 70 [274]
Monawet MB-45 [599]

Sodium diisobutyl sulfosuccinate (45% conc.).
Alcopol OB [37]

Sodium diisooctyl sulfosuccinate.
Geropon CYA/DEP [550]
Rewopol SB-DO 70 [274]
Steinapol SB-DO 70 [274]

Sodium diisooctyl sulfosuccinate (75% conc.).
Geropon CYA/DEP [550]

Sodium diisopropyl naphthalene sulfonate.
Aerosol OS [233]

Sodium diisopropyl naphthalene sulfonate (75% conc.).
Aerosol OS [233]
Morwet IP [688]

Sodium {4-(dimethylamino) phenyl} diazene sulfonate.
Bayer 5072 [597a]
Lesan [112], [597a]

Sodium di-(methylamyl) sulfosuccinate.
Manoxol MA [544]

Sodium dimethyl dithiocarbamate.
Vulnopol NM [27]
Wing-Stop [362]

Sodium dimethyl dithiocarbamate/ sodium-2-mercapto-benzothiazole 27.6%:2.4%.
Vancide 51 [894]

Sodium dinaphthyl methane sulfonate.
Geropon RM/77, RM/100 [550]

Sodium dinaphthyl methane sulfonate (80% conc.).
Geropon RM/77 [550]
Geropon RM/100 [788]

Sodium dinonyl naphthalene sulfonate in light min. oil.
Na-Sul SS [894]

Sodium dinonyl sulfosuccinate.
Manoxol N [544]

Sodium dioctyl sulfosuccinate.
Aerosol GPG, OT-70 PG, OT-75, OT-100, OTB, OTS [233]
Aerosol OT-100% [50]
Alcopol O 60%, O 70PG, O 100% [37]
Alkasurf SS-O-60, SS-O-75 [33]
Alrowet D-65 [190]
Astrowet O-75 [28]
Avirol SO-70P [400]
Cyanosol OT 100% Surface Active Agent [50]
Denwet CM [365]
Detersil [401]
Drewfax 0007 [264]
Emcol 4500, 4560 [935]
Empimin OP45, OP70, OT, OT75 [23]
Hipochem EK-18 [410]
Monawet MO-70, MO-70E, MO-70R, MO-75E, MO-84R2W [599]
Nekal WT-27 [338]
Penetron Conc. [391]
Tex-Wet 1001 [460]

Sodium dioctyl sulfosuccinate (20% conc.).
Empimin OP45, OP70 [23]
Empimin OT [33]

Sodium dioctyl sulfosuccinate (30% conc.).
Nissan Rapisol B-30 [636]

Sodium dioctyl sulfosuccinate (35% conc.).
Alkasurf SS-O-75 [34]

Sodium dioctyl sulfosuccinate (40% conc.).
Alkasurf SS-O-40 [33]

Sodium dioctyl sulfosuccinate (42-45% conc.).
Thorowet G-40 [198]

Sodium dioctyl sulfosuccinate (60% conc.).
Alcopol O 60% [37]
Alkasurf SS-O-60 [33], [34]
Tex Wet 1001 [460]

Sodium dioctyl sulfosuccinate (65% conc.).
Santex T-14 [399], [400]

Sodium dioctyl sulfosuccinate (70% conc.).
Aerosol GPG, OT70PG, OTS [233]
Alcopol O 70PG [37]
Progawet DOSS [109]
Sanmorin OT70 [764]
Schercowet DOS-70 [771]
Secosol DOS/70 [825a]

Sodium dioctyl sulfosuccinate (72% conc.).
Empimin OP70 [25]

Sodium dioctyl sulfosuccinate (73-76% conc.).
Thorowet G-75 [198]

Sodium dioctyl sulfosuccinate (75% conc.).
Aerosol OT-75 [233]
Alkasurf SS-O-75 [33]
Hipochem EK-18 [410]
Insidol 75 [77]
Pentex 99 [200]

Sodium dioctyl sulfosuccinate (80% conc.).
Aerosol OT-80 PG [50a]
Nissan Rapisol B-80 [636]

Sodium dioctyl sulfosuccinate (85% conc.).
Aerosol OTB [233]

Sodium dioctyl sulfosuccinate (contains ethanol).
Lankropol KO 2 [253]

Sodium dioctyl sulfosuccinate (contains ethanol) (60% conc.).
Lankropol KO 2 [509]

Sodium dioctyl sulfosuccinate (contains mineral oil).
Lankropol KO Special [253]

Sodium dioctyl sulfosuccinate (contains mineral oil) (60% conc.).
Lankropol KO Special [509]

Sodium dioctyl sulfosuccinate, modified.
Emcol 4560 [935]

Sodium dioctyl sulfosuccinate and sodium benzoate.
Aerosol OT-B [50]

Sodium diphenyl methane sulfonate (70% conc.).
Geropon FMS [550]

Sodium diprotrizoate.
Miokon Sodium [542]

Sodium ditridecyl sulfosuccinate.
Alcopol OD [37]
Monawet MT-70, MT-70E, MT-80H2W [599]

Sodium ditridecyl sulfosuccinate (60% conc.).
Alcopol OD [37]

Sodium dodecyl benzene sulfonate.
Arylan SC 30, S Flake [509]
Bio Soft D-35X, D-40, D-60, D-62 [825]
Calsoft F-90, L-40, L-60 [697]
Conco AAS-35, AAS-40, AAS-40G, AAS-45S, AAS-65, AAS-90 [209]
Cycloryl DDB 40 [234]
Di-Aqua [251]
Elfan WA, WA Powder [17]
Hetsulf 40, 40X, 60S, Acid [407]
Kadif 50 Flakes [937]
Manro BA 30, DL 32, SDBS 30 [545]
Marlon A 350, A 360, A 365, A 375, A 390, A 396, AM 70, AM 80, AM 90, ARL [178]
Mercol 25, 30 [147]
Nacconol 35SL, 40F, 40G, 90G [825]
Nacconol 90F [26]
Nansa HS 55, HS 80 [26]
Nansa HS80/S [25]
Nansa HS 80/S, HS 85/S [548]
Nansa SB 25, SB 62 [23]
Nansa SL 30, SS 30, SS 60 [548]
Nansa 1106, 1169 [23]
Parnol 40 [643]
Reworyl NKS 50 [25]
Reworyl NKS 50, NKS 100 [725a]
Richonate 1850U, 40 B, 45 B, 45BX, 50 B D, 60B, 1850, C-50H [732]
Siponate DS-4, DS-10 [29]
Sul-fon-ate AA-9, AA-10, LA-10 [852]
Ultrawet 60K, 68 KN, K [72]
Witconate 90 Flakes, 1238 Slurry [935]
Witconate 1250 Slurry [934], [935]

Sodium dodecyl benzene sulfonate (23% conc.).
Siponate DS-4 [29]

Sodium dodecyl benzene sulfonate (25% conc.).
Nansa SB 25 [25]

Sodium dodecyl benzene sulfonate (30% conc.).
Manro BA 30, SDBS 30 [545]
Nansa 1106, 1169 [25]

Sodium dodecyl benzene sulfonate (31.5% conc.).
Conco AAS-35 [209]

Sodium dodecyl benzene sulfonate (32% conc.).
Manro DL 32 [545]

Sodium dodecyl benzene sulfonate (40% conc.).
Conco AAS-40F, AAS-40S [209]
Norfox 40 [650]
Witconate 1850 [935]

Sodium dodecyl benzene sulfonate (42% conc.).
Witconate 45BX [935]

Sodium dodecyl benzene sulfonate (44% conc.).
Witconate C-50H [935]

Sodium dodecyl benzene sulfonate (50% conc.).
Elfan WA 50 [764]
Marlon A 350 [178]
Reworyl NKS 50 [725]

Sodium dodecyl benzene sulfonate (60% conc.).
Conco AAS-65S [209]
Witconate 60B [935]

Sodium dodecyl benzene sulfonate (62% conc.).
Nansa SB 62 [25]

Sodium dodecyl benzene sulfonate (65% conc.).
Marlon A 365 [178]

Sodium dodecyl benzene sulfonate (70% conc.).
Soitem 5C/70 [807]

Sodium dodecyl benzene sulfonate (75% conc.).
Marlon A 375 [178]
Serdet DMK 75 [788]

Sodium dodecyl benzene sulfonate (80% conc.).
Elfan WA Powder [764]

Sodium dodecyl benzene sulfonate (85% conc.).
Conco AAS-90S [209]
Norfox 85 [650]

Sodium dodecyl benzene sulfonate (90% conc.).
Conco AAS-90S [209]
Marlon A 390 [178]
Norfox 90 [650]

Sodium dodecyl benzene sulfonate (96% conc.).
Marlon A 396 [178]

Sodium dodecyl benzene sulfonate (98% conc.).
Siponate DS-10 [29]
Sul-fon-ate AA-10 [194]

Sodium dodecyl benzene sulfonate, branched (hard) (79-82% conc.).
Ufaryl DB80 [874]

Sodium dodecyl benzene sulfonate, flake, branched (chain).
Arylan S90 Flake [509]

Sodium dodecyl benzene sulfonate, hard.
Emulsifier 30 [697]

Sodium dodecyl benzene sulfonate, linear.
Merpisap AE 50, AE 60, AE 70, AP 80W, AP 85W, AP 90, AP 90P, KH 30 [285]
Sterling AB-40, AB-80, LA Paste 55%, LA Paste 60% [145]
Tairygent CB-1 [335]

Sodium dodecyl benzene sulphonate, linear (30% conc.).
Merpisap KH 30 [285]

Sodium dodecyl benzene sulfonate, linear (50% conc.).
Merpisap AE 50 [285]

Sodium dodecyl benzene sulfonate, linear (55% conc.).
Sterling LA Paste 55% [145]

Sodium dodecyl benzene sulfonate, linear (57% conc.).
Sterling LA Paste 60% [145]

Sodium dodecyl benzene sulfonate, linear (60% conc.).
Merpisap AE 60 [285]

Sodium dodecyl benzene sulfonate, linear (70% conc.).
Merpisap AE 70 [285]

Sodium dodecyl benzene sulfonate, linear (80% conc.).
Merpisap AP 80W [285]

Sodium dodecyl benzene sulfonate, linear (85% conc.).
Merpisap AP 85W, AP 90P [285]

Sodium dodecyl benzene sulfonate, linear (88% conc.).
Merpisap AP 90 [285]

Sodium dodecyl benzene sulfonate, modified.
Marlon AFR [178]

Sodium dodecyl benzene sulfonate, modified (30% conc.).
Marlon AFR [178]

Sodium dodecyl benzene sulfonate, nonionic blends (70% conc.).
Marlon AM 70 [178]

Sodium dodecyl benzene sulfonate, nonionic blends (80% conc.).
Marlon AM 80 [178]

Sodium dodecyl benzene sulfonate, nonionic blends (90% conc.).
Marlon AM 90 [178]

Sodium dodecyl benzene sulfonic acid.
Sandet 60 [764]

Sodium dodecyl diphenyl ether disulfonate.
Dowfax 2A1 [262]
Eleminol MON-7 [764]

Sodium dodecyl diphenyl oxide disulfonate (45% conc.).
Conco Sulfate 2A1 [209]
Dowfax 2A1, XDS-30237 [261]

Sodium dodecyl diphenyl oxide disulfonate (50% conc.).
Pelex SS [478]

Sodium ether-alcohol sulfate.
Duponol RA [269]

Sodium ether lauryl sulfate.
Sterling Super [145]

Sodium ether sulfate, ethoxylated (high) (28% conc.).
Texapon 130E [399]

Sodium ether sulfate, ethoxylated (high) (58% conc.).
Texapon 125E Conc. [399]

Sodium (2)-ethylhexyl polyglycol ether phosphate (3 EO) (85% conc.).
Servoxyl VPT 3/85 [788]

Sodium (2)-ethylhexyl sulfate.
Avirol SA-4106 [400]
Carsonol SHS [155], [529]
Merpinal EH 40 [285]
Rewopol NEHS 40 [725a]
Sipex BOS [29]
Sordet DS K 40 [788]
Sulfotex OA [400]
Witcolate D-510 [934], [935]

Sodium (2) ethylhexyl sulfate (5% conc.).
Emersal 6465 [289]

Sodium (2)-ethylhexyl sulfate (35% conc.).
Sole-Terge TS-2-S [415]

Sodium (2)-ethylhexyl sulfate (40% conc.).
Carsonol SHS [529]
Merpinal EH 40 [285]
Niaproof 08 [627]
Serdet DSK 40 [788]
Sipex BOS [29]
Sulfotex OA [399], [400]

Sodium (2) ethylhexyl sulfate (45% conc.).
Avirol SA-4106 [399], [400]

Sodium (2) ethylhexyl sulfosuccinate.
Chimin DOS [514]

Sodium (2) ethylhexyl sulfosuccinate (68% conc.).
Chimin DOS [514]

Sodium fatty acid sarcosinate (35% conc.).
Sarkosine KA Liquid [55]

Sodium fatty acid sarcosinate (65% conc.).
Sarkosine KA Conc. [55]

Sodium fatty acid soaps.
Doittol 891 [253]

Sodium fatty alcohol phosphate.
Forlanit F 452 [401]

Sodium fatty alcohol phosphate (14-16% conc.).
Forlanit F 452 [401]

Sodium fatty alcohol triethoxy sulphate.
Empimin SQ25 [23], [24]
Empimin SQ70 [23]

Sodium ferric ethylene diamine tetraacetate, trihydrate.
Hamp-Ene NaFe Purified Grade [363]

Sodium fluoride.
Fluoritab Tablets & Fluoritab Liquid [331]
Luride Drops, Lozi-Tabs Tablets [427]
Pediaflor Drops [747]
Phos-Flur Oral Rinse Supplement [427]
Point-Two Dental Rinse [427]
Poly-Vi-Flor 1.0 mg Vitamins w/ Fluoride Chewable Tablets, 0.5 mg Vitamins w/ Fluoride Drops, 1.0 mg Vitamins w/ Iron & Fluoride Chewable Tablets [571]
Polyvitamin-Fluoride Drops & Tablets [769]
Thera-Flur Gel-Drops [427]
Tri-Vi-Flor 1.0 mg Vitamins w/ Fluoride Chewable Tablets, 0.25 mg Vitamins w/ Fluoride Drops, 0.5 mg Vitamins w/ Fluoride Drops [571]

Sodium fluosilicate.
Safsan [471a]

Sodium formaldehyde hydrosulfite.
Formopon [742]

Sodium formaldehyde sulfoxylate.
Hydrosulfite AWC [252]

Sodium-α-d-glucoheptonate dihydrate.
Seqlene 540 [689]

Sodium D-gluconate.
PMP Sodium Gluconate Tech [707]

Sodium glyceryl oleate phosphate.
Emphos D70-30C [934], [935]

Sodium heptadecyl sulfate.
Niaproof 7 [627]

Sodium heptadecyl sulfate (26% conc.).
Niaproof 7 [627]

Sodium n-hexadecyl diphenyl oxide disulfonate (35% conc.).
Dowfax XDS-8390 [261]

Sodium hexametaphosphate.
Calgon [577]
Fosfodril [449]
Hexaphos [332]
Polyphos [659]

Sodium hexametaphosphate, sodium phosphate glass.
Metafos [755]

Sodium hyaluronate.
Healon [693]

Sodium hydrocarbon sulfonate.
Charlab Leveler AT Special, Leveler DSL [159]

Sodium hydrosulfite.
Lykopon [742]
Vatrolite [753]

Sodium hydrosulfite, buffered.
Clalite [753]

Sodium hydrosulfite, reducing agent.
Zepar BP [270]

Sodium hydroxide.
Acculog 81-0650-00, 81-0650-10 [172]
Caustic Soda Beads [261]

Sodium (2) hydroxy ethane sulfonate.
Rewopol NI 56 [725]

Sodium (2) hydroxy ethane sulfonate (56% conc.).
Rewopol NI 56 [725]

Sodium hydroxyethyl cocamidoethyl glycinate (50% conc.).
Emery 5412 [290]

Sodium hydroxyethyl octamido ethyl glycinate (30% conc.).
Emery 5418 [290]

Sodium (2) hydroxyethyl sulfate.
Rewopol NI 56 [725a]

Sodium hypochlorite (15% conc.).
Hipochem Bleach [410]

Sodium (iron EDTA).
Sequestrene NAFe 13% Fe [190]

Sodium isethionate.
Emery 5440 [291]

Sodium isethionate (56% conc.).
Emery 5440 [290]

Sodium isethionate, coco base (85% conc.).
Jordapon CI [474]

Sodium isethionate, coconut oil acid ester (83% conc.).
Igepon® AC-78 [338]

Sodium isethionate, fatty acid ester.
Hostapon 50, KA Pdr. Hi. Conc. [55]

Sodium isoascorbate.
Curona [913]

Sodium isodecyl sulfate.
Sipex CAV [29]

Sodium isopropyl naphthalene sulfonate.
Aerosol OS [50]
Geropon IN [550]

Sodium isopropyl naphthalene sulfonate (65% conc.).
Geropon IN [550]

Sodium isosteanoyl-21 actylate.
Pationic ISL [674a]

Sodium isostearoyl lactylate.
Crodactil SISL [223]
Pationic ISL [676], [735]

Sodium isostearoyl (2) lactylate.
Pationic ISL [674a], [735]

Sodium lactate.
Patlac NAL 60% [735]
Sodium Lactate [223]
Sodium Lactate (60% conc.) [674a], [675]

Sodium laureth (13) carboxylate.
Miranate LEC [592]

Sodium laureth-(4) phosphate.
Gafac MC-470F [338]

Sodium laureth sulfate.
Alkasurf ES-60 [33]
Calfoam ES-30 [697]
Carsonol SES-S, SLES, SLES-2, SLES-4 [529]
Conco Sulfate 219, 219 Special, WE [209]
Cycloryl NA [939]
Cycloryl NA 2, NA 61, NA 61 CG, NA 305 [234]
Emersal 6452, 6453 [290]
Empicol ESB3, ESB30, ESB50, ESB70, ESC3 [548]
Empimin KSN27, KSN60, KSN70 [548]
Genapol LRO Liquid, LRO Paste [55]
Maprofix 60S, ES, ES-2, ESY, LES60C [663]
Montelane CL 2288, LAU [785]
Rewopol NL3 [725]
Richonol OME, S-1285, S-1285C, S-5260 [732]
Serdet DCK 2/30, DCK 2/65, DCK 3/60, DCK 30, DFK 40 [175]
Sipon ES, ES-2, ESY [29]
Standapol ES-1, ES-2, ES-3, ES-7099, SL-60 [400]
Steol 4N, CS-125, CS-460 [825]
Sterling Super A, IE [145]
Texapon ES-1, ES-2, ES-3, ES-40, N25, N40, N70, N70LS [400]
Witcolate SE-5 [935]
Witcolate S1285C [934]

Sodium laureth sulfate (24% conc.).
Witcolate™ OME [935]

Sodium laureth sulfate (25% conc.).
Standapol ES-1 [399], [400]

Sodium laureth sulfate (26% conc.).
Standapol ES2 [399], [400]

Sodium laureth sulfate (28% conc.).
Standapol ES-3 [399], [400]

Sodium laureth sulfate (60% conc.).
Witcolate S-1285 C [935]

Sodium laureth (2) sulfate.
Sipon ES 2 [29]
Sterling Super 2E-25%, 2E-60% [145]

Sodium laureth (3) sulfate.
Sipon 201-10, ES [29]

Sodium laureth (3.5) sulfate.
Sterling Super 3.5E [145]

Sodium laureth (5) sulfate.
Carsonol SLES-5 [529]

Sodium laureth (7) sulfate.
Sipon ES-7 [29]
Steol CS-760 [825]

Sodium laureth (12) sulfate.
Sipon ES-12 [29]
Standapol 125-E, 130-E [400]

Sodium laureth (12) sulfate (30% conc.).
Standamul LC [399], [400]

Sodium lauriminodipropionate.
Deriphat 160C [399]
Mirataine H2C-HA [592]
Velvetex 610-L [400]

Sodium lauriminodipropionate (30% conc.).
Velvetex [400]

Sodium lauroyl glutamate.
Acylglutamate LS-11 [14]

Sodium lauroyl lactylate.
Pationic 138C [674a], [675], [735]
Pationic R-138C [676]

N-Sodium N-lauroyl-N-methyl alaninate.
Nikkol Alaninate LN-30 [631]

Sodium N-lauroyl methyl taurate (92% conc.).
Nikkol LMT [631]

Sodium lauroyl sarcosinate.
Hamposyl L-30, L-95 [363]
Maprosyl 30 [663]
Nikkol Sarcosinate LN [631]
Sarkosine LD [56]
Sarkosyl NL-30 [190]

Sodium lauroyl sarcosinate (30% conc.).
Hamposyl L-30 [363]
Maprosyl 30 [663]

Sodium N-lauroyl sarcosinate.
Crodasinic LS30, LS 35 [223]

Sodium lauroyl sarcosine.
Hamposyl L-30, L-95 [363]

Sodium lauryl alkylolamide sulfosuccinate.
Alkasurf SS-L9DE, SS-L9ME [33]

Sodium lauryl diethoxysulfate.
Empicol ESB, ESB3, ESB3GA, ESB30 [23]

Sodium lauryl diethoxysulfate (20% conc.).
Empicol ESB50, ESB70 [24]
Empicol ESB70-AU [23]

Sodium lauryl ether phosphate.
Forlanit P [401]
Forlanon [401]

Sodium lauryl ether phosphate (10% conc.).
Forlanon [401]

Sodium lauryl ether phosphate (30% conc.).
Forlanit P [401]

Sodium lauryl ether sulfate.
Akyposal DS 28, DS 56, EO 20, EO 20 MW, EO 20 PA, EO 20 PA/TS, EO 20 SF, MS, MS Conc., 23 ST 70 [182]
Alkasurf ES-60 [33]
Alscoap TAP-30 [864]
Berol 452 [116]
Calfoam ES-30 [697]
Carsonol SES-S, SLES, SLES-2 [155], [529]
Condanol NL 3, 28% [274]
Cosmopon LE 50 [514]
Cycloryl JB-8, MI, NA, NA-2, NA 61, NA61 CG, NWC, TS30 [234]
Drewpon EBO, EG, ELB, ETH, EVE [265]
Elfan NS 242, NS 242 Conc., NS 243 S, NS 243 S Conc., NS 252 S, NS 252 S Conc. [16]
Empicol ESB-70 [23]
Empicol ESC3 [25]
Empicol ESC 70 [23]
Manro NEC 70 [545]
Merpoxal LM 3028, LM 3070 [285]
Polyfac AES-60S [918]
Rewopol NL 3-28 [274], [725a]
Rewopol NL 3, 28% [725]
Rewopol NL 3 S-70 [725a]
Steinapol NL 3-28 [274]
Steol CS-130, CS-230, CS-330 [825]
Sulfatol E3, KE3 [1]
Sulfotex LMS-E [400]
Syntol N77 [1]
Texapon N 40 [398]

Sodium lauryl ether sulfate (22% conc.).
Texapon SG [401]

Sodium lauryl ether (2) sulfate (22% conc.).
Tylorol BS [858]

Sodium lauryl ether sulfate (24.5-25.5% conc.).
Nutrapon ESY [650]

Sodium lauryl ether sulfate (25% conc.).
Cedepal SS-203, SS-303, SS-403 [256]
Cycloryl NA2 [234]
Polystep B-16 [825a]
Sactipon 2 OS [521]
Sactol 2 OS [521]
Sipon ESY [29]
Sterling Super IE, 2E 25% [145]
Tylorol S [858]

Sodium lauryl ether (2) sulfate (25% conc.).
Emersal 6452 [290]
Tylorol 2S/25 [858]

Sodium lauryl ether sulfate (3.5 EO) (25% conc.).
Maprofix ES [663]

Sodium lauryl ether sulfate (26% conc.).
Polystep B-19 [825]
Sipon ES-2 [29]
Texapon ES-2 [399]

Sodium lauryl ether sulfate (26-28% conc.).
Drewpon EVE [266]

Sodium lauryl ether sulfate (3 EO. synthetic) (26-28% conc.).
Ungerol CG27 [874]

Sodium lauryl ether sulfate (27% conc.).
Sipon ES [29]
Swascol L-327 [837]
Zoharpon ETA 27 [946]

Sodium lauryl ether sulfate (27-28% conc.).
Texapon N 25 [401]

Sodium lauryl ether sulfate (3 EO) (27-28% conc.).
Texapon N 103 [401]

Sodium lauryl ether sulfate (2 EO. natural) (27-29% conc.).
Ungerol N2-28 [874]

Sodium lauryl ether (3 EO. natural) (27-29% conc.).
Ungerol N3-28 [874]

Sodium lauryl ether sulfate (28% conc.).
Akyposal DS 28, EO 20 [182]
Berol 452 [116]
Elfan NS 242, NS 243S, NS 252S [764]
Merpoxal LM 3028 [285]
Norfox SLES-02, SLES-03 [650]
Rewopol NL 3, 28% [725]
Sactipon 2 OS 28 [521]
Sactol 2 OS 28 [521]
Sterling Super 3.5E [145]
Texapon ES-3 [399]
Texapon N 40, NSO [401]

Sodium lauryl ether (2) sulfate (28% conc.).
Tylorol 2S [858]

Sodium lauryl ether (3) sulfate (28% conc.).
Emersal 6453 [290]
Tylorol 3S [858]

Sodium lauryl ether sulfate (30% conc.).
Alscoap TAP-30 [864]
Avirol SE-3001, SE-3002, SE-3003 [399], [400]
Carsonol SLES, SLES-2 [529]
Conco Sulfate WE [209]
Cycloryl NA [234]
Norfox SLES-30 [650]

Sodium lauryl ether (2) sulfate (30% conc.).
Maprofix ES2 [663]

Sodium lauryl ether sulfate (35-37% conc.).
Texapon EVR [401]

Sodium lauryl ether sulfate (50% conc.).
Sactol 2 OS 2 [521]

Sodium lauryl ether sulfate (56% conc.).
Akyposal DS 56 [182]

Sodium lauryl ether (3) sulfate (56-59% conc.).
Nutrapon ES 60 [650]

Sodium lauryl ether sulfate (57% conc.).
Texapon N Conc. [401]

Sodium lauryl ether sulfate (58% conc.).
Alkasurf ES-60 [33]

Sodium lauryl ether sulfate (60% conc.).
Carsonol SES-S [529]
Cedepal SS-306, SS-406 [256]
Cycloryl NA 61, NA 61 CG [234]
Hartenol LES 60 [391]
Norfox SLES-60 [650]
Polystep B-12 [825]

Sodium lauryl ether sulfate (3.5 EO) (60% conc.).
Maprofix LES-60C [663]

Sodium lauryl ether sulfate (60 ± 1% conc.).
Mars ES-60 [554]

Sodium lauryl ether sulfate (2 EO. natural) (68-72% conc.).
Ungerol N2-70 [874]

Sodium lauryl ether sulfate (3 EO. natural) (68-72% conc.).
Ungerol N3-70 [874]

Sodium lauryl ether sulfate (69% conc.).
Empicol ESC70 [25]

Sodium lauryl ether sulfate (70% conc.).
Akyposal 23 ST70 [182]
Elfan NS 242 Conc., NS 243 S Conc., NS 252 S Conc. [764]
Empicol ESB-70 [25]
Manro NEC 70 [545]
Merpoxal LM 3070 [285]
Texapon N 70, N 70 LS, N 70 N [401]

Sodium lauryl ether sulfate (1.0 ETO).
Maprofix ESY [663]

Sodium lauryl ether sulfate (2 EO) (70% conc.).
Zoharpon ETA 70 [946]

Sodium lauryl ether sulfate (3 EO).
Maprofix ES [663]

Sodium lauryl ether sulfate (3.0 EO).
Maprofix ES, LES-60C [663]

Sodium lauryl ether sulfate (3 EO) (70% conc.).
Zoharpon ETA 703 [946]

Sodium lauryl ether sulfate, amide.
Cosmopon ESC [514]

Sodium lauryl ether sulfate, built, pearlized.
Cycloryl MI [234]

Sodium lauryl ether sulfate, industrial grade.
Empimin KSN27, KSN 60 [25]
Empimin KSN 70 [23], [25]

Sodium lauryl ether sulfate, sol'n.
Polyfac LES-60S, SD-60S [918]

Sodium lauryl ether sulfonate (3 EO. synthetic) (53-55% conc.).
Ungerol Les3-54 [874]

Sodium lauryl ether sulfonate (3 EO. synthetic) (68-72% conc.).
Ungerol Les3-70 [874]

Sodium lauryl ether sulfosuccinate.
Sulfatol 88.22 [1]

Sodium lauryl monoether sulfate (1 EO) (25% conc.).
Maprofix ESY [663]

Sodium lauryl myristyl ether sulfate (30% conc.).
Texapon K 14S Special [401]

Sodium lauryl myristyl ether sulfate (70% conc.).
Texapon K 14S 70 Special [401]

Sodium lauryl oleyl sulfate.
Duponol D Paste [269]

Sodium lauryl poly (oxyethylene sulfate) (30% conc.).
Sulfotex EL [399]

Sodium lauryl polyether sulfate.
Lakeway 201-10, 201-11 [126]

Sodium lauryl polyglycol ether sulfate (3 ethylene oxide).
Condanol NL3 [274]

Sodium lauryl sarcosinate (30% conc.).
NL-30 [190]

Sodium lauryl sulfate.
Alkasurf WAQ [33]
Alscoap LN-40, LN-90 [864]
Arsul WAQ [76]
Berol 474 [116]
Calfoam SLS-30 [697]
Carsonol SLS [529]
Carsonol SLS Paste B [155], [529]
Carsonol SLS Special [155]
Carsonol SLS Special (Lonzol LS-300) [529]
Conco Sulfate WA, WA-1200, WA-1245, WA Dry, WA Special, WAG, WAN, WAS, WB-45, WBS-45, WN, WR, WX [209]
Condanol NLS 28, NLS 30, NLS 90 [274]
Cycloryl 21 [234], [934]
Cycloryl 21 LS, 21 SP [234]
Cycloryl 31, 580 [939]
Cycloryl 580, 585N [938]
Cycloryl 585N [939]
Drewpon 100, AS, CN [265]
Drewpon 100 [266]
Duponol C, ME Dry, QC, WA Dry Surface Active Agent, WA Paste, WAQ, WAQE [269]
Elfan 240 [16]
Emal O, 10 [478]
Emersal 6400, 6403 [290]
Empicol 0045 [23], [25], [548]
Empicol 0048 [548]
Empicol 0185 [23], [25], [548]
Empicol 0919 [23]
Empicol LM [23], [548]
Empicol LM40 [548]
Empicol LM45 [23], [548]
Empicol LM/T [23]
Empicol LMV [23], [548]
Empicol LMV/T [23]
Empicol LS 30, LS30B, LS30P [23]
Empicol LX [548]
Empicol LX28 [23], [548]
Empicol LXS95, LXSV [23]
Empicol LXV [23], [548]
Empicol LY28/S [23]
Empicol LZ, LZ34 [23], [548]
Empicol LZ41 [23]
Empicol LZ41/S, LZ/D [548]
Empicol LZ/E, LZG 30, LZGV [23], [548]
Empicol LZGV/C [23]
Empicol LZP, LZV [23], [548]
Empicol LZV/D [548]
Empicol LZV/E [23], [548]
Empicol WAK, WAQ [23]
Empimin LR28 [23]
Equex S, SP, SW [709]
Gardinol WA Paste [745]
Hartenol LAS-30, LES-60 [391]
Hoe S 2627 [55]
Incronol SLS [223b]
Lakeway 101-10, 101-11, 101-19 [126]
Lonzol LS-300 [529]
Manro SLS 28 [545]
Maprofix 563, 563-SD, LCP, LK, LK USP, MM, WAC, WAC-LA, WAC-LV, WAM, WA Paste, WAQ [663]
Mars SLS-A95 [554]
Melanol CL 30 [785]
Merpinal LM 50, LS 40, LS 50 [285]
Montopol LA Paste [711]
Neopon LS [635]
Nikkol SLS [631]
Orvus WA Paste [709]
Peri-Wash [839]
Polystep B-3, B-5, B-24 [825]
Rewopol NLS [725]
Rewopol NLS 28 [274], [725]
Rewopol NLS 15/L, NLS 30L [725a]
Rewopol NLS 90 [725]
Rewopol NLS 90-Powder [274]
Richonol 6522, A, AF, A Powder, C [732]
Sipex LCP, SB, UB [29]
Sipon LS, LSB [29]
Standapol WA-AC, WAS-100, WAQ-115, WAQ-LC [400]
Standapol WAQ Special [399], [400]

Sodium lauryl sulfate *(cont'd).*
Steinapol NLS 28, NLS 90-Powder [274]
Stepanol ME Dry, WA 100, WAC, WA Extra, WA Paste, WA Special, WAQ [825]
Sterling PA Paste, WA Paste, WAQ, WAQ-CH, WAQ-Cosmetic, WAQ-LO [145]
Sulfatol 33 Pasta, 33 Pasta 13, 33 Paste [1]
Sulfopon 101, 101 Special, 103 [400]
Sulfopon WA-3 [399]
Sulfopon WA 30 [398]
Sulfopon WAQ Special [399]
Sulfotex LCX [400]
Surgi-Kleen [839]
Swascol 3L [837]
Texapon K-12 [400]
Texapon K-1296 [399], [400]
Texapon L-100, V HC Needles, V Highly Conc. Needles, ZHC Needles, ZHC Powder, Z Highly Conc. Needles, Z Highly Conc. Powder [400]
Trichotine Liquid, Vaginal Douche, Powder, Vaginal Douche [719]
Witcolate A [934], [935]

Sodium lauryl sulfate (5% conc.).
Emersal 6400 [289]

Sodium lauryl sulfate (20% conc.).
Empicol LM45, LM/T, LMV/T, LX, LX28, LX34, LXV, LZ/E, LZG 30, LZGV, LZGV/C, LZP, LZV, LZV/E [23]

Sodium lauryl sulfate (20% ± 1 conc.).
Mars SLS-30 [554]

Sodium lauryl sulfate (28% conc.).
Empicol LX 28, LY 28S [25]
Manro SLS 28 [545]
Polystep B-24 [825]
Rewopol NLS 28 [725]
Sterling WA-Paste [145]
Sulfotex LCX [399]
Sulfotex WA, WA-6576 [399], [400]

Sodium lauryl sulfate (28-30% conc.).
Drewpon CN [266]
Nutrapon W, WAC, WAQ [198]
Standapol WA-AC, WAQ-LC, WAQ-LCX [399], [400]
Stepanol WA-100 [825a]
Stepanol WA-C [825], [825a]
Stepanol WA Paste [825]
Stepanol WA Paste, WA Special, WAQ [825a]
Sulfotex WA Paste [399], [400]

Sodium lauryl sulfate (29% conc.).
Alkasurf WAQ [33]
Carsonol SLS-Paste B [529]
Emersal 6403 [290]
Neopon LS [937a], [938]
Norfox SLS, SLS-03 [650]
Polystep B-5 [825], [825a]
Sipex SB [29]
Sipon LSB [29]
Sterling WAQ-CH [145]

Sodium lauryl sulfate (29.5% conc.).
Sterling WAQ-Cosmetic, WAQ-LO [145]

Sodium lauryl sulfate (30% conc.).
Arsul WAQ [76]
Avirol SL-2010 [399], [400]
Calfoam SLS-30 [697]
Carsonol SLS, SLS-300, SLS-Special [529]
Cedepon LS-30, LS-30P [256]
Conco Sulfate WA [209]
Cycloryl 21LS [234]
Emersal 6400, 6402, 6410 [290]
Empicol LS30/P [24]
Empicol LZG 30 [25]
Hartenol LAS-30 [391]
Jordanol SL 300 [474]
Maprofix LCP, WAC, WAC-LA, WA Paste, WAQ [663]
Polyfac SLS-30 [918]
Sipex LCP, UB [29]
Stepanol WA-Extra [825]
Sulfopon WAQ LCX [399]
Witcolate 6522, A [935]
Zoharpon LAS, LAS Special [946]

Sodium lauryl sulfate (C_{10}-C_{16}) (30% conc.).
Sulfopon 102 [401]

Sodium lauryl sulfate (C_{12}-C_{16}) (30% conc.).
Sulfopon 101, 101 Special, 103 [401]

Sodium lauryl sulfate (31% conc.).
Akyposal NLS [182]

Sodium lauryl sulfate (33% conc.).
Zoharpon LAS Special [946]

Sodium lauryl sulfate (34% conc.).
Empicol LZ 34 [25]

Sodium lauryl sulfate (35% conc.).
Empicol 9019 [24]
Empicol LS30B, LS30P [34]
Empicol WAK [25]

Sodium lauryl sulfate (C_{12}-C_{18}) (35% conc.).
Sulfopon K 35 [401]

Sodium lauryl sulfate (40% conc.).
Alscoap LN-40 [864]
Berol 474 [116]
Gardinol WA Paste [745]
Merpinal LS 40 [285]
Suphonated 'Lorol' Paste [745]

Sodium lauryl sulfate (41% conc.).
Sermul EA 150 [788]

Sodium lauryl sulfate (41.5% conc.).
Maprofix Paste MM [663]

Sodium lauryl sulfate (43% conc.).
Sactipon 2 S3 [521]

Sodium lauryl sulfate (45% conc.).
Monotpol LA Paste [711]

Sodium lauryl sulfate (47% conc.).
Polystep B-3 [825]

Sodium lauryl sulfate (49% conc.).
Sactol 2 S 3 [521]

Sodium lauryl sulfate (50% conc.).
Merpinal LM 50 [285]

Sodium lauryl sulfate (C_{12}-C_{14}) (50% conc.).
Texapon Z granules [401]

Sodium lauryl sulfate (C_{12}-C_{14}) (58-62% conc.).
Texapon Z [401]

Sodium lauryl sulfate (82% conc.).
Empicol LMV [25]

Sodium lauryl sulfate (84% conc.).
Empicol LXV, LZGV, LZV [25]

Sodium lauryl sulfate (85% conc.).
Texapon 2H C needles [399], [400]

Sodium lauryl sulfate (C_{12}) (85% conc.).
Texapon K 12 granules [401]

Sodium lauryl sulfate (86% conc.).
Empicol LM [25]

Sodium lauryl sulfate (88% conc.).
Empicol LZ [25]

Sodium lauryl sulfate (C_{12}-C_{16}) (88% conc.).
Texapon Z High Conc. Needles. [401]

Sodium lauryl sulfate (C_{12}-C_{18}) (88% conc.).
Texapon Z High Conc. Needles [401]

Sodium lauryl sulfate (C_{12}-C_{16}) (88-92% conc.).
Texapon V High Conc. Needles [401]

Sodium lauryl sulfate (89% conc.).
Empicol LX, LZP [25]

Sodium lauryl sulfate (90% conc.).
Alscoap LN-90 [864]
Drewpon 100 [266]
Emal 10 [478]
Empicol LZ/E, LZV/E [25]
Maprofix LK, LK (USP) [663]
Michelene LS-90 [581]
Texapon K-12 [399], [400], [401]
Texapon VH C needles [399], [400]
Witcolate A Powder [935]

Sodium lauryl sulfate (C_{12}-C_{14}) (90% conc.).
Texapon LS Highly Conc. Needles [401]

Sodium lauryl sulfate (C_{12}-C_{18}) (90% conc.).
Texapon Z High Conc. [401]

Sodium lauryl sulfate (90-96% conc.).
Conco Sulfate WX [209]

Sodium lauryl sulfate (C_{12}-C_{18}) (91% conc.).
Texapon OT H.C Needles [401]

Sodium lauryl sulfate (93% conc.).
Stepanol ME Dry [825]

Sodium lauryl sulfate (94% conc.).
Empicol 0045, 0185 [25]
Empicol 0266, 0303 [5]
Empicol LXS 95 [25]

Sodium lauryl sulfate (C_{12}) (94% conc.).
Texapon K 12-94 [401]

Sodium lauryl sulfate (95% conc.).
Swascol IP [837]

Sodium lauryl sulfate (96% conc.).
Texapon K 1296 [399], [400]
Texapon VH.C-B [400]

Sodium lauryl sulfate (C_{12}) (96% conc.).
Texapon K 12-96 [401]

Sodium lauryl sulfate (97% conc.).
Nikkol SLS [631]

Sodium lauryl sulfate (97% min. C12 content).
Empicol 0045 [25]

Sodium lauryl sulfate (98.5% conc.).
Emal O [478]

Sodium lauryl sulfate (99% conc.).
Conco Sulfate WR [209]
Maprofix 563 [663]
Stepanol WA-100 [825], [825a]
Sterling WA Powder [145]
Texapon L-100 [399], [400]

Sodium lauryl sulfate, (aq. sol'n.).
Empicol LY28/S [23]

Sodium lauryl sulfate (B.P. grade).
Empicol LZ/D, LZV/D [23]

Sodium lauryl sulfate (C_{12}-C_{14}, 70:30).
Swascol 1 PC [837]

Sodium lauryl sulfate, ethoxylated.
Empicol 0342, 0352, ESA, ESA70, ESB, ESB3 [23]
Empicol ESB-3 [25]
Empicol ESB3/CP [23]
Empicol ESB-30 [25]
Empicol ESB50 [23], [25]
Empicol ESB70, ESB 70-AU, ESC3 [23]
Empicol ESC-3 [25]
Empicol ESC 70, MD [23]
Empimin KSN 27/T, KSN 50, KSN 60, KSN 70 [23]
Polystep B-16 [825]

Sodium lauryl sulfate, ethoxylated (4 EO).
Polystep B-10, B-12 [825]

Sodium lauryl sulfate, ethoxylated (12 EO).
Polystep B-21, B-23 [825]

Sodium lauryl sulfate, ethoxylated (30 EO).
Polystep B-19 [825]

Sodium lauryl sulfate, high purity (20% conc.).
Empicol LXS95 [23]

Sodium lauryl sulfate, modified.
Manro DL 28 [545]
Wetanol [359]

Sodium lauryl sulfate, modified (28% conc.).
Monro DL 28 [545]

Sodium lauryl sulfate, natural (29-31% conc.).
Ungerol LSN [874]

Sodium lauryl sulfate, paste (28-30% conc.).
Sulfopon WA-3 [399]

Sodium lauryl sulfate, pearlescent.
Sterling WAQ-P [145]

Sodium lauryl sulfate, pure (99% conc.).
Texapon L 100 [401]

Sodium lauryl sulfate/sodium sarcosinate.
Nutrapon RS 1147 [198]

Sodium lauryl sulfate, synthetic (39-41% conc.).
Ungerol LS [874]

Sodium lauryl sulfate, USP.
Duponol C [269]
Empicol LZ/E, LZV/E [23]

Sodium lauryl sulfoacetate.
Lanthanol LAL [35]
Lathanol LAL, LAL Flake, LAL Powder [825]
Lathanol LAL 70 Powder [825a]

Sodium lauryl sulfoacetate (70% conc.).
Lathanol LAL Powder [825]
Lathanol LAL 70 Powder [825a]

Sodium lauryl sulfoacetate (92.5% conc.).
Nikkol LSA [631]

Sodium lauryl sulfonate (92% conc.).
Zoharpon Las Spray Dried [946]

Sodium lauryl triethoxysulfate.
Empicol ESC-AU, ESC 70-AU [23]

Sodium levothyroxine.
Levothroid [891]
Synthroid [327]
Thyrolar Tablets [891]

Sodium lignosulfate.
Amasperse N [190]
Darvan #2 [894]
Dyqex [349]
Lignosite 458, 823, 854 [349]
Lignosol D-10, D-30, FTA, NSX 120, SFX, XD [720]
Maracell E [47]
Marasperse CB, N-22 [47]
Maratan 22, SN [47]
Norlig 12 [47]
Orzan S [229]
Polyfon [919]

Sodium lignosulfonate, desulfonated.
Maracell-E [47]

Sodium liothyronine.
Thyrolar Tablets [891]

Sodium magnesium fluorosilicate.
Laponite B, S [511]

Sodium magnesium lauryl sulfate, ethoxylated.
Empicol BSD [23]

Sodium magnesium lauryl sulfonate, ethoxylated (26% conc.).
Empicol BSD [25]

Sodium magnesium silicate.
Laponite® RD, RDS, XLG, XLS [511]

Sodium mannuronate methylsilanol.
Algisium [469]

Sodium mercaptobenzothiazole.
Troysan 28 [869]

Sodium (2) mercaptobenzothiazole.
Nacap [894]

Sodium metaborate/sodium chlorate 68:30.
Monobor-Chlorate [802a]

Sodium metasilicate.
Drymet [821]
Metso 20, Beads 2048, Pentabead 20 [706]

Sodium metasilicate, anhydrous.
Drymet [219], [821]

Sodium metasilicate pentahydrate.
Crystamet [219]

Sodium methane naphthalene sulfonate.
Soitem 207 [807]

Sodium-N-methyl-N-acyl taurate.
Igepon TN-74 [338]

Sodium-N-methyl-N-"coconut oil acid" taurate.
Igepon TC-42 [338]

Sodium methyl cocoyl taurate.
Chempon 42 [174]
Hostapon CT [55]
Igepon TC-42, TC-99 [338]
Protapon 42 [713]
Tauranol WS, WS-Powder [320]

Sodium N-methyl-N-cocoyl taurate.
Adinol CT95 [223]

Sodium N-methyl-N cocoyl taurate (24% conc.).
Tauranol WS [320]

Sodium N-methyl-N cocoyl taurate (31% conc.).
Tauranol WS Conc. [320]

Sodium N-methyl-N cocoyl taurate (70% conc.).
Tauranol WS Powder [320]

Sodium methyl cocoyl taurate (95% min. conc.).
Tauranol WS H.P. [320]

Sodium N-methyl-N-cocoyl taurate (95% min. conc.).
Adinol CT 95 [223]

Sodium N-methyl dithiocarbamate.
Metam-Fluid BASF [107]
Trimaton [683]
Vapam [821]

Sodium N-methyl-N-myristoyl taurate.
Igepon TM-43 [338]

Sodium methyl oleoyl taurate.
Chempon 33 [174]
Concogel 2 Conc. [209]
Hostapon T [55]
Igepon T-33, T-43, T-51, T-77 [338]
Protapon 33 [713]
Tauranol MS, MS Powder [320]
Tergenol G [391]

Sodium methyl-N-oleoyl taurate.
Amasperse N [190]
Amaterg T [190]

Sodium N methyl N oleoyl taurate.
Adinol T, T35, T Gel, T High Conc., T Paste [223]
Consamine K-Gel [207]
Igepon T-51, TC-42, TM-43 [338]
Tergenol G, S Liquid, Slurry [391]

Sodium N-methyl-N-oleoyl-taurate (14% conc.).
Concogel 2 Conc. Dry [209]
Igepon T-51 [338]
Tauranol T-Gel [320]

Sodium N-methyl-N-oleoyl taurate (16% conc. min.).
Adinol T35 [223]

Sodium N-methyl-N oleoyl taurate (23% conc.).
Tauranol M-35 [320]

Sodium methyl oleoyl taurate (32% conc.).
Tergenol S Liquid [391]

Sodium N-methyl-N oleoyl taurate (32% conc.).
Cresterge L [222]
Igepon T-33 [338]

Sodium N-methyl-N-oleoyl taurate (33% conc.).
Igepon T-43 [338]
Tauranol ML, MS [320]

Sodium methyl oleoyl taurate (40% conc.).
Tergenol Slurry [391]

Sodium N-methyl-N-oleoyl taurate (67% conc.).
Igepon T-77 [338]

Sodium N-methyl-N oleoyl taurate (75% conc.).
Tauranol M Powder [320]

Sodium N-methyl-N-palmitoyl taurate.
Igepon TN-74 [338]

Sodium N-methyl-N-palmitoyl taurate (44% conc.).
Igepon TN-74 [338]

Sodium-N-methyl-N-"tall oil acid" taurate.
Igepon TK-32 [338]

Sodium N-methyl-N-tall oil-acid taurate (20% conc.).
Igepon TK-32 [338]

Sodium mono alkyl ethanol amide sulfo-succinate (40% conc.).
Secosol EXP/40 [825a]

Sodium mono alkyl ether sulfosuccinate (40% conc.).
Secosol ALL [825a]

Sodium mono alkyl polyglycol ether sulfosuccinate (40% conc.).
Servoxyl VLB 1123 [788]

Sodium mono alkyl sulfosuccinate (33% conc.).
Secosol AL/959 [825a]

Sodium mono ethanolomine alkyl ethoxy sulfosuccinate.
Empicol SHH [23]

Sodium mono ethanolomine alkyl ethoxy sulfosuccinate (32% conc.).
Empicol SHH [25]

Sodium mono nonylphenol polyglycol ether sulfosuccinate (10 EO) (35% conc.).
Sermul EA 176 [788]
Serrvoxyl VLE 1159 [788]

Sodium monobasic.
Uro-KP-Neutral [820]

Sodium myreth sulfate.
Cycloryl ME 60 [234]
Genapol CRO Liquid, CRO Paste [55]
Rewopol NM [725]
Standapol ES-40 Conc. [400]
Texapon K 14S Special [400]

Sodium myreth sulfate (58% conc.).
Standapol ES-40 Conc. [399], [400]

Sodium N-myristoyl methyl taurate (92% conc.).
Nikkol MMT [631]

Sodium myristoyl sarcosinate.
Hamposyl M-30 [363]
Nikkol Sarcosinate MN [631]

Sodium myristoyl sarcosine.
Hamposyl M-30 [363]

Sodium myristyl ether sulfate.
Cycloryl ME 60 [234]

Sodium myristyl ether sulfate (60% conc.).
Cycloryl ME60 [234]
Texapon ES-40 Conc. [399]

Sodium myristyl sulfate.
Carsonol SMS [529]
Cycloryl MS [234]
Maprofix MSP 90 [663]
Nikkol SMS [631]

Sodium myristyl sulfate (27% conc.).
Cycloryl MS [234]

Sodium naphthalene sulfonate.
Lomar LS [252]

Sodium naphthalene sulfonate-formaldehyde condensate.
Blancol, N [338]

Sodium naphthalene sulfonic acid, condensation product, sodium salt.
Tamol NNO, NNOK [106]

Sodium naphthalene sulfonic acid formaldehyde.
Arylan SNS [252]

Sodium n-1-naphthyl phthalamate.
Alanap [881]
Naptalam [881]

Sodium o-nitro-benzene sulfonate.
Matexil PA-L [442]

Sodium nitroprusside.
Nipride Injectable [738]
Nitropress [3]
Sodium Nitroprusside Injection [287]

Sodium nonoxynol (6) phosphate.
Emphos CS-1361 [934], [935]
Gafac LO-529 [338]

Sodium nonoxynol (9) phosphate.
Emphos CS-1361 [935]

Sodium nonoxynol (4) sulfate.
Alipal CO-433 [338]

Sodium nonyl phenol ether sulfate (4 EO) (31% conc.).
Serdet DNK 30 [788]

Sodium nonyl phenol polyglycol ether sulfate (4 EO) (31% conc.).
Sermul EA 54 [788]

Sodium nonyl phenol polyglycol ether sulfate (10 EO) (35% conc.).
Sermul EA 151 [788]

Sodium nonyl phenol polyglycol ether sulfate (15 EO) (35% conc.).
Sermul EA 146 [788]

Sodium octadecyl sulfate.
Merpinal Q 101 [285]

Sodium octadecyl sulfate (30% conc.).
Merpinal Q 101 [285]

Sodium N-octadecyl sulfosuccinamate.
Alkasurf SS-TA-125 [33]

Sodium N-octadecyl sulfosuccinamate (35% conc.).
Alkasurf SS-TA, SS-TA 125 [33]
Alkasurf SS-TA-125 [34]

Sodium N-octadecyl sulfosuccinate.
Alkasurf SS-TA [33]

Sodium N-octadecyl sulfosuccinate (35% conc.).
Alkasurf SS-TA [34]
Protowet 4678 [710]

Sodium octoxynol-2 ethane sulfonate.
Triton X-200 [742]

Sodium octyl phenoxy ethyl sulfonate, ethoxylated.
Newcol 861SE [635]

Sodium octyl phenoxy ethyl sulfonate, ethoxylated (30% conc.).
861S [635]

Sodium octyl sulfate.
Carsonol SHS [529]
Sipex OLS, OS [29]
Sulfotex OA [400]

Sodium n-octyl sulfate.
Duponol 80 [269]

Sodium octyl sulfate (33% conc.).
Sipex OLS [29]

Sodium oleate.
Witco Sodium Oleate [935]

Sodium oleate (90% conc.).
Norfox Oleic Flakes [650]

Sodium α -olefin sulfonate.
Bio Terge AS-90 Beads [825]
Bio-Terge AS-90F [825a]
Surco AOS [663]
Witconate AOS [934]

Sodium α -olefin (C_{14}-C_{16}) sulfonate.
Siponate 246-L, 301-10F, A 246 LX [29]
Sterling AOS [145]
Ultrawet AOK [72]
Witconate AOS [934], [935]

Sodium α -olefin (C_{16}-C_{18}) sulfonate.
Siponate A-167 [29]

Sodium α -olefin (C_{16}-C_{18}) sulfonate (30% conc.).
Siponate A-168 [29]

Sodium α -olefin sulfonate (39% conc.).
Sterling AOS [145]

Sodium α -olefin sulfonate (40% conc.).
Bio Terge AS-40 [825]
Conco AOS-40 [209]
Polystep A-18 [825]

Sodium α -olefin (C_{14}-C_{16}) sulfonate (40% conc.).
Siponate A-246, A-246L [29]

Sodium α -olefin sulfonate (90% conc.).
Bio-Terge AS-90F [825a]
Conco AOS-90F [209]
Ultrawet AOK [72]

Sodium α -olefin (C_{14}-C_{16}) sulfonate (90% conc.).
Siponate 301-1OF [29]

Sodium oleoyl lactylate.
Pationic OL [676]

Sodium N-oleoyl sarcosinate.
Crodasinic OS 35 [223]

Sodium oleth (7) phosphate.
Gafac RB-520B [338]

Sodium oleth (8) phosphate.
Hostaphat KO280 [417]

Sodium oleyl sulfate.
Duponol LS Paste [269]
Sipex OS [29]

Sodium oleyl sulfate (26% conc.).
Sipex OS [29]

Sodium oleyl sulfate (30% conc.).
Conco Sulfate O [209]

Sodium orthosilicate.
Orthosil [682]

Sodium oxalate.
Natox [821]

Sodium oxychlorosene.
Clorpactin WCS-90 [376]

Sodium N-palmitoyl methyl taurate (92% conc.).
Nikkol PMT [631]

Sodium pareth (15-7) carboxylate.
Surfine WCT Gel, WCT Liquid, WCT-LS [320]

Sodium pareth (25-7) carboxylate.
Surfine WNT Gel, WNT Liquid, WNT-LS [320]

Sodium pareth (23) sulfate.
Akyposal DS56 [182]

Sodium pareth (25) sulfate.
Neodol 25-3S [791]
Serdet DPK 30 [175]
Standapol SP-60 [400]
Sterling Super A [145]
Witcolate 1050 [935]

Sodium pentachloro dihydroxy triphenyl methane sulfonate.
Eulan CN [338]

Sodium pentachloro phenate, technical.
Santobrite [602a]

Sodium pentobarbital.
Nembutal Sodium Capsules, Solution, Suppositories [3]
Pentobarbital Sodium Injection [28]
Pentobarbital Sodium in Tubex [942]
WANS (Webcon Anti-Nausea Supprettes) [915]

Sodium perborate.
Trichotine Powder, Vaginal Douche [719]

Sodium perchlorate.
KM [490]

Sodium petroleum sulfonate.
Petronate CR, HL [936]
Petrosul 545, 550, 742, 744CL, 745, 750 [684]
Pyronate 40 [936]

Sodium petroleum sulfonate (50% conc.).
Petrosul 545, 550 [684]

Sodium petroleum sulfonate (60% conc.).
Petrosul 742, 745, 750 [684]
Soi Mul 440 [807]

Sodium petroleum sulfonate (62% conc.).
Aristonate 430, 460, 500 [697]
Petronate CR, HL, L [936]

Sodium petroleum sulfonate, natural (70% conc.).
Morco H-70, M-70 [547]

Sodium o-phenyl phenol tetrahydrate.
Dowicide A [261]

Sodium o-phenylphenate.
Dowicide A [261]

Sodium phosphate.
Fleet Enema, Phospho-Soda, Prep Kits [325]
Sodium Phosphates Oral Solution [751]

Sodium phosphate, anhydrous glassy.
Hy-Phos [421]

Sodium phosphate, dibasic.
Fleete Enema, Phospho-Soda, Prep Kits [325]
K-Phos Neutral Tablets [113]
Neutra-Phos Powder & Capsules [929]

Sodium phosphate, glassy.
Calgon [142]
Oilfos [602], [602a]

Sodium phosphate, monobasic.
K-Phos M.F. (Modified Formula) Tablets [113]
Neutra-Phos Powder & Capsules [929]
Uroqid-Acid Tablets, No. 2 Tablets [113]

Sodium phosphate, tribasic.
Dri-Tri [821]

Sodium phosphate, water-soluble glassy.
Polyphos [659]

Sodium POE (2) alkyl ether carboxylic acid (85% conc.).
Nikkol ECCO-2N [631]

Sodium POE (3) alkyl ether carboxylic acid (85% conc.).
Nikkol ECT-3N [631]

Sodium POE (3) lauryl sulfate (5% conc.).
Emersal 6453 [289]

Sodium polyacrylate.
Acrysol GS [742]
Alcogum 6625, 6940, 6945, 9445, 9635, 9636 [27]
Antiprex A [37]
Densol 1010 [365]
Empicryl PAS 140, PAS 240 [23]
Tanamer [50]

Sodium-polyacrylate solution.
Thickener L [416]

Sodium polyglycol ether sulfosuccinate.
Alkasurf SS-LA-3 [33]

Sodium polymethacrylate.
Darvan #7 [894]

Sodium polynaphthalene sulfonate.
Darvan #1 [894]
Lomar PWC [252]

Sodium polyoxyethylene alkylaryl ether sulfate.
Emal 20C, NC [478]

Sodium polyoxyethylene alkyl ether sulfate.
Emal 20C, E-25C, E-70C, NC [478]

Sodium polyoxyethylene alkyl ether sulfate (25% conc.).
Emal 20C [478]

Sodium polyoxyethylene alkyl ether sulfate (70% conc.).
Emal E-25C, E-70C [478]

Sodium polyoxyethylene alkyl sulfosuccinate (32% conc.).
Protowet 4951 [710]

Sodium polyoxyethylene nonylphenyl ether sulfate.
Newcol 560SN [635]

Sodium polyoxyethylene nonylphenyl ether sulfate (30% conc.).
560SN [635]

Sodium polyoxyethylene tridecyl ether sulfate (30% conc.).
1305SN, 1310SN [635]

Sodium polystyrene sulfonate.
Flexan 130 [619]
Kayexalate [133]
Sodium Polystyrene Sulfonate Suspension [751]

Sodium propionate.
Amino-Cerv [586]
Impedex [269]

Sodium pyrithione.
Sodium Omadine [659]

Sodium riboflavin phosphate.
Riboflavin-5′-Phosphate Sodium, No. 60295 [738]

Sodium ricinoleate.
NL Sodium Ricinoleate [640a], [641a]
Solcrin 535 [640a], [641]

Sodium salicylate.
Pabalate Tablets [737]

Sodium sesquicarbonate.
Snow Fine [35]
Snowflake [35]
Snowflake Crystals [450]

Sodium silicate.
Britesil [706]
D, N Sodium Silicate, O Silicate [706]

Sodium silicate, ferrosilicon.
Exobond [251]

Sodium silicates, hydrous.
Britesil [706]

Sodium silicoaluminate.
Zeolex 7, 23A, 35 [428]

Sodium soap, aq. sol'n.
Dresinate 81 Liquid Soap [406]

Sodium soya hydrolyzed animal protein.
Maypon K [825]

Sodium starch glycolate.
Primojel [343a]

Sodium stearate.
Witco Sodium Stearate C-1, C-6, C-7 [935]

Sodium stearate (96% conc.).
Norfox B [650]

Sodium stearoyl lactylate.
Admul SSL 2003, SSL 2004 [704]
Artodan [373]
Crodactil SSL [223]
Emplex [675]
Grindtek FAL 1 [373]
Lamegin NSL [176], [375]
Pationic 145A [674a], [675]
Pationic SSL [676], [735]
Stearolac S [672]

Sodium stearoyl (2) lactylate.
Lisat N [360], [856]
Pationic SSL [674a], [675]
Radiamuls SSL [658]

Sodium N-stearoyl methyl taurate (92% conc.).
Nikkol SMT [631]
SMT [631]

Sodium stearyl (2) lactylate.
Artodan [373]

Sodium sulfacetamide.
Metimyd Ophthalmic Ointment-Sterile, Suspension [772]
Sulfacet-R Acne Lotion [246]

Sodium sulfacetamide (10% conc.).
Ocu-Sul-10 [655a]

Sodium sulfacetamide (15% conc.).
Ocu-Sul-15 [655a]

Sodium sulfacetamide (30% conc.).
Ocu-Sul-30 [655a]

Sodium sulfate.
Pedi-Bath Salts [679]

Sodium sulfate, anhydrous.
Trona [490]

Sodium sulfate, synthetic sperm oil (75% conc.).
Value-MS-1 [556]

Sodium sulfite.
Oxygen Scavenger K-91 [142]

Sodium sulfite, anhydrous.
Santosite [602a]

Sodium sulfonate, copolymerizable (35% conc.).
Polystep RA-35 S [825a]

Sodium sulfonate, α olefin.
Sulframin AOS, AOS 90 [935]

Sodium sulfonate, olefin.
Bio Soft LD-70, LD-80 [825]

Sodium sulfonate, olefin linear.
Stepantan 29N, 39N [825]

Sodium sulfonate, α olefin (C_{14}-C_{16}) sodium salt.
Elfan OS 46 [16]

Sodium sulfonate, organic (50% conc.).
Karasperse DDL-12 [537]

Sodium sulfosuccinate.
Geropon T/18/2 [550]

Sodium sulfosuccinate, amido ester (30% conc.).
Schercopol CMS-Na [771]

Sodium sulfosuccinate, amido ester, ethoxylated (40% conc.).
Schercopol OMES-Na [771]

Sodium sulfosuccinate, asymmetrical diester.
Emcol 4930 [934]

Sodium sulfosuccinate, diester.
Emcol 4500, 4580PG, 4600 [934]

Sodium sulfosuccinate, dioctyl ester.
Drewfax 528 [266]

Sodium sulfosuccinic acid, bis tridecyl ester.
Aerosol TR-70 [50a]

Sodium sulfosuccinic acid, dialkyl ester.
Aerosol A-196, A196-40 [50a]

Sodium sulfosuccinic acid, diamyl ester.
Aerosol AY-65, AY-100 [50a]

Sodium sulfosuccinic acid, dihexyl ester.
Aerosol MA-80 [50a]

Sodium sulfosuccinic acid, di-isobutyl ester.
Aerosol IB-45 [50a]

Sodium sulfosuccinic acid, dioctyl ester.
Aerosol GPG, OT-70-OG, OT-75, OT-100, OT-B, OT-S [50a]
Marlinat DF 8 [178]
Solusol 75%, 84%, 85%, 100% [50]

Sodium sulfosuccinic acid, dioctyl ester (60% conc.).
Crestopen 5X [222]

Sodium sulfosuccinic acid, dioctyl ester (70% conc.).
Marlinat DF 8 [178]

Sodium sulfosuccinic acid, dioctyl ester (75% conc.).
Mackanate DOS-75 [567]

Sodium sulfoxylate formaldehyde, modified.
Stripolite [753]

Sodium tallow alcohol sulfate.
Empicol TAS30 [25]

Sodium tallow alcohol sulfate (30% conc.).
Maprofix TAS [663]

Sodium tallow dimethyl glycinate, hydrogenated.
Monateric 1203 [599]

Sodium tallow fatty alcohol sulfate.
Merpinal T 35 [285]

Sodium tallow fatty alcohol sulfate (35% conc.).
Merpinal T 35 [285]

Sodium tallow sulfate.
Maprofix TAS [663]

Sodium tallow sulfate (30% conc.).
Conco Sulfate T [209]

Sodium/TEA-lauroyl animal protein, hydrolyzed.
Lipoproteol LCO, LK [730]

Sodium/TEA-undecylenoyl animal protein, hydrolyzed.
Lipoproteol UCO [730]

Sodium tetraborate, anhydrous.
Dehybor [887]

Sodium tetraborate decahydrate.
Borax [887]

Sodium tetradecyl sulfate.
Niaproof 4 [627]
Sotradecol Injection [287]

Sodium tetradecyl sulfate (28% conc.).
Niaproof 4 [627]

Sodium tetrahydro naphthalene sulfonate.
Alkanol S [269]

Sodium tetraphosphate.
Quadrafos [755]
Sodaphos [449]

Sodium tetrapropylene benzene sulfonate.
Merpisap DE 70, DP 82 [285]

Sodium tetrapropylene benzene sulfonate (70% conc.).
Merpisap DE 70 [285]

Sodium tetrapropylene benzene sulfonate (82% conc.).
Merpisap DP 82 [285]

Sodium thiopental.
Pentothal [3]

Sodium thiosulfate.
Acculog 81-6380-00 [172]
Komed Acne Lotion, HC Lotion [103]
Tineasol Lotion [78]
Tinver Lotion [103]

Sodium, thyroxine.
Choloxin [327]
Synthroid [327]

Sodium toluene sulfonate.
Eltesol ST Pellets [23], [25]
Eltesol ST33 [23]
Eltesol ST 34 [26], [548]
Eltesol ST40 [23], [25], [548]
Eltesol ST 90 [23], [25], [548]
Eltesol STV [23]
Manro STS 40, STS 90 [545]
Nacconol 90F [178]
Na Toluene Sulfonate 30, 40 [178]
Reworyl NTS 30, 40 [725]
Reworyl NTS 40 [725a]
Stepanate T [825]
Witconate STS [934], [935]

Sodium toluene sulfonate (30% conc.).
Na-Toluene Sulfonate 30, 40 [178]

Sodium toluene sulfonate (40% conc.).
Eltesol ST 40 [25]
Manro STS 40 [545]
Naxonate 4ST [754]
Witconate STS Liquid [935]

Sodium toluene sulfonate (88% conc.).
Eltesol ST 90 [25]

Sodium toluene sulfonate (90% conc.).
Manro STS 90 [545]
Witconate STS Powder [935]

Sodium toluene sulfonate (93% conc.).
Naxonate ST [754]

Sodium toluene sulfonate, aq. sol'n.
Eltesol ST40 [25]

Sodium toluene sulfonate, 40% aq. sol'n.
Manro STS 40 [545]

Sodium toluene xylene sulfonate.
STXS [210]

Sodium toluene xylene sulfonate (40% conc.).
Conoco STXS [206]

Sodium trichloroacetate.
NaTA [423]

Sodium trichlorophenate.
Preventol I [338]

Sodium trideceth (7) carboxylate.
Rewopol CTN [725]
Sandopan DTC [762]

Sodium trideceth sulfate.
DV-674C [29]
Liposurf EST-30, EST-40 [525]
Simpex EST-75 [29]
Sipex EST-30 [29]

Sodium tridecyl benzene sulfonate.
Richonate TDB [732]

Sodium tridecyl benzene sulfonate (44% conc.).
Witconate TDB [935]

Sodium tridecyl ether sulfate.
Merpoxal TR 4060 [285]
Sipex EST-30, EST-60, EST-75 [29]

Sodium tridecyl ether sulfate (29% conc.).
Sipex EST-30 [29]

Sodium tridecyl ether sulfate (30% conc.).
Avirol SE-3004 [399], [400]
Maprofix OS [663]

Sodium tridecyl ether sulfate (35% conc.).
Cedepal TD-404 [256]

Sodium tridecyl ether sulfate (60% conc.).
Merpoxal TR 4060 [285]

Sodium tridecyl ether sulfate (75% conc.).
Cedepal TD-407 [256]
Sipex EST-75 [29]

Sodium tridecyl sulfate.
Liposurf EST [525]
Sipex TDS [29]

Sodium tridecyl sulfate (25% conc.).
Sipex TDS [29]

Sodium tridecyl sulfate (30% conc.).
Avirol SA-4113 [399], [400]
Sulfotex 113 [399]

Sodium triethoxy sulphate.
Empimin SQ70 [23]

Sodium trimetaphosphate (STMP).
IP-61 [602]

Sodium tripolyphosphate.
Empiphos STP [25]

Sodium tripolyphosphate (STP).
Nutrifos [602]

Sodium tripolyphosphate, anhydrous.
Armofos [821]

Sodium tyropanoate.
Bilopaque Sodium [932]

Sodium valproate.
Depakene [729]

Sodium xylene sulfate.
Mars SXS [554]

Sodium xylene sulfonate.
Alkatrope SX-40 [33]
Carsosulf SXS [155], [529]
Conco SXS [209]
Eltesol SX30 [23], [26], [548]
Eltesol SX33, SX40, SX 93 [23]
Eltesol SX 93 [26], [548]
Eltesol SX Pellets [23], [25]
Eltesol SXV [23]
Lakeway SXS [126]
Lankrosol SXS-30 [509]
Manro SXS 30, SXS 40, SXS 93 [545]
Ninex 303 [762]
Pilot SXS-40, SXS-96 [697]
Polystep A-2 [825]
Reworyl NXS 30, NXS 40 [725]
Reworyl NXS 40 [725a]
Richonate SXS [732]
Stepanate X [825], [825a]
Sterling 2XS [145]
Sulfotex SXS [400]
Surco SXS [663]
SXS [210]
Ultra SXS Liquid, SXS Powder [935]
Ultrawet 40SX [72]
Witconate SXS [934], [935]

Sodium xylene sulfonate, (30% conc.).
Eltesol SX 30 [25]
Manro SXS 30 [545]

Sodium xylene sulfonate (40% conc.).
Alkatrope SX-40 [33]
Conoco SXS [206]
ESI-Terge SXS [294]

Sodium xylene sulfonate (40% conc.) *(cont'd).*
Manro SXS 40 [545]
Naxonate 4L [754]
Ninex 303 [825a]
Norfox SXS-40 [650]
Siponate SXS [29]
Surco SXS [663]
Witconate SXS Liquid [935]

Sodium xylene sulfonate (41% conc.).
Ultrawet 40SX [72]

Sodium xylene sulfonate (90% conc.).
Witconate SXS Powder [935]

Sodium xylene sulfonate (93% conc.).
Eltesol SX 93 [25]
Manro SXS 93 [545]
Naxonate G [754]

Sodium xylene sulfonate (96% conc.).
Norfox SXS-96 [650]

Sodium zinc sulfoxylate formaldehyde.
Reducolite [753]

Somatropin.
Asellacrin (somatropin) [787]
Crescormon [693]

Sorbate (20)/sorbitan monolaurate.
ML-33-F [397]

Sorbate (40)/sorbitan monopalmitate.
MP-33-F [397]

Sorbate (60)/sorbitan monostearate.
MS-33-F [397]

Sorbate (85)/sorbitan trioleate.
TO-33-F [397]

Sorbate (65)/sorbitan tristearate.
TS-33-F [397]

Sorbeth-(20).
Ethosperse SL-20 [358]
Liponic SO-20 [525]

Sorbic acid.
Aflaban DF [602]
Guard Sorbic Acid [65]
Preservastat [867]
Sorbistat [691]

Sorbide dioleate.
Atlas G-950 [445]

Sorbitan acid ester.
Sorbax SMS [170]

Sorbitan dioleate.
DO-33-F [397], [416]

Sorbitan distearate.
Emasol S-20 [478]

Sorbitan ester.
Crill [223]
Drewmulse 75 [266]
Glycomul [358]
Sorban AL, AO, AO1, AP, AST, CO [937a], [938]
Sorbanox AST [937a], [938]

Sorbitan esters, ethoxylated.
Glycosperse [358]
Protasorbs [713]
Sorbanox AL, AOM, AP [937a], [938]
Sorbanox AST [938]
Sorbanox CO [937a], [938]
Sorbax PML-20, PMO-5, PMO-20, PTO-20, PTS-20 [170]
T-MAZ [564]

Sorbitan fatty acid ester.
Atpet 100, 200 [445]
Emsorb [290]
Radiamuls SORB [658]
Sorbax SML, SMO, SMP, STO, STS [170]

Sorbitan fatty ester.
Industrol 68, 1186, L-4-S, L-20-S, O-5-S, P-20-S, S-20-S, TO-16, TO-20-S [109]

Sorbitan isostearate.
Montane 70 [785]

Sorbitan laurate.
Arlacel 20 [445]
Armotan ML [79]
Emsorb 2515 [290]
Glycomul L, LC [358]
Grindtek SML [373]
Kuplur SML [923]
Liposorb L [525]
Lonzest SML [529]
Montane 20 [785]
Newcol 20 [635]
Protachem SML [713]
S-Maz 20, 20R [564]
Span 20 [445]

Sorbitan laurate, ethoxylated.
Radiamuls SORB 2137 [658]

Sorbitan mono fatty acid ester.
Arlacel 987 [445]

Sorbitan monocaprylate.
Nissan Nonion CP-08R [636]

Sorbitan monoisostearate.
Arlacel 987 [92]
Crill 6 [223]
Montane 60 [711]
Nikkol SI-10R, SI-10T, SI-15R, SI-15T [631]
S-Maz 67 [564]

Sorbitan monoisostearate, polyethoxylated (20).
Montanox 70 [711]

Sorbitan monolaurate.
Ahco 759 [438]
Alkamuls SML [33], [34]
Arlacel 20 [92], [438], [445]
Armotan ML [16]
Crill 1 [223]
Drewmulse SML [715]
Durtan 20 [272]
Emasol L-10 [478], [478a]
Emsorb 2515 [289], [290]
Glycomul L, LC [358]
Hodag SML [415]
Ionet S-20 [764]
Kuplur SML [109]
Liposorb L [525]
Lonzest SML [525], [529]
Montane 20 [711]
Nikkol SL-10 [631]
Nissan Nonion LP-20R, LP-20RS [636]
Radiamuls 125, SORB 2125 [658]
Radiasurf 7125 [658]
S-Maz 20 [564]
Soprofor S/20 [550]
SORB 2125 [658]
Sorbax SML [170]
Sorbon S-20 [864]
Sorgen 90 [237]
Span 20 [92], [93, [438], [445]

Sorbitan monolaurate (97% conc.).
Esterol 1090 [390]

Sorbitan monolaurate, cosmetic grade.
Glycomul LC [358]

Sorbitan monolaurate ester, ethoxylated.
Alkamuls PSML-20 [33]

Sorbitan monolaurate, ethoxylated.
Soprofor T/20 [550]

Sorbitan (20) monolaurate, ethoxylated.
Alkamuls PSML-20 [33]

Sorbitan monolaurate, polyethoxylated (5).
Montanox 21 [711]

Sorbitan monolaurate, polyethoxylated (20).
Montanox 20 [711]

Sorbitan monomyristate.
Nissan Nonion MP-30R [636]

Sorbitan monooleate.
Ahco 832, 944 [438]
Alkamuls SMO [33], [34]
Arlacel 80 [92], [438], [445]
Armotan MO [16]
Atpet 80 [92], [445]
Atpet 100, 200 [92]
Capmul O [829]
CPH-367-N [379]
Crill 4 [223]
Drewmulse SMO [715]
Durtan 80 [272]
Emasol O-10 [478], [478a]
Emsorb 2500 [289], [290]
Flo Mo SMO [247]
G-4884 [92]
Glycomul O [358]
Hodag SMO [415]
Ionet S-80 [764]
Kuplur SMO [109]
Liposorb O [525]
Lonzest SMO [525], [529]
Montane 80 [711]
Nikkol SO-10 [631]
Nissan Nonion OP-80R [636]
Radiamuls 155, SORB 2155 [658]
Radiasurf 7155 [658]
S-Maz 80 [564]
Soprofor S/80 [550]
SORB 2155 [658]
Sorbax SMO [170]
Sorbon S-80 [864]
Sorgen HO, S-40-H [237]
Span 80 [92], [93], [445]

Sorbitan monooleate, ethoxylated.
Soprofor T/20, T/60, T/65, T/80, T/85 [550]

Sorbitan monooleate, N.E.
Span 80 [438]

Sorbitan monooleate, polyethoxylated (5).
Montanox 81 [711]

Sorbitan monooleate, polyethoxylated (20).
Montanox 80 [711]

Sorbitan monooleate, technical.
Atlas G-4884 [92]
Crill 50 [223]

Sorbitan monooleate ester, ethoxylated.
Alkamuls PSMO-5 [33]

Sorbitan monopalmitate.
Ahco FP-67 [438]
Arlacel 40 [92], [438], [445]
Armotan MP [16]
Crill 2 [223]
Emasol P-10 [478], [478a]
Emsorb 2510 [289], [290]
Glycomul P [358], [529]
Hodag SMP [415]
Kuplur SMP [109]
Liposorb P [525]
Lonzest SMP [525], [529]
Montane 40 [711]

Sorbitan monopalmitate *(cont'd).*
Nikkol SP-10 [631]
Nissan Nonion PP-40R [636]
Radiamuls 135, SORB 2135 [658]
Radiasurf 7135 [658]
S-Maz 40 [564]
Soprofor S/40 [550]
SORB 2135 [658]
Sorbax SMP [170]
Sorbon S-40 [864]
Span 40 [92], [93], [438], [445]

Sorbitan monopalmitate, ethoxylated.
Soprofor T/40 [550]

Sorbitan monopalmitate, polyethoxylated (20).
Montanox 40 [711]

Sorbitan monostearate.
Ahco 909 [438]
Alkamuls SMS [33], [34]
Arlacel 60 [92], [438], [445]
Armotan MS [16]
Capmul S [829]
Crill 3 [223]
Drewmulse SMS [715]
Drewsorb 60 [715]
Durtan 60, 60 K [272]
Emasol S-10 [478], [478a]
Emsorb 2505 [289], [290]
Emultex SMS [514]
Glycomul S [358], [529]
Hodag SMS [415]
Ionet S-60 C [764]
Kuplur SMS [109]
Liposorb S [525]
Lonzest SMS [525], [529]
Montane 60 [711]
Nikkol SS-10 [631]
Nissan Nonion SP-60 R [636]
Radiamuls 145, SORB 2145 [658]
Radiasurf 7145 [658]
S-Maz 60 [564]
Soprofor S/60 [550]
SORB 2145 [658]
Sorbax SMS [170]
Sorbon S-60 [864]
Sorgen 50 [237]
Span 60 [92], [93], [438], [445]

Sorbitan monostearate, polyethoxylated (4).
Montanox 61 [711]

Sorbitan monostearate, polyethoxylated (20).
Montanox 60 [711]

Sorbitan monostearate, vegetable.
Glycomul S Veg [358]

Sorbitan monostearate ester, ethoxylated.
Alkamuls PSMS-4, PSMS-20 [33]

Sorbitan monotallate.
Flo Mo SMT [247]
Glycomul TAO [358]
S-Maz 90 [564]
Witcomul 78 [934], [935]

Sorbitan oleate.
Alkamuls SMO [33]
Arlacel 80 [438]
Armotan MO [79]
Capmul O [147]
Emsorb 2500 [290]
Glycomul O [358]
Grindtek SMO [373]
Kuplur SMO [923]
Liposorb O [525]
Lonzest SMO [529]
Montane 80 [785]
Newcol 3-80, 3-85, 80 [635]
Nikkol SO-10 [631]
Protachem SMO [713]
S-Maz 80, 80R [564]
Span 80 [438]

Sorbitan oleate, ethoxylated.
Radiamuls SORB 2157 [658]

Sorbitan palmitate.
Arlacel 40 [438]
Emsorb 2510 [290]
Glycomul P [358]
Kuplur SMP [923]
Liposorb P [525]
Lonzest SMP [529]
Montane 40 [785]
Newcol 40 [635]
Protachem SMP [713]
S-Maz 40, 40R [564]
Span 40 [438]

Sorbitan-POE-monolaurate.
Radiasurf 7137 [658]

Sorbitan-POE (20)-monolaurate.
Radiamuls 137, SORB 2137 [658]
Radiasurf 7137 [658]
SORB 2137 [658]

Sorbitan POE monooleate.
Radiasurf 7157 [658]

Sorbitan POE (20) monooleate.
Radiamuls 157, SORB 2157 [658]
Radiasurf 7157 [658]
SORB 2157 [658]

Sorbitan-POE-monopalmitate.
Radiasurf 7136 [658]

Sorbitan POE monostearate.
Radiasurf 7147 [658]

Sorbitan POE (20) monostearate.
Radiamuls 147, SORB 2147 [658]
Radiasurf 7147 [658]
SORB 2147 [658]

Sorbitan sesquiisostearate.
Emsorb 2518 [290]
Montane 73 [785]

Sorbitan sesquioleate.
Arlacel 83 [92], [438], [445]
Arlacel C [92], [438], [445]
CPH-390-N [379]
Crill 43 [223]
Emasol O-15 [478]
Emsorb 2502 [289], [290]
Glycomul SOC, SOC Special [358]
Hodag SSO [415]
Liposorb SQO [525]
Montane 83 [711], [785]
Nikkol SO-15 [631]
Nissan Nonion OP-83-RAT [636]
Protachem SOC [713]
QO-33-F [397]
S-Maz 83, 83R [564]
Sorgen 30, S-30H [237]

Sorbitan sesquistearate.
Nikkol SS-15 [631]

Sorbitan stearate.
Alkamuls SMS [33]
Arlacel 60 [438]
Armotan MS [79]
Capmul S [147]
Emsorb 2505 [290]
Glycomul S [358]
Grindtek SMS [373]
Hodag SMS [415]
Kuplur SMS [923]
Liposorb S, SC [525]
Lonzest SMS [529]
Montane 60 [785]
Newcol 60 [635]
Nikkol SS-10 [631]
Protachem SMS [713]
S-Maz 60, 60R [564]
Span 60 [438]

Sorbitan stearate, ethoxylated.
Radiamuls SORB 2147 [658]

Sorbitan triisostearate.
Emsorb 2519 [290]

Sorbitan trioleate.
Ahco FO-18 [438]
Alkamuls STO [33]
Arlacel 85 [92], [445]
Atlox 4885 [92]
Crill 45 [223]
Emasol O-30 [478], [478a]
Emsorb 2503 [289], [290]
Glycomul TO [358]
Hodag STO [415]
Ionet S-85 [764]
Kuplur STO [923]
Liposorb TO [525]
Lonzest STO [525], [529]
Montane 85 [711], [785]
Nikkol SO-30 [631]
Nissan Nonion OP-85-R [636]
Protachem STO [713]
Radiamuls SORB 2355 [658]
S-Maz 85, 85R [564]
Soprofor S/85 [550]
SORB 2355 [658]
Sorbax STO [170]
Span 85 [92], [93], [438], [445]

Sorbitan trioleate, ethoxylated.
Soprofor T/85 [550]

Sorbitan trioleate, polyethoxylated (20).
Montanox 85 [711]

Sorbitan trioleate, technical.
Atlas G-4885 [92]

Sorbitan trioleate ester.
Alkamuls STO [33]

Sorbitan trioleate ester, ethoxylated.
Alkamuls PSTO-20 [33]

Sorbitan tristearate.
Ahco FS-21 [438]
Alkamuls STS [33]
Crill 35 [223]
Drewmulse STS [715]
Emasol S-30 [478], [478a]
Emsorb 2507 [289], [290]
Glycomul TS [358]
Grindtek STS [373]
Hodag STS [415]
Kuplur STS [109], [923]
Liposorb TS [525]
Lonzest STS [525], [529]
Montane 65 [711], [785]
Nikkol SS-30 [631]
Protachem STS [713]
Radiamuls 345, SORB 2345 [658]
Radiasurf 7345 [658]
S-Maz 65, 65R [564]
Soprofor S-65 [550]
SORB 2345 [658]
Sorbax STS [170]
Span 65 [92], [93], [438], [445]]

Sorbitan tristearate, ethoxylated.
Soprofor T/65 [550]

Sorbitan tristearate, polyethoxylated (20).
Montanox 65 [711]

Sorbitan tristearate ester, ethoxylated.
Alkamuls PSTS-20 [33]

Sorbitan tritallate.
Glycomul T-TAO [358]
S-Maz 95 [564]

Sorbitol.
A-641 [445]
Arlex [445]
Liposorb 70 [525]
Sorbitol U.S.P. Solution [529]

Sorbitol (20 EO).
Polyol 20 [109]

Sorbitol (70% conc.).
Sorbitol USP [529]

Sorbitol ester.
Sorbitan Monostearate [222]

Sorbitol fatty acid ester.
Glycomuls [358]

Sorbitol fatty ester.
S-MAZ [564]

Sorbitol oleate, ethoxylated.
Industrol 1186 [109]

Sorbitol stearate, ethoxylated.
Industrol 68 [109]

Sorbitol trioleate, ethoxylated.
Twix 20TO [109]

Sorbitol USP (70% conc.).
SORBO [438]

Soy glyceride, hydrogenated.
Monomuls 90-45 [375]
Myverol 18-06 Distilled Monoglycerides, 18-06K Distilled Monoglycerides [278]

Soy phosphatides.
Centrolex® Series [167]

Soy phosphatides (95% conc.).
Alcolec Granules, Powder [57]

Soy sterol.
Seboside [901]

Soy trimonium chloride.
Kemamine Q-9973B [429]

Soya acid, distilled.
Industrene 226 FG [429]

Soya, epoxidized.
Peroxidol 780 [722]

N-Soya-(3, amidopropyl), N-N dimethyl, N-ethyl ammonium ethyl sulfate; cationic.
Schercoquat SOAS [771]

Soya amine (2 EO).
Teric 16M2, 16M Series [439]

Soya amine (5 EO).
Icomeen S-5 [109]
Teric 16M5, 16M Series [439]

Soya amine (10 EO).
Teric 16M10, 16M Series [439]

Soya amine (15 EO).
Teric 16M15, 16M Series [439]

Soya amine, ethoxylated.
Accomeen S2 [81]
Accomeen S2, S5, S10 [829]
Accomeen S10 [81]
Accomeen T2, T5, T15 [829]
Alkaminox SO-5 [33]
Chemeen S-2, S-5, S-30, S-30/80 [170]
Kostat P997/2, P997/5, P997/10, P997/15 [429a]

Soya amine, ethoxylated (2 EO on primary soya amine).
Accomeen S2 [81]

Soya amine, ethoxylated (5 EO on primary soya amine).
Accomeen S5 [81]

Soya amine, ethoxylated (10 EO on primary soya amine).
Accomeen S10 [81]

Soya amine, ethoxylated (80% conc.).
Varonic L230 [796]

Soya amine, primary.
Armeen SD [79]
Kemamine P-997 (Distilled), P-997 D (Distilled) [429]

Soya bean, sulfated.
Actrasol OY75 [811]

Soya bean diethanolamide.
Empilan 2125-AU [23], [24]

Soya bean oil, epoxidized.
Paraplex G-62 [379]

Soya bean oil, epoxidized unsaturated.
Plastoflex 2307 [459]

Soya bean oil, sulfonated.
Actrasol CS75 [811]

Soya DEA.
Alkamide SDO [33]

N-Soya-1,3-diaminopropane.
Duomeen S [79]

Soya diethanolamide.
Mackamide S [567]

Soya dimethyl ethyl ammonium ethosulfate.
Larostat 264 A [474]

Soya ethyl morpholinium ethosulfate.
Barquat SME-35 [529]
G-271 [438]

N-Soya N-ethyl morpholinium ethosulfate (35%).
Atlas G-271 [92]

N-Soya-N-ethyl morpholinium ethosulfate (35% conc.) (aq. sol'n.).
Atlas G-271 [438]

Soya fatty acid.
Industrene 226 FG [429]

Soya fatty acid, imidazoline, substituted.
Miramine SC [592]

Soya fatty alkyl acid, distilled.
Industrene 226 [429]

Soya glyceride, hydrogenated.
Neustrene 064 [429]

Soya imidazoline.
Miramine SC [592]

Soya imidazoline, quaternary.
Alkaquat S [33]

Soya lecithin.
Clearate, Special Extra [195]

Soya lecithin (60% conc.).
Clearate WDF [195]

Soya primary amine.
Kemamine P-997 D [429a]

Soya primary amine, distilled.
Kemamine P-997 [429a]

Soya stearine.
Neobee 62 [715]

Soya sterol.
Generol 122 [399], [400]

Soya sterol, (10 EO).
Generol 122 E10 [399], [400]

Soya sterol, (16 EO).
Generol 122 E16 [399]

Soya sterol (25 EO).
Generol™ 122 E25 [218], [399]

Soya sterol, adduct (5 EO).
Generol 122 E5 [399], [400]

Soya sterol, refined (90% conc.).
Generol 122 (Cosmetic Grade) [399], [400]

Soya trimethyl ammonium chloride.
Adogen 415 [711]
Jet Quat S-50 [471]
Tomah Q-S [866]

Soyamide DEA.
Mackamide S [567]
Mazamide SS-10 [564]

Soyamidopropyl benzyl dimonium chloride.
Schercoquat SOAB [771]

Soyamidopropyl ethyl dimonium ethosulfate.
Schercoquat SOAS [771]

Soyamine.
Kemamine P-997 [429]

Soybean alkylamine, primary.
Nissan Amine SB [636]

Soybean lecithin.
Centrolex P [166a]

Soybean lecithin, de-oiled.
Dulectin [268]

Soybean lecithin, modified.
Centrolene Series [167]

Soybean lecithin, natural.
Centrol Series [167]

Soybean lecithin, refined.
Centrophil Series [167]

Soybean oil.
Edsoy [818]

Soybean oil, epoxidized.
Admex 710, 711 [711]
Drapex 6.8 [74]
Epoxol 7-4 [840]
Flexol Plasticizer EPO [878]
Paraplex G-62 [379]
Plas-Chek 775 [315]
Polycizer ESO [392]

Soybean oil, partially hydrogenated.
Durkee 17 [272]

Soybean oil alkanolamide.
Mazamide SS-10 [564]

Soybean oil compound, epoxidized.
Estabex 138-A [459]

Soybean oil distilled monoglyceride, saturated.
Dimodan PV, PV 300 [373]

Soybean oil mono-diglyceride.
Tegomuls SO [856]

Soyo dicoco quaternary ammonium chloride.
Jet Quat S-2C-50 [471]

Spectinomycin hydrochloride.
Trobicin, -Sterile Powder [884]

Sperm oil sodium sulfate, synthetic (75% conc.).
Value-MS-1 [556]

Spermaceti, synthetic.
Cyclochem SPS [234]
Liponate SPS [525]

Spermaceti wax, synthetic.
Kessco 653 [79]

Spironolactone.
Aldactazide [782]
Aldactone [510], [782]
Spironazide Tablets [769]
Spironolactone/Hydrochlorothiazide, Tablets [240]
Spironolactone, Tablets [344], [769]
Spironolactone, Tablets, USP [674]
Spironolactone w/ Hydrochlorothiazide Tablets [344], [674]

β -Spodumene, ceramic material.
Lithafrax [364]

Squalane.
Nikkol Squalane [631]
Robane [736]
Supraene [736]

Squalane NF.
Robane [736]

Stannous chloride.
Stannochlor [610]

Stannous chloride, anhydrous.
Stannochlor [919]

Stannous chloride/P 75/25.
Kenmix [488]

Stanozolol.
Winstrol [932]

Staphylococcus bacterial antigen.
Staphage Lysate (SPL) [245]

Staphylococcus vaccine.
Staphage Lysate (SPL) [245]

Starch diethylaminoethyl ether.
78-1567, 78-1712 [619]

Starch ether.
Nikkagum A [630]

Starch hydrolysate, hydrogenated.
Polyol HM-75 [529]

Steapyrium chloride.
Emcol E-607S [924], [925]

Stearalkonium chloride.
Barquat SB-25 [529]
Dehyquart STC-25 [400]
Mackernium SDC-25, SDC-85 [567]
Maquart SC18-25% [558]
Standamul STC-25 [400]
Stedbac [408]
Triton CG-400, CG-500 [742]
Varisoft SDC [771]

Stearalkonium hectorite.
Bentone 27 [640a]
Bentone 27 Rheological Additive [641a]

Stearamide.
Armid 18 [79]
Armoslip 18 [16], [653]
Crodamide S, SR [223]
Isochem Wax 6 [462]
Kemamide S [429]
Petrac Vyn-Eze [688]

Stearamide DEA.
Aminol N-1918, STD [320]
Clindrol Superamide 100S, 200-S, 868 [197]
Condensate N-1918 [174]
Condensate Q [209]
Condensate S-100 [174]
Cyclomide DS 280/S [234]
Cyclomide SD [939]
Hetamide DS [407]
Keilamide 75 [483]
Lipamide S [525]
Monamid 15-65, 150-CEW-C, 718, 718A [599]
Onyxol 42 [663]
Protamide N-1918, SA [713]
Rewo-Amid DS 280 [725]
Unamide S [529]

Stearamide DEA distearate.
Schercomid SD-DS [771]

Stearamide DIBA stearate.
Polytex 10 [494]

Stearamide ester.
Isochem Wax L [462]

Stearamide MEA.
Carsamide SMEA [529]
Clindrol 200MS [197]
Comperlan HS [400]
CPH-380-N [379]
Cyclomide S280 [234]
Cyclomide SM [939]
Emid 6507 [290]
Hetamide MS [407]
Mazamide SME-M [564]
Monamid S [599]
Product 2-M-19, LT 14-8 [197]
Rewo-Amid S 280 [725]
Schercomid SME, SME-A [771]
Witcamide 70 [934]

Stearamide MEA stearate.
Cerasynt D [895]
Schercomid SME-S [771]
Witcamide MAS [934], [935]

Stearamide MEA superamide (98-100% conc.).
Aminol MS Flakes [320]

Stearamide MIPA.
Cyclomide SP [939]

Stearamido amine.
Mackine 301 [567]

Stearamido ethyl DEA.
Chemical Base 6532 [761], [762]
Lexamine 22 [453]
Sapamine COB-ST [190]

Stearamidoethyl DEA (97-100% conc.).
Chemical Base 6532 [761]

Stearamidoethyl ethanolamine.
Chemical 39 Base [762]

Stearamido morpholine.
Mackine 321 [567]

Stearamidopropyl dimethylamine.
Cyclomide SODI [234]
Incromine SB [223b]
Lexamine S13 [453]
Mackine 301 [567]
Mazeen SA [564]
Schercodine S [771]

Stearamidopropyl dimethylamine lactate.
Emcol 3780 [934], [935]
Hetamine 5L25 [407]
Incromate SDL [223b]
Lexamine S-13 Lactate [453]
Mackalene 316 [567]
Richamate 3780 [732]

Stearamidopropyl-dimethyl-β-hydroxyethyl ammonium dihydrogen nitrate.
Cyastat SN [52]

Stearamidopropyl-dimethyl-β-hydroxyethyl ammonium dihydrogen phosphate.
Cyastat SP [52]
Cyastat SP Antistatic Agent [50]

Stearamidopropyl-dimethyl-β-hydroxyethyl-ammonium nitrate.
Cyastat SN Antitstatic Agent [50]

Stearamidopropyl dimethyl myristyl acetate ammonium chloride.
Ceraphyl 70 [895]

Stearamidopropyl morpholine lactate.
Richamate SML [732]

Stearamine.
Armeen 18, 18D [79]
Kemamine P-990 [429]

Stearamine oxide.
Ammonyx SO [663]
Barlox 18S [529]
Carsamine Oxide S [529]
Conco XAS [209]
Jordamox SDA [474]
Rewominoxid S 300 [725]
Schercamox DMS [771]

Steareth-(2).
Alkasurf SA-2 [33]
Brij 72 [438]
Hetoxol STA-2 [407]
Lipocol S-2 [525]
Macol SA-2 [564]
Procol SA-2 [713]
Simulsol 72 [785]
Volpo S.2 [223]

Steareth-(4).
Nikkol BS-4 [631]
Procol SA-4 [713]

Steareth-(7).
Lamercreme SAY7 [375]

Steareth-(10).
Brij 76, 700 [438]
Lipocol S-10 [525]
Macol SA-10 [564]
Simulsol 76 [785]
Volpo S.10 [223]

Steareth-(11).
Empilan KM11 [548]

Steareth-(13).
Macol SA-13 [564]

Steareth-(15).
Macol SA-15 [564]

Steareth-(20).
Alkasurf SA-20 [33]
Brij 78 [438]
Hetoxol STA-20 [407]
Lipal 20SA [715]
Lipocol S-20 [525]
Macol SA-20 [564]
Procol SA-20 [713]
Simulsol 78 [785]
Volpo S.20 [223]

Steareth-(40).
Macol SA-40 [564]

Steareth-(50).
Empilan KM50 [548]

Steareth-(100).
Volpo S-100 [223]

Stearic acid.
Acme 210, 220, 230, 270, 280, 290 [6]
Akrochem Stearic Acid - Rubber Grade [15]
Century 1240 [876]
Crosterene [223]
Dar-Chem 14 [241]
Emersol 120, 132, 150 [288]
Emery 400 [289]
Formula 300 [715]
Groco 54, 55L, 58, 59 [374]
Hydrofol Acid 1855, 1895 [796]
Hy-Phi 1199, 1303, 1401 [241]

Stearic acid *(cont'd).*
Hystrene 4516, 5016, 5016 NF FG, 7018, 7018 FG, 8718 FG, 9718, 9718 NF FG [429]
Industrene 5016, 7018 FG, 8718 FG [429]
Loxiol G 20 [401]
Neo-Fat 18, 18-54, 18-55, 18-59, 18-61, 18-S [79]
Pearl Stearic [241]
Petrac 270 [688]
Pristerene 4900, 4901 [875]
Radiacid 408, 411, 420, 423 [658]
WGS 1135 T.P.S. [916]
Wochem #737 [941]

Stearic acid (9 EO).
Teric SF9, SF Series [439]

Stearic acid (15 EO).
Teric SF15, SF Series [439]

Stearic acid (55% conc.).
Hydrofol Acid 1855 [796]

Stearic acid (70% conc.).
Industrene 7018 [429]

Stearic acid (80% conc.).
Hystrene 8018 [429]

Stearic acid (90% conc.).
Industrene 9018 [429]

Stearic acid (95% conc.).
Hydrofol Acid 1895 [796]

Stearic acid, alkyl amidoamine.
Witcamine 20 [938]

Stearic acid, double pressed.
Dar Chem-12 [241]
Emersol Stearic Acid [289]

Stearic acid DEA.
Lipamide S [525]

Stearic acid, rubber grade.
Petrac 250 [688]

Stearic acid, single pressed.
Dar-Chem-11 [241]
Emersol Stearic Acid [289]

Stearic acid, tertiary amine, ethoxylated.
Ethomeen 18/12, 18/15, 18/20, 18/25 [79]

Stearic acid, triple pressed.
Dar Chem-13 [241]
Emersol Stearic Acid [289]
Groco 55L [374]
Hystrene 5016 [429]

Stearic acid, triple pressed, low I.V.
Dar Chem 14 [241]

Stearic acid, triple pressed, U.S.P./N.F.
Emersol 132 [289]

Stearic acid ester amide.
CPH-3800-N [379]

Stearic acid, ethoxylated.
Alkasurf S-2, S-5, S-8, S-14, S-40 [33]
Emulphor VT-650 [338]

Stearic acid, ethylene oxide condensate.
Ethofat 60/15, 60/20, 60/25 [79]

Stearic acid imidazoline, substituted.
Schercozoline S [771]

Stearic acid MEA.
Aminol N-1918 [320]
Clindrol LT 14-8 [197]
Comperlan HS [401]
Loramine S 280 [274]
Rewomid S 280 [274]
Schercomid SME, SME-A, SME-M [771]
Schercopearl EA-100 [771]
Steinamid S 280 [274]

Stearic acid POE (4) (4 ethylene oxide).
Crodet S4 [223]

Stearic acid POE (8).
Crodet S8 [223]
Simulsol M45 [711]

Stearic acid, POE (9).
Emulphor VT-650 [338]

Stearic acid POE (12).
Crodet S12 [223]

Stearic acid, POE (20).
Simulsol M49 [711]

Stearic acid POE (24).
Crodet S24 [223]

Stearic acid, POE (30).
Simulsol M51 [711]

Stearic acid POE (40).
Crodet S40 [223]
Simulsol M 52 [711]

Stearic acid, POE (50).
Simulsol M 53 [711]

Stearic acid POE (100).
Crodet S100 [223]
Simulsol M 59 [711]

Stearic acid polyglycol ester.
Eumulgin ST8 [401]

Stearic alkylolamine condensate.
Nopcogen 14-S [252], [253]

Stearic amido alkyl dimethyl amine.
Schercodine S [771]

Stearic DEA.
Alkamide HTDE [33]
Clindrol Superamide 100S (formerly Clindrol 868) [147]

Stearic DEA *(cont'd).*
Cyclomide DS280S [234]
Monamid 718 [599]
Unamide S [529]

Stearic DEA, 2:1 type.
Felsamid-SPD [823]

Stearic dimethyl amine oxide.
Schercamox DMS [771]

Stearic ester amide.
Schercomid SME-S [771]

Stearic ethyl amino hydroxyethyl amide.
Unamide SI [529]

Stearic fatty acid (1:1) alkanolamide.
Onyxol 42 [663]

Stearic fatty acid, ethoxylated.
Alkasurf S-1, S-2, S-5, S-8, S-14, S-40 [33]
Chemax E200MS, E400S, E600MS, E1000MS [170]

Stearic fatty acid imidazoline, substituted.
Miramine GS [592]

Stearic fatty acid polyglycol ester (6 EO).
Merpoxen, SFS 60 [285]

Stearic fatty acid polyglycol ester (8 EO).
Merpoxen SFS 80 [285]

Stearic hydroxyethyl imidazoline.
Alkazine ST [33], [34]

Stearic imidazoline.
Schercozoline S [771]

Stearic isopropanolamide.
Felsamid-SI [823]

Stearic MEA.
Alkamide HTME [33]
Ardet WB [73]
Cyclomide S280 [234]
Emid 6507 [290]
Felsamid-SM [823]
Monamid S [599]
Ninol SNR [762], [825a]
Rewomid S 280 [725]
Witcamide 70 [937a], [938]

Stearic/palmitic acid, 48:52%.
Hydrofol Acid 1655-CG-NF [796]

Stearo amphocarboxyglycinate.
Schercoteric STS [771]

Stearoamphoglycinate.
Amphoterge S [529]
Miranol DM, DM Conc. 45% [592]
Rewoteric AM-2S [725]

Stearoamphopropionate.
Miranol DM-SF [592]

Stearoamphopropylsulfonate.
Miranol DS [592]

Stearoxy dimethicone.
Silicone Copolymer F-755 [841]

Stearoyl alkanolamide.
Monamid 718, S [599]

Stearoyl colamino formyl methyl pyridinium chloride.
Emcol E 607 S [934]

N (Stearoyl colamino formyl methyl) pyridinium chloride.
Emcol E-607S [935]

Stearoyl colamino formyl methyl pyridinium chloride (90% conc.).
Emcol E 607S [938]

Stearoyl lactylate.
Artodan [373]

Stearoyl (2) lactylate.
Admul CSL, SSL [704]

Stearoyl lactylic acid.
Grindtek FAL 3 [373]

Stearoyl (2) lactylic acid, calcium salt.
Radiamuls CSL 2980 [658]

Stearoyl (2) lactylic acid, sodium salt.
Radiamuls SSL 2990 [658]

Stearoyl sarcosine.
Hamposyl S [363]

Steartrimonium animal protein, hydrolyzed.
Quat-Pro S [562]

Steartrimonium chloride.
Kemamine Q-9903B [429]
M-Quat 1850 [564]

Stearyl, amphoteric.
Sipoteric HDS [29]

Stearyl, amphoteric (65% conc.).
Sipon GSC [29]

Stearyl alcohol.
Adol 62, 63, 64 [796]
Alfol 18 [206]
Cachalot S-53, S-56, S-54 [581]
CO-1895 [709]
Crodacol S, S-95 [223]
Cyclal Stearyl Alcohol [939]
Cyclogol Stearyl Alcohol [234]
Dehydag Wax 18 [400]
Kalcohl 68, 80 [478]
Lanol S [711], [785]
Laurex 18 [548]
Lipocol S [525]
Lorol [270]
Stearal [44]

Stearyl alcohol (97% C_{18} content).
Adol 61 NF [796]

Stearyl alcohol, ethoxylated.
Alkasurf SA-2, SA-10, SA-20 [33]
Volpo S2, -S10, S20 [223]

Stearyl alcohol, polyethoxylated (2).
Simulsol 72 [711]

Stearyl alcohol, polyethoxylated (10).
Simulsol 76 [711]

Stearyl alcohol, polyethoxylated (20).
Simulsol 78 [711]

Stearyl alcohol polyglycol ether.
Genapol S-020, S-050, S-080 [55]

Stearyl alcohol, polypropoxylated.
Witconol APS [937]

Stearyl alcohol, technical.
Adol 64 [796]

N (Stearyl amido propyl) N,N dimethylamine.
Servamine KES 350 S [788]

N-Stearyl-(3-amido propyl)-N,N-dimethyl-N-benzyl ammonium chloride.
Schercoquat SAB [771]

N (Stearyl amido propyl) N,N dimethyl N benzyl ammonium chloride (80% conc.).
Servamine KES 4523 [788]

N-Stearyl-(3-amidopropyl)-N-N dimethyl-N-ethyl ammonium ethyl sulfate.
Schercoquat SAS [771]

Stearyl amine.
Crodamine 1.18D [223]
Kemamine P-990D [429]

Stearyl amine (5 EO).
Icomeen 18-5 [109]

Stearyl amine (50 EO).
Icomeen 18-50 [109]

Stearyl amine acetate.
Crodamac 1.18A [223]

Stearyl amine oxethylate.
Genamine S-020, S-050, S-080, S-100, S-150, S-200, S-250 [55]

Stearyl amino acid, complex, potassium salt (70% conc.).
Mafo 13 [564]

n-Stearyl amino acid, potassium salt, amphoteric.
Mafo 13 [564]

N-Stearyl amino dicarboxylic acid, complex, dipotassium salt (70% conc.).
Mafo 213 [564]

Stearyl ammonium chloride, substituted (26% conc.).
Miramine TA-26 [592]

Stearyl amphocarboxy glycinate, modified (65% conc.).
Sipoteric HDS [29]

Stearyl betaine.
Alkateric STB [33]
Chimin BX [514]
Lonzaine 18S [529]

Stearyl betaine (40% conc.).
Chimin BX [514]

Stearyl betaine (50% conc.).
Alkateric STB [33]
Varion SDG [796]

Stearyl cetyl alcohol.
Adol 63 [796]
Kalcohl 68, 86 [478]

Stearyl cetyl alcohol (10 EO).
Siponic E5 [29]

Stearyl cetyl alcohol, ethoxylated.
Siponic E7, E15 [29]

Stearyl dimethyl amine.
Barlene 18S [529]

Stearyl dimethyl amine oxide.
Ammonyx SO [663]
Barlox 18S [529]
Incromine Oxide S [223b]

Stearyl dimethyl amine oxide (25% conc.).
Ammonyx SO [663]
Mazox SO [564]

Stearyl dimethyl benzyl ammonium chloride.
Alkaquat DMB-ST [33]
Ammonyx 4, 4B, 485, 490, 4002 [663]
Barquat SB-25 [529]
Carsoquat SDQ-25, SDQ-85 [155]
Carsoquat SDQ-25 (Barquat SB-25), Carsoquat SDQ-85 [529]
Catinal OB-80E [864]
Cation S [764]
Incroquat S-85, SDQ-25 [223b]
Stedbac [317], [408], [408a]
Varisoft SDC [796]

Stearyl dimethyl tertiary amine, distilled.
Lilamin 342 D [522]

Stearyl erucamide.
Kemamide E-180 [429]

Stearyl erucate.
Schercemol SE [771]

Stearyl ether phosphate, acid form (4 EO).
Servoxyl VPSZ 4/100 [788]

Stearyl glyceryl ether.
Batyl Alcohol EX [631]
Nikko Batyl Alcohol 100 [631]

Stearyl glycyrrhetinate.
Co-Grhetinol [557]

Stearyl heptanoate.
Crodamol W [223]
Purcellin Solid 2/066220 [263]

Stearyl hydroxyethyl imidazoline.
Unamine S [529]

Stearyl imidazoline.
Amine S [190]
Schercozoline S [771]

Stearyl imidazoline amphoteric.
Amphoterge S [529]

Stearyl imidazoline monocarboxylate.
Amphoterge S [529]

Stearyl nitrile.
Arneel 18D [79]

Stearyl polyglycol ester (9 EO).
Serdox NSG 400 [788]

Stearyl primary amine.
Kemamine P-990D [430]

Stearyl stearamide.
Kemamide S-180 [429]

Stearyl stearate.
Cyclochem SS [234]
Lexol SS [453]
Liponate SS [525]
Rilanit STS-T [401]
Schercemol SS [771]

Streptokinase.
Kabikinase [693]
Streptase [418]

Streptomycin sulfate.
Streptomycin Sulfate Injection [690]

Streptozocin.
Zanosar [884]

Strontium carbonate.
Ultralog 82-2460-00 [172]

Strontium chloride.
Ultralog 82-2485-00 [172]

Strontium nitrate.
Ultralog 82-2600-00 [172]

Styrene acrylate latex.
Darex X442 [363]

Styrene acrylonitrile latex.
Darex 165L [363]

Styrene butadiene.
Kraton 101 [791]

Styrene butadiene latex.
Bayer SBR Latex 200 C, SBR Latex 210 C, SBR Latex 310 C, SBR Latex 320 C, SBR Latex 410 C, SBR Latex 420 C, SBR Latex 603 C [112]
Darex, 5281L, 632L, 636L, 637L, 643L [363]

Styrene-butadiene latex, carboxylated.
Darex 508L, 510L, 526L, 537L [363]

Styrene butadiene latex, modified.
Darex 3333 [363]

Styrene butadiene rubber.
Ameripol SBR [361]

Styrene maleic anhydride.
Scripset 520 [602]

Styrene maleic anhydride, esterfied.
Scripset 540, 550 [602]

Styrene maleic anhydride amide/NH_4OH acid salt.
Scripset 808 [602]

Styrene maleic anhydride amide/NH_4OH basic salt (25% conc.).
Sripset 720 [602]

Styrene maleic anhydride, sodium salt.
Scripset 500 [602]

Styrene maleic anhydride, sodium salt (30% conc.).
Scripset 700 [602]

Succinic acid 2,2-dimethyl hydrazide.
Alar [881]

Succinylcholine chloride.
Anectine [139]

Sucralfate.
Carafate Tablets [551]

Sucrose cocoate.
Crodesta SL-40 [223]

Sucrose distearate.
Crodesta F-10, F-50, F110 [223]
Ryoto Sugar Ester S-570, S-770 [758]

Sucrose distearate, acetylated.
Crodesta A-10, A-20 [223]

Sucrose ditristearate.
Ryoto Sugar Ester S-270, S-370 [758]

Sucrose fatty acid ester.
DK Ester F-10, F-20, F-50, F-70, F-90, F-110, F-140, F-160 [237]

Sulfate fatty acid ester (45% conc.).
Doittol 14 [253]

Sucrose fatty acid ester compound.
Sunny Safe [237]

Sucrose mono cocoate.
Crodesta SL-40 [223]

Sucrose monococoate (40% conc.).
Crodesta SL-40 [223]

Sucrose monodistearate.
Ryoto Sugar Ester A-970, S-1170 [758]

Sucrose monolaurate (40% conc.).
Ryoto Sugar Ester LWA-1540 [758]

Sucrose monooleate (40% conc.).
Ryoto Sugar Ester OWA-1570 [758]

Sucrose monopalmitate.
Ryoto Sugar Ester P-1570, P-1670 [758]

Sucrose monostearate.
Crodesta F-110 [223]
Ryoto Sugar Ester S-1570, S-1670 [758]

Sucrose polystearate.
Ryoto Sugar Ester S-170 [758]

Sucrose stearate.
Crodesta F-110, F-160 [223]

Sulfabenzamide.
Sultrin Triple Sulfa Cream, Triple Sulfa Vaginal Tablets [666]
Triple Sulfa Vaginal Cream [769]
Trysul [767]

Sulfacetamide.
Sultrin Triple Sulfa Cream, Triple Sulfa Vaginal Tablets [666]
Triple Sulfa Vaginal Cream [769]
Trysul [767]

Sulfacetamide sodium.
Sodium Sulamyd Ophthalmic Solutions & Ointment-Sterile [772]

Sulfacetamide sodium/prednisolone acetate, 100 mg:5 mg.
Ocu-Lone-C [655a]

Sulfacytine.
Renoquid [356]

Sulfadoxine.
Fansidar Tablets [738]

Sulfamethizole.
Azotrex Capsules [134]
Sul-Azo Tablets [512]
Thiosulfil Duo-Pak, Forte, -A Forte [97]
Urobiotic-250 [739]

Sulfamethoxazole.
Azo Gantanol Tablets [738]
Bactrim DS Tablets, I.V. Infusion, Pediatric Suspension, Suspension, Tablets [738]
Cotrim, D.S. [519]
Gantanol DS Tablets, Suspension, Tablets [738]
Septra DS Tablets, I.V. Infusion, Suspension, Tablets [139]
Sulfamethoxazole Tablets [344]
Sulfamethoxazole & Trimethoprim Pediatric Suspension, Tablets [119]
Sulfamethoxazole with Trimethoprim Tablets, Tablets Double Strength [240]
Sulfamethoxazole w/ Trimethoprim DS, SS [344]
Sulfatrim & Sulfatrim D/S Tablets [769]

Sulfanilamide.
AVC Cream, Suppositories [580]
Vagimide Cream [518]
Vaginal Sulfa Suppositories [769]

Sulfasalazine.
Azulfidine Tablets, EN-tabs, Oral Suspension [693]
Sulfasalazine Tablets [240], [344], [769]

Sulfate fatty acid ester (45% conc.).
Doittol 14 [252]

Sulfathiazole.
Sultrin Triple Sulfa Cream, Triple Sulfa Vaginal Tablets [666]
Triple Sulfa Vaginal Cream [769]
Trysul [767]

Sulfinpyrazone.
Anturane Tablets & Capsules [189]
Sulfinpyrazone Tablets [240], [769]

Sulfinpyrazone tablets.
Anturane Tablets [510]

Sulfisoxazole.
Azo Gantrisin Tablets [738]
Gantrisin Tablets [738]
Koro-Sulf Vaginal Cream [945]
Pediazole [747]
SK-Soxazole Tablets [805]
Sulfisoxazole Tablets [344], [751], [769]
Vagilia Cream, Suppositories [519]

Sulfisoxazole diolamine.
Gantrisin Injectable, Ophthalmic Ointment/Solution [738]

Sulfo betaine.
Rewoteric AM-CAS [725a]

Sulfo betaine, modified (50% conc.).
Varion CAS [796]

Sulfo-coco amido betaine (35% conc.).
Schercotaine SCAB [771]

Sulfo dicarboxylic acid ester (25% conc.).
Insidol [77]

Sulfo dicarboxylic acid ester, wetting agent.
Insidol [77]

Sulfonate.
Dilasoft TF [762]
Richonate [732]

Sulfonate, complex aliphatic.
Emcol K-8300 [937]

Sulfonate, creosol-free, containing solvents.
Mercerol QW [762]

Sulfonate fatty acid.
Witconate 1840-X [935]

Sulfonate, fatty acid (55% conc.).
Lankropol OPA [509]

Sulfonate, liquid.
Witconate P10-59 [938]

Sulfonate, organic.
Cordon COT [320]

Sulfonate, organic (aq. sol'n.).
Mercerol SM [762]

Sulfonate, synthetic.
Barium Petronate 50-S Neutral [936]

Sulfonic acid.
Polystep RA-90 [825]
Richonic Acid, Acid B, 94H, 97H [732]

Sulfonic acid, aromatic (76% conc.).
Manro FCM 130 [545]

Sulfonic acid, aromatic (85% conc.).
Manro FCM 140 [545]

Sulfonic acid, aromatic (90% conc.).
Manro FCM 100 [545]

Sulfonic acid, aromatic (91% conc.).
Manro FCM 95 LV [545]

Sulfonic acid, aromatic (92% conc.).
Manro FCM 150 [545]

Sulfonic acid, aromatic, condensated salt.
Demol AS, MS [478]

Sulfonic acid, high molecular.
Osimol Grunau DP [176]

Sulfonic acid, high molecular (40% conc.).
Lamepon 287 SF, N, RE [176]

Sulfonic acid, naphthalene, sodium salt.
Tamol L. Conc., SN [742]

Sulfonic acid, organic.
Wetting Agent 611, CG 16 [176]

Sulfonic acid salt.
Primasol FP [109]

3-(Sulfonyl)-O-O ξ (methylamino) carbonyl ξ oxime-2-butanone.
Plant pin [910a]

Sulfosuccinate.
Amawet SS [190]
Anionyx [663]
Condanol TS 25 [274]
Cyclopol [234]
Emcol 4100M, 4300, 4600 [935]
Gemtex; Fizul [320]
Jordawet [474]
Rewopol B 2003, SBV [725a]
Schercopol [771]
Siponate [29]

Sulfosuccinate, asymmetrical.
Emcol 4930, 4940 [935]

Sulfosuccinate, complex aliphatic.
Emcol K 8300 [937a], [938]

Sulfosuccinate, ethoxylated.
Schercopon 2WD [771]

Sulfosuccinate, fatty alcohol.
Rewopol SBF 12, SBF 18 [725]

Sulfosuccinate, fatty alcohol (40% conc.).
Rewopol SBF 12 [725]

Sulfosuccinate, fatty alcohol (95% conc.).
Rewopol SBF 18 [725]

Sulfosuccinate, fatty alcohol ether.
Rewopol SBFA 30, SB-FA 50 [725]

Sulfosucciante, fatty alcohol ether (40% conc.).
Rewopol SBFA 30, SB-FA 50 [725]
Varsulf SBFA 30 [796]

Sulfosuccinate, half-ester.
Emcol 4161L, 4300 [937a], [938]

Sulfosuccinates, half ester (25% conc.).
Fizul 10-127, 12-25, 30-25, K-127 [320]

Sulfosuccinates, half ester (35% conc.).
Fizul O-653 [320]

Sulfosuccinates, half ester (40% conc.).
Fisul 10-126 Conc. [320]

Sulfosuccinate, isopropylamine salt.
Pestilizer B [825]

Sulfosuccinate, lauric acid monoethanol amide, sodium salt.
Marlimat HA 12 [178]

Sulfosuccinate, modified.
Condanol SBV [274]
Cyclonate K8300 [938]
Lankropol ADF [509]

Sulfosuccinate, special alkanolamine salt (40-42% conc.).
Setarin M [948]

Sulfosuccinic acid, alcohol half ester, disodium ethoxylated.
Aerosol A-102 [50], [50a]

Sulfosuccinic acid, dioctyl ester, sodium salt.
Elfanol 883 [16]

Sulfosuccinic acid ester, sodium salt.
Marlinat, DF 8, HA 12, SRN 30 [178]

Sulfosuccinic acid, half ester, disodium salt (31% conc.).
Aerosol A-102 [233]

Sulfosuccinic acid, half ester, disodium salt (50% conc.).
Aerosol [223]

Sulfosuccinic acid, nonyl phenol half ester, disodium ethoxylated.
Aerosol A-103 [50a]

Sulfosuccinic acid, nonyl phenol half ester, disodium ethoxylated (34% conc.).
Aerosol A 103 [233]

Sulfur.
Aveenobar Medicated [214]
Code 104 Rubbermaker's, 209 Oil Treated RM, 296 Oil Treated Conditioner RM, 338 Superfine RM, 793 Conditioned RM [392]
Hill Cortac [412]
Pedi-Bath Salts [679]
Royal RM98 D Sulfur [752]
Sulfacet-R Acne Lotion [246]
Transact [920]

Sulfur; added oil, polymerized.
Crystex 95 OT 20 Insoluble Sulfur [821]

Sulfur, agricultural dusting.
Palmetto [659]

Sulfur, colloidal.
Acne-Dome Medicated Cleanser [585]

Sulfur, polymerized.
Crystex Insoluble Sulfur [821]

Sulfurated lime.
Vlemasque [246]

Sulfuric acid.
Acculog 82-8200-00, 82-8200-10 [172]

Sulfuric acid ester, high saturated fatty alcohol, sodium salt (90% conc.).
Dehydag WAX E [101]

Sulfur & 5-nitro-benzene-1,3-dicarboxylic acid bis (1-methylethyl) ester.
Kumulan [107]

Sulfur-nitrogen compound, heterocyclic.
Vanlube 601 [894]

Sulfur preparations.
Fostex Medicated Cleansing Bar, Medicated Cleansing Cream [920]
Fostril [920]
Pernox Lotion, Medicated Lathering Scrub [920]
Sebulex & Sebulex Cream Shampoo, Shampoo with Conditioners [920]
Sebutone & Sebutone Cream Shampoo [920]
Xerac [686]

Sulfuryl fluoride.
Vikane Gas Fumigant [261]

Sulindac.
Clinoril Tablets [578]

Sunflower oil mono-diglyceride.
Tegomuls SB [856]

Sunflower oil monoglyceride, unsaturated, distilled.
Dimodan LS [373]

Sunflower seed oil.
Lipovol Sun [525]

Sunflower seed oil glyceride.
Monomuls 90-40 [375]

Super-phosphoric acid.
Phospholeum [602a]

Sutilains.
Travase Ointment [327]

Sweet almond oil.
Lipovol ALM [525]

Talbutal.
Lotusate Caplets [932]

Talc.
#907 Micro Talc [188]
AGI Talc, BC 1615 [924]
Alpine Talc USP, BC 127, BC 662 [924]
Ceramitalc No.1, 10A [894]
Korman [236]
Lo-Micron Brown Oxide 7172 [924]
Lo-Micron Suntan B.C. 34-2500, B.C. 34-2509 [414]
Lo Micron Talc 1, Talc, BC 1621, Talc USP, BC 2755 [924]
Metro Talc 4609 [924]
Mistron Spray [236]
Monoblend [236]
Nytal [894]
Olympic [236]
Orefino [236]
Oreon, Oreon EF [236]
Purtalc USP [188]
Talc, Heavy USP, Bacteria Controlled 1745 [924]
Talc OOS, OXO [760]

Tallamide DEA.
Schercomid SO-T, TO-2 [771]

Tallamphopropionate.
Miranol LM-SF Conc. [592]

Tall, ethoxylated.
Isoplast D [461]

Tall oil alkanolamide.
Mazamide 70, 75, TO-10, TO-20 [564]
Monamine T-100 [599]

N (Tall oil amidopropyl) N,N dimethyl amine.
Servamine KET 350 [788]

Tall oil amidopropyl dimethylethyl ammonium ethosulfate.
Servamine KET 4542 [788]

Tall oil amino ethyl imidazoline.
Alkazine TO-A [33]

Tall oil diethanolamide.
Monamine T-100 [599]
Schercomid SO-T [771]

Tall oil dioleates.
Polyfac TDO 5, TDO 9, TDO 14 [918]

Tall oil, distilled.
Acosix 700 [722]

Tall oil & EO, 5 moles.
Teric T Series [439]

Tall oil & EO, 7 moles.
Teric T Series [439]

Tall oil & EO, 10 moles.
Teric T Series [439]

Tall oil, epoxidized.
Peroxidol 781 [722]

Tall oil ethoxylate.
Aconol X10 [390]

Tall oil, ethoxylated (EO 16).
Industrol TO-16 [109]

Tall oil fatty acid diethanolamide.
Schercomid TO-2 [771]

Tall oil fatty acid, ethoxylated.
Chemax TO-10, TO-16 [170]
Trydet 22 [289]

Tall oil fatty acid, ethoxylated (EO 10).
Industrol TO-10 [109]

Tall oil fatty acid, 2-ethyl hexyl alcohol ester.
Croplas EH [226]

Tall oil fatty acid imidazoline.
Amine T [190]

Tall oil fatty acids, imidazoline, substituted.
Miramine GT, TOC [592]

Tall oil fatty dimer acid.
Dimer C 36-18, C 36-20D, C 36-22, C 36-24, C 36-27D, C 36-35, CD-1100 [181]
Monomer C18 [181]

Tall oil; high pH, sulfated.
Cordon PB-870 [320]

Tall oil hydroxyethyl imidazoline.
Alkazine TO [33], [34]
Unamine T [529]

Tall oil imidazoline.
Textamine T-1 [399], [400]

Tall oil, imidazoline, substituted.
Mazoline TA [564]
Monazoline T [564]

Tall oil; low pH, sulfated.
Cordon LB-870 [320]

Tall oil monooleate series.
Polyfac TMO 5, 7, 9, 14 [918]

Tall oil, sulfated.
Actrasol SP175K [811]

Tallow acetate, n-hyd.
Acetamin HT [803]
Crodamac 1.HTA [223]

Tallow alcohol ethoxylate.
Alkasurf TA-20 [33]
Elfapur T 110, T 250 [16], [764]

Tallow alcohol polyglycolether.
Rewopal TA 11, TA 25 [725]

Tallow alkyl dimethylamine, hydrogenate.
Nissan Tertiary Amine ABT [636]

Tallow-alkyl propylenediamine.
Nissan Amine DT [636]

Tallow alkyl propylenediamine, hydrogenated.
Nissan Amine DTH [636]

Tallow alkyl trimethyl ammonium chloride, hydrogenated.
Nissan Cation ABT-350, ABT-500 [636]

Tallow amide.
Kemamide S-65 [429a]

Tallowamide DEA.
Mackamide T [567]
Schercomid SO-T, TO-2 [771]

Tallow amide, ethoxylated hydrogenated (EO 50).
Icomid HT-50 [109]

Tallow amide, hyd.
Kemamide S-65 [429]

Tallow amidoalkylamine oxide (50% conc.).
Textamine Oxide TA [399], [400]

Tallow amidopropylamine oxide.
Jordamox TAPA [474]

Tallowamidopropyl betaine.
Mackam TA [567]

Tallowamidopropyl dimethylamine.
Incromine TB [223b]
Schercodine T [771]

Tallow amido propyl dimethylamine oxide.
Ammonyx TDO [663]

Tallowamidopropyl hydroxysultaine.
Mirataine® TABS [592]

Tallowamidopropyl sultaine.
Mirataine TABS [592]

Tallow amine.
Amine BGD [487]
Crodamine 1.TD [223]
Kemamine P-974 [429a]
Lilamin BGD [487]
Noram S [162]
Radiamine 6170, 6171 [658]

n-Tallow amine acetate.
Acetamin T [803]
Armac T [16], [79]
Crodamac 1.TA [223]
Kemamine A974 [429a]
Noramac S [162]
Radiamac 6179 [658]

Tallow amine acetate, hydrogenated.
Armac HT [79]
Kemamine A970 [429a]
Radiamac 6148, 6149 [658]

Tallow amine, dimethyl hydrogenated.
Amine 2MHBG [487]
Lilamin 2MHBG [487]

Tallow amine, distilled, ethoxylate 20-mole.
Polyfac TA-20 [917a]

Tallow amine (distilled), hydrogenated.
Radiamine 6141 [658]

Tallow amine (distilled), polyoxyethylated (15).
Katapol PN-730 [338]

Tallow amine (distilled), polyoxyethylated (20).
Katapol PN-810 [338]

Tallow amine E.O. + 2 moles.
Teric 17M Series [439]

Tallow amine E.O. + 5 moles.
Teric 17M Series [439]

Tallow amine E.O. + 15 moles.
Teric 17M Series [439]

Tallow amine, ethoxylated.
Accomeen T15 [81]
Alkaminox T-5, T-10, T-12, T-15. T-25, T-30 [33]
Chemeen T-2, T-5, T-10, T-15, T-20 [170]
Ethylan TT-15 [509]
Icomeen T-25 CWS [109]
Varonic T205, T210, T215 [796]

Tallow amine, ethoxylated hydrogenated.
Chemeen HT-2, HT-50 [170]
Varonic U202, U205, U215, U230 [796]

Tallow amine, ethoxylated, EO 2 moles.
Accomeen T2 [81]
Icomeen T-2 [109]

Tallow amine ethoxylate (EO 5).
Icomeen T-5 [109]

Tallow amine ethoxylate (EO 7).
Icomeen T-7 [109]

Tallow amine ethoxylate (EO 15).
Icomeen T-15 [109]

Tallow amine ethoxylate (EO 20).
Icomeen T-20 [109]

Tallow amine ethoxylate (EO 25).
Icomeen T-25 [109]

Tallow amine ethoxylate (EO 40).
Icomeen T-40, T-40-80% [109]

Tallow amine, hydrogenated.
Amine HBGD [487]
Crodamine 1.HTD, 2.HT [223]
Kemamine P-970 [429a]
Lilamin HBGD [487]
Noram SH [162]

Tallow amine, hydrogenated *(cont'd).*
Radiamine 6140 [658]

Tallow amine, hydrogenated primary, ethoxylated.
Alkaminox HT-12, HT-25, HT-30 [34]

Tallow amine oxide.
Mackamine TAO [567]

Tallow amine polyglycol ether, distilled, EO 15.
Serdox NJAD 15 [788]

Tallow amine polyglycol ether, distilled, EO 20.
Serdox NJAD 20 [788]

Tallow amine polyglycol ether, distilled, EO 30.
Serdox NJAD 30 [788]

Tallow amine, polyoxyethylated (5).
Katapol PN-430 [338]

Tallow amine, polyoxyethylated (15).
Katapol PN-730 [338]

Tallow amine, polyoxyethylated (20).
Katapol PN-810 [338]

Tallow amine primary.
Kemamine P-974 [429a]

Tallow amine primary, distilled.
Kemamine P-974 D [429a]

Tallow amine primary, distilled hyd.
Kemamine P-970D [429], [429a]

Tallow amine primary, hydrogenated.
Kemamine P-970 [429a]
Nissan Amine ABT [636]

Tallow amine primary, technical.
Kemamine P-974 [429]
Tomah Tallow Amine [866]

Tallow amine primary, technical hyd.
Kemamine P-970 [429]

Tallow amine secondary.
Crodamine LTD [223]

Tallow amine, secondary hydrogenated.
Crodamine 2.HT [223]

Tallow amine, N-tallow ethoxylated.
Noramox S2 to 11 [162]

Tallow ammonium carboxylate, complex.
Ampho T-35 [829]

Tallowamphoglycinate.
Rewoteric AM-2T [725]

Tallow betaine.
Alkateric TB [33]

Tallow betaine, hydrogenated.
Mirataine TDMB-H [592]

Tallow bis-hydroxyethyl amine oxide.
Schercamox T-12 [771]

Tallow diamine.
Dicrodamine 1.T [223]
Radiamine 6570 [658]
Tomah Tallow Diamine [866]

Tallow diamine diacetate.
Dicrodamac 1.TDA [223]

Tallow diamine dioleate.
Adogen 570-DO [796]
Dicrodamac 1.TDO [223]

Tallow diamine, ethylene oxide adduct 30 mole.
Icodimeen T-30 [109]

Tallow diamine, hydrogenated.
Dicrodamine 1.HT [223]
Radiamine 6540 [658]

N-Tallow-1,3-diaminopropane.
Duomeen T, T Special [79]
Duomeen TX [16]

Tallow dicoco am. chloride, quaternary.
Jet Quat T-2C-50 [471]

Tallow diethanolamide.
Mackamide T [567]
Product LT 19-14 [197]

N-tallow dimethyl amine.
Noram DMS [162]

Tallow-dimethyl-t-amine (distilled), hydrogenated.
Lilamin 345 D [522]

Tallow dimethyl amine, hydrogenated.
Armeen DMHTD [16]
Noram DMSH [162]

Tallow dimethyl amine oxide, hydrogenated.
Aromox DMHTD [16]

Tallow dimethyl ammonium chloride, dihydrogenated.
Adogen 442 H [796]
Arquad 2HT-75 [79]
Noramium M2SH [162]
Radiaquat 6442 [658]

Tallow dimethyl benzyl ammonium chloride.
Ammonyx 856 [663]
Noramac S 75 [162]
Noramium S 75 [162]

Tallow dimethyl benzyl ammonium chloride, hydrogenated.
Arquad DMHTB-75 [16]
Empigen BCM75, BCM75/A [25]

Tallow dimethyl trimethylpropylene diammonium chloride.
Tomah Q-D-T [866]

N-Tallow dipropylene triamine.
Trinoram S [162]

Tallow ether sulfate.
Elfan NS 682 KS [16]

Tallow fatty acid.
Emery 540 [289]

Tallow fatty acid, imidazoline, substituted.
Miramine® TC [592]

Tallow fatty acid monoglyceride polyglycol ether.
Rewoderm LI 48, LI 420 [725a]

Tallow fatty alcohol (EO 11).
Dehydol TA 11 [401]

Tallow fatty alcohol ethoxylate.
Alkasurf TA-20, TA-30, TA-40, TA-50 [33]

Tallow fatty alcohol polyglycol ether (EO 25).
Rewopal TA 25/S [725a]

Tallow fatty alcohol sulfate.
Sulfopon T 35 [401]

Tallow glyceride.
Myverol 18-30 Distilled Monoglycerides [278]

Tallow glyceride, acetylated hydrogenated.
Lamegin EE-50, EE-70, EE-90 [375]
Tegin E41 [360]

Tallow glyceride citrate, hydrogenated.
Lamegin ZE-30, ZE-60 [375]
Tegin C 61, C 62, C 64 [360]

Tallow glyceride ethyloxylate, complex.
Acconon TGH [81], [829]

Tallow glycerides, hydrogenated.
Neustrene 059 [429]

Tallow glyceride lactate, hydrogenated.
Lamegin GLP-10, GLP-20 [375]
Tegin L61, L62 [360]

Tallow glyceride, refined hydrogenated.
Neustrene 060 [429]

Tallow imidazoline quaternary.
Accosoft 808 [829]

Tallow imidazolinium methosulfate.
Carsosoft S-75, S-90, S-90M [529]
Incrosoft S-75, S-75 CG, S-90, S-90M [223b]

N-Tallow β imino dipropionate, disodium salt.
Deriphat 154 [365]

Tallow methylamine, N-dihydrogenated.
Noram M2 SH [162]

Tallow monoglyceride distilled, saturated.
Dimodan TH [373]

Tallow nitrile.
Arneel T [79]

N-Tallow pentamethyl propane diammonium dichloride.
Adogen 477 [796]

N-Tallow polypropylene polyamine.
Polyram S [162]

Tallow, primary amine, hydrogenated.
Radiamine 6140 [658]

N-Tallow-1,3-propanediamine.
Duomeen T, T Special [79]

N-Tallow-1,3-propanediamine diacetate.
Duomac T [79]

N-Tallow-1,3-propanediamine dioleate.
Duomeen TDO [79]

Tallow-1,3-propylene diamine.
Diamine BG [487]
Dinoram S [162]
Kemamine D 974 [429a], [430]
Lilamin BG [487]

N-Tallow propylene diamine acetate.
Dinoramac S [162]

N-Tallow 1,3 propylene diamine diacetate.
Kemamine AD 974 [429a]

N-Tallow 1,3 propylene diamine dioleate.
Kemamine OD 974 [429a]

Tallow propylene diamine, N-hydrogenated.
Dinoram SH [162]
Kemamine D-970 [429a], [430]

N-Tallow propylene diamine, polyethoxylated.
Dinoramox S 3 to 12 [162]

Tallow propylene diamine, triethoxylated N-hydrogenated.
Dinoramox SH 3 [162]

Tallow proylene diammonium caprylate.
Rewocor TPAC 100 [725a]

Tallow sulfated.
Nopcosulf TA-30 [252]
Nopcosulf TA-45V [252], [253]

Tallow sulfosuccinamate.
Arylene TA [390]

Tallow triamine.
Tomah Tallow Triamine [866]

Tallow trimethyl ammonium chloride.
Adogen 471 [796]
Jet Quat T-50 [471]

Tallow trimethyl ammonium chloride *(cont'd).*
Radiaquat 6471 [658]

Tallow trimethyl ammonium chloride, hydrogenated.
Noramium MSH 50 [162]

N-Tallow trimethylene diamine diacetate.
Duomac T [79]

Tallow trimonium chloride.
Kemamine Q-9703B [429a]

Tamoxifen citrate.
Nolvadex Tablets [831]

Tannic acid.
Dalidyne [239]

Tar preparations.
Alphosyl Lotion, Cream [719]
Bainetar [920]
DHS Tar Shampoo [686]
Estar Gel [920]
Fototar Cream 1.6%, Stik 5% [284]
Sebutone & Sebutone Cream Shampoo [920]
Zetar Emulsion, Shampoo [246]

Taurate.
Jer-Gel [842]
J-Pongeler [842]

Tauride, sodium salt, fatty acid.
Hostapon KTW New [416]

TBHO; citric acid; propylene glycol, 20%:3%:77%.
Sustane 20-3 [883]

TBHO; citric acid; propylene glycol, 20%:10%:70%.
Sustane 20 [883]

TBHO; propyl gallate; citric acid; veg. oil; glycerol monooleate, 20%:15%:3%:30%:32%.
Sustane 20A [883]

TEA animal protein, abietoyl hydrolyzed.
Lamepon PA-TR [375]
Lexein A-520 [453]

TEA animal protein, coco hydrolyzed.
Lamepon S2-TR, S-TR, ST-40 [375]
Maypon 4CT [825]
May-Tein CT [562]

TEA animal protein, myristoyl hydrolyzed.
Lexein A-220 [453]

TEA animal protein, oleoyl hydrolyzed.
Lamepon PO-TR [375]

TEA animal protein, undecylenoyl hydrolyzed.
Maypon UDT [825]

TEA-alkylaryl sulfonate.
Witconate 60T, TAB [934]

TEA alkyl ether sulfate.
Steol 7T [825a]

TEA-cocoyl glutamate.
Acylglutamate CS-12 [14]

TEA-dodecylbenzenesulfonate.
Alkasurf T [33]
Bio Soft N-300 [825]
Carosulf T-60-L [529]
Conco AAS-60 [209]
Cycloryl DDB-60T [234]
Cycloryl TABS [939]
Marlopon AT 50 [178]
Nansa TS 60 [548]
Richonate 5725, S-1280, TAB [732]
Rueterg 97T [320]
Sterling LA Oil [145]
Surco 60T [663]
Ultrawet 60L [72]
Witconate 60T, TAB [934], [935]
Witconate 79S [935]

TEA-laureth sulfate.
Alscoap TA-40 [864]
Cycloryl TD [939]
Empicol ETB [548]
Montelane LT 4088 [785]
Texapon NT [400]

TEA-lauroyl glutamate.
Acylglutamate LT-12 [14]

TEA lauryl ether sulfate.
Neopon LOT [937a], [938]

TEA-lauryl sulfate.
Alkasurf TLS [33]
Avirol 300 [400]
Carsonol TLS [529]
Conco Sulfate TL [209]
Cycloryl TA [939]
Cycloryl WAT, WAT-1 [234]
Emersal 6434 [291]
Empicol TDL75, TL40 [548]
Equex T [709]
Genapol LRT 40 [55]
Lonzol LT-300 [529]
Maprofix TLS 65 [663]
Maprofix TLS-106, TLS 500, TLS 513 [663]
Melanol LP 20 T [785]
Nikkol TEALS [631]
Rewopol TLS [725]
Richonol T [732]
Serdet DFL 40 [175]
Sipon LT 6, TLS-65 [29]
Standapol T, TS-100 [400]
Stepanol WAT [825], [825a]
Sterling WAT [145]

TEA-lauryl sulfate *(cont'd)*.
Texapon T [399]
Texapon T42, TH [400]
Witcolate T [934], [935]

TEA lauryl sulfate fatty alcohol, sulfonated.
Neopon LT [635], [937a], [938]

TEA monooleamido PEG-2 sulfosuccinate.
Standapol SH-300 [400]

TEA-monooleamide, sulfosuccinate.
Cyclopol SBG 280/T [234]

TEA-oleamido PEG-2 sulfosuccinate.
Mackanate ODT [567]
Standapol SH-300 [400]

TEA oleic sulfosuccinate.
Mackanate ODT [567]

TEA-palm kernel sarcosinate.
Sarkosine KF [55]

TEA-PEG-3 cocamide sulfate.
Genapol AMS [55]

TEA-salicylate.
Fitrosol B [647]
Sunarome W [312]

Tellurium.
Tellow [894]

Tellurium diethyldithiocarbamate.
Ethyl Tellurac [894]
Ethyl Tellurium [708]
Tellurac [894]

Temazepam.
Restoril Capsules [762]

Terbutaline sulfate.
Brethine, Ampuls [340]
Bricanyl Injection, Tablets [580]

Terpene amide.
Isomul Extra, Extra Type B [462]

Terpene hydrocarbon.
Dipanol [227]

Terpene laurate.
IsoPent 21 [462]

Terpene liquid.
Terpineol 318, 318 prime [406]

Terpene solvent.
Gensol No. 6 [850]

Terphenyl, partially hydrogenated.
HB-40 [602]

Terpineol.
Terpineol 318 [406]

Terpin hydrate.
SK-Terpin Hydrate and Codeine Elixir [805]
Terpin Hydrate & Codeine Elixir [751]

Testolactone.
Teslac Tablets [817]

Testosterone.
BayTestone-50, -100 [111]

Testosterone cypionate.
Depo-Testosterone [884]
T-Cypionate 200 [518]

Testosterone enanthate.
Ditate-DS [767]
Testaval 90/4 [518]

Tetanus antitoxin.
Tetanus Antitoxin (equine), refined [778]
Tetanus Immune Globulin (Human) Hyper-Tet [232]

Tetanus & diphtheria toxoids adsorbed (for adult use).
Tetanus & Diphtheria Toxoids, Adsorbed Purogenated [516]
Tetanus & Diphtheria Toxoids, Adsorbed (For Adult Use) [778]

Tetanus & diphtheria toxoids combined, aluminum phosphate adsorbed (for adult use).
Tetanus & Diphtheria Toxoids Adsorbed (Adult) [942]
Tetanus & Diphtheria Toxoids Adsorbed (Adult) in Tubex [942]

Tetanus & diphtheria toxoids combined, aluminum potassium sulfate adsorbed (for adult use).
Tetanus & Diphtheria Toxoids Adsorbed (For Adult Use) [815]

Tetanus immune globulin (human).
Homo-Tet [767]
Tetanus Immune Globulin (Human) Hyper-Tet [232]
Tetanus Immune Globulin (Human), in Tubex [942]

Tetanus toxoid, aluminum hydroxide adsorbed.
Tetanus Toxoid, Adsorbed [778]

Tetanus toxoid, aluminum phosphate adsorbed.
Tetanus Toxoid Adsorbed, Aluminum Phosphate Adsorbed, Ultrafined, Ultrafined in Tubex [942]

Tetanus toxoid, aluminum potassium sulfate adsorbed.
Tetanus Toxoid Adsorbed [815]

Tetanus toxoid, fluid.
Tetanus Toxoid [815]

Tetanus toxoid, fluid *(cont'd).*
Tetanus Toxoid Fluid, Purified, Ultrafined, Ultrafined in Tubex [942]

Tetraammonium EDTA.
Sequestrene Tetraammonium [190]

Tetrabromobisphenol A.
BA-59, BA-59P [369]
Emery 9350 [289]

Tetrabromobisphenol A, bis (allyl ether).
BE-51 [369]

Tetrabromobisphenol A, bis (2,3-dibromopropyl ether).
Great Lakes PE-68 [369]

Tetrabromobisphenol A, bis (2-hydroxyethyl ether).
BA-50, BA-50P [369]

Tetrabromobisphenol A, ethoxylated, bis acrylate.
BA-43 [369]

Tetrabromophthalate diol.
PHT-4 Diol [369]

Tetrabromophthalic anhydride.
Great Lakes PHT4 [369]

Tetrabromoxylene.
Emery 9345 [289]

Tetrabutylthiuram disulfide.
Butyl Tuads [894]

Tetrabutylthiuram monosulfide (q.v.).
Pentex [881]

Tetrabutyl titanate.
Tyzor TBT [269]

Tetracaine hydrochloride.
Cetacaine Topical Anesthetic [168]
Pontocaine Hydrochloride for Spinal Anesthesia [133]

1,2,4,5-Tetrachloro-3-nitrobenzene.
Fusarex [444]

2,3,5,6-Tetrachloro-1,4-benzoquinone.
Spergon [881]

2,3,5,6-Tetrachloro-4-pyridinol.
Daxtron [261]

Tetrachlorobutyl alcohol.
AA-81C [259]

cis-N-((1,1,2,2,-Tetrachloroethyl) thio) 4-cyclohexene-1,2-dicarboximide.
Difolatan [183]
Haipen [183]
Mycodifol [809a]
Pillartan [696a]
Sanspor [444]

2,4,5,6-Tetrachloro-isophthalonitrile.
Bravo [252], [780]
Bravo 500 [780]
Daconil 2787 [780]
Exotherm Termil [732a]
Nopcocide N-96 [252]

Tetrachlorophthalic anhydride.
Tetrathal® [602]

Tetracycline.
Achromycin V Oral Suspension [516]
Mysteclin-F Capsules, Syrup [817]
SK-Tetracycline Syrup [805]
Sumycin Syrup [817]
Tetracycline HCl Capsules (Cyclopar), (Cyclopar 500) [674]

Tetracycline hydrochloride.
Achromycin [51]
Achromycin Intramuscular, Intravenous, Ophthalmic Ointment, Ophthalmic Suspension 1%, 3% Ointment, V Capsules [516]
Mysteclin-F Capsules, Syrup [817]
Panmycin [884]
Robitet '250' Hydrochloride Robicaps, '500' Hydrochloride Robicaps [737]
SK-Tetracycline Capsules [805]
Sumycin Capsules, Tablets [817]
T-500 [884]
Tetracycline HCl Capsules [240]
Tetracycline HCl Capsules & Syrup [769]
Topicycline [652]

Tetracycline phosphate complex.
Azotrex Capsules [134]
Tetrex Capsules & bid CAPS [134]

n-Tetradecylamine.
Amine 14D [487]
Lilamin 14D [487]

Tetradecyl amine, primary.
Kemamine P-790 [429a]

Tetradecyl amine, primary, distilled.
Kemamine P-790 D [429a]

Tetradecyl benzyl dimethyl ammonium chloride dihydrate.
Roccal MC-14 [413]

Tetradecyl dimethyl amine.
Adma 4 [302]
Armeen DMI4D [16]
Nissan Tertiary Amine MB [636]
Onamine 14 [663]

Tetradecyl dimethyl amine oxide.
Aromox DM14D-W [16]

Tetradecyl dimethyl benzyl ammonium chloride.
Arquad DM14B-90 [16]
Nissan Cation M_2-100 [636]

Tetradecyl dimethyl benzyl ammonium chloride monohydrate.
BTC 824 P100 [663]

n-Tetradecyl ether, ethoxylated.
Nikkol BM-1SY thru BM-8SY [631]

Tetra(diphenylphosphito)pentaerythritol.
Pentite [421]

Tetraethylammonium chloride (q.v.).
Etamon [674]

O,O,O,O-Tetraethyl dithiopyrophosphate.
Bladafum [112]

Tetraethylene glycol diacrylate.
SR-268 [766]

Tetraethylene glycol di-2-ethylhexoate.
TegMeR 804, 804 Special [379]

Tetraethylene glycol diheptanoate.
TegMeR 704 [379]

Tetraethylene glycol dimethacrylate.
SR-209 [766]

Tetraethylene glycol dimethyl ether.
Tetraglyme [367]

Tetraethylene glycol dodecyl ether.
Nikkol BL-4SY [631]

Tetraethylenepentamine.
O.E.H. 26 [261]

Tetraethyl orthosilicate.
Extrema Grade [774]

Tetraethyl pyrophosphate.
Tetron [59]

Tetraethylthiuram disulfide.
Akrochem TETD [15]
Ethyl Thiurad [392], [602]
Ethyl Tuads [894]
Etiurac [682]
TETD [154], [402]
TETD Vulcanization Accelerator [602]

Tetrahydro-3,5-dimethyl-4H-1,3,5-oxadiazine-4-thione.
R-240 DATU #1 [318]

Tetrahydro-5,5-dimethyl-2(1H)-pyrimidinone ξ3-ξ4-(trifluoromethyl)phenylʒ-1-ξ2-ξ4-trifluoromethyl) phenyl ʒ ethen-ylʒ-2-propenyl-idene ʒ hydrazone.
Amdro [50]

Tetrahydro-3,5-dimethyl-2H-1,3,5-thiadiazine-2-thione.
Basamid-Granular [107]
Crag Fungicide 974, Crag Nemacide [878]
Dazomet-Powder BASF [107]
Microfume [587]
Mylone [423]

Tetrahydrofuran.
THF [338]

Tetrahydrofurfuryl acrylate.
SR-285 [766]

Tetrahydrofurfuryl methacrylate.
SR-203 [766]

Tetra-hydrogen EDTA.
Versene Acid [261]

Tetrahydronaphthalene.
Tetralin [269], [270]

3,4,5,6-Tetrahydrophthalimidomethyl chrysanthemate.
Multicide [832]
Neo-Pynamin [832], [832a]
Tetralate [832]

2,2′,4,4′-Tetrahydroxybenzophenone.
Uvinul D-50 [338]

Tetrahydroxypropyl ethylenediamine.
Mazeen 173, 174-75 [564]
Quadrol [109]

Tetrahydrozoline HCl 0.05%.
Eye-Zine [655a]
Murine Plus Eye Drops [3]
Tyzine [491]

Tetraisopropyl titanate.
Tyzor TPT [269]

[Tetrakis(2,4-di-t-butyl-phenyl)4,4′-biphenylenediphosphonite].
Sandostab P—EPQ [761]

Tetrakis (2-ethylhexyl) titanate.
Tyzor TOT [269]

N,N,N′,N′-Tetrakis (2-hydroxypropyl) ethylene diamine.
Quadrol [109]

Tetrakis[methylene (3,5-di-t-butyl-4-hydroxyhydrocinnamate)] methane.
Irganox 1010 [190]

1,1,3,3-Tetramethyl butyl hydroperoxide.
Lupersol 215 [536]

Tetramethyl decynediol.
Surfynol 104 [13], [854]

Tetramethylthiuram disulfide.
Accelerator Thiuram [5]
Akrochem TMTD [15]
Methyl Tuads [894]
Metiwrac-O [682]
Royal TMTD [752]
Thiurad [392], [602]
TMTD [110], [154], [183], [402], [708]
Tuex [881]
Vulcafor TMTD [212], [910]
Vulkacit Thiuram, Thiuram/C [597]

Tetramethylthiuram disulfide, oil treated.
Vulkacit Thiuram/C [597]

Tetramethylthiuramdisulfide pellets.
Cyuram [50]

Tetramethylthiuram monosulfide.
Akrochem TMTM [15]
Monex [881]
Mono Thiurad® Vulcanization Accelerator [392], [602]
Thionex [270]
TMTM [283], [402], [708], [752]
Unads [894]
Vulcafor TMTM (G) [212], [910]
Vulkacit Thiuram MS/C [597]

Tetramethylthiuram monosulfide, oil treated.
Vulkacit Thiuram MS/C [597]

Tetraoctanoate.
Caprylate [736]

Tetrapeptide abietic acid condensate, triethanolamine salt.
Lamepon PA-TR [176]

Tetrapeptide coconut fatty acid condensate, potassium salt.
Lamepon S [176]

Tetrapeptide coconut fatty acid condensate, triethanolamine salt.
Lamepon ST40, S-TR [176]

Tetrapeptide oleic acid condensate, triethanolamine salt.
Lamepon PO-TR [176]

Tetrapeptide undecylenic acid condensate, potassium salt.
Lamepon UD [176]

Tetraphenyl dipropylene glycol diphosphite.
Weston THOP [129]

Tetrapotassium pyrophosphate.
Empiphos 4KP [26], [548]

O,O,O,O-Tetra-n-propyl dithiopyrophosphate.
Aspon [821]

Tetrapropylenebenzene sulfonic acid.
Merpisap TS 98 [285]

Tetrasodium N-alkyl, N-(1,2 dicarboxyethyl) sulfosuccinamate.
Astromid 22 [28]

Tetrasodium N (1,2 dicarboxyethyl).
Alcopol FL [37]

Tetrasodium (1,2-dicarboxyethyl)-N-alkyl sulfosuccinamide.
Cardanol B 2003 [274]
Rewomat B 2003 [274]
Steinamat B 2003 [274]

Tetrasodium N-(1,2-dicarboxyethyl)-N-octadecyl sulfosuccinamate.
Aerosol 22 [50], [50a]
Aerosol 22 N [233]
Cyanasol 22 Surface Active Agent [50]
Lankropol ATE [509]
Monawet SNO-35 [599]
Rewopol B 2003 [725]

Tetrasodium dicarboxyethyl stearyl sulfosuccinamate.
Aerosol 22 [50]
Monawet SNO-35 [599]
Rewopol B2003 [725]

Tetrasodium EDTA.
Alkaquest EDTA [33]
Cheelox BF-12, BF-13, BF-78 [338]
Chelon 100 [821]
Hamp-ene 100, 220, Na4 [363]
Kalex Liquid 50% [390]
Perma Kleer 100, Tetra CP [663]
Questex 4SW [821]
Sequestrene 30A, Na 4 [190]
Versene 100, 220 Crystals [261]

Tetrasodium ethylenediaminetetraacetate, dihydrate.
Hamp-Ene Na_4 [363]
Kalex Powder FC [390]
Sequestrene 220 [190]

Tetrasodium ethylenediaminetetraacetate, tetrahydrate.
Hamp-Ene 220 [363]
Questrex 4SW Crystals [821]

Tetrasodium etidronate.
Turpinal 4NL [400]

Thallium (I) nitrate.
Ultralog 85-3700-00 [172]

Thallium (III) oxide.
Ultralog 85-3760-00 [172]

Thallium sulfate.
Ratox [112]
Zelio [112]

Theophyllinate.
Eclabron Elixir [921]

Theophylline.
Accurbron [580]
Aerolate Liquid, Sr. & Jr. & III Capsules [326]
Aquaphyllin Syrup & Sugarbeads Capsules [314]
Bronkodyl, Bronkodyl S-R, Bronkolixir, Bronkotabs [133]
Constant-T Tablets [340]
Duraphyl [570]
Elixicon Suspension [115]

Theophylline *(cont'd).*
Elixophyllin Capsules, Elixir, -GG, SR Capsules [115]
Isuprel Hydrochloride Compound Elixir [133]
LABID 250 mg Tablets [652]
Lodrane Capsules-130 & 260 [703]
Marax Tablets & DF Syrup [739]
Mudrane GG Elixir [703]
Priamatene Tablets-M Formula, Tablets-P Formula [922]
Pulm 100 mg. TD, 200 mg. TD, 300 mg. TD [765]
Quibron & Quibron-300, Plus, -T & Quibron-T/SR [572]
Respbid [125]
Slo-bid Gyrocaps [746]
Slo-Phyllin 80 Syrup, GG Capsules, Syrup, Gyrocaps, Tablets [746]
Somophyllin-CRT Capsules, -T Capsules [324]
Sustaire Tablets [739]
Synophylate -GG Tablets/Syrup, Tablets/Elixir [166]
T.E.H. Tablets [344]
T-E-P Tablets [769]
T.E.P. Tablets [344]
Tedral -25 Tablets, Elixir & Suspension, Expectorant, SA Tablets, Tablets [674]
Theobid, Theobid Jr. Duracap [354]
Theoclear L.A.-65, -130 & -260 Capsules, -80 Syrup [166]
Theo-Dur Sprinkle, Tablets [491]
Theofedral Tablets [240]
Theolair & Theolair-SR, -Plus Tablets & Liquid [734]
Theon Elixir [124]
Theo-Organidin Elixir [911a]
Theophyl Chewable Tablets [570]
Theophylline Elixir [344]
Theophylline Elixir & KI Elixir [769]
Theophylline Oral Solution [751]
Theophylline S.R. Tablets [344]
Theophyl-SR, -225 Elixir, -225 Tablets [570]
Theospan-SR Capsules [513]
Theostat 80 Syrup, Tablets [513]
Theovent Long-Acting Capsules [772]
Theozine Syrup & Tablets - Dye-Free [769]

Theophylline anhydrous.
Theo-24 [782]
Theophylline Anhydrous Tablets [769]

Theophylline calcium salicylate.
Quadrinal Tablets & Suspension [495]

Theophylline glyceryl guaicolate.
Lanophyllin GG Capsules [510]

Theophylline sodium glycinate.
Asbron G Inlay-Tabs/Elixir [258]
Synophylate -GG Tablets/Syrup, Tablets/Elixir [166]

Thiabendazole.
Mintezol Chewable Tablets & Suspension [578]

Thiadiazine.
Vanax NP [894]

Thiamine hydrochloride.
Albafort Injectable [105]
Alba-Lybe [105]
BayBee-1 [111]
Besta Capsules [394]
Eldercaps [563]
Eldertonic [563]
Geravite Elixir [394]
Glutofac Tablets [489]
I.L.X. B_{12} Elixir Crystalline, Tablets [489]
Neuro B-12 Forte Injectable, B-12 Injectable [505]
Thiamine HCl Injection [287]
Thiamine HCl in Tubex [942]
Tia-Doce Injectable Monovial [105]

Thiamine mononitrate.
Mega-B [72]
Megadose [72]
Nu-Iron-V Tablets [563]
Prenate 90 Tablets [124]
Vicon-C Capsules [354]

Thiamine nitrate.
Thiamine Mononitrate, No. 60132 [738]

Thiamylal sodium.
Surital [674]

2-(4′-Thiazolyl)-benzimidazole.
Ap1-Luster [682]
Arbotect [577]
Mertect [577]
RPH [748a]
TBZ [577]
Tecto [577]
Thibenzole [577]

2-(4-Thiazolyl) benzimidazole, hypophosphite salt.
Arbotect S [423]
Arbotect 20-S [423], [577]

Thiethylperazine.
Torecan Injection (Ampuls), Suppositories, Tablets [125]

Thimerosal.
Merthiolate [286]

4,4′-Thiobis(6-t-butyl-m-cresol).
Santonox® and Santonox® R [602]
Santowhite® Crystals Antioxidant [602]

4,4′-Thiobis(6-t-butyl-o-cresol).
Ethanox 736 [302]

Thiocarbamyl sulfenamide.
Cure-rite Accelerator [361]
Cure-Rite 18 [15], [361], [752]

2-Thiocyanoethyl laurate.
Lethane [742]

2-(Thiocyanomethylthio) benzothiazole.
Busan 30A [137]

O,O′-(Thiodi-4,1-phenylene) bis (O,O-dimethyl phosphorothioate.
Abate [50]
Abathion [50]
Biothion [50]
Ecopro [50]
Lypor [50]

Thiodiethylene bis-(3,5-di-t-butyl-4-hydroxy)hydrocinnamate.
Irganox 1035 [190]

Thiodiethylene glycol.
Glyecine A [338]
Kromfax [29]

Thiodipropionate polyester.
TDP 2000 [278]

Thiodipropionic acid.
Thiodipropionic Acid [303]

Thioglycerin.
Thiovanol [303]

Thioglycerol.
Thiovanol [303]

Thioglycolic acid.
Thiovanic Acid [303]

Thioguanine.
Tabloid Brand Thioguanine [139]

Thiohydropyrimidine.
Thiate A [894]

Thionazin.
Zinophos [50]

Thiophosphoric acid-tris-(p-isocyanatophenylester).
Desmodur RF [112]

Thioridazine.
Mellaril, -S [762]
Thioridazine Tablets [344], [769]

Thioridazine hydrochloride.
Thioridazine HCl Tablets [240]

Thiosulfate, organic.
Lankrolan SHR-3 [252], [253]

Thiotepa.
Thiotepa [516]

Thiothixene.
Navane Capsules and Concentrate, Intramuscular [739]

Thiourea.
Vulcafor 322 [212], [910]

Thioxanthene derivatives.
Navane Capsules and Concentrate, Intramuscular [739]

Thiphenamil hydrochloride.
Trocinate [703]

Thrombin.
Thrombinar [80]
Thrombostat [674]

Thyroglobulin.
Proloid Tablets [674]

Thyroid.
Armour Thyroid Tablets [891]
Cytomel Tablets [805]
Euthroid [674]
Levothroid [891]
Proloid Tablets [674]
S-P-T [326]
Synthroid [327]
Thyroid Strong Tablets, Tablets [551]
Thyroid Tablets [891]

Thyrotropic hormone.
Thytropar [80]

Thyroxine.
Choloxin [327]
Euthroid [674]
Levothroid [891]
L-Thyroxine Tablets [769]
Synthroid [327]
Thyrolar Tablets [891]

Ticarcillin disodium, sterile.
Ticar® 1 Gm, 3 Gm, 6 Gm, 20 Gm, 30 Gm [114]

Timolol maleate.
Blocadren Tablets [578]
Timolide Tablets [578]
Timoptic Sterile Ophthalmic Solution [578]

Tin.
Baum® [856]

Tin catalyst.
Fomrez UL-1 [935]
Kosmos® 19, 20, 29 [856]

Tin (II) chloride dihydrate.
Ultralog 86-5820-00 [172]

Tin (IV) chloride.
Ultralog 86-5830-00 [172]

Tin (IV) oxide.
Ultralog 86-5890-00 [172]

Tin mercaptide.
Mark 292 [74]

Tin stabilizer.
Pennmax Four, Four-A, Five [682]

TIPA-lauryl sulfate.
Carsonol TILS [529]

Titania catalyst.
TI-0720 T 1/8″ [389]

Titanium carbide.
Kentanium [492]

Titanium diboride.
TiB_2 [8]

Titanium dioxide.
Atlas Sup-R-White Titanium Dioxide, White Titanium Dioxide [496]
C-Weiss 7 [351]
Cosmetic Titanium Dioxide 300309 [930]
Cosmetic White C47-5175, C47-9623 [835]
Horse Head A-410, A-420, R-710 [626]
Kowet Titanium Dioxide [496]
Tiebright [543]
Titandioxid [838]
Titanium Dioxide P 25 [244]
Titanium Dioxide, USP, CTFA, Bacteria Controlled [924]
Titanox 1025, 2010 [640a]

Titanium dioxide, anatase.
Titanox 1000, 1050, 1072 [641a]

Titanium dioxide, nonpigmentary.
Titanox 3020, 3030 [641a]

Titanium dioxide, surface modified.
Titankup [543]

Titanium oxide.
Rutile, Ceramic [615a]
Ultralog 86-6720-00 [172]

Tobramycin sulfate.
Nebcin, Sterile [254]

Tocopherol.
Vitamin E No. 60524 [738]

Tocopherols (d-alpha, d-gamma and d-beta, d-delta), 7-9%, 32-38%, and 10-15%.
Coviox T-50 [400]

d- α Tocopherol concentrate FCC.
Covitol F-600, F-1000 [400]

Tocopheryl acetate.
Protavit E Acetate [713]

d- α -Tocopheryl acetate.
Covitol 700C [400]

d- α Tocopheryl acetate concentrate FCC.
Covitol 544, 1100 [400]

d- α Tocopheryl acetate FCC.
Covitol 1360 [400]

d- α Tocopheryl acid succinate.
Covitol 1185 [400]

d- α Tocopheryl acid succinate FCC.
Covitol 1210 [400]

TOFA, sulfonated.
Actrasol SP [811]

Tolazamide.
Tolinase® [884]

Tolazoline hydrochloride.
Priscoline Hydrochloride Multiple-Dose Vials [189]

Tolbutamide.
Orinase® [884]
SK-Tolbutamide Tablets [805]
Tolbutamide Tablets [240], [344], [769]

Tolmetin sodium.
Tolectin Tablets & DS Capsules [570]

Toluene-2,5-diamine sulfate.
Rodol BLFX [531]

Toluene-2,4-diisocyanate.
Hylene T [270]

Toluene-2,4-diisocyanate/toluene-2,6-diisocyanate, 80:20.
Hylene TM [270]

Toluene diisocyanate-polyol reaction product.
Desmodur L 75 [112]

o,p-Toluene sulfonamide.
Santicizer 9 [502]

Toluene sulfonates.
Para Toluene Sulfonic Acid, Sulfonic Acid solution [194]

Toluene sulfonate, potassium salt.
Hartotrope KTS 50 [390]

Toluene sulfonate, sodium salt.
Hartotrope STS 40, STS Powder [390]

Toluene sulfonic acid.
Eltesol TA [25]
Eltesol TA65 [23], [25], [26]
Eltesol TA96 [23], [25], [26], [548]
Eltesol TA/B, TA/E [23]
Eltesol TA/F [23], [25]
Eltesol TA/H, TA/K [23]
Eltesol TPA [25]
Eltesol TSE, TSX [23]
Manro PTSA 65 E, PTSA 65 H, PTSA 65 LS [545]
Toluenesulfonic Acid [754]
Witconate TX Acid [937a], [938]

p-Toluene sulfonic acid.
Condasol T65 [725]

p-Toluene sulfonic acid *(cont'd)*.
Cyclophil P.T.S.A. [938]
Reworyl T 65 [725]

p-Toluene sulfonic acid monohydrate.
Manro PTSA Crystals [545]

p-(P-Toluene-sulfonylamido)-diphenylamine.
Aranox [881]

p-Toluene sulfonyl chloride, technical grade.
Manro PTSC [545]

p-Toluene sulfonyl semicarbazide.
Celogen RA [881]

Toluhydroquinone.
THQ [278]

m-Tolyl-n-methylcarbamate.
Kumiai [832a]
Metacrate [832a]
Tsumacide [594]

N-m-tolylphthalamic acid.
Tomaset [540]

Tolyltriazole.
Cobratec® TT-100 [797]

p-(p-Tolysulfonyl-amido)-diphenylamine.
Aranox [881]

Tomato extract.
Phytelene of Tomato EN 310 Powder [900a]

Toxaphene.
Estonox [59]

Tranylcypromine sulfate.
Parnate [805]

Trazodone hydrochloride.
Desytel [572]

Tretinoin.
Retin-A (tretinoin) [666]

Triacetin.
Enzactin Cream [97]
Estol 1580 [875]
Fungoid Creme & Solution, Tincture [679]
Kessco Triacetin [79]

Trialkanolammonium alkenyl succinate.
Produkt B 3032 [223]

Trialkanolammonium dodecyl benzene sulfonate.
Reworyl TKS 90/F [725a]

Trialkanolammonium lauryl sulfate.
Rewopol TLS 90/L [725a]

Triallate.
Avadex® BW [602]
Far-Go® [602]

Triallyl cyanurate.
TAC [50]

Triallyl trimellitate.
Natro-Cure TATM [392]

Triamcinolone acetate 0.01%.
Cinolar [655a]

Triamcinolone acetate 0.025%.
Cinolar [655a]

Triamcinolone acetate 0.5%.
Cinolar [655a]

Triamine dodecylbenzene sulfonate.
Reworyl TKS 90/L [725]

Triamine lauryl sulfate.
Cycloryl WAT-I [234]
Rewopol TLS 90L [725]

Triammonium dodecylbenzenesulfonate.
Condasol TKS 90/F [274]
Reworyl TKS 90/F [274]
Steinaryl TKS 90/F [274]

Triamterene.
Dyazide [805]
Dyrenium [805]

Triaryl phosphate.
Kronitex 100 [379]
Santicizer® 143, 154 [602]

Triaryl phosphate, t-butyl-substituted.
Phosflex 61B [392]

Triaryl phosphate ester.
Santicizer 143 [602]

Triaryl phosphate, isopropyl-substituted.
Phosflex 41P [392], [821]

Triazolam.
Halcion [884]

Tribasic lead sulfate.
Epistatic Stabilizers [15]
Tribase, Tribase XL, E Special, EXL Special [88]
Vanstay 5996 [894]

2,4,6-Tribromoaniline.
AN-73 [369]

2,4,6-Tribromophenol.
PH-73 [369]

Tribromophenol.
Emery 9332 [289]

Tribromo silane.
Extrema Grade [774]

Tributoxyethyl phosphate.
KP-140 [332], [379]
(Phosflex) TBEP [392]
TBEP [379]

Tributyl citrate.
Citroflex 4 [691]

Tributyl citrate *(cont'd).*
TBC [223]

Tributyl-2,4-dichloro-benzylphosphonium chloride (q.v.).
Phosfon [897]

Tributyl phenol polyglycol ether (EO 4 to 50).
Sapogenat T Brands [416]

Tributyl phosphate.
Celluphos 4 [821]
Kroniflex TBP [379]
TBP [332]

S,S,S-Tributylphosphorotrithioate.
DEF [597]
De-Green [597a], [821]
Fos-Fall "A" [597a]
Ortho Phosphate Defoliant [597a]

Tributyl phosphorotrithioite.
Deleaf Defoliant [730a]
Easy Off-D [730a]
Folex [730a]

Tributyltin compound.
Intercide N-628, TMP, T-0, FC [459]

Tributyltin oxide.
Keycide® X-10 [935]

Tricapryl trimellitate.
Uniflex TCTM [876]

Tri (caprylyl capryl)-t-amine.
Lilamin 364 [522]

Tricaprylyl methyl ammonium chloride.
Aliquat 336-PTC [400]

Tricarboxymethyl diamino alkyl amide.
Rewopol CHT 12 [725a]

Triceteth (5) phosphate.
Nikkol TCP-5 [631]

Trichlorbenzene.
Anthrapole DW [77]

Trichlorfon (q.v.).
Dipterex [184]
Dylox [184]

2,2,2-Trichloro-1-(3,4-dichlorophenyl)-ethyl acetate.
Baygon MEB [112]

2,4,4′-Trichloro-2′-hydrozydiphenyl ether.
Irgasan DP300 [190]

3,5,6-Trichloro-2-pyridinyl-oxyacetic acid.
Garlon 3A, 4 [261]

2,3,5-Trichloro-4-propylsulfonyl pyridine.
Dowicil A-40 [261]

Trichloroacetic acid.
Konesta [18a]
NaTA [416]
Varitox [561a]

S-(2,3,3-Trichloroalkyl) diisopropyl thiocarbamate.
Avadex BW [601]

S-(2,3,3-Trichloroallyl) diisopropylthiocarbamate.
Far-Go [601]

1,2,4-Trichlorobenzene, emulsified.
Hipochem GM, Jet Dye T, NTB [410]
Marvanol Carrier OTC [552]

2,3,6-Trichlorobenzoic acid.
Benzac [877]

2,3,6-Trichlorobenzyloxypropanol.
Tritac [421]

3,4,4′ Trichlorocarbanilide.
TCC [602]

Trichloroethane.
Aerothene TT [261]
Chlorothene NU [261]
Solvent 111 [909]

1,1,1-Trichloroethane.
Chlorothene NU [379]
Extrema Grade [774]

1,1,1-Trichloroethane solvent.
Tri-ethane [705]

Trichloroethyl phosphate.
Disflamoll TCA [597]

Trichlorofluoromethane.
Isotron II [682]
Isotron M Solvent [682]

Trichloroisocyanuric acid.
ACL 85 [602]

Trichloromethylphenylcarbinyl acetate.
Crystarose [337]
Rosetone [452]

cis-N-((Trichloromethyl)thio)-4-cyclohexene-1,2-dicarboximide.
Merpan [540]
Orthocide [183]
Pillarcap [696a], [945a]
Vancide 89, 89RE [894]
Vond-captan [683]

N-(Trichloromethylthio)-phthalimide.
Folpan [540]
Phaltan [183]
Troysan Anti-Mildew O [869]

2,4,5-Trichlorophenoxy acetic acid.
Amine 2,4,5-T for Rice [878]
Brush Rhap [900a]
Esteron [900a]
Tormona [162a]
Transamine [900a]
Trinoxol [878]

2,4,5-Trichlorophenoxy acetic acid *(cont'd).*
U46 Brushkiller HV, LV [107]
Weedar [878]
Weedone [878]

2-(2,4,5,-Trichlorophenoxy)-propionic acid.
AquaVex [682]
Double Strength [878]
Kuron [261]
Silvi-Rhap [900a]
Weed-B-Gon [183]

(2,3,6-Trichlorophenyl) acetic acid.
Fenatrol [877]
Trifene [338]

2,4,6-Trichlorophenyl-4′-nitrophenyl ether.
MO [595], [798a]

Trichlorotrifluoroethane.
Isotron 113, T Solvent [682]

Tricresyl phosphate.
Disflamoll TKP [597]
Kronitex TCP [332]
Lindol [392], [821]
Phosflex 179C [392]

Tricyclohexyl hydroxystannone.
Plictran [261]

1-(Tricyclohexylstannyl)-1H-1,2,4-triazole.
Peropal [112]

(E), (Z)-4-Tridecen-1-yl acetate.
NoMate Gusano A.T., Pinworm Suppressant [21]

Trideceth-3.
Emulphogene BC-420 [338]
Hetoxol TD-3 [407]
Lipal 3TD [715]
Lipocol TD-3 [525]
Procol TDA-3 [713]

Trideceth-6.
Emulphogene BC-610 [338]
Ethosperse TDA-6 [358]
Hetoxol TD-6 [407]
Lipal 6TD [715]
Lipocol TD-6 [525]
Procol TDA-6 [713]

Trideceth-6 phosphate.
Gafac RS-610 [338]

Trideceth-7.
Carsodet TD-7C [155]

Trideceth-7 carboxylic acid.
Carsodet TD-7C [529]
Emery 5340 [291]
Rewopol CT [725]
Sandopan DTC-Acid [762]
Surfine T-Acid [320]

Trideceth-10.
Emulphogene BC-720 [338]
Lipal 10TD [715]

Trideceth-11.
Carsonon TD-11 [529]

Trideceth-12.
Hetoxol TD-12 [407]
Lipocol PT-400, TD-12 [525]
Procol TDA-12 [713]

Trideceth-15.
Emulphogene BC-840 [338]
Macol TD-15 [564]
Procol TDA-15 [713]

(Tridecylalcohol).
Iconol™ TDA-3, TDA-6, TDA-8, TDA-9, TDA-10, TDA-12, TDA-15 [109]

Tridecyl alcohol, EO, 9 moles.
Teric 13A9 [439]

Tridecyl alcohol, ethoxylated.
Alkasurf TDA-5, TDA-6, TDA-7, TDA-7.5, TDA-8, TDA-12, TDA-15 [33]
Carsonon TD-7, TD-11 [529]
Soprofor TR/Series [550]
Trycol TDA-11 [290]

Tridecyl alcohol, ethoxylated, (EO 3).
Iconol TDA-3 [109]

Tridecyl alcohol, ethoxylated, (EO 6).
Iconol TDA-6 [109]

Tridecyl alcohol, ethoxylated, (EO 8).
Iconol TDA-8, TDA-8-90% [109]

Tridecyl alcohol, ethoxylated, (EO 9).
Iconol TDA-9 [109]

Tridecyl alcohol, ethoxylated, (EO 10).
Iconol TDA-10 [109]

Tridecyl alcohol, ethoxylated, (EO 18).
Iconol TDA-18-80% [109]

Tridecyl alcohol, ethoxylated, (EO 29).
Iconol TDA-29-80% [109]

Tridecyl alcohol ethylene oxide adduct, 60%.
T-DET TDA-60 [859]

Tridecyl alcohol ethylene oxide adduct, 65%.
T-DET TDA-65 [859]

Tridecyl alcohol ethylene oxide adduct, 70%.
T-DET TDA-70 [859]

Tridecyl alcohol ethylene oxide adduct, 150 mole.
T-DET TDA-150 [859]

Tridecyl alcohol, phosphate ester.
Actrafos T [811]

Tridecyl alcohol polyglycolether.
Produkt RT 38 [948]

(Tridecyl benzene).
Nalkylene 600 Detergent Alkylate [206]

Tridecyl benzene sulfonic acid.
Conco ATR-98, ATR-98S [209]
Nansa TDB [25], [548]

Tridecyl benzene sulfonic acid, branched chain.
Arylan SO60 Acid [509]

Tridecyl benzene sulphonic acid, branched (hard).
Ufacid A60 [874]

Tridecyl ether carboxylic acid.
Carsodet TD-7C [529]

Tridecyl ethers, ethoxylated.
Trycol TDA-6, TDA-9 [290]

Tridecyl ether, phosphated (EO 2 moles).
Crodafos T2 Acid [223]

Tridecyl ether phosphate (EO 3).
Servoxyl VPDZ 3/100 [788]

Tridecyl ether, phosphated (EO 5 moles).
Crodafos T5 Acid [223]

Tridecyl ether, phosphated (EO 10 moles).
Crodafos T10 Acid [223]

Tridecyl ethoxylate.
Siponic TD-3, TD-12 [29]

Tridecyl phosphite.
Mark TDP [75], [934]

Tridecyl polyglycolether.
Merpemul 7080, 7100 [285]

Tridecyl salicylate.
BTN [901]

Tridecyl stearate.
Emerest 2308 [289]
Grocor 5721 [374]
Rilanit ITS [401]

Tridecyloxypoly (ethyleneoxy) ethanol.
Emulphogene BC-420, BC-610, BC-630, BC-720, BC-840, TB-970 [338]

N-Tridecyloxypropyl-1,3-diaminopropane.
Tomah DA-17 [866]

Tridecyloxypropyl dihydroxyethyl methyl ammonium chloride.
Tomah Q-17-2 [866]

Tridecyloxypropylamine.
Tomah PA-17 [866]

Tridihexethyl chloride.
Milpath [911a]
Pathibamate [516]
Pathilon [516]

Tri (2-ethylhexyl) trimellitate.
Hatcol TOTM [393]
Palatinol TOTM [99]

Triethanolamide lauryl sulfate.
Empicol TL40/T [23]

Triethanolamine, activated.
Polymel Actisil [701]

Triethanolamine alkyl aryl sulfonate.
Calsoft T-60 [697]
Cycloryl DDB60T [234]
Sulfamin [116]
Surco 60T, 170-60T [663]
Witconate™ TAB [935]

Triethanolamine alkyl aryl sulfonate, linear.
Carsofaom T-60L [529]

Triethanolamine alkyl benzene sulfonate.
Louryl T-50 [858]
Nansa TS50, TS60 [23]

Triethanolamine alkyl benzene sulfonate, linear.
Cedepon T [256]

Triethanolamine alkyl benzene sulfonate, linear, (C_{12}).
Polystep A-9 [825]

Triethanolamine alkyl benzene sulfonate, linear (soft).
Ufasan Tea [874]

Triethanolamine alkyl ether sulfate.
Alscoap LE-240 [864]

Triethanolamine alkyl sulfate.
Manro TL 40S [545]

Triethanolamine alkylate sulfonate, linear.
Bio Soft N-300 [825a]

Triethanolamine/ammonium lauryl sulfate, 20:30.
Empicol TCR/T [25]
Sulphonated 'Lorol' Liquid TN [745]

Triethanolamine ammonium lauryl sulfate, built.
Sulphonated 'Lorol' Liquid TNR [745]

Triethanolamine chelate.
Tyzor TE [269]

Triethanolamine dioleate.
Rilanit TDO [401]

Triethanolamine dodecyl benzene sulphonate.
Alkasurf T [33]
Calsoft T-60 [697]
Carsofoam T60L [529]
Conco AAS-60S [209]

Triethanolamine dodecyl benzene sulphonate *(cont'd).*
Elfan WAT [16], [764]
ESI-Terge T-60 [294]
Marlopon AT 50, CA [178]
Nansa TS 60 [23], [25]

Triethanolamine dodecylbenzene sulfonate, linear.
Merpisap AT 50, AT 90 [285]
Sterling LA Oil [145]

Triethanolamine lauroyl sarcosinate.
Hamposyl TL-40 [363]

Triethanolamine lauryl ether sulfate.
Cycloryl TD [234]
Drewpon EDT [266]
Empicol ETB [25]
Sactipon 2 OT [521]
Sactol 2 OT [521]
Texapon NT [401]
Tylorol LT 50 [858]

Triethanolamine lauryl ethoxy sulfate.
Empicol ETB [23]

Triethanolamine lauryl sarcosinate.
Cycloryl TEALS [234]

Triethanolamine lauryl sulfate.
Akyposal TLS 42 [182]
Alkasurf TLS [33]
Berol 480 [116]
Calfoam TLS-40 [697]
Carsonol TLS [155]
Cedepon LT-40 [256]
Conco Sulfate TL [209]
Condanol TLS 40 [274]
Cosmopon TR [514]
Cycloryl WAT [234]
Drewpon 40 [265], [266]
Drewpon 40LS [265]
Emal TD [478]
Emersal 6434 [289]
Empicol TA40 [23], [24]
Empicol TA40A, TL 40/T, TLP/T, TLR/T, XC35, XM17, XT45 [23], [25]
Empicol 0308 [23]
Equex T [709]
Incronol TLS [223b]
Jordanol TEAL-400 [474]
Lakeway 101-30 [126]
Lonzol LT400 [529]
Manro TL 40 [545]
Maprofix TLS, TLS-65, TLS-106, TLS-500, TLS-513 [663]
Mars TLS-40 [554]
Merpinal LMT 40 [285]
Nutrapon TLS-500 [198]
Polyfac TLS-40 [917a]
Polystep B-6 [825]
Rewopol TLS 40 [725]
Sactipon 2 T [521]
Sactol 2 T [521]
Sipex TEA [29]
Sipon LT-6 [29]
Standapol T [399], [400]
Sterling WAT [145]
Sulfatol 33 TR [1]
Sulfotex WAT [399], [400]
Sulphonated 'Lorol' Liquid TA [745]
Swascol 4L [837]
Texapon N 42 [398]
Texapon T [400]
Texapon T 35, T-42, TH [401]
Witcolate™ T [935]
Zoharpon LAT [946]

Triethanolamine monooleic acid ester.
Emulan FM [106]

Triethanolamine oleoyl-cocoyl sarcosinate (70-30).
Hamposyl TOC-30 [363]

Triethanolamine salicylate.
Aspercreme [860]
Myoflex Creme [7]

Triethanolamine salt.
Genapol CRT 40 [416]

Triethanolammonium alkyl sulfate (C_{12}-C_{14}).
Serdet DFL 40 [788]

N,N,N-Triethanolammonium dodecylbenzene sulfonate.
Norfox T-60 [650]
Serdet DML 45 [788]

Triethanolammonium lauryl sulfate.
Rewopol TLS 40, TLS 45/B [725a]

Triethonium animal protein ethosulfate, hydrolyzed.
Quat-Pro E [562]

Triethyl citrate.
Citroflex 2 [691]

Triethylene glycol caprate-caprylate.
Plasthall 4141 [379]

Triethylene glycol diacrylate.
SR-272 [766]

Triethylene glycol dicaprylate/caprate.
Tricap [223]

Triethylene glycol di-2-ethylhexanoate.
Kodaflex TEG-EH [278]

Triethylene glycol di-2-ethylhexoate.
TegMeR 803, 803P [379]

Triethylene glycol diheptanoate.
TegMeR 703 [379]

Triethylene glycol dimethacrylate.
SR-205 [766]

Triethylene glycol dimethyl ether.
Triglyme [367]

Triethylene glycol dipelargonate.
Plastolein 9404 TGP [289]
TegMeR 903P [379]

Triethylene glycol distearate.
Nikkol Estepearl 30, 35 [631]

Triethylene glycol dodecyl ether.
Nikkol BL-3SY [631]

Triethylenediamine.
DABCO [13]

Triethylenetetramine.
D.E.H. 24 [261]

Triethylenethiophosphoramide.
Thiotepa [516]

Trifluoperazine hydrochloride.
Stelazine [805]
Trifluoperazine Tablets [344], [769]

a,a,a-Trifluoro-2,6-dinitro-N,N-dipropyl-p-toluidine.
Crisalina [230a]
Digermin [310]
Elancolan [282]
Ipersan [716a]
Trefanocide [282]
Treflan [282]
Triflurex [540]

Trifluoroacetic acid.
Sequalog Grade 88-6720-00 [172]

Trifluoromethanesulfonic acid.
FC-24 [862]

3-Trifluoromethyl-4-nitrophenol, sodium salt.
Lamprecid [416]

Trifluridine.
Viroptic Ophthalmic Solution [139]

Triglyceride, epoxidized.
Epoxol 8-2B, 9-5 [840]

Triglyceride, ethoxylated.
Cirrasol ALN-WY [93]
Mulsifan RT 7, RT 69, RT 163 [948]

Triglyceride, fatty acid (C_6-C_{10}).
Softenol 3819 [275]

Triglyceride, fatty acid (C_7).
Softenol 3107 [275]

Triglyceride, fatty acid (C_8-C_{10}).
Softenol 3108 [275]

Triglyceride, fatty acid (C_8-C_{18}).
Softenol 3178 [275]

Triglyceride, fatty acid (C_{12}-C_{18}).
Softenol 3100 [275]

Triglyceride, fatty acid (C_{14}).
Softenol 3114 [275]

Triglyceride, fatty acid (C_{18}).
Softenol 3118 [275]

Triglyceride, medium chain.
Miglyol 812 [275]
Nissan Panacete 810 [636]

Triglyceride, medium chain (capric acid base).
Coconad SK [478a]

Triglyceride, medium chain (caprylic acid base).
Coconad MT [478a]

Triglyceride, polyethoxylated.
Chemcol CO-30 [171]

Triglycerol diisostearate.
Emerest® 2452 [290]

Triglycerol mono shortening.
Caprol 3GVS [829]

Triglycerol monooleate.
Caprol 3GO [829]
Drewpol 3-1-O [715]
Mazol PG031 [564]

Triglycerol monostearate.
Aldo TGMS [358]
Caprol 3GS [829]
Drewpol 3-1-S [715]
Polyaldo® TGMS [358]
Santone 3-1-S [272]

Trihexyphenidyl hydrochloride.
Trihexyphenidyl HCl Tablets [240]

Trihexyphenidyl preparations.
Artane Elixir, Sequels, Tablets [516]

2,4a,7-Trihydroxy-1-methyl-8-methylenegibb-3-ene 1, 10-carboxylic acid 1 ⇒ 4-lactone.
Activol [444]
Berelex [444]
Cekugib [167a]
Gibrel [577]
Grocel [444]
Pro-Gibb, Plus [3]

Trihydroxystearin.
Thixcin E [640a]
Thixcin R [640a], [641a]

Triiodothyronine.
Thyrolar Tablets [891]

Triisodecyl phosphite.
Weston TDP [129]

Triisononanoin.
Isodragil 2/050300 [263]

Triisooctyl phosphite.
Weston TIOP [129]

Triisooctyl trimellitate.
Plasthall TIOTM [379]
Staflex TIOTM [722]

Triisopropanolamine lauryl sulfate.
Carsonol TILS [529]

Triisopropyl trimerate.
Schercemol TT [771]

Triisostearin.
Sun Espol G-318 [847]

Triisostearyl trimerate.
Schercemol TIST [771]

Trilaneth-4 phosphate.
Hostaphat KW340N [55]

Trilaureth-4 phosphate.
Hostaphat KL340N [55]
Nikkol TLP-4 [631]

Trilaurin.
Cyclochem GTL [234]
Lipo 320 [525]
Softisan 100 [275], [482]
Witepsol E-75 [275]

Trilauryl t-amine.
Lilamin 363 [522]

Trilauryl citrate.
Citrest LT [234]

Trilauryl phosphite.
Weston TLP [129]

Trilauryl trithio phosphite.
Weston TLTTP [129]

Trimeprazine tartrate.
Temaril [805]

Trimer acid.
Hystrene 5460 [429]

Trimethadione.
Tridione [3]

Trimethaphan camsylate.
Arfonad Ampuls [738]

Trimethobenzamide hydrochloride.
Stemetic [518]
Tigan® Capsules, 100 mg., 250 mg. [114]
Trimethobenzamide HCl Suppositories [769]

Trimethoprim.
Bactrim DS Tablets, I.V. Infusion, Pediatric Suspension, Suspension, Tablets [738]
Cotrim, D.S. [519]
Proloprim [139]
Septra DS Tablets, I.V. Infusion, Suspension, Tablets [139]
Sulfamethoxazole & Trimethoprim Pediatric Suspension, Tablets [119]
Sulfamethoxazole with Trimethoprim Tablets, Tablets (Double Strength) [240]
Sulfamethoxazole w/ Trimethoprim DS, SS [344]
Sulfatrim & Sulfatrim D/S Tablets [769]
Trimethoprim Tablets [119]
Trimpex Tablets [738]

Trimethyl ammonium chloride, C_8 C_{10} quaternary.
Adogen 464 [796]

Trimethyl benzyl ammonium chloride.
Variquat B200 [796]

Trimethyl cetyl ammonium chloride, quaternary.
Adogen 444 [796]

Trimethyl coco ammonium chloride.
Arquad C-33, C-50 [79]

Trimethyl coco ammonium chloride, quaternary.
Adogen 461 [796]

Trimethyl coconut ammonium chloride.
Kemamine Q 6503 B [429a]

Trimethyl cyclohexyl-ethyl ether.
Herbavert [399], [400]

2,2,4-Trimethyl-1,2-dihydroquinoline, polymerized.
Akrochem Antioxidant DQ [15]
Permanax TQ [212], [910]
Vulkanox HS/LG, HS/Powder [597]

Trimethyl dodecyl ammonium chloride.
Arquad 12-33, 12-50 [79]
Kemamine Q 6903 B [429a]
Nissan Cation BB [636]

Trimethyl hexadecyl ammonium chloride.
Arquad 16-29, 16-50 [79]
Kemamine Q 8803 B [429a]
Nissan Cation PB—40 [636]

Trimethyl nonyl polyethylene glycol ether.
Tergitol® TMN-3, TMN-6, TMN-10 [879]

Trimethyl octadecyl ammonium chloride.
Arquad 18-50 [79]
Kemamine Q 9903 B [429a]
Nissan Cation AB [636]

2,2,4-Trimethyl-1,3-pentanediol, diisobutyrate.
Kodaflex TXIB [278]

2,2,4-Trimethyl-1,3-pentanediol monoisobutyrate.
Texanol Ester-Alcohol [278]

Trimethyl soya ammonium chloride.
Arquad S-50 [79]
Kemamine Q 9973 B [429a]

Trimethyl soya ammonium chloride, quaternary.
Adogen 415 [796]

Trimethyl tallow ammonium chloride.
Adogen 471 [796]
Arquad T-27W, T-50 [79]
Kemamine Q 9743 B [429a]
Querton BGCL50 [487]

Trimethyl tallow ammonium chloride/ dimethyl dicoco ammonium chloride, 1:1.
Arquad T-2C-50 [79]

Trimethyl tallow ammonium chloride, hydrogenated.
Kemamine Q 9703 B [429a]

Trimethyl tetradecyl ammonium chloride.
Kemamine Q 7903 B [429a]

1,3,5-Trimethyl-2,4,6-tris (3,5-di-t-butyl-4-hydroxybenzyl) benzene.
Ethanox 330 [302]

Trimethylene phosphonic acid.
Unihib 305 [529]

Trimethylene phosphonic acid, penta sodium salt.
Unihib 314 [529]

Tri-methylol-propane complex ester.
Rilanit TMNC [401]

Trimethylolpropane monooleate.
Radiasurf 7172, 7372 [658]

Trimethylolpropane triacrylate.
SR-351 [766]

Trimethylolpropane triacrylate, ethoxylatd.
SR-454 [766]

Trimethylolpropane tri-C_7.
Radiasyn 7367 [658]

Tri-methylol-propane tri-fatty acid ester.
Rilanit TMTC [401]

Tri-methylolpropane trilaurate.
Kemamine TMP-12 [429]

Trimethylolpropane trimethacrylate.
SR-350 [766]

Trimethylolpropane trimethacrylate/ hydroquinone inhibitor 80 ± 20 ppm.
SR-350 [766]

Trimethylolpropane trioctanoate.
Kessco X-330 [79]

Trimethylolpropane-tris- (B-(N-aziridinyl) propionate).
XAMA®-2 [215]

Trimethylpropane trioleate.
Radia 7370 [658]

Trimethylthiourea.
Thiate E [894]

Trimipramine maleate.
Surmontil Capsules 25 mg., 50 mg., & 100 mg. [463]

Trimyristin.
Dynasan 114 [275]

Tri-(nonylated phenyl) phosphite.
Polygard [881]

Trioctyl trimellitate.
Kodaflex TOTM [278]
Plasthall TOTM [379]
Polycizer TOTM [392]
Staflex TOTM [722]
Uniflex TOTM [876]

Triolein.
Aldo TO [358]
CPH-399-N [379]
Emerest 2423 [291]
Emery 2423 [288]
Grocor 1000, 1200 [374]

Trioleth-8 phosphate.
Hostaphat KO380 [55]

Trioleyl phosphate.
Hostaphat KO300 [55]

Trioxsalen.
Trisoralen Tablets [284]

Tripelennamine hydrochloride.
PBZ Hydrochloride Cream, -SR Tablets [340]
Tripelennamine HCl Tablets [240]

Tripelennamine preparations.
PBZ Tablets & Elixir [340]

Triprolidinen hydrochloride.
Actifed-C Expectorant [139]
Pseudoephedrine Hydrochloride & Triprolidine Hydrochloride Syrup, Tablets [751]
Triafed-C Expectorant [769]
Trifed Tablets & Syrup [344]
Tripodrine Tablets [240]

Triprolidine preparations.
Triafed Syrup & Tablets [769]

Triphenyl phosphate (q.v.).
Celluflux TPP [821]
Disflamoll TP [597]

Triphenyl phosphite.
Mark TPP [75], [934]

Triphenyl phosphite *(cont'd).*
Weston TPP [129]

Triphenyl phosphite (contains ≈ triisopropanol amine, 0.5% wt.).
Weston EGTPP [129]

Triphenyl phosphite (contains ≈ triisopropanol amine, 1.0% wt.).
Weston 310 [129]

Triphenylmethane acid blues.
Alphazurine [35]

Triphenylmethane triisocyanate.
Desmodur R [112]

Triphenyltin acetate.
Batasan [416]
Brestan [416]
Suzu [629]

Triphenyltin hydroxide.
Duter [268]
Tubotin [561a]

Tri-POE (2) alkyl ether phosphate.
Nikkol TDP-2 [631]

Tri-POE (4) alkyl ether phosphate.
Nikkol TDP-4 [631]

Tri-POE (6) alkyl ether phosphate.
Nikkol TDP-6 [631]

Tri-POE (8) alkyl ether phosphate.
Nikkol TDP-8 [631]

Tri-POE (10) alkyl ether phosphate.
Nikkol TDP-10 [631]

Tripropylene glycol diacrylate.
SR-306 [766]

Tris alkylamido tri quaternary.
Monaquat P-TC, P-TD, P-TL, P-TS [599]

Tris imidazoline tri quaternary.
Monaquat P-TZ [599]

Tris phosphite 25.
Weston 474 [129]

Tris {ammine {ethylene bis (dithiocarbamato)} zinc (2+)} {tetrahydro-1,2,4,7-dithiadiazocine-3,8 diathione}, polymer.
Ariso G [107]
NIA 9102 [331a]
Pallinal [107]
Pallitop [107]
Polyram [331a]
Polyram-Combi [107]

Tris {ammine {propylene-1, 2-bis (dithiocarbamato)} zinc (2+)} {5-methyl-tetrahydro-1,2,4,7-dithiadiazocine-3,8-dithione}, polymer.
Basfungin [107]

1,3,5-Tris (4-t-butyl-3-hydroxy-2,6-dimethyl benzyl)-1,3,5-triazine-2,4,6-(1H, 3H, 5H)-trione.
Cyanox® 1790 [52]

Tri (β -chloroethyl) phosphite.
Tyrol CEF [392]

Tris (3,5-di-t-butyl-4-hydroxybenzyl.
Good-rite 3114 [361]

Tris (3,5-di-t-butyl-4-hydroxy benzyl) isocyanurate.
Vanox GT [894]

Tris (2,4-di-butylphenyl) phosphite.
Mark 2112 [934]

Tris (2,4-dichlorophenoxyethyl) phosphite.
Falone [881]

Tris (dichloropropyl) phosphate.
Tyrol FR-2 [392]

Tris (dipropyleneglycol) phosphite.
Weston 430 [129]

N,N′,N′-Tris (2-hydroxyethyl)-N-tallow-1,3 diaminopropane.
Ethoduomeen T/13 [16]

1,3,5-Tris (2-hydroxyethyl)-s-triazine-2,4,6 (1H, 3H, 5H)-trione, 3,5-di-t-butyl-4-hydroxyhydrocinnamic acid triester.
Agerite SKT [894]

Tris (hydroxymethyl) aminomethane.
Chemalog Grade 91-0160-00 [172]
Tris Amino [449], [451]
Ultralog 91-0160-20 [172]

Tris (hydroxymethyl) nitromethane.
Tris Nitro [456]

Tris [1-(2-methyl)-aziridinyl]-phosphine oxide (q.v.).
MAPO [452]

Tris [1-(2-methyl)-aziridinyl]-phosphine sulfide.
MAPS [452]

1,1,3-Tris (2-methyl-4-hydroxy-5-t-butyl phenyl) butane.
Topanol CA [438]

Trisnonyl phenyl phosphite.
Mark TNPP [74], [75], [934]
Stabilizer Mark 1178 [74]
Weston TNPP [129]
Wytox® 312 [659]

Trisnonylphenyl phosphite (contains ≈ triisopropanol amine, 0.75% wt.).
Weston 399 [129]

Trisnonylphenyl phosphite (contains ≈ triisopropanol amine, 1.0% wt.).
Weston 399B [129]

Tris-nonyl phenyl polyphosphite.
Stave TNPP (Non-toxic) [822]

Trisodium EDTA.
Hamp-ene Na3T [363]
Sequestrene Na 3 [190]

Trisodium EDTA sol'n.
Hamp-Ene Na_3 Liquid [363]

Trisodium EDTA, trihydrate.
Hamp-Ene Na_3T [363]

Trisodium HEDTA.
Cheelox HE-24 [338]
Chel DM-41 [190]
Hamp-ol 120, Crystals [363]
Perma Kleer 80, 80 Crystals [663]
Versenol 120 [261]

Trisodium n-hydroxyethylenediaminetriacetate aq. sol'n., tech. grade.
Chelon 120 [821]

Trisodium nitrilotracetate monohydrate.
Hampshire NTA Na_3 Crystals [363]

Trisodium sulfosuccinate.
T.S.S. [390]

Tristearate.
Dynasan 118 [275]

Tristearin.
Lipocerite Standard [901]
Neobee 62 [715]

Tristearyl citrate.
Citrest ST [234]

Tristearyl phosphite.
Weston TSP [129]

Trixylenyl phosphate.
Kronitex TXP [332]

Trixylyl phosphate.
Phosflex 179A [392]

Troleandomycin.
Tao Capsules & Oral Suspension [739]

Tromethamine.
Tris Amino [456]

Tromethamine magnesium aluminum silicate.
Veegum PRO [894]

Tropicamide (0.5% conc.).
Ocu-Tropic [655a]

Tropicamide (1.0%).
Ocu-Tropic [655a]

Trypsin.
Granulex [409]
Orenzyme, Bitabs [580]
Stimuzyme Plus [615]
Zypan Tablets [819]

Tuberculin, old.
Mono-Vacc Test (O.T) [579]
Tuberculin, Mono-Vacc Test (O.T.)
Tuberculin, Old, Tine Test (Rosenthal) [516]

Tungsten carbide.
Haystellite [878]

Tungsten (6) oxide.
Ultralog 91-5105-00 [172]

Turpentine.
Terpex [357]

Typhoid vaccine.
Typhoid Vaccine [942]

Valproic acid.
Depakene Capsules & Syrup [3]

Vanadia catalyst.
V-0701 T 1/8″ [389]

Vanadium.
Total Formula [907]

Vanadium (V) oxide.
Ultralog 92-5005-00 [172]

Vanadium pentoxide.
KM® [490]

Vancomycin.
Vancocin HCl, for Oral Solution, Vials [523]

Vanillin.
Vanillin Monsanto, N.F., F.C.C. [602]

Vasopressin.
Pitressin [674]

Vegetable fat, sulfated.
Softol Servo TLB [788]

Vegetable glycerol monostearate.
Radiamuls MG 2143 [658]

Vegetable Oil (cottonseed, soybean), partially hydrogenated.
Kaorich Beads [272]
KLX Flakes [272]

Vegetable oil (cottonseed, soybean, palm), partially hydrogenated.
Aratex [272]

Vegetable oil, epoxidized.
Agrilan FS 101 [509]

Vegetable oil, epoxilated.
Agrilan FS 112 [509]

Vegetable oil ethoxylate.
Cedepal EL-719 [256]

Vegetable oil, high sulfated.
Nopco 1471 [252]
Nopco 1471-M [253]

Vegetable oil mono-diglyceride.
Emuldan HV 40 K, HV 52 K, HVF 52 K [373]

Vegetable oil monoglyceride, distilled, saturated.
Dimodan PVP [373]

Vegetable oil monoglyceride, distilled, unsaturated.
Dimodan O [373]

Vegetable oil, oxalkylated.
Wettol EM 3 [106]

Vegetable oil, oxyethylated.
Uniperol EL [109]

Vegetable oil, polyethoxylated.
Emulgane E [253]

Vegetable oil, polyoxyethlated.
Emulphor® El-719, EL-620 [338]

Vegetable oil, sulfated.
Ahco AJ110 [438]

Verapamil hydrochloride.
Calan for IV Injection, Tablets [783]
Isoptin Ampules, for Intravenous Injection, Oral Tablets [495]

Versenate, calcium disodium.
Calcium Disodium Versenate Injection [734]

Verxite.
Zonolite [364a]

Vidarabine.
Vira-A Ophthalmic Ointment, 3% [674]

Vidarabine monohydrate.
Vira-A for Infusion [674]

Vinblastine sulfate.
Velban [523]

Vincristine sulfate.
Oncovin [523]

Vinyl.
Geon 2042, 8610, 8720, 8730, 8737, 8745, 8800, 8801-021, 8801-025, 8803, 8804, 8806, 8825, 8830, 8841, 8843, 8852, 8863, 8870, 8880, 8884, 8891, 8896, 8898, 8918, 82461, 82718, 84059, 84203, 84295 [361]

Vinyl acrylic emulsion.
Everflex E, T [363]

Vinyl comp..
Geon 8714, 8761, 8812, 8813, 8814, 8815, 8857, 8883, 83457, 83718, 83741, 83794, 85483, 85555, 85672, 85690, 85707, 86100, 86101, 86103 [361]
Kohinor 101, 102, 103, 104, 105, 106, 107, 108, 142, 144, 145, 148, 149, 150, 152, 154, 157 [673]

Vinyl comp., rigid.
Geon 7082, 7084, 8700A, 8702, 8750, 82662, 87242, 87255, 87256, 87262 [361]

Vinyl film.
Krene [878]

Vinylether polymer.
BASF Wax V [106]

Vinylidene chloride acrylic latex emulsion.
Versaflex 6 [363]

Vinylpyridine latex.
Pyratex [881]

Vitamin B_{12} *(cont'd).*
- Geravite Elixir [394]
- Hemo-Vite, Liquid [267]
- Heptuna Plus [739]
- I.L.X. B_{12} Elixir Crystalline, Tablets [489]
- Mega-B, Megadose [72]
- Natacomp-FA Tablets [868]
- Neuro B-12 Forte Injectable, B-12 Injectable [505]
- Nu-Iron-Plus Elixir, Nu-Iron-V Tablets [563]
- Orexin Softab Tablets [831]
- Perihemin [516]
- Prenate 90 Tablets [124]
- Pronemia Capsules [516]
- Rovite Tonic [748]
- The Stuart Formula Tablets [831]
- Therabid [593]
- Tia-Doce Injectable Monovial [105]
- Vicon Forte Capsules [354]
- Viopan-T Tablets [868]

Vitamin B_{12} feed supplements.
- Proferm [204]

Vitamin B complex.
- Added Protection III Multi-Vitamin & Multi-Mineral Supplement [589]
- BayBee Complex [350]
- Becotin [254]
- Besta Capsules [394]
- Cetol Filmtab Tablets [3]
- Dayalets Filmtab, plus Iron Filmtab [3]
- Hemo-Vite, Liquid [267]
- Iberet-500 [3]
- Iromin-G [593]
- Mega-B, Megadose [72]
- Natalins Rx, Tablets [572]
- Orexin Softab Tablets [831]
- Prenate 90 Tablets [124]
- Probec-T Tablets [831]
- The Stuart Formula Tablets [831]
- Vicon Forte Capsules, -C Capsules, -Plus Capsules [354]
- Vi-Penta F Chewables, F Multivitamin Drops, Multivitamin Drops [738]

Vitamin B complex with vitamin C.
- Added Protection III Multi-Vitamin & Multi-Mineral Supplement [589]
- Adeflor Chewable Tablets [884]
- Al-Vite [267]
- B-C-Bid Capsules [350]
- Becotin with Vitamin C, -T [254]
- Beminal-500, Forte w/ Vitamin C, Stess "Plus" [97]
- Berocca Tablets, -C, -C 500, -WS Injectable [738]
- Besta Capsules [394]
- Eldercaps [563]
- Enviro-Stress with Zinc & Selenium [907]
- Hemocyte Plus Tabules [888]
- Hemo-Vite [267]
- Heptuna Plus [739]
- Iberet, -500 Liquid, -Folic-500, Liquid [3]
- Iromin-G [593]
- Larobec Tablets [738]
- M.V.I., -12, -12 Lyophilized, Pediatric [80]
- Mediplex Tabules [888]
- Nu-Iron-V Tablets [563]
- Orabex-TF [512]
- Prenate 90 Tablets [124]
- Probec-T Tablets [831]
- Sigtab Tablets [884]
- Surbex Filmtab, w/ C Filmtab [3]
- Tabron Filmseal [674]
- Thera-Combex H-P [674]
- Vicon Forte Capsules, -C Capsules, -Plus Capsules [354]
- Vio-Bec, Forte [750]
- Viopan-T Tablets [868]

Vitamin C.
- ACE + Z Tablets [518]
- Al-Vite [267]
- Ascorbic Acid Tablets [751]
- B-C-Bid Capsules [350]
- Besta Capsules [394]
- C-Span Capsules [281]
- Cee-500, Cee-1000 T.D. Tablets [518]
- Cefol Filmtab Tablets [3]
- Cevi-Bid Capsules, -Fer Capsules (sustained release) [350]
- Dayalets Filmtab, plus Iron Filmtab [3]
- ECEE Plus Capsules & Tablets [281]
- Eldercaps [563]
- Enviro-Stress with Zinc & Selenium [907]
- Ferancee Chewable Tablets, -HP Tablets [831]
- Fergon Plus [133]
- Fetrin [512]
- Glutofac Tablets [489]
- Hemo-Vite [267]
- I.L.X. B_{12} Tablets [488]
- Iromin-G [593]
- Mevanin-C Capsules [117]
- Natacomp-FA Tablets [868]
- Natalins Rx, Tablets [572]
- Niferex w/ Vitamin C [166]
- Nu-Iron-V Tablets [563]
- Peridin-C [117]
- Prenate 90 Tablets [124]
- Probec-T Tablets [831]
- The Stuart Formula Tablets [831]

Walnut oil.
Lipovol W [525]

Wareflex plasticizer.
SR-650 [766]

Warfarin sodium.
Coumadin [269], [510]
Panwarfin [3]
Warfarin Sodium Tablets [769]

Watercress extract.
Phytelene of Cresson EG 224 Liquid, of Watercress EG 224 Liquid [900a]

Wheat germ oil.
Emcon W [308]
Lipovol WGO [525]

Wheat germamido PEG-L sulfosuccinate.
Mackanate WGD [567]

Wheat germamidopropy betaine.
Mackam WGB [567]

Wheat germamidopropyl dimethylamine lactate.
Mackalene 716 [567]

Wheat germamidopropylamine oxide.
Mackamine WGO [567]
Mackanate WGO [567]

White nettle extract.
Phytelene of Ortie Blanche EG 157 Liquid, of Ortie Blanche EN 103 Powder, of White Dead Nettle EG 157 Liquid, of White Dead Nettle EN 103 Powder [900a]

Witch hazel.
Medicated Cleansing Pads By the Makers of Preparation H Hemorrhoidal Remedies [922]
Tucks Cream, Ointment, Premoistened Pads [674]

Witch hazel extract.
Phytelene of Hamamelis EG 138 Liquid, of Hamamelis EN 092 Powder, of Witch Hazel EG 138 Liquid, of Witch Hazel EN 092 Powder [900a]

Wood cellulose.
Solka Floc [392]

Wool fat, refined.
Golden Fleece Anhydrous Lanolin P80, P95, R.A. [916a]

Wool wax alcohols.
Hartolite [223]

Wool wax alcohols BP.
Hartolan [223]

Wool wax alcohols BP and DAB, distilled.
Super Hartolan [223]

Wool wax alcohol sulfosuccinate.
Rewolan E LAN/E, 5 [725]

Yarrow extract.
Milfoil HS, LS [20]

Yellow fever vaccine.
YF-VAX (Yellow Fever Vaccine)(Live, 17 Virus, Avian Leukosis-Free, Stabilized) [815]

Yohimbine hydrochloride.
Yohimex Tablets [498]

Ytterbium (III) chloride.
Ultralog 93-8920-00 [172]

Ytterbium (III) oxide.
Ultralog 93-9280-00 [172]

Yttrium chloride hexahydrate.
Ultralog 94-0100-00 [172]

Yttrium oxide.
Ultralog 94-0180-00 [172]

Zinc.
Beminal Stress Plus [97]
Besta Capsules [394]
Eldercaps [563]
Enviro-Stress with Zinc & Selenium [907]
Megadose [72]
Prenate 90 Tablets [124]
Total Formula [907]
Vicon Forte Capsules, -C Capsules, -Plus Capsules [354]
Vi-Zac Capsules [354]
Zinc-220 Capsules [38]

Zinc ammonium chloride.
Preact [591]

Zinc-bis (undecylenic amidopropyl dimethyl glycinate).
Schercotaine UAB-Z [771]

Zinc borate.
Firebrake ZB [110], [392], [887]
ZB-112, ZB-112R, ZB-237, ZB-325, ZB-X511 [431]

Zinc carbonate.
Zinc Oxide Transparent [597]

Zinc chloride.
Zinctrace [80]

Zinc chloride 1 mg./ml..
Zinctrace™ [80]

Zinc chromate primer, one-pack.
Fertegol® ZA 40, ZA 80 [856]

Zinc chromate primer, two-pack.
Fertegol® ZM [856]

Zinc chrome.
Zn-0602 T 1/8″ [389]

Zinc chromite.
Zn-0312 T 1/4″ [389]

Zinc diamyldithiocarbamate.
Amyl Zimate [894]
Vanlube AZ [894]

Zinc dibenzyl dithiocarbamate.
Arazate [881]

Zinc dibenzyl dithiocarbonate.
Akrochem ZBED [15]

Zinc dibutyl dithiocarbamate.
Akrochem Accelerator BZ [15]
Mastermix 2DB DTC 4685 MB [392]
Vulcafor ZDBC [212], [910]
ZDBC [402]

Zinc di-n-butyl dithiocarbamate.
Butyl Zimate [894]

Zinc-N-dibutyl dithiocarbamate.
Vulkacit LDB/C [597]

Zinc-o,o-di-n-butylphosphorodithioate.
Vocol Vulcanization Accelerator [602]

Zinc diethyl dithiocarbamate.
Akrochem Accierator EZ [15]
Ethyl Zimate [894]
Vulcafor ZDEC [212], [910]
Vulkacit LDA [597]
ZDEC [402]

Zinc-N-diethyl dithiocarbamate.
Vulkacit LDA [597]

Zinc dimethyldithiocarbamate.
Corozate [705]
Cuman [190]
Methasan [602a]
Methasan® Vulcanization Accelerator [602]
Mezene [310]
Pomarsol Z Forte [112]
Tricarbamix Z [683]
Triscabol [683]
Vancide MZ-96 [894]
Z-C-Spray [332]
Zerlate [269]
Ziram Technical [682]

Zinc-N-dimethyl dithiocarbamate.
Vulkacit L [597a]

Zinc dimethyl dithiocarbamate/zinc 2-mercaptobenzothiazole, 87:7.7%.
Vancide 512 [894]

Zinc dimethyl dithiocarbonate.
Akrochem MZ [15]
Metazin-O [682]
Methasan [392], [602]
Methazate [881]
Methyl Zimate [894]
Methyl Ziram [708]
Vulcafor ZDMC [212], [910]
Vulkacit L [597]
ZM [154]

Zinc ethyl phenyl-dithiocarbamate.
Vulkacit P extra N [597]

Zinc ethylenebisdithiocarbamate.
Aspor [310]
Dithane Z-78 [742]
Lonacol [112]
Mancozan [730]
Parzate [269]
Polyram Z [106]
Tiezene [310]
Tritoftorol [683]
Zinosan [730]

Zinc 2-ethylhexoate.
Octoate Z [894]

Zinc fluosilicate.
Hornstone [364]

Zinc formaldehyde sulfoxylate.
Hydrozin [643]

Zinc gluconate.
Libidinal Capsules [304]
Megadose [72]

Zinc isopropylxanthate.
Propyl Zithate [894]

Zinc laurate.
Witco Zinc Soap #26 [935]

Zinc mercaptobenzothiazole.
Vulcafor ZMBT [212], [910]
Zetax [894]
ZMBT [708]

Zinc 2-mercaptobenzothiazole.
Zetax [894]

Zinc 2-mercaptotolylimidazole.
Vanox ZMTI [894]

Zinc methacrylate.
Processing Agent 748 [881]

Zinc neodecanoate.
Zinc Ten-Cem [605]

Zinc oxide.
Actox-14 [626]
Allersone [541]
Anusol-HC, Ointment, Suppositories [674]
Azo 11, 55, 55-LO, 66, 77 [84]
AZO-33, -55, -55TT, -66, -66TT, -77, -77TT [60]
Azodox-55, -55TT [60]
B Lead Free Zinc Oxide [804]
CO-064 [626]
Denlox [797]
Derma Medicone-HC Ointment [575]
Dome-Paste Bandage [585]
Hill Cortac [412]
Horse Head, U.S.P.-12, XX [626]
Kadox-15, -72 [626]
Osti-Derm Lotion [679]
Ozide 10-R, 10-RS, 101R [797]
Protox 166, 167, 168, 169 zinc oxide [626]
RR-Regular Grade [906]
RVPaque Ointment [284]
Rectal Medicone-HC Suppositories [575]
Saint Joe No. 500, No. 911, No. 920, No. 922, USP-111, USP-120 [828]
Ultralog 94-2705-00 [172]
Wyanoids HC Rectal Suppositories, Hemorrhoidal Suppositories [942]
Zinc Oxide USP 66 [924]
Ziradryl Lotion [674]
Zn-0401 E 3/16″ [389]

Zinc oxide, doped.
XX-85 [626]

Zinc oxide, fluorescent.
Ottalume 2100 [315]

Zinc oxide, lead-free.
Denzox [276]
Kadox [626]

Zinc oxide, precipitated.
Zinkoxyd Activ [597]

Zinc oxide, surface treated.
Saint Joe No. 500-36, No. 911-36, No. 920-36 [828]

Zinc oxide, untreated.
Akrochem RR Zinc Oxide [15]

Zinc-N-pentamethylene dithiocarbamate.
Vulkacit ZP [597]
ZPMC [402]

Zinc phosphates, crystalline.
Irco Bond [457]

Zinc phosphate primer, one-pack.
Fertegol® ZP 40, ZP 80 [856]

Zinc phosphate primer, two-pack.
Fertegol® ZM P [856]

Zinc phosphide.
Billy [799a]
Ratol [799a]
ZP [114a]

Zinc-ξN,N′-propylene-1,2-bis-(dithiocarbamate)ξ.
Airone [310]
Antracol [112]

Zinc pyrithione.
DHS Zinc Dandruff Shampoo [686]
Sebulon Dandruff Shampoo [920]
Zinc Omadine [659]

Zinc pyrithione 48% conc.
ZNP [19a]

Zinc selenide (q.v.).
Irtran 4 [279]

Zinc stearate.
Interstab ZN-18-1 [459]
Lubrazinc W, Superfine [935]
Petrac ZN-41, ZN-42 [688]
Plymouth Zinc Stearate, USP [680]
Witco Zinc Stearate U.S.P. [935]

Zinc stearate, heat stable.
Petrac ZN-44HS [688]

Zinc sulfate.
ACE + Z Tablets [518]
Besta Capsules [394]
ECEE Plus Capsules & Tablets [281]
Eldercaps, Eldertonic [563]
Glutofac Tablets [489]

Zinc sulfate *(cont'd)*.
Hemocyte Plus Tabules [888]
Mediplex Tabules [888]
Medizinc Scored Tablets [888]
Vicon Forte Capsules, -C Capsules, -Plus Capsules [354]
Vio-Bec Forte [750]
Vi-Zac Capsules [354]
Zinc-220 Capsules [38]
Zinckel-220 Tablets [498]

Zinc sulfate, basic.
BSZ [941a]

Zinc sulfoxylate formaldehyde.
Decroline [338]
Parolite [753]

Zinc undecylenate.
Pedi-Dri Foot Powder [679]

Zirconium (IV) oxide.
Ultralog 94-5700-00 [172]

Zirconium oxide/silicon dioxide, 88% min.:7% max..
Opax [615a]

Zirconium oxide/silicon dioxide, 90% min.: purity = 5% max..
Treopax [615a]

Zirconium oxide/silicon dioxide, min. 94%:max. 5%.
Insuloxide [615a]

Zirconium silicate, particle size 92 to 94.5%.
Superpax [615a]

Zirconyl chloride octahydrate.
Ultralog 94-5825-00 [172]

APPENDIX

Chemical Manufacturers

1	Aarhus Oliefabrik	Bruunsgade, Arhus C
2	Abbott Diagnostics Div.	North Chicago, IL
3	Abbott Laboratories	North Chicago, IL
4	AC & C	Greenwich, CT
5	Aceto Chem. Co.	Flushing, NY
6	Acme-Hardesty, Inc.	Jenkitown, PA
7	Adria Labs, Inc.	Dublin, OH
8	Advanced Refractory Technologies	New York, NY
9	Aeropres	Sheveport, LA
10	Age-Rite Geltrol	New York, NY
11	Agrashell	Bethlehem, PA
11a	Agri Products Marketing Corp.	Bakersfield, CA
12	Agvar Chemicals Inc.	New York, NY
12a	Agway Inc.	Syracuse, NY
13	Air Products & Chemicals, Inc.	Allentown, PA
14	Ajinomoto U.S.A., Inc.	New York, NY
15	Akron Chemical	Akron, OH
16	Akzo Chemie	Burt, NY
17	Akzo Chemie GmbH	Netherlands
18	Akzo Chemie Italia	Netherlands
19	Alan Protection Division	Vancouver, BC
19a	Alba Chemical	Tenafly, NJ
20	Alban Muller, Int'l	Paris, France
21	Albany International Corp.	Needham Heights, MA
22	Albermar Co.	New York, NY
23	Albright & Wilson	Norwood, NJ
24	Albright & Wilson (Australia)	Melbourne, Australia

25	Albright & Wilson Detergent Div.	Cumberland, UK
26	Albright & Wilson/Marchon Div.	Cumbria, UK
27	Alcoa (Closure Systems Int'l.)	Richmond, IN
28	Alco Chemical	Chattanooga, TN
29	Alcolac Inc.	Baltimore, MD
30	Alconox Inc.	New York, NY
31	Alcon (P.R.) Inc.	Fort Worth, TX
32	Aldrich Chemical Co. Inc.	Milwaukee, WI
33	Alkaril Chemicals (Inc.)	Winder, GA
33a	Alkaril Chemicals (Ltd.)	Mississauga, Ont., Canada
34	Alkaril Chemicals U.K.	Wilmslow Cheshire, UK
35	Allied Chemical Corp.	Morristown, NJ
36	Allied Chem., Fibers & Plastics Co.	Morristown, NJ
37	Allied Colloids Ltd.	Yorkshire, UK
38	Alto Pharmaceuticals, Inc.	Tampa, FL
39	Aluminum Co. of America	Pittsburgh, PA
40	Alza Corp.	Palo Alto, CA
41	Alzo, Inc.	Matawan, NJ
42	Amchem Products Inc.	Ambler, PA
43	Amerace Corp.	New York, NY
44	Amerchol Corp.	Englewood Cliffs, NJ
45	Amerchol Europe	Englewood Cliffs, NJ
46	American Agricultural Chemical Co.	Tulsa, OK
47	American Can Co.	Greenwich, CT
47a	American Colloid Co.	Skokie, IL
48	American Color & Chem. Corp.	New York, NY
49	American Critical Care	McGraw Park, IL
50	Amer. Cyanamid Co.	Wayne, NJ
50a	American Cyanamid Chemical Prod. Div.	Wayne, NJ
51	American Cyanamid Co. Fine Chemicals Dept.	Wayne, NJ
52	American Cyanamid Plastic Add. & Textile Chem. Dept.	Wayne, NY
52a	American Cyanamid Co./Resins Dept.	Wayne, NJ
53	American Dermal Corp.	Somerset, NJ
54	American Enzyme Corp.	Brown Deer, WI
55	American Hoechst Corp.	Somerville, NJ
56	Amer. Hoechst Corp./Plastics Div.	Somerville, NJ
57	American Lecithin Co.	Atlanta, GA
58	American Metal Climax, Inc.	Greenwich, CT
58a	American Pelletizing Corp.	Des Moines, IA
59	American Potash & Chemical Corp. Div. of Kerr-McGee	Oklahoma City, OK
59a	American Rubber and Chem.	Louisville, KY
60	Amer. Smelting & Refining Co.	New York, NY
61	American Synthetic Rubber	Cambridge, MA
62	Ameripol Inc.	Cleveland, OH
63	Ames Laboratories Inc.	Elkhart, IN
64	Amoco Chemicals	Chicago, IL
65	Anheuser-Busch	St. Louis, MO
66	Ansul Co.	Australia
67	Applied Science Laboratories	St. Petersburg, FL
68	Arapahoe Chemicals Inc.	Englewood, CO
69	ARC Chem. Division Balchem Corp.	State Hill, NY
70	Archer Daniels Midland Co.	Decatur, IL
71	Acheson Colloids Co.	Port Huron, MI
72	Arco Chem. Co.	Los Angeles, CA
73	Ardmore Chemical Co.	Newark, NJ

74	Argus	Brooklyn, NY
75	Argus Chem. Div./Witco Chem. Corp.	New York, NY
76	Arjay, Inc.	Houston, TX
77	Arkansas Co. Inc.	Newark, NJ
78	Arlo Corp.	Hato Rey, PR
78a	Arm & Hammer	Princeton, NJ
79	Armak Ind. Chem. Div.	Asheville, NC
80	Armour	Tarrytown, NY
81	Armstrong Chem.	Janesville, WI
82	Arol Chem. Prod.	Newark, NJ
83	Arthur H. Thomas Co.	Philadelphia, PA
84	Asarco	New York, NY
85	Asbury Graphite Mills	Asbury, NJ
86	B.F. Ascher & Co.	Lenexa, KS
87	Ashland Chem. Co.	Ashland, KY
88	Associated Lead Inc.	London, UK
89	Astor Chemical Durachem	Jacksonville, FL
90	Astra Pharmaceutical	Westboro, MA
91	Atlantic Powdered Metals	New York, NY
92	Atlas Chemical Industries, N.V.	London, UK
93	Atlas Chemical Industries (UK) Ltd.	London, UK
94	Atlas Refining Inc.	Newark, NJ
95	Atramax Inc.	Paterson, NJ
96	Australia Ltd.	Alhambra, CA
97	Ayerst Labs.	New York, NY
98	Babcock & Wilcox Co.	New Orleans, LA
99	Badische Corp.	Williamsburg, VA
100	Baker Castor Oil Co.	New York, NY
100a	Baker Chemical	Houston, TX
101	Bareco Div. Petrolite	Tulsa, OK
102	Barium Chemicals	Steubenville, OH
103	Barnes-Hind/Hydrocurve	New York, NY
104	Barry Labs, Inc.	Pompano Beach, FL
105	A.J. Bart, Inc.	Gurabo, PR
106	BASF AG	West Germany
107	BASF Aktiengeselschaft	Ludwigshafen, Germany
108	BASF India	Bombay, India
109	BASF Wyandotte Corp.	Parsippany, NJ
110	Bate Chemical	Ont., Canada
111	Bay Pharmaceuticals, Inc.	San Rafael, CA
112	Bayer AG	Leuerkusen, West Germany
113	Beach	Tampa, FL
114	Beecham Labs	Bristol, TN
115	Berlex Labs, Inc.	Cedar Knolls, NJ
116	Berol Kemi A.B.	Sweden
117	Beutlich, Inc.	Chicago, IL
118	BFC Chemicals, Inc.	Wilmington, DE
119	Biocraft Labs, Inc.	Elmwood Park, NJ
120	Bio-Lab Inc.	Decatur, GA
121	Biophil Chemical Fine S.A.S.	Milano, Italy
122	Bio Systems Research Inc.	Salida, CO
122a	Black Leaf Products Co.	Elgin, IL
123	Blew Chemical Co.	Palos Heights, IL
124	Bock	Burbank, CA
125	Boehringer Ingelheim	Ridgefield, CT

126 Bofors Lakeway .... Muskegon, MI
127 Boots .... Nottingham, UK
128 Borden Chemical .... Columbus, OH
129 Borg-Warner Chem. Inc. .... Parkersburgh, WV
130 Bostiksouth .... Greenville, SC
131 Bowman .... Herndon, VA
132 Boyle .... Cerritos, CA
133 Breon Laboratories .... New York, NY
134 Bristol-Myers .... New York, NY
135 Bromine Producers Div., Drug Research Inc. .... Adrian, MI
136 R.J. Brown Co. .... St. Louis, MO
137 Buckman Laboratories Inc. .... Memphis, TN
138 Burgess Pigment .... Macon, GA
139 Burroughs Wellcome Co. .... RTC, NC
140 Bush Boake Allen .... Montvale, NJ
141 Cabot Corp. .... Tuscola, IL
142 Calgon Corp., Pittsburgh Activated Carbon Division .... Pittsburgh, PA
143 California Industrial Minerals Co. .... Fresno, CA
144 Campbell .... New York, NY
145 Canada Packers, Inc. .... Mississauga, Ont., Canada
146 Cancarb .... Calgary, Alta
147 Capital City .... Janesville, WI
148 Cardiovascular Research .... Concord, CA
149 Cargill Inc. .... Longview, TX
150 Carl Becker .... Hamburg, West Germany
151 Carlisle Chemical Works Inc. .... Cincinnati, OH
151a Carlisle Chemical Works Inc., Advance Division .... Cincinnati, OH
152 Carlton .... Boston, MA
153 Carnrick .... Cedar Knolls, NJ
154 R.E. Carroll .... Harrisburg, PA
155 Carson Chem. .... Long Beach, CA
156 Carstab Corp. .... Reading, OH
157 Carus Chemical Co. .... La Salle, IL
158 Cas Chem Inc. .... Bayonne, NJ
159 Catawba-Charlab .... Charlotte, NC
160 Caf Chimie .... Cleveland, UK
161 C.E. Basic Chemicals .... Cleveland, OH
162 Ceca S.A. .... Cedex, France
162a Celamerck GmbH & Co. .... Phem, F.R.G.
163 Celanese Chemical Co. .... New York, NY
164 C.E. Minerals .... Stamford, CT
165 Center Division of E.M. Industries, Inc. .... Port Washington, NY
166 Central Pharmaceuticals .... Seymar, IN
166a Central Soya/Chemurgy Div. .... Fort Wayne, IN
167 Central Soya Co., Inc. .... Fort Wayne, IN
168 Cetylite .... Pennsauken, NJ
169 Charles Taylor & Sons Co. .... Todmorden, UK
170 Chemax Inc. .... Greenville, SC
171 Chemform Corp. .... W. Memphis, AR
172 Chemical Dynamics Corp. .... S. Plainfield, NJ
173 Chemical Mfg. Co. Inc. .... New York, NY
174 Chemical Products Corp. .... Elmwood Park, NJ
175 Chemische Fabriek Servo B.V. .... Delden, Netherlands
176 Chemische Fabrik Grunau GmbH .... West Germany
177 Chemische Fabrik Wibarco GmbH .... West Germany

178 Chemische Werke Huls AG .... Marl, West Germany
179 Chemithon Corp. .... Seattle, WA
180 Chemplast, Inc. .... Wayne, NJ
181 Chemtall Inc. .... Riceboro, GA
182 Chem-y .... Bodengraven, Holland
183 Chevron Chemical Co. .... San Francisco, CA
184 Chimagro Agricultural Div. Mobare Chem. Corp. .... Bayerwerk, West Germany
185 Chimex .... Gonesse, France
186 Chipman Chemical Co. .... River Rouge, MI
187 Chromex Chem. Corp. .... Brooklyn, NY
188 Chrystal .... New York, NY
189 CIBA .... Basel, Switzerland
190 Ciba-Geigy Co. .... Greensboro, NC
190a Ciba Geigy Corp. Agricultural Div. .... Basie, Switzerland
191 Ciba Pharmaceutical Co. .... Summit, NJ
192 Cincinnati Milacron Chem. .... Reading, OH
193 Cindet Chemicals .... Greenboro, NC
194 Cities Service Co. .... Atlanta, GA
195 W.A. Cleary Corp. .... Somerset, NJ
196 Climax Molybdenum Co. .... Greenwich, CT
197 Clintwood Chem. Co. .... Chicago, IL
198 Clough Chemical .... St. Jean, Quebec
199 C & M .... Louisville, KY
199a Colloides Naturels .... Far Hills, NJ
200 Colloids Inc. .... Newark, NY
201 Colorado Chemical .... Golden, CO
202 Columbian Chemicals .... Tulsa, OK
203 Commercial Minerals .... Tulsa, OK
204 Commercial Solvents Corp. .... Northbrook, IL
205 Compagnie Parento .... Ossining, NY
206 Conoco Chemicals .... Houston, TX
207 Consos Inc. .... Charlotte, NC
208 Continental Carbon .... Wilmington, DE
209 Continental Chemical Co. .... Clifton, NJ
210 Continental Oil .... Saddle Brook, NJ
211 Contour Chemical .... Windsor Loeks, CT
212 Cook Chemical .... Palm Beach, FL
213 Cooke-Waite .... New York, NY
214 Cooper Care .... Palo Alto, CA
215 Cordova Chem. Co. .... N. Muskegon, MI
216 Corning Glass Works .... Corning, NY
216a Corn Products Unit CPC N. Amer. .... Englewood Cliffs, NJ
217 Cosden Oil & Chem Co. .... Dallas, TX
218 Cosmedial Dept. of Gen'l. Mills Chem. .... Minneapolis, MN
218a Costec Inc. .... Palatine, IL
219 Cowles Chemical Co. .... Kent, UK
220 C.P. Chemicals Inc. .... Sewaren, NJ
220a CPC International .... Englewood Cliffs, NJ
222 Crest Chem. Corp. .... Stamford, CT
223 Croda Inc. .... New York, NY
223a Croda Italiana S.R.C. .... Mortara, Italy
223b Croda Surfactants Inc. .... North Humberside, UK
224 Crodet Chem. .... New York, NY
225 Crompton & Knowles Dyes & Chemicals Division .... Charlotte, NC
226 Crosby Chemical .... De Ridder, LA

227	Crowley Chemical	New York, NY
228	Crown Metro Inc.	Greenville, SC
229	Crown Zellerbach	Vancouver (Orchards), WA
230	Crucible Chemical Co.	Greenville, SC
230a	Crystal Chemical Inter-America	Huntington, IN
231	Crystal Soap	Lansdale, PA
232	Cutter Biological	Elkhart, IN
233	Cyanamid B.V.	Rotterdam, The Netherlands
235	see 674	Morris Plains, NJ
234	Cyclo Chemicals Corp.	Miami, FL
236	Cyprus Ind. Minerals Co.	Los Angeles, CA
237	Dai-ichi Kogyo Seiyaku Co., Ltd.	Tokyo, Japan
238	Dainippon	Tokyo Japan
239	Dalin	Farmingdale, NY
240	Danbury	Danburg, CT
241	Darling & Co.	Chicago, IL
242	Davison Chem. Div.	Baltimore, MD
243	Degesch America Inc.	Weyers Cave, VA
244	Degussa Corp.	Teterboro, NJ
245	Delmont	Swarthmore, PA
246	Dermik	Bluebell, PA
247	DeSoto, Inc. Sellers Chemical Div.	Harahan, LA
248	Detergents, Inc.	Santa Fe Springs, CA
249	Devan Chemicals S.A.	Ronse-Renaix, Belgium
250	Dexter Chem. Corp.	Bronx, NY
251	Diamond Alkali Co.	Dallas, TX
251a	Diamond Crystal	Salto, CO
252	Dimond Shamrock (Process Chemicals Div.)	Morristown, NJ
253	Diamond Shamrock Europe	Brussels, Belgium
254	Dista	Indianapolis, IN
255	Devcon Corp.	Danvers, MA
256	Domtar, Inc./CDC Div.	Mississauga, Ont., Canada
257	Dorsey Laboratories (Division of Sandoz, Inc.)	Lincoln, NE
258	Dorsey Pharmaceuticals (Div. of Sandoz, Inc.)	East Hanover, NJ
259	Dover Chemical	Dover, OH
260	Don B. Hickam Inc.	Houston, TX
261	Dow Chemical Co.	Midland, MI
262	Dow Corning Corp.	Midland, MI
263	Dragoco	Totowa, NJ
264	Drew Chemical Corp.	Ashland, KY
265	Drew Produtos Quimicos Ltda.	Sao Paulo, Brazil
266	Drew Produitos Quimicos S.A.	Sao Paulo, Brazil
267	Drug Industries	Ferndale, MI
268	Duphar B.V.	The Netherlands
269	Du Pont	Wilmington, DE
270	E.I. dupont de Nemours & Co. Inc.	Wilmington, DE
271	Dura Commodities Corporation	Harrison, NY
272	Durkee Industrial Foods Group/SCM Corp.	Cleveland, OH
273	Dusa	Camas, WA
274	Dutton & Reinisch Ltd.	Surrey, UK
275	Dynamit-Nobel AG	West Germany
	Dynamit-Nobel America	Ruckleigh, NJ
276	Eagle Pichar Industries Inc.	Joplin, MO
277	Eastern Color & Chem. Co.	Providence, RI
278	Eastman Chemicals	Kingsport, TN

279 Eastman Kodak Co. .... Rochester, NY
280 Edward Mendell .... Carmel, NY
281 Edwards Pharmacal .... Osceola, AR
282 Elanco Products Co. .... Indianapolis, IN
283 Elastochem .... West Germany
284 Elder .... Bryan, OH
285 Elektrochem Fabrik Kempen GmbH .... Kempen, West Germany
286 Eli Lilly .... Indianapolis, IN
287 Elkins-Sinn .... Cherry Hill, NJ
288 Emery FDAG .... Cincinnati, OH
288a Emery Ind., Fatty & Dibasic Acids Group .... Cincinnati, OH
289 Emery Industries, Inc. .... Linden, NJ
290 Emery Ind./Personal Care Products .... Linden, NJ
291 Emery PCSG .... Mauldin, SC
291a EM Industries Plant Protection Div. .... Hawthorne, NY
292 Emkay Chem Co. .... Elizabeth, NJ
293 Emulan .... Kenosha, WI
294 Emulsion Systems Inc. .... Brooklyn, NY
295 Engelhard Min. & Chem. Corp. .... Edison, NJ
296 Enjay Chem. Co. .... New York, NY
297 Enthone, Inc. .... New Haven, CT
297a Enzyme Development Corporation .... New York, NY
298 Escambia Chemical Co. .... Allentown, PA
299 Eschem .... Chicago, IL
300 Essential Chemicals Corp. .... Merton, WI
301 ETC-Euarmes Textiel Centrum .... Eurag-Shoes, Belgium
302 Ethyl Corp. .... Baton Rouge, LA
303 Evans Chemetics Inc. .... Darien, CT
304 Everett .... East Orange, NJ
304a Excel Industries .... Bombay, India
305 Exxon Chemical Americas .... Houston, TX
306 Exxon Corp .... Houston, TX
307 Fairmont Chemical Co. .... Newark, NJ
307a Fairmount Granite Ltd. .... Cranston, RI
308 Fanning .... Chicago, IL
309 Farben Fabriken Bayer AG .... Leverkusen, Germany
310 Farmoplant S.P.A. .... Milano, Italy
311 Feldspar Corp. .... Spruce Pine, NC
312 Felton .... Brooklyn, NY
313 Fermco Biochemists .... Elk Grove Village, IL
314 Ferndale .... Ferndale, MI
315 Ferro Corp. Chem. Div. .... Bedford, OH
316 Fiberfil, Inc. .... Evansville, IN
317 Fine Organics Inc. .... Lodi, NJ
318 Fike Chemicals Inc. .... Nitro, WV
319 Fine Organics LaPorte Industries (Holdings) PLC .... London, UK
320 Finetex Inc. .... Elmwood Park, NJ
321 Firestone Synthetic Fibers Co. .... Akron, OH
322 Firestone Synthetic Rubber and Latex Co. .... Akron, OH
323 Firmenich .... Princeton, NJ
324 Fisons, Pharmaceutical Div. .... Bedferd, MA
325 Fleet .... Lynchburg, VA
326 Fleming .... Fenton, MO
327 Flint, Division of Travenol Lab. Inc. .... Deerfield, IL
328 Flintkote .... Hunt Valley, MD

329 Florida Department of Citrus .... Lakeland, FL
330 Floridin Co. .... Pittsburgh, PA
331 Fluoritab .... Flint, MI
331a FMC Agricultural Chemical Group .... Philadelphia, PA
332 F.M.C. Corp. .... Philadelphia, PA
333 Foote Mineral Co. .... New York, NY
334 Forest .... New York, NY
335 Formosa Chem. & Fibre Corp. .... Taipei, Taiwan
335a Freeman .... Port Washington, WI
336 Freeport Kaolin .... Gordon, GA
337 Fritzsche Dodge & Olcutt Inc. .... New York, NY
338 GAF Corp, Chemical Products .... New York, NY
339 Gatte Fosse Ets. .... St. Priest, France
340 Geigy, Div. of CIBA-GEIGY Corp. .... Ardsley, NY
341 General Electric .... Waterferd, NY
342 General Electric Copolymers Product Dep. .... Fairfield, CT
343 General Tire & Rubber Co. .... Akron, OH
343a Generichem Corp. .... New York, NY
344 Geneva (Generics) .... Broomfield, CO
345 Genstar Stone Products .... Hunt Valley, MD
346 Geoliquids Inc. .... Chicago, IL
347 Georgia Kaolin Co. .... Elizabeth, NJ
348 Georgia Marble .... Atlanta, GA
349 Georgia-Pacific/Bellingham Div. .... Bellingham, WA
350 Geriatric .... New Hyde Park, NY
351 Germany Exploration Co. .... Dallas, TX
352 Gilbert .... Chester, NJ
353 Givaudan Corp. .... Clifton, NJ
354 Glaxo .... Research Triangle Park, NC
355 Glenbrook, Div. of Sterling Drug .... New York, NY
356 Glenwood .... Tenafly, NJ
357 Glidden/Durkee Div. SCM Corp .... New York, NY
358 Glyco Inc. .... Greenwich, CT
359 Glycol Chem. .... Greenwich, CT
360 Goldschmidt Chem. Corp. .... Hopewell, VA
361 B.F. Goodrich Chem. Co. .... Cleveland, OH
362 Goodyear Tire & Rubber Co. .... Akron, OH
363 W.R. Grace & Co.-Organic Chem. Div. .... Lexington, MA
364 W.R. Grace/Construction Products Div. .... New York, NY
364a W.R. Grace & Co./Davison Chemical Division .... Baltimore, MD
364b W.R. Grace-Organic Chem. Div. .... Lexington, MA
365 Graden Chemical .... Havertown, PA
366 Grain Processing .... Muscaline, IA
367 Grant Chem. Div./Ferro Corp. .... Bedford, OH
368 Gray, Affiliate the Purdue Frederick Co. .... Norwalk, CT
369 Great Lakes Chem. Corp. .... W. Lafayette, IN
370 Greeff .... Old Greenwich, CT
370a Greenwood Chemical Co. .... Greenwood, VA
371 Grefco .... Torrance, CA
372 Grindtek .... New York, NY
373 Grinsted Prod. .... Brabrand, Denmark
374 A. Gross & Co., Millmaster Onyx Group .... Newark, NJ
375 Grunau .... Federal Republic of Germany
376 Guardian .... Smithtown, NY
376a Gujarat Pesti-Chem Industries .... Baroda, India

377	Gulf Oil Chemicals Co.	Houston, TX
377a	G. & W. Laboratories, Inc.	South Plainfield, NJ
378	Haarman & Reimer	Springfield, NJ
379	The C.P. Hall Co.	Chicago, IL
380	Hamblet & Hayes Co.	Salem, MA
381	Hammil & Gillespie	Livingston, NJ
382	Handy & Harman	Totowa, NJ
383	Harbison Carborundum Corp.	Levittown, PA
384	Harbison-Walker Refractories Co.	Pittsburgh, PA
385	Hardman Inc.	Belleville, NJ
386	A. Harrison & Co.	Pawtucket, RI
386a	Harrisons & Crosfield (Canada)	London, UK
387	Harrisons & Crosfield (Pacific)	London, UK
388	Harrison Enterprises	Dayton, OH
389	Harshaw Chem. Co.	Cleveland, OH
390	Hart Chemical	Guelph, Canada
391	Hart Products Corp.	Jersey City, NJ
392	Harvick	Akron, OH
393	Hatco Chemical	Fords, NJ
394	W.E. Hauck	Roswell, GA
395	Haveg Industries Inc., Sub. Hercules Indus.	Wilmington, DE
396	Hayashibara Biochemical	Okayamo, Japan
397	Hefti Ltd.	Zurich, Switzerland
398	Henkel Argentina	Capital Federal, Argentina
399	Henkel Chemicals (Canada)	Montreal, Canada
400	Henkel Corp.	Maywood, NJ
401	Henkel KGaA	Dusseldorf, West Germany
402	Henley	New York, NY
403	Herbert	Westbury, NY
404	Hercon, Div. Health Chem. Corp.	New York, NY
404a	Hercules BV	Rijswijk, The Netherlands
405	Hercules France	Wilmington, DE
406	Hercules Inc.	Wilmington, DE
407	Heterene Chem.	Paterson, NJ
408	Hexcel Chem. Prod. Div.	Zeiland, MI
408a	Hexcel Corp. Spec. Chem. Div.	Maywood, NJ
409	Dow B. Hickam	Sugar Land, TX
410	High Point Chemical	High Point, NC
411	High Temperature Materials, Inc.	San Fernando, CA
412	Hill Dermaceuticals	Orlando, FL
413	Hilton-Davis	Cincinnati, OH
414	Hilton-Davis-Thomasset	New York, NY
414a	Himont	Wilmington, DE
415	Hodag Chem. Corp.	Stokie, IL
416	Hoechst AG	Frankfurt, West Germany
417	Hoechst Japan	Tokyo, Japan
418	Hoechst-Roussel	Somerville, NJ
419	Hoffman-La Roache Inc.	Nutley, NJ
420	Holloway	Birmingham, AL
421	Hooker Chemicals & Plastics Corp.	Niagara Falls, NY
422	Hoover-Hanes Rubber Corporation	Tallopoosa, GA
423	Hopkins Agricultural Chem. Co.	Madison, WI
424	Hormmel	Pittsburgh, PA
425	E.F. Houghton & Co.	Valley Forge, PA
426	Howard Hall & Co.	Cos Cob, CT

427 Hoyt, Division of Colgate-Palmolive Co. .... Norwood, MA
428 J.M. Huber Corp. .... Havre de Grace, MD
429 Humko Chem./Div. Witco Chem Corp. .... Memphis, TN
429a Humko Products Chem. Div./Witco Chem. .... Memphis, TN
430 Humko Sheffield Chem. .... Memphis, TN
431 Humphrey Chemical .... North Haven, CT
432 Hung Ltd. .... Hong Kong, UK
433 Huntington Alloy Products Div. International Nickel Co. Inc. .... Toronto, Ont., Canada
434 Hyland Therapeutics, Travenol Laboratories, Inc. .... Glendale, CA
435 Hynson, Westcott, & Dunning, Inc. .... Baltimore, MD
436 Hyrex .... Memphis, TN
437 Hysan .... Chicago, IL
438 ICI Americas Inc. .... Wilmington, DE
439 ICI Australia Ltd. .... Melbourne, Australia
440 ICI England .... Millbank, London, UK
441 ICI England Petrochemicals & Plastics Div. .... Cleveland, UK
442 ICI Organics Div. .... Manchester, UK
443 ICI Petrochemicals .... Wilmington, DE
444 ICI Plant Protection Division .... Surrey, UK
445 ICI United States .... Wilmington, DE
446 ICN Pharmaceuticals .... Covina, CA
447 Ikeda Bussan Kaisha .... Tokyo, Japan
448 Illinois Minerals .... Cairo, IL
449 IMC Chem. Group .... Des Plaines, IL
449a IMC Chem-NP Div. .... Northbrook, IL
450 Independent Chemical .... Glendale, NY
451 Industrial Minerals, Industrial Mineral Ventures, Inc. .... Las Vegas, NV
452 Inmont Corp. .... Clifton, NJ
453 Inolex Chem. Co. .... Philadelphia, PA
454 International Minerals & Chem. Corp. NV Div. .... Northbrook, IL
455 International Dioxide, Inc. .... Clark, NJ
456 International Minerals & Chemicals .... Terre Haute, IN
.... Des Plaines, IL
457 International Rustproof Co. .... Cleveland, OH
458 International Wax .... Valley Stream, NY
459 Interstab Chem. Inc. .... New Brunswick, NJ
460 Intex Products .... Greenville, SC
461 Isochem Corp. .... Lincoln, RI
462 Isochem Resins .... Lincoln, RI
463 Ives .... New York, NY
464 Jacobus .... Princeton, NJ
465 Jahres Fabrikker .... Sandefijord, Norway
466 James River Limestone .... Buchanan, VA
467 Janssen .... Piscataway, NJ
468 Jayco .... Camp Hill, PA
469 Jean D'Avez .... France
470 Jefferson Chemical Co., Inc. .... Austin, TX
471 Jetco Chemicals .... Cursicana, TX
471a Jewmin-Jofte Chemicals, Ltd. .... Israel
472 Johns-Manville .... Denver, CO
472a Jojoba Commodities Group .... North Hollywood, CA
473 Jojoba Growers .... New York, NY
474 Jordan Chemical .... Folcroft, PA
475 Kaiser Chemicals Div. Kaiser Aluminum .... Oakland, CA

476	Kaken Chemical Co.	Tokyo, Japan
476a	Kalama Chemical, Inc.	Seattle, WA
477	Kalo Agricultural Chem. Inc.	Overland Park, KS
478	Kao Corp.	Tokyo, Japan
478a	Kao Food	Tokyo, Japan
479	Kaopolite	Elizabeth, NJ
480	Kao Soap	Tokyo, Japan
481	Katsura	Tokyo, Japan
482	Kay-Fries Inc.	Rockleigh, NJ
483	Keil Chemical	Hammond, IN
484	Kelco Co.	Rahway, NJ
485	Kelco/Div. of Merck & Co. Inc.	San Diego, CA
485a	Kemira Oy	Helsinki, Finland
486	KenoGard VT AB	Stockholm, Sweden
487	Keno Gard S.A.	Bruxelles, Belgium
488	Kenrich Petrochemicals	Bayonne, NJ
489	Kenwood	New Rochelle, NY
490	Kerr-McGee	Oklahoma City, OK
491	Key Pharmaceuticals	Miami, FL
492	Kennametal Inc.	Latrobe, PA
493	Klinger Engineering Plastics Div.	UK
494	Knapp Products, Inc.	Lodi, NJ
495	Knoll Pharmaceutical Co.	Whippany, NJ
496	H. Kohnstamm & Co.	New York, NY
497	Koppers, Co. Inc.	Pittsburgh, PA
498	Kramer	Miami, FL
499	Kremers-Urban	Milwaukee, WI
500	Kumiai Chem. Ind.	Tokyo, Japan
501	Laboratoires Serobiologiques	Nancy, France
501a	Laboratori Vevy, Inc.	New York, NY
502	LactAid	Pleasantville, NJ
503	Lafayette	Fort Worth, TX
504	Lake States Chemical, Div. St. Regis Paper Co.	New York, NY
505	Lambda	Gurabo, P.R.
506	La Motte Chemical Products, Inc.	Chesterton, MD
507	Lanaetex	Elizabeth, NJ
508	Lancaster	Lakeland, FL
509	Lankro Chem. Ltd.	Manchester
510	Lannett	Philadelphia, PA
511	Laporte Industries Ltd.	London, UK
512	LaSalle	Detroit, MI
513	Laser	Crown Point, IN
514	La Tassilchimka	Bergamo, Italy
515	Laur Silicone Rubber Compounding, Inc.	Beaverton, MI
516	Lederle	Wayne, NJ
517	Leeming Pfizer Inc.	Parsippany, NJ
518	Legere	Scottsdale, AZ
519	Lemmon	Sellersville, PA
520	Lensfield	Bedford, UK
521	Lever Industriel	Noisy-le-Sec, France
522	Lilachim S.A.	Brussels, Belgium
523	Eli Lilly & Co.	Indianapolis, IN
524	Lion Corp.	Tokyo, Japan
525	Lipo Chemicals	Paterson, NJ
526	Liquid Nitrogen Processing Corp.	Malvern, PA

527	LNP Corp.	Malvern, PA
528	Lomar PWA	New York, NY
529	Lonza Inc.	Fairlawn, NJ
530	Los Angeles Chemical Co.	South Gate, CA
531	Lowenstein Dyes & Cosmetics, Inc.	Brooklyn, NY
532	Lubrisol Corp.	Wickliffe, OH
533	Lucasmeyer	West Germany
534	Lucas Meyer GmbH & Co.	West Germany
535	Luchem Pharmaceuticals Inc.	New York, NY
536	Luciool Div./Pennwalt Corp.	Buffalo, NY
536a	Lucidol Div., Ram Div.	Buffalo, NY
537	Lyndal Chemical	Dalton, GA
537a	Maag Agrochemicals (Switzerland)	Dielsdorf, Switzerland
538	MacFarlan Smith	Edinburgh, Scotland
539	Magna Corp.	Houston, TX
540	Makhteshim Agan	Beer-sheva, Israel
541	Mallard	Detroit, MI
542	I. Mallinckrodt Inc.	St. Louis, MO
543	Malvern Minerals	Hot Springs Natl. Park, AR
544	Manchem Ltd.	Manchester, UK
545	Manro Products Ltd.	Cheshire, UK
546	See 353	Clifton, NJ
547	Marathon Morco Co.	Dickinson, TX
548	Marchon	London, UK
549	Marine Colloids	Springfield, NJ
550	Mario Geronazzo Ind. Chim SpA	Italy
551	Marion Labs	Kansas City, MO
552	Marlowe-Van Loan Corp.	Hight Point, NC
553	Marlyn	Escondido, CA
554	Mars Chemical Corp.	Atlanta, GA
555	Martin Marietta	Hunt Valley, MD
556	Marubishi Oil Chemical Co.	Osaka, Japan
557	Maruzen Fine Chemicals	Moorestown, NJ
558	Mason Pharmaceutical	Newport Beach, CA
559	Masonite Corp.	Chicago, IL
560	Mastar Pharmaceutical	Bethlehem, PA
561	Matsumoto Yushi-Seiyaku Co.	Osaka, Japan
561a	May & Baker	Essex, UK
562	Maybrook	Lawrence, MA
563	Mayrand	Greensboro, NC
564	Mazer Chemicals Inc.	Gurnee, IL
565	McDanel Refractory Porcelain Co.	Beaver Falls, PA
566	McGregor Pharmaceutical	Farmington, MI
567	McIntyre Chem. Co. Ltd.	Chicago, IL
568	McLauglin Gormley King Co.	Minneapolis, MN
569	McNeil Consumer Products, McNEILAB, Inc.	Fort Washington, PA
570	McNeil Pharmaceutical, McNEILAB, Inc.	Spring House, PA
571	Mead Johnson Nutritional, Mead Johnson & Company	Evansville, IN
572	Mead Johnson Pharmaceutical, Mead Johnson & Company	Evansville, IN
573	Mearl Corp.	Ossining, NY
574	Med-Chem Labs	Monroe, MI
575	Medicone Comp.	New York, NY
576	Meer Corp.	North Bergen, NJ
577	Merck & Co. Inc.	Rahway, NJ
578	Merck Sharp & Dohme, Div. of Merch & Comp., Inc.	West Point, PA

579 Merieux Inst., Inc. .... Miami, FL
580 Merrell Dow Phar., Inc., Subsid. of Dow Chemical Co. .... Cincinnati, OH
581 M. Michel & Co. Inc. .... New York, NY
582 Michigan Chemical Corp. .... Chicago, IL
583 Midox Ltd. .... Leicestershire, UK
584 Miles Laboratories Inc. .... Elkhart, IN
585 Miles Pharmaceuticals, Div. of Miles Laboratories, Inc. .... West Haven, CT
586 Milex Products, Inc. .... Chicago, IL
587 Miller Chemical & Fertilzer Corp. .... Hanover, PA
588 Milliken Chemical .... Spartenburg, SC
589 MineraLab Inc. .... Hayward, CA
590 Minerals & Chemicals/Div. Engelhard Minerals & Chemicals Corp. .... Edison, NJ
591 Mineral Research & Development Corp. .... Charlotte, NC
592 Miranol Chem. Co. Inc. .... Dayton, NJ
593 Mission Phar. Co. .... San Antonio, TX
594 Mitsubishi Chem. Ind. Ltd. .... Tokyo, Japan
595 Mitsui Toatsu Chemicals Inc. .... Tokyo, Japan
596 M & J Chemicals .... Houston, TX
597 Mobay Chem. Corp. .... Pittsburgh, PA
597a Mobay Chemical Corp., Agricultural Chemicals Division .... Kansas City, MO
598 Mobil Chem. Co., Chem. Prod. Div. .... Richmond, VA
599 Mona Industries .... Paterson, NJ
600 Mono-chem. Corp. .... Atlanta, TX
601 Monsanto Agricultural Products Co. .... St. Louis, MO
602 Monsanto Co. .... St. Louis, MO
602a Monsanto Industrial Chemicals Co. .... St. Louis, MO
603 Monsanto Plastics & Resins .... St. Louis, MO
604 Montedison USA, Inc. .... New York, NY
605 Mooney .... Cleveland, OH
606 Moores Lime Co. .... Springfield, OH
607 Morton Chemical, Div. Morton-Norwich .... Chicago, IL
608 Morton Thiokol, Inc. .... Princeton, NJ
609 Motomco, Inc. .... Clearwater, FL
610 M & T Chemicals .... Rahway, NJ
611 H. Muehlstein .... Greenwich, CT
611a Multiform Dessicant, Inc. .... Buffalo, NY
612 Muro Phar. Inc. .... Tewksbury, MA
612a Murphy-Phoenix Co. .... Beachwood, OH
613 See 548 .... London, UK
614 See 548 .... London, UK
615 National Dermaceutical .... Maineville, OH
615a National Lead Co. .... New York, NY
616 National Lead Co. Baroid Div. .... New York, NY
617 National Lead Co., De Lore Division .... New York, NY
617a National Lead Co. Evans Lead Div. .... New York, NY
618 National Poly-Chemicals Inc. .... Wilmington, MA
619 National Starch & Chem. Corp. .... Bridgewater, NJ
620 Natrochem .... Savannah, GA
621 Nease Chemical Co. Inc. .... State College, PA
622 Neutrogena Dermatologies, Div. of Neutrogena Corp. .... Los Angeles, CA
623 Neville Chemicals .... Pittsburgh, PA
624 Neville-Synthese Organics, Inc. .... Pittsburgh, PA
625 New Japan Chemical Co. Ltd. .... Osaka, Japan
626 New Jersey Zinc .... Nashville, TN
627 Niacet Corp. .... Niagara Fallss, NY

628	Nihon Emulsion Co. Ltd.	Tokyo, Japan
629	Nihon Nohyaku Co.	Tokyo, Japan
630	Nikka Chem. Ind. Co., Ltd.	Fukui City, Japan
631	Nikko Chemicals Co. Ltd.	Tokyo, Japan
632	Nipa Laboratories Ltd.	Mid-Glamorgan, UK
633	Nippon Chemical	Tokyo, Japan
634	Nippon Kayaku Co.	Tokyo, Japan
635	Nippon Nyukazai Co.	Tokyo, Japan
636	Nippon Oil & Fats Co.	Tokyo, Japan
637	Nippon Senka Chemical Industries, Ltd.	Osaka, Japan
638	Nippon Soda Co.	Tokyo, Japan
639	Nissan Chem. Ind.	Tokyo, Japan
640	Nisshin Oil Mills	Tokyo, Japan
640a	NL Chemicals	Hightstown, NJ
641	NL Ind./Baroid Div.	Houston, TX
641a	NL Industries Inc.	Hightstown, NJ
642	N.L. Ind./Ind. Chem. Div.	Hightstown, NJ
643	Nopco Division Shamrock Corp.	London, Ont., Canada
644	Norac Co. Inc.	Azusa, CA
645	Nor-Am Agricultural Products Inc.	Naperville, IL
646	Norcliff Thayer	Tarrytown, NY
647	Norda	East Hanover, NJ
648	Nordisk-USA	Bethesda, MD
649	Norgine Laboratories, Inc.	Scarsdale, NY
650	Norman, Fox & Co.	Vernon, CA
651	Norsolor/Co. of C.D.F. Chemie	New York, NY
652	Norwich Eaton	Norwich, NY
653	Noury Chem. Corp.	Burt, NY
654	Novo Industri A/S	Bagsvoerd, Denmark
655	Nyacol Products Inc.	Ashland, MA
656	Ohio Lime	Hamilton, Ont., Canada
655a	Ocumed	Freeport, NY
657	OLC Laboratories	Miami, FL
658	Oleofina S.A.	Brussels, Belgium
659	Olin Corp.	Little Rock, AR
660	Omya	Proctor, VT
661	O'Neal, Jones & Feldman	Maryland Heights, MO
662	See 663	Jersey City, NJ
663	Onyx Chemical Co., Millmaster Onyx Group	Jersey City, NJ
664	Organon	West Orange, NJ
665	Ortho Diagnostic Systems	Raritan, NJ
666	Ortho Pharmaceutical (Dematological Div.)	Raritan, NJ
667	Osaka Kasei Co., Ltd.	Osaka, Japan
668	Otsuka Chem. Co.	New York, NY
669	Ottawa Chem. Co.	Toledo, Ohio
670	Ozark Mahoning Co., A Pennwalt Subsidiary	Tulsa, OK
671a	Pamol Ltd. Arad, Luxemburg Chemicals	Tel Aviv, Israel
671	Paisley Products	Edison, NJ
672	Paniplus	Olathe, KS
673	Pantasote, Inc.	Passaic, NJ
674	Parke Davis & Co.	Morris Plains, NJ
674a	Patco Cosmetic Products	Kansas City, MO
675	Patco Products Div. C.J.P.	Kansas City, MO
676	C.J. Patterson Co.	Kansas City, MO
677	PBI/Gordon Corp.	Kansas City, MO

678	Pearsall Chem. Div./Witco Chem. Corp.	Houston, TX
678a	Pecten Chemical Inc.	Houston, TX
679	Pedinol Phar.	W. Babylon, NY
680	S.B. Penick Div.	Newark, NJ
681	Pennsylvania Glass Sand Corp.	Pittsburgh, PA
682	Pennwalt Corp.	Philadelphia, PA
684	Penreco Div. of Pennzoil	Bulter, PA
683	Pennualt Holland B.V.	Rotterdam, Holland
685	Permutit Co.	Paramus, NJ
686	Person & Covey	Glendale, CA
687	Pestcon Systems Inc.	Alhambra, CA
688	Petrochemicals Co. Inc.	Ft. Worth, TX
689	Pfanstiehl Laboratories	Waukegan, IL
690	Pfipharmecs Div., Pfizer Inc.	New York, NY
690a	Pfister	New York, NY
691	Pfizer Inc.	Greensboro, NC
692	PFW	Middletown, NY
693	Pharmacia Fine Chemicals	Piscataway, NJ
694	Philadelphia Quarts Co.	Augusta, GA
695	Phillips Chem. Co.	Stow, OH
696	Pigment Dispersions Inc.	Edison, NJ
696a	Pillar International Co.	Taipei, Taiwan
697	Pilot Chem. Co.	Sante Fe Springs, CA
698	See 680	Newark, NJ
699	Poco Graphite Inc.	Houston, TX
699a	Polaroid Corp.	Cambridge, MA
700	Polyfil DL/Sub. Reed International PLC	London, UK
701	Polymel	Baltimore, MD
701a	Polymer Applications	Tonawanda, NY
702	Polyvinyl Chem. Industries	Wilmington, MA
703	WM. P. Poythress & Co., Inc.	Richmond, VA
704	PPF International	The Netherlands
705	P.P.G. Industries Inc.	Pittsburgh, PA
706	The PQ Corp.	Valley Forge, PA
707	Premier Malt Products Inc.	Milwaukee, WI
708	Prochinire Avenches SA	Basel, Switzerland
709	Procter & Gamble Co.	Cincinnati, OH
710	Proctor Chem. Co.	Salisbury, NC
711	Produits Chimiques de la Montagne Noire	Cedex, France
712	Produits Chimiques Ugine Kullmann (France)	Paris, France
713	Protameen Chem., Inc.	Totowa, NJ
714	Purdue Frederick	Norwalk, CT
715	PVO Int'l.	Boonton, NJ
716	Quaker Oats Co.	Chicago, IL
717	RAM Chem., Div. Whittaker Corp.	Gardena, CA
718	RAM Laboratories	Miami, FL
719	Reed & Carnick	Piscataway, NJ
719a	Reed Lignin, Inc.	Greenwich, CT
720	Reed Ltd. Chemical Div.	Quebec, Canada
721	Reheis Chem. Co.	Berkeley Heights, NJ
722	Reichhold Chem. Inc.	White Plains, NY
723	Reid-Provident, Direct Div.	Atlanta, GA
723a	Reid Provident Labs.	Atlanta, GA
724	Reilly-Whiteman Inc.	Conshohocken, PA
725	Rewo Chemical Group	Morden Surrey, UK

725a Rewo Chemische Werke GmbH .... West Germany
726 Rexar Phar. Corp. .... Valley Stream, NY
727 Rexolin Chem. AB .... Helsingborg, Sweden
728 Rhein Chemie .... West Germany
730a Rhone-Poulenc Inc., Agrochemical Div. .... Monmouth Junction, NJ
729 Rhode Island Laboratories Inc. .... West Warwick, RI
730 Rhone-Poulenc Inc. .... Paris, France
731 Riceland Foods, Sub. Witco Chem. Corp., Organics Div. .... New York, NY
732 The Richardson Co. .... Des Plains, IL
733 Riken Vitamin Oil .... Tokyo, Japan
734 Riker Lab., Inc. .... Northridge, CA
735 R.I.T.A. Corp. .... Crystal Lake, IL
736 Robeco Chem., Inc. .... New York, NY
737 A.H. Robins Co., Pharmaceutical Division .... Richmond, VA
738 Roche Labs., Div. of Hoffmann-La Roche Inc. .... Nutley, NJ
739 Roerig, Div. of Pfizer Phar. .... New York, NY
740 Rogers Corp. .... Rogers, CT
741 Rohm & Haas (Australia) Pty. Ltd. .... Victoria, Australia
742 Rohm & Haas Co. .... Philadelphia, PA
743 Rohm & Haas France SA .... Philadelphia, PA
744 Rond Pearl .... Bayonne, NJ
745 Ronsheim & Moore .... West Yorkshire, UK
746 W.H. Rorer, Inc. .... Fort Washington, PA
747 Ross Labs., Div. Abbott Labs. .... Columbus, OH
748 Rotex Phar., Inc. .... Orlando, FL
749 Roussel Uclaf. .... Paris, France
750 Rowell Labs. Inc. .... Baudette, MN
751 Roxane Labs., Inc. .... Columbus, OH
752 H.M. Royal, Inc. .... Trenton, NJ
753 Royce Chem. Co. .... E. Rutherford, NJ
754 Ruetgers Nease Chem. Co. .... State College, PA
755 Rumford Chemical Works, Essex Chem. Corp. .... Clifton, NJ
756 Ryco, Inc. .... Conshohocken, PA
757 Rydelle Labs., Inc. .... Racine, WI
758 Ryoto Co., Ltd. .... Tokyo, Japan
759 Rystan Co., Inc. .... Little Falls, NJ
760 Salomon & Bros. .... Port Washington, NY
760a Sandoz, Agro Div. .... San Diego, CA
761 Sandoz Colors & Chemicals .... East Hanover, NJ
762 Sandoz Ltd. .... Basel, Switzerland
762a Sandoz Ltd., Agro Division .... Basle, Switzerland
763 Sankyo Co. Ltd., Agrochemicals Planning Dept. .... Tokyo, Japan
764 Sanyo Chem. Ind. Ltd. .... Kyoto, Japan
764a Sardesia Brothers Ltd. .... India
765 Saron Phar. Corp. .... St. Petersburg, FL
766 Sartomer Co. .... West Chester, PA
767 Savage labs. .... Melville, NY
767a Saykeck Inc. .... East Brunswick, NJ
768 Santech Inc. .... Philadelphia, PA
769 Henry Schein, Inc. .... Port Washington, NY
770 Schenectady Chem. Inc. .... Schenectady, NY
771 Scher Chemicals, Inc. .... Clifton, NJ
772 Schering AG, Agrochemical Div. .... Berlin, West Germany
773 Schmid Products Co. .... Little Falls, NJ
774 J.C. Schumacher Co. .... West Germany

775	Schuylkill Chem. Co.	Philadelphia, PA
776	Schwarz/Mann	Orangeburg, NY
777	Scientific Chemicals, Inc.	Huntington Beach, CA
778	Sclavo Inc.	Wayne, NJ
779	Scot-Tussin Phar. Co., Inc.	Cranston, RI
780	SDS BioTech Corp.	Painesville, OH
781	Seamless Hospital Products Co.	Wallingford, CT
782	Searle & Co.	San Juan, Puerto Rico
783	Searle Pharmaceuticals, Inc.	Chicago, IL
783a	Secol, Inc.	Newark, NJ
784	Sederma	Meudon, France
785	SEPPIC	Paris, France
786	Seres Labs., Inc.	Santa Rosa, CA
787	Serono Labs., Inc.	Randolph, MA
788	Servo BV	Delden, Holland
789	Shamokin Filler	Shamokin, PA
790	Shawinigan Products, Sub. Lavalin Inc.	Montreal, Canada
791	Shell Chemicals Co.	Houston, TX
792	Shell Chemicals UK Ltd., Shell Chimie France	London, UK
793	Shell Chimie France	Paris, France
794	Shell Int'l Chem. Co. Ltd.	London, UK
795	Shell Nederland Chemie N.V	The Hague, The Netherlands
796	Sherex Chemical Co.	Dublin, OH
797	Sherwin Williams Chemicals	Cleveland, OH
798	Shin-Ei Chem. Co., Ltd.	Osaka, Japan
798a	Shinung Corp	Taichung, Taiwan
799	See 800	Wayne, NJ
799a	Shroffs Industrial Chemicals Pvt. Ltd.	Bombay, India
800	Shulton Inc.	Wayne, NJ
801	Richardson Co.	Des Plains, IL
802	Silas Flotronics	New York, NY
802a	J.R. Simplo Co.	Pocatellow, ID
803	Siror Kao, S.A.	Barcelona, Spain
804	E.W. Smith Chemical Co.	Industry, CA
804a	Smith Kline Corp.	Philadelphia, PA
805	Smith Kline & French	Philadelphia, PA
806	Smith Labs Inc.	Northbrook, IL
807	Soitem-Societa Ital. Emulsionanti	Milano, Italy
808	Solem Industries Inc.	Norcross, GA
809	Solvay	Brussels, Belgium
810	Southeastern Clay	Aiken, SC
811	Southland Corp. Chem. Div.	Argo, IL
812	Spaulding Fibre Co., Inc.	Buffalo, NY
813	Spencer-Kellogg	Buffalo, NY
814	Springbok	Houston, TX
815	Squibb/Connaught	Princeton, NJ
816	Squibb-Novo	Princeton, NJ
817	E.R. Squibb & Sons Inc.	Princeton, NJ
818	Staley Chemical Co.	Decatur, IL
819	Standard Process Labs., Inc.	Milwaukee, WI
820	Star Pharmaceut., Inc.	Pompano Beach, FL
821	Stauffer Chemical Co.	Weston, CT
822	Staveley Chemical	Derbyshire, UK
823	F. Steinfels AG	Zurich, Switzerland
824	Stein Hall & Co. Inc., Sub. Celanese Corp.	New York, NY

825 Stepan Chemical Co. .... Northfield, IL
825a Stepan Europe .... Voreppe, France
826 Stevenson Bros. & Co. .... Philadelphia, PA
827 Stiefel Labs., Inc. .... Coral Gables, FL
828 St. Joe Resources .... New York, NY
829 Stokely-Van Camp, Inc. .... Janesville, WI
830 Strahl & Pitsch .... West Babylon, NY
831 Stuart Pharmaceut. Div., of ICI Americas Inc. .... Wilmington, DE
832 Sumitom Chemical Americal Inc. .... New York, NY
833 Summit Chemical .... Osaka, Japan
834 Sunkist Growers .... Ontario, CA
835 Sun Petroleum .... Philadelphia, PA
836 Sutton Labs .... Chathan, NJ
837 Swastik Household & Industrial Products, Div. of Ambalal Sara Bnai Ent. Ltd. .... Bombay, India
838 Sweden Inc. .... New York, NY
840 Swift Chemical .... Guelph, Ont., Canada
839 Sween .... Lake Crystal, MN
841 SWS Silicones .... Adrian, MI
842 Sybron Corp. Chem. Div. .... Wellford, SC
843 Sylacauga Calcium Prod. .... Sylacauga, AL
844 Syntex Labs., Inc. .... Palo Alto, CA
845 Synthetic Products Co. .... Racine, WI
846 Tac Industries .... Dalton, GA
847 Taiyo Kagaku .... Yokkaichi, Mie-Pref., Japan
848 Takeda Chemicals Industries .... Tokyo, Japan
849 Takemoto Oil & Fat Co. .... Gamagori Aichi, Japan
850 Tenneco Chemicals Inc. .... Saddle Brook, NJ
851 Tenneco Chemicals Inc. Intermediates Division .... Piscataway, NJ
852 Tennessee Corp. .... Atlanta, GA
853 Terry Corp. .... Satellite Beach, FL
854 Texaco Chemical Co. .... Austin, TX
855 Texo Corp. .... Cincinnati, OH
855a Theodor Leonhard Wax Co. .... Haledon, NJ
856 Th. Goldschmidt AG .... West Germany
857 Thioleol Chemical Corp. .... Chicago, IL
857a Thomas Alabama Kaolin Co. .... New York, NY
858 Thomas Triantaphyllou S.A. .... Athens, Greece
859 Thompson-Hayward Chemical Co. .... Kansas City, KS
860 Thompson Medical .... New York, NY
861 Thompson, Weinman & Co. .... Cartersville, GA
862 3 (Three) M/ Commercial Chem. Div. .... St. Paul, MN
862a (3) Three V Chem. .... St. Paul, MN
863 Titan Chem. Prod. Inc. .... Short Hills, NJ
864 Toho Chem. Industry Co. .... Tokyo, Japan
865 Tokai Seiyu Ltd. .... Nagoya, Japan
866 Toman Products Inc. .... Milton, WI
867 Tri-K Ind. .... Westwood, NJ
868 Trimen Labs., Inc. .... Pittsburgh, PA
869 Troy Chemical Corp. .... Newark, NJ
870 Tuco/Div. of Upjohn Co. .... Kalamazoo, MI
871 Tulco, Inc. .... No. Billerica, MA
872 Tyson and Assoc., Inc. .... Santa Monica, CA
873 U.C.T. Inc. .... Louisville, KY
874 Unger Fabrikker AS .... Fredrikstad, Norway

875 Unichema Internat. .... West Germany
876 Union Camp Corp. .... Wayne, NJ
877 Union Carbide Agricultural Products Co. .... Research Triangle Park, NC
878 Union Carbide Corp. .... Danbury, CT
879 Union Carbide Corp. Ethylene Oxide Derivatives Div. .... Danbury, CT
880 Union Carbide Corp., Linde Division .... Danbury, CT
881 Uniroyal Chemical Div. of Uniroyal Inc. .... Naugatuck, CT
882 Universal Chemicals .... Cincinnati, OH
882a Universal Oil Products Co. .... Des Plaines, IL
883 UOP .... Des Plaines, IL
884 Upjohn Co. .... Kalamazoo, MI
885 Upjohn Polymer Chem. .... LaPorte, TX
887 U.S. Borax & Chemical Corp. .... Los Angeles, CA
888 U.S. Chemical Marketing Group, Inc. .... Decatur, GA
889 U.S. Industrial Chemicals .... New York, NY
890 U.S. Peroxygen Div./Witco .... Richmond, CA
891 USV Pharmaceutical .... Tarrytown, NY
886 Upsher-Smith Labs., Inc. .... Minneapolis, MN
892 Valchem (Australia) Pty. Ltd. .... Victoria, Australia
893 Valchem Chem. Div. of United Merchants & Manufacturers Inc. .... Langley, SC
894 B.T. Vanderbilt Co. Inc. .... Norwalk, CT
895 Van Dyk & Co. .... Belleville, NJ
896 Van Schuppen Chemie .... The Netherlands
897 V-C Chemical Co. .... Richmond, VA
898 Veisicol Chemical Corp. .... Chicago, IL
899 Ventron Corp. .... Danvers, MA
900 Verex Labs., Inc. .... Englewood, CO
901 Vevy .... New York, NY
902 Vicks Pharmacy Products, Richardson-Vicks Inc. .... Wilton, CT
903 Vikon Chemical Co. Inc. .... Burlington, NC
904 Vineland Chem. Co. .... Vineland, NJ
905 Viobin Corp. .... Monticello, IL
906 Virginia Chemical .... Portsmoth, VA
907 Vitaline Formulas .... Incline Village, NV
908 Vitamins, Inc. .... Chicago, IL
909 Vulcan Chemicals .... Birmingham, AL
910a Wacker-Chemie GmbH .... Munchen, West Germany
910 Vulnax, Sub. Imperial Chemical Ind. PLC .... London, UK
911 Walker, Corp. .... Syracuse, NY
911a Wallace Labs. .... Cranbury, NJ
912 Wallace & Tierman Turnan Inc., Harchem Div. .... Belleville, NJ
913 Wallerstein Co. .... Kingstree, SC
914 Washington Penn Plastic Co. Inc. .... Washington, PA
915 Webcon Phar. .... Humacao, Puerto Rico
916 Werner G. Smith, Inc. .... Cleveland, OH
917 Westinghouse Electric Corp., Micarta Div. .... Hampton, SC
917a Westvaco-Oleochemical Div. .... N. Charleston Heights, SC
918 Westvaco Polychemicals .... Charleston Heights, SC
919 West Virginia Pulp & Paper Co. .... Covington, VA
920 Westwood Phar. .... Buffalo, NY
921 Wharton Labs., Inc. .... Long Island City, NY
922 Whitehall Labs., Inc. .... New York, NY
923 Whitestone Chemical .... Spartanburg, SC
924 Whittaker/Clark & Daniels .... South Plainfield, NJ
925 See 926 .... Huguenot, NY

926	Wickhen Products Inc.	Huguenot, NY
927	Wilbur B. Driver Co.	Newark, NJ
928	Wilkinson Minerals	Gordon, GA
929	Willen Drug Co.	Baltimore, MD
930	Williams Ltd.	Middlesex, UK
931	Wilmington Chem. Corp.	Wilmington, DE
932	Winthrop Laboraties Sterling Drug Co.	New York, NY
933	Wisconsin Alumni Research Fdn.	Madison, WI
934	Witco Chem.	Brooklyn, NY
935	Witco Chem. Co. Organics Division	New York, NY
936	Witco Chem. Corp. Sonneborn Division	New York, NY
937	Witco Chem. Ltd., Israel	Haifa, Israel
937a	Witco Chemical S.A.	Paris, France
938	Witco Chem. UK	Worcester, UK
939	Witcolcyclo Div.	Paris, France
940	Witco Golden Bear Div.	Los Angeles, CA
941	Woburn Chem. Corp.	Paterson, NJ
942	Wyeth Labs.	Philadelphia, PA
943	XCEL Corp.	Newark, NJ
944	Yoshimura Oil Chem. Co. Ltd.	Osaka, Japan
945	Youngs Drug Products Corp.	Piscataway, NJ
945a	Yuen Fa Chemical Co.	Taipei, Taiwan
946	Zohar Detergent Factory	Kibbutz Dalia, Israel
947	Zocon Corp.	Palo Alto, CA
948	Zschimmer & Schwarz	Lahnstein, West Germany

Chemical Manufacturers' Addresses

Aarhus Oliefabrik
Als, M.P. Bruunsgade 27
DK-8100 Arhus C
West Germany

Abbott Diagnostics Div.
1400 Sheridan Rd.
North Chicago, IL 60064

Abbott Laboratories
14th & Sheridan
North Chicago, IL 60064

AC & C
American Lane
Greenwich, CT 06830

Aceto Chem. Co.
126-02 Northern Blvd.
Flushing, NY 11368

Acme-Hardesty, Inc.
Suite 910 Benjamme Fox Pavillion
Jenkitown, PA 19046

Adria Labs, Inc.
5000 Post Rd.
Dublin, OH 43017

Advanced Refractory Technologies
40 W. 57th St.
New York, NY 10019

Aeropres
1108 Petroleum Tower
Shreveport, LA 71106

Agrashell
275 Keystone Dr.
Bethlehem, PA 18017

Agri Products Marketing Corp.
2813 Hangar Way
Bakersfield, CA 93308

Agvar Chemicals Inc.
1 Lincoln Plaza
New York, NY 10023

Agway Inc.
P.O. Box 4741
Syracuse, NY 13721

Air Products & Chemicals, Inc.
P.O. Box 538
Allentown, PA 18105

Ajinomoto U.S.A., Inc.
9 West 57th St. #4625
New York, NY 10019

Akron Chemical
255 Fantain St.
Akron, OH 44304

Akzo Chemie
2153 Lockport-Olcott Rd.
Burt, NY 14028

Akzo Chemie GmbH
6800 LS Arnhem
Velperweg 76, Postbus 186
Netherlands

Akzo Chemie Italia
6800 LS Arnhem
Velperweg 76, Postbus 186
Netherlands

Alna Protection Division
86 SE Marine Dr.
Vancouver, BC V5X 4P8

Alba Chemical Co.
285 Canty Rd.
Tenafly, NJ 07670

Alban Muller Int'l.
19 Rue St. Just
93100 Montrevill
Paris, France

Albany International Corp.
110A St.
Needham Heights, MA 02194

Albermar Co.
51 Madison Ave.
New York, NY 10011

Albright & Wilson
475 Walnut St.
Norwood, NJ 07648

Albright & Wilson (Australia)
610 St. Kilda Rd.
Box 4544
Melbourne 3001, Australia

Albright & Wilson Detergent Div.
P.O. Box 15
White Haven, Cumberland
CA28 9QQ, UK

Albright & Wilson/Marchon Div.
Whitehaven, Cumbria
CA28 9QQ, UK

Alcoa (Closure Systems Int'l.)
809 Dillion Dr.
Richmond, IN 47374

Alco Chemical
909 Mueller Dr.
Chattanooga, TN 37406

Alcolac Inc.
3440 Fairfield Rd.
Baltimore, MD 21226

Alconox Inc.
215 Park Ave. So.
New York, NY 10003

Alcon (P.R.) Inc.
P.O. Box 1959
6201 South Free Way
Fort Worth, TX 76134

Aldrich Chemical Co. Inc.
940 W St. Paul Ave.
Milwaukee, WI 53233

Alkaril Chemicals Inc.
P.O. Box 1010
Winder, GA 30680

Alkaril Chemicals Ltd.
Mississauga, Ontario, Canada

Alkaril Chemicals U.K.
Southbank Daveylands
Winslow Cheshire
SK9 2AG, UK

Allied Chemical Corp.
P.O. Box 2064R
Morristown, NJ 07960

Allied Chem., Fibers & Plastics Co.
P.O. Box 2332 R
Morristown, NJ 07960

Allied Colloids Ltd.
P.O. Box 38
Low Moor, Bradford
BD12 0JZ, Yorkshire, UK

Alto Pharmaceuticals, Inc.
15509 Cassey Rd Ext.
Tampa, FL 33624

Aluminum Co. of America
1501 Alcoa Bldg.
Pittsburgh, PA 15219

Alza Corp.
950 Page Mill Rd.
Palo Alto, CA 94304

Alzo, Inc.
6 Gulfstream Blvd.
Matawan, NJ 07744

Amchem Products Inc.
300 Brookside Ave.
Ambler, PA 19002

Amerace Corp.
555 Fifth Ave.
New York, NY 10017

Amerchol Corp.
International Plaza
Englewood Cliffs, NJ 07632

Amerchol Europe
International Plaza
Englewood Cliffs, NJ 07632

American Agricultural Chemical Co.
One Williams Center
Tulsa, OK 74103

American Can Co.
American Lane
Greenwich, CT 06830

American Colloid Co.
5100 Suffield Ct.
Skokie, IL 60077

American Color & Chem. Corp.
100 E 42nd St.
New York, NY 10017

American Critical Care
Div. Am. Hosp. Supply
McGraw Park, IL 60085

Amer. Cyanamid Co.
One Cyanamid Plz.
Wayne, NJ 07470

American Cyanamid Chemical Prod. Div.
One Cyanamid Plz.
Wayne, NJ 07470

American Cyanamid Co. Fine Chemicals Dept.
One Cyanamid Plz.
Wayne, NJ 07470

American Cyanamid Plastic Add. & Textile Chem. Dept.
One Cyanamid Plz.
Wayne, NJ 07470

American Cyanamid Co./Resins Dept.
One Cyanamid Plz.
Wayne, NJ 07470

American Dermal Corp.
12 Worlds Fair Dr.
Somerset, NJ 08873

American Enzyme Corp.
P.O. Box 23441
Brown Deer, WI 53223

American Hoechst Corp.
1041 Rt 202-206 N
Somerville, NJ 08876

Amer. Hoechst Corp./Plastics Div.
1041 Rt 202-206 N
Somerville, NJ 08876

American Lecithin Co.
451-67 Stephens St. SW
Atlanta, GA 30310

American Metal Climax, Inc.
AMAX Center
Greenwich, CT 06836

American Pelletizing Corp.
P.O. Box 3628
Des Moines, IA 50334

American Potash & Chemical Corp.
Div. of Kerr-McGee
P.O. Box 25861
Oklahoma City, OK 73125

American Rubber and Chemical Co.
Box 1034
Louisville, KY 40201

Amer. Smelting & Refining Co.
120 Broadway
New York, NY 10271

American Synthetic Rubber
575 Technology Sq.
Cambridge, MA 02139

Ameripol Inc.
3135 Euclid Ave.
Cleveland, OH 44115

Ames Laboratories Inc.
1127 Myrtle St.
P.O. Box 70
Elkhart, IN 46515

Amoco Chemicals
200 E Randolph Dr.
P.O. Box 5910a
Chicago, IL 60680

Anheuser-Busch
One Busch Pl.
St. Louis, MO 63118

Ansul Co.
1578 Sydney, NSW
2001, Australia

Applied Science Laboratories
One Plaza Pl. N.E.
St. Petersburg, FL 33701

Arapahoe Chemicals Inc.
7409 S Alton Ct.
Englewood, CO 80155

ARC Chem. Division Balchem Corp.
Box 180
State Hill, NY 10973

Archer Daniels Midland Co.
4666 Faries Pky
Decatur, IL 62526/5

Acheson Colloids Co.
Port Huron, MI

Arco Chem. Co.
515 S. Flower St.
Los Angeles, CA 90051

Ardmore Chemical Co.
29 Riverside Ave.
Newark, NJ 07102

Argus
366 Cart St.
Brooklyn, NY 11231

Argus Chem. Div./Witco Chem. Corp.
520 Madison Ave.
New York, NY 10022

Arjay, Inc.
P.O. Box 33387
Houston, TX 77233

Arkansas Co. Inc.
185 Fandry St. Box 210
Newark, NJ 07101

Arlo Corp.
212 Mayaguez St.
Hato Rey, PR 00919

Armak Ind. Chem. Div.
Pack Sq.
Asheville, NC 28802

Armour
303 S. Broadway
Tarrytown, NY 10591

Armstrong Chem.
1530 S. Jackson St.
Janesville, WI 53545

Arol Chem. Prod.
649 Ferry St.
Newark, NJ 07105

Arthur H. Thomas Co.
Philadelphia, PA

Asarco
120 Broadway
New York, NY 10271

Asbury Graphite Mills
Main St. & RD #1
Asbury, NJ 08802

B.F. Ascher & Co.
15501 W. 109 St.
Lenexa, KS 66219

Ashland Chem. Co.
1409 Winchester Ave.
Ashland, KY 41101

Associated Lead Inc.
14 Gresham St.
London, EC2V 7AT, UK

Astor Chemical Durachem
5050 Edgewood CT.
Jacksonville, FL 32203

Astra Pharmaceutical
50 Otis St.
Westboro, MA 01581-4428

Atlantic Powdered Metals
225 Broadway
New York, NY 10007

Atlas Chemical Industries, N.V.
Imperial Chemical House
Milbank, London, SW1P 3JF, UK

Atlas Chemical Industries (UK) Ltd.
Imperial Chemical House
Milbank, London, SW1P 3JF, UK

Atlas Refinery Inc.
142 Lockwood St.
Newark, NJ 07105

Atramax Inc.
35 N. Straight St.
Paterson, NJ 07824

Australia Ltd.
1000 S. Fremont Ave.
P.O. Box 400C
Alhambra, CA 91802

Ayerst Labs.
685 Third Ave.
New York, NY 10017

Babcock & Wilcox Co.
P.O. Box 60035
New Orleans, LA 70160

Badische Corp.
Off Hwy #60 E
Williamsburg, VA 23185

Baker Castor Oil Co.
1230 Ave. of the Americas
New York, NY 10020

Baker Chemical
2200 Post Oak Blvd.
Houston, TX 77056

Bareco Div. Petrolite
6910 E. 14th St.
Tulsa, OK 74112

Barium Chemicals
P.O. Box 218
City Rd. 44
Steubenville, OH 43952

Barnes-Hind/Hydrocurve
Sub. Revlon, Inc.
767 Fifth Ave.
New York, NY 10153

Barry Labs, Inc.
461 NE 27th St.
P.O. Box 1967
Pompano Beach, FL 33061

A.J. Bart, Inc.
P.O. Box 813
Gurabo, PR 00658

BASF AG
D-6700 Ludwigshafen/Rhein
West Germany

BASF Aktiengeselschaft
Ludwigshafen, West Germany

BASF India
Maybaker House
Sudam Kalu Ahire Marg
Bombay 400 025 India

BASF Wyandotte Corp.
100 Cherry Hill Rd.
Parsippany, NJ 07054

Bate Chemical
1210 Shappard Ave. East
Suite 113
Willowdale, Ontario
M1K 1E3 Canada

Bay Pharmaceuticals, Inc.
1111 Francisco Rd.
San Rafael, CA 94901

Bayer AG
Div. KL D-5090
Leuerkusen, West Germany

Beach
5220 S. Manhattan Ave.
Tampa, FL 33611

Beecham Labs
501 5th St.
Bristol, TN 37620

Berlex Labs, Inc.
Cedar Knolls, NJ 07927

Berol Kemi A.B.
S-44401 Stenungsund
Sweden

Beutlich, Inc.
7006 N. Western Ave.
Chicago, IL 60645

BFC Chemicals, Inc.
4311 Lancaster Pike
Wilmington, DE 19805

Biocraft Labs, Inc.
92 Route 46
Elmwood Park, NJ 07407

Bio-Lab Inc.
P.O. Box 1489
Decatur, GA 30030

Biophil Chemical Fine S.A.S.
Via Oslavia 7
Milano, Italy

Bio Systems Research Inc.
P.O. Box 1037
Salida, CO 81201

Black Leaf Products Co.
Elgin, IL

Blew Chemical Co.
P.O. Box 501
Palos Heights, IL 60463

Bock
132 W. Providencia
Burbank, CA 91502

Boehringer Ingelheim
90 East Ridge
P.O. Box 368
Ridgefield, CT 06877

Bofors Lakeway
5025 Evanston Ave.
Muskegon, MI 49443

Boots
1 Thane Rd.
Nottingham NG2 3AA, UK

Borden Chemical
180 E. Broad St.
Columbus, OH 43215

Borg-Warner Chem. Inc.
International Center
Parkersburgh, WV 26101

Bostik South
P.O. Box 5695
Greenville, SC 29606

Bowman
11710 Sunset Hills Rd.
Herndon, VA 22090

Boyle
13260 Moore St.
Cerritos, CA 90701

Breon Laboratories
90 Park Ave.
New York, NY 10016

Bristol-Myers
345 Park Ave.
New York, NY 10022

Bromine Producers Div./Drug Research Inc.
Adrian, MI

R.J. Brown Co.
St. Louis, MO

Buckman Laboratories Inc.
1256 N. McLean Blvd.
Memphis, TN 38108

Burgess Pigment
P.O. Box 4146
Macon, GA 31208

Burroughs Wellcome Co.
3030 Cornwallis Rd.
RTC, NC 27709

Bush Boake Allen
7 Mercedes Dr.
Montvale, NJ 07645

Cabot Corp.
Box 188
Tuscola, IL 61953

Calgon Corp.
Pittsburgh Activated Carbon Division
P.O. Box 1346
Pittsburgh, PA 15230

California Industrial Minerals Co.
2728 S. Cherry
Fresno, CA 93706

Campbell
300 E 51st St.
New York, NY 10022

Canada Packers, Inc.
5100 Timberlea Blvd.
Mississauga, Ontario
L4W 2S5 Canada

Cancarb
P.O. Box 500
407 8th Ave. SW
Calgary, Alta T2P 2M7

Capital City
1530 S. Jackson St.
Janesville, WI 53545

Cardiovascular Research, Ltd.
1061-B Shary Circle
Concord, CA 94518

Cargill Inc.
208 N. Green St.
Longview, TX 75606-0992

Carl Becker
Hamburg, West Germany

Carlisle Chemical Works Inc.
1700 Dubois Tower
Cincinnati, OH 45262

Carlisle Chemical Works Inc.
Advance Division
1700 Dubois Tower
Cincinnati, OH 45262

Carlton
44 Bromfield St.
Boston, MA 02108

Carnrick
65 Herse Hill Rd.
Cedar Knolls, NJ 07927

R.E. Carroll
Eisenhower Blvd.
Harrisburg, PA 17105

Carson Chem.
2779 East El Presido
Long Beach, CA 90810

Carstab Corp.
West Street
Reading, OH 45215

Carus Chemical Co.
1500 Eighth St.
La Salle, IL 61301

Cas Chem Inc.
40 Avenue A
Bayonne, NJ 07002

Catawba-Charlab
P.O. Box 240497
Charlotte, NC 28224

CaF Chimie
Eaglescliffe Industrial Estate
Stockton-on-Tees
Cleveland TS16 0PN, UK

C.E. Basic Chemicals
845 Hanna Bldg.
Cleveland, OH 44115

Ceca S.A.
11 Avenue Morane Saulnier
BP 66 78141 Velizy Villacablay
Cedex, France

Celamerck GmbH & Co.
P.O. Box 202
6501 Ingelheim
Phem, F.R.G.

Celanese Chemical Co.
1211 Ave. of the Americas
New York, NY 10036

C.E. Minerals
P.O. Box 9308
Stamford, CT 06904

Center
Division of E.M. Industries, Inc.
35 Channel Drive
Port Washington, NY 11050

Central Pharmaceuticals
112-128 East Third St.
Seymour, IN 47274

Central Soya/Chemurgy Div.
Fort Wayne, IN 46802

Central Soya Co., Inc.
P.O. Box 1400
Fort Wayne, IN 46801

Cetylite
P.O. Box CN6
9051 River Rd.
Pennsauken, NJ 08110

Charles Taylor & Sons Co.
Todmorden, UK

Chemax Inc.
P.O. Box 6067
Greenville, SC 29606

Chemform Corp.
Vertac Bridgeport Ind. Park
P.O. Box 1745
W. Memphis, AR 72301

Chemical Dynamics Corp.
3001 Hadley Rd.
S. Plainfield, NJ 07080

Chemical Mfg. Co. Inc.
20 Pine St.
New York, NY 10005

Chemical Products Corp.
Elmwood Park, NJ 07407

Chemische Fabriek Servo B.V.
Delden, Netherlands

Chemische Fabrik Grunau GmbH
Postfach, F.R.G.

Chemische Fabrik Wibarco GmbH
6700 Ludwigshafen/Rhein
West Germany

Chemische Werke Huls AG
Marl, West Germany

Chemithon Corp.
5430 Marginal Way, SW
Seattle, WA 98106

Chemplast, Inc.
150 Dey Rd.
Wayne, NJ 07476

Chemtall Inc.
P.O. Box 250
Riceboro, GA 31323

Chem-y
Fabrik van Chemische Producten BV
P.O. Box 50
2410 AB Bodengraven, Holland
Netherlands

Chevron Chemical Co.
575 Market St.
San Francisco, CA 94105

Chimagro Agricultural
Div. Mobare Chem. Corp.
D-5090
Leverkusen, Bayerwerk
West Germany

Chimex
Gonesse, France

Chipman Chemical Co.
800 Marion
P.O. Box 718
River Rouge, MI 48218

Chromex Chem. Corp.
19 Clay St.
Brooklyn, NY 11222

Chrystal
53 Park Place
New York, NY 10007

CIBA
141, CH-4002
Basel, Switzerland

Ciba-Geigy Co.
P.O. Box 18300
Greensboro, NC 27409

Ciba Geigy Corp.
Agricultural Div.
CH-4002
Basie, Switzerland

Ciba Pharmaceutical Co.
556 Morris Ave.
Summit, NJ 07901

Cincinnati Milacron Chem.
(CARSTAB)
West Street
Reading, OH 45215

Cindet Chemicals
Greensboro, NC

Cities Service Co.
3475 Lenox Rd., NE
Atlanta, GA 30326

W.A. Cleary Corp.
1049 Somerset St.
Somerset, NJ 08873

Climax Molybdenum Co.
1 Greenwich Plaza
Greenwich, CT 06836

Clintwood Chem. Co.
4342 S. Wolcott Ave.
Chicago, IL 60609

Clough Chemical
178 St. Pierre
St. Jean, Quebec J3B 5W4

C & M
207 E. Broadway
Louisville, KY 40202

Colloides Naturels
P.O. Box 561
Route 202 & Dumont Rd.
Far Hills, NJ 07931

Colloids Inc.
394 Frelinghuysen Ave.
Newark, NY 07114

Colorado Chemical
4295 McIntyre St.
Golden, CO 80401

Columbian Chemicals
P.O. Box 47
Tulsa, OK 74102

Commercial Minerals
2808 E 26th St.
Tulsa, OK 74114

Commercial Solvents Corp.
2315 Sanders Rd.
Northbrook, IL 60062

Compagnie Parento
Weslerly Rd.
Ossining, NY 10562

Conoco Chemicals
15990 N. Barkers Landing Rd.
P.O. Box 19029
Houston, TX 77224

Consos Inc.
P.O. Box 34186
Charlotte, NC 28234

Continental Carbon
1007 Market Street
Wilmington, DE 19898

Continental Chemical Co.
270 Clifton Blvd.
Clifton, NJ 07015

Continental Oil
Park-Eighty Plaza East
Saddle Brook, NJ 07662

Contour Chemical
1 Elm Street
Windsor Loeks, CT 06096

Cook Chemical
322 Roayl Poinciania Plaza
Palm Beach, FL 33480

Cooke-Waite
90 Park Ave.
New York, NY 10016

Cooper Care
Dermatology Prods.
3145 Porter Dr.
Palo Alto, CA 94304

Cordova Chem. Co.
500 Agard Rd.
P.O. Box 5150
N. Muskegon, MI 49445

Corning Glass Works
Corning Medical & Scientific
MS-2158
Corning, NY 14831

Corn Products Unit CPC N. Amer.
Englewood Cliffs, NJ 07632

Cosden Oil & Chem. Co.
P.O. Box 2159
3350 N. Central Expressway
Dallas, TX 75221

Cosmedial Dept. of Gen'l. Mills Chem.
4620 W 77th St.
Minneapolis, MN 55435

Costec Inc.
800 E. Northwest Highway
Suite 314
Palatine, IL 60067

Cowles Chemical Co.
Springfield Mill
Maidstone, Kent ME14 2LE, UK

C.P. Chemicals Inc.
Arbor Street
P.O. Box 158
Sewaren, NJ 07077

CPC International
International Plaza
Englewood Cliffs, NJ 07632

Crest Chem. Corp.
One Harbor Plaza
Stamford, CT 06902

Croda Inc.
51 Madison Ave.
New York, NY 10010

Croda Italiana S.R.C.
Mortara (PV) Italy

Croda Surfactants Inc.
Cowich Hall, Snaith
Goole, North Humberside
PN14 9AA, UK

Crodet Chem.

Crompton & Knowles
Dyes & Chemicals Division
P.O. Box 33188
Charlotte, NC 28233

Crosby Chemical
De Ridder, LA

Crowley Chemical
261 Madison Ave.
New York NY 10016

Crown Metro Inc.
P.O. Box 5695
Greenville, SC 29606

Crown Zellerbach
Chemical Products Div.
P.O. Box 4266
Vancouver (Orchards), WA 98662

Crucible Chemical Co.
P.O. Box 6786
Greenville, SC 29606

Crystal Chemical Inter-America
970 Tipton St.
Huntington, IN 46750

Crystal Soap
8th St. & Moyers Rd.
Lansdale, PA 19446

Cutter Biological
1127 Myrtle St.
Elkhart, IN 46514

Cyanamid B.V.
Postbus 1523, 3000 BM Rotterdam
Rotterdam, The Netherlands

Cyclo Chemicals Corp.
7500 N.W. 66th St.
Miami, FL 33166

Cyprus Ind. Minerals Co.
Los Angeles, CA 90017

Dai-ichi Kogyo Seiyaku Co., Ltd.
Miki Building 3-12-1
Nihombashi, Chou-ku
Tokyo 103, Japan

Dainippon
3-3 Nihonbashi-Dori
Chuo-ku
Tokyo, Japan

Dalin
74-80 Marine St.
Farmingdale, NY 11735

Danbury
131 West Street
P.O. Box 296
Danburg, CT 06810

Darling & Co.
1251 West 46th St.
Chicago, IL 60609

Davison Chem. Div.
P.O. Box 2117
Baltimore, MD 21203

Degesch America Inc.
P.O. Box 116
Weyers Cave, VA 24486

Degussa Corp.
Route 46 at Hollister Rd.
Teterboro, NJ 07680

Delmont
P.O. Box AA
Swarthmore, PA 19081

Dermik
1777 Walton Rd.
Bluebell, PA 19422

De Soto, Inc.
Sellers Chemical Div.
P.O. Box 23523
Harahan, LA 70183

Detergents, Inc.
12143 Altamar Pl.
Santa Fe Springs, CA 90670

Devan Chemicals S.A.
34 St. Sauvar Str.
B-9600
Ronse-Renaix, Belgium

Devcon Corp.
Danvers, MA 01923

Dexter Chem. Corp.
845 Edgewater Rd.
Bronx, NY 10474

Diamond Alkali Co.
717 North Harwood St.
Dallas, TX 75201

Diamond Shamrock
(Process Chemicals Div.)
350 Mt. Kemble Ave.
Morristown, NJ 07960

Diamond Shamrock Europe
Chaussee dela Hulpe
185, B1170
Brussels, Belgium

Dista
307 East McCarty St.
Indianapolis, IN 46285

Domtar, Inc./CDC Div.
1136 Matheson Blvd.
Mississauga, Ontario
Canada L4W 2V4

Dorsey Laboratories
(Division of Sandoz, Inc.)
P.O. Box 83288
Lincoln, NE 68501

Dorsey Pharmaceuticals
(Div. of Sandoz, Inc.)
East Hanover, NJ 07936

Dover Chemical
P.O. Box 40
Dover, OH 44622

Don B Hickam Inc.
Houston, TX

Dow Chemical Co.
2020 Bldg.
Dow Center
Midland, MI 48640

Dow Corning Corp.
P.O. Box 0994
Midland, MI 48640

Dragoco
Gordon Drive
Box 261
Totowa, NJ 07511

Drew Chemical Corp.
1409 Winchester Ave.
Ashland, KY 41101

Drew Produtos Quimicos Ltda.
Rua Sampaio Viana, 425
Sao Paulo, 04004
Brazil-CP4885

Drew Produitos Quimicos S.A.
Rua Sampaio Viana, 425
Sao Paulo, 04004
Brazil-CP4885

Drug Industries
3237 Hilton Rd.
Ferndale, MI 48220

Duphar B.V.
3900 AB Veenendaal
(P.O. Box 70)
The Netherlands

Du Pont
Elastomer Chem. Dept.
Wilmington, DE 19898

E.I. Dupont de Nemours & Co. Inc.
1007 Market Street
Wilmington, DE 19898

Dura Commodities Corporation
111 Calvert Street
P.O. Box 618
Harrison, NY 10528

Durkee Industrial Foods Group/SCM Corp.
900 Union Commerce Bld.
Cleveland, OH 44115

Dusa
P.O. Box 1007
Camas, WA 98607

Dutton & Reinisch Ltd.
Crown House, London Rd.
Morden, Surrey
SM4 5DU, UK

Dynamit-Nobel-AG
Postfach 1269
D-5810 Witten (Ruhr)
West Germany

Dynamit-Nobel America
10 Link Dr.
Rockleigh, NJ 07647

Eagle Picner Industries Inc.
P.O. Box 550
Joplin, MO 64802

Eastern Color & Chem. Co.
35 Livingston St.
Providence, RI 02904

Eastman Chemicals
P.O. Box 431
Kingsport, TN 37662

Eastman Kodak Co.
343 St. Street
Rochester, NY 14650

Edward Mendell
Carmel, NY

Edwards Pharmacal
100 East Hale
Osceola, AR 72370-9990

Elanco Products Co.
740 S. Alabama St.
Indianapolis, IN 46285

Elastochem
6700 Ludwigshafen/Rhein
West Germany

Elder
705 E. Mulberry St.
Bryan, OH 43506

Elektrochem Fabrik Kempen GmbH
P.O. Box 21
4152 Kempen, West Germany

Eli Lilly
Indianapolis, IN 46206

Elkins-Sinn
2 Esterbrook Lane
Cherry Hill, NJ 08034

Emery FDAG
1300 Carew Tower
Cincinnati, OH 45202

Emery Ind., Fatty & Dibasic Acids Group
1300 Carew Tower
Cincinnati, OH 45202

Emery Industries, Inc.
Malmstrom Div.
1501 W. Elizabeth Ave.
Linden, NJ 07036

Emery Ind./Personal Care Products
1501 West Elizabeth Ave.
Linden, NJ 07036

Emery PCSG
P.O. Box 628
Mauldin, SC 29662

EM Industries
Plant Protection Div.
5 Skyline Dr.
Hawthorne, NY 10532

Emkay Chem. Co.
319-325 Second St.
Elizabeth, NJ 07206

Emulan
P.O. Box 582
3726 Roosevelt Rd.
Kenosha, WI 53141

Emulsion Systems, Inc.
215 Kent Ave.
Brooklyn, NY 11211

Engelhard Min. & Chem. Corp.
Menlo Park
Edison, NJ 08837

Enjay Chem. Co.
Div. of Humble Oil & Refining Co.
60 W. 49th St.
New York, NY

Enthone, Inc.
P.O. Box 1900
New Haven, CT 06508

Enzyme Development Corp.
New York, NY

Escambia Chemical Co.
Box 538
Allentown, PA 18105

Eschem
30 North LaSalle Street
Chicago, IL 60602

Essential Chemicals Corp.
2839 Essential Rd.
Merton, WI 53056-0012

ETC-Eurames Textiel Centrum
Subsidiary of Ames Textile Corp.
Heerde, Hagestraat 5
Eurag-Shoes, Belgium NV, SA

Ethyl Corp.
451 Florida Blvd.
Baton Rouge, LA 70801

Evans Chemetics Inc.
90 Tokeneke Rd.
Darien, CT 06820

Everett
76 Franklin Street
East Orange, NJ 07017

Excel Industries
184-87 S.V. Road
Jogeshwari, Bombay 400102
India

Exxon Chemical Americas
P.O. Box 3272
Houston, TX 77001

Exxon Corp.
P.O. Box 2180
Houston, TX 77001

Fairmont Chemical Co.
117 Blanchard St.
Newark, NJ 07105

Fairmount Granite Ltd.
Sub. of Nortek Inc.
815 Reservoir Ave.
Cranston, RI 02910

Fanning
3117 No. Clybourn Ave.
Chicago, IL 60618

Farben Fabriken Bayer Ag
Div. KL
D-5090 Leverkusen
West Germany

Farmoplant S.P.A.
Piazza Della Republical 6
20124 Milano, Italy (02 63331)

Feldspar Corp.
P.O. Box 99
Spruce Pine, NC 28777

Felton
599 Johnson Ave.
Brooklyn, NY 11237

Fermco Biochemists
2020 Lunt Ave.
Elk Grove Village, IL

Ferndale
780 W. Eight Mile Rd.
Ferndale, MI 48220

Ferro Corp. Chem. Div.
7050 Crick Rd.
Bedford, OH 44146

Fiberfil, Inc.
2267 West Mills Rd.
Evansviille, IN

Fine Organics Inc.
205 Main St.
Lodi, NJ 07644

Fike Chemicals Inc.
P.O. Box 546
Nitro, WV 25143

Fine Organics
La Porte Industries (Holdings) PLC
Hanover House
14 Hanover Square
London W1R 0BE, UK

Finetex Inc.
418 Falmouth Ave.
P.O. Box 216
Elmwood Park, NJ 07407

Firestone Synthetic Fibers Co.
1200 Firestone Pky.
Akron, OH 44317

Firestone Synthetic Rubber and Latex Co.
P.O. Box 2786
Firestone Park Sta.
Akron, OH 44301

Firmenich
P.O. Box 5880
Princeton, NJ 08540

Fisons Pharmaceutical Div.
Two Preston Court
Bedferd, MA 01730

Fleet
4615 Murray Pl.
Lynchburg, VA 24506

Fleming
1600 Fenpark Dr.
Fenton, MO 63026

Flint
Division of Travenol Lab. Inc.
Deerfield, IL 60015

Flintkote
Executive Plaza IV
11350 McCormick Rd.
Hunt Valley, MD 21031

Florida Department of Citrus
P.O. Box 148
Lakeland, FL 33802

Floridin Co.
Three Penn Center
Pittsburgh, PA 15235

Fluoritab
P.O. Box 381
Flint, MI 48501

FMC Agricultural Chemical Group
2000 Market St.
Philadelphia, PA 19103

F.M.C. Corp.
2000 Market Street
Philadelphia, PA 19103

Foote Mineral Co.
Sub. Newmont Mining Corp.
200 Park Ave.
New York, NY 10166

Forest
919 Third Ave.
New York, NY 10022

Formosa Chem. & Fibre Corp.
1 Nanking East Road
Sec. 2, Taipei
Taiwan, China

Freeman
Sub. of H.H. Robertson Co.
222 E. Main Street
Port Washington, WI 53074

Freeport Kaolin
Gordon, GA 31031

Fritzsche Dodge & Olcutt Inc.
76 Ninth Ave.
New York, NY 10011

GAF Corp., Chemical Products
140 W. 51st St.
New York, NY 10020

GatteFosse Ets.
36 Chemm De Genas
BP 603
69800 St. Priest, France

Geigy
Div. of CIBA-GEIGY Corp.
Ardsley, NY 10502

General Electric
Silicone Prod. Dept.
Waterferd, NY 12188

General Electric
Copolymers Product Dept.
3135 Easton Tpk.
Fairfield, CT 06431

General Tire & Rubber Co.
Sub. Gencorp Inc.
1 General St.
Akron, OH 44329

Geneva (Generics)
2599 W. Midway Blvd.
Broomfield, CO 80020

Genstar Stone Products
Executive Plaza IV
Hunt Valley, MD 21031

Geoliquids Inc.
3127 W. Lake St.
Chicago, IL 60612

Georgia Kaolin Co.
1185 Mary St.
Elizabeth, NJ 07207

Georgia Marble
Industrial Sales Div.
2575 Cumberland Parkway, NW
Atlanta, GA 30339

Georgia-Pacific/Bellingham Div.
P.O. Box 1236
Bellingham, WA 98225

Geriatric
149 Covert Ave.
P.O. Box 1098
New Hyde Park, NY 11040

Germany Exploration Co.
8111 Preston Rd.
Dallas, TX 75225

Gilbert
31 Fairmount Ave.
Chester, NJ 07930

Givaudan Corp.
100 Delawanna Ave.
Clifton, NJ 07014

Glaxo
Five Moore Drive
Research Triangle Park, NC 27709

Glenbrook
Div. of Sterling Drug
90 Park Ave.
New York, NY 10016

Glenwood
83 North Summit St.
Tenafly, NJ 07670

Glidden/Durkee
Div. SCM Corp.
299 Park Ave.
New York, NY 10171

Glyco Inc.
P.O. Box 700
51 Weaver St.
Greenwich, CT 06830

Glycol Chem.
51 Weaver St.
Greenwich, CT 06830

Goldschmidt Chem. Corp.
P.O. Box 1299
Hopewell, VA 23860

B.F. Goodrich Chem. Co.
6100 Oak St. Blvd.
Cleveland, OH 44131

Goodyear Tire & Rubber Co.
1144 East Market St.
Akron, OH 44316

W.R. Grace & Co.
Organic Chem. Div.
55 Hayden Ave.
Lexington, MA 02173

W.R. Grace/Construction Products Div.
Grace Plaza
1114 Ave. of the Americas
New York, NY 10036

W.R. Grace & Co./Davison Chemical Division
P.O. Box 2117
Baltimore, MD 21203

W.R. Grace
Organic Chem. Div.
55 Hayden Ave.
Lexington, MA 02173

Graden Chemical
426 Bryan St.
Havertown, PA 19083

Grain Processing
1600 Oregon Street
Muscaline, IA 52761

Grant Chem. Div./Ferro Corp.
7050 Crick Rd.
Bedford, OH 44146

Gray
Affiliate the Purdue Frederick Co.
100 Connecticut Ave.
Norwalk, CT 06854

Great Lakes Chem. Corp.
P.O. Box 2200
W. Lafayette, IN 47906

Greeff
1445 East Putnam Ave.
Old Greenwich, CT 06870

Greenwood Chemical Co.
P.O. Box 26
Greenwood, VA 22943

Grefco
3435 W. Lomita Blvd.
Torrance, CA 90509

Grinsted Prod.
Grindastedvaerket A/S
38 Edwin Rahrs Vej
DK-8220 Brabrand
Denmark

A. Gross & Co.
Millmaster Onyx Group
P.O. Box 818
Newark, NJ 07101

Grunau
7918 Jllertissen
Postfach 120
F.R.G. 07303/13-1

Guardian
A Div. of United Guardian, Inc.
230 Marcus Boulevard
P.O. Box 2500
Smithtown, NY 11787

Gujarat Pesti-Chem Industries
111 Nandesari Industrial Estate
Nandesari, Baroda, India

Gulf Oil Chemicals Co.
P.O. Box 3766
Houston, TX 77001

G & W Laboratories, Inc.
111-T Coolidge St.
South Plainfield, NJ 07080

Haarman & Reimer
11 U.S. Highway 22
Springfield, NJ 07081

The C.P. Hall Co.
7300 S. Central Ave.
Chicago, IL 60638

Hamblet & Hayes Co.
Colonial Rd.
P.O. Box 730
Salem, MA 01970

Hammil & Gillespie
154 South Livingston Ave.
Livingston, NJ

Handy & Harman
Metalsmiths Systems Div. 21
Campus Rd.
Totowa, NJ 07512

Harbison Carborundum Corp.
Sub. Harbison Ind. Prod.
7200 Hibbs Lane
Levittown, PA 19057

Harbison-Walker Refractories Co.
Two Gateway Ctr.
Pittsburgh, PA 15222

Hardman Inc.
600 Cortlandt St.
Belleville, NJ 07109

A. Harrison & Co.
P.O. Box 494
Pawtucket,RI 02862

Harrisons & Crosfield (Canada)
1-4 Great Tower Street
London EC3R 5AB, UK

Harrisons & Crosfield (Pacific)
1-4 Great Tower St.
London EC3R 5AB, UK

Harrison Enterprises
115 Columbia St.
Dayton, OH 45401

Harshaw Chem. Co.
1945 E. 97 St.
Cleveland, OH 44106

Hart Chemical
Guelph, Canada N1H 6K8

Hart Products Corp.
173 Sussex St.
Jersey City, NJ 07302

Harwick
P.O. Box 9360
60 S. Selberling St.
Akron, OH 44305

Hatco Chemical
King George Post Rd.
Fords, NJ 08863

W.E. Hauck
P.O. Box 1065
Roswell, GA 30075

Haveg Industries Inc.
Sub. Hercules Indus.
900 Greenbank Rd.
Wilmington, DE 19898

Hayashibara Biochemical
2-3, I-Chome, Shimuishii
Okayamo, Japan
0862-24-4311
Mabit

Heftt Ltd.
P.O. Box 1623
CH 8048 Zurich, Switzerland

Henkel Argentina
Avda. E. Madero Piso 14
1106 Capital Federal, Argentina

Henkel Chemicals (Canada)
9550 Blvd. Ray Lawson
Ville d'Anjou (Montreal)
P.Q. Canada H1J 1L3

Henkel Corp.
255 West Spring Valley Ave.
Maywood, NJ 07607

Henkel KGaA
Postfach 1100
D-4000
Dusseldorf 1, West Germany

Henley
750 Third Ave.
New York, NY 10017

Herbert
180 Linden Ave.
Westbury, NY 11590

Hercon, Div. Health Chem. Corp.
1107 Broadway
New York, NY 10010

Hercules BV
8 Veraartlaan
P.O. Box 5822
2280 HV Rijswijk, The Netherlands

Hercules France
910 Market Street
Wilmington, DE 19899

Hercules, Inc.
910 Market Street
Wilmington, DE 19898

Heterene Chem.
792 21st Avenue
Paterson, NJ 07513

Hexcel Chem. Prod. Div.
315 N. Centennial St.
Zeiland, MI 49494

Hexcel Corp. Spec. Chem. Div.
255 W. Spring Valley Ave.
Maywood, NJ 07607

Dow B. Hickam
P.O. Box 2006
Sugar Land, TX 77478

High Point Chemical
609 Taylor Street
P.O. Box 2316
High Point, NC 27261

High Temperature Materials, Inc.
11541 Bradley Ave.
San Fernando, CA 91340

Hill Dermaceuticals
P.O. Box 19283
Orlando, FL 32814

Hilton-Davis
2235 Langdon Farm Rd.
Cincinnati, OH 45237

Hilton-Davis-Thomasset
90 Park Ave.
New York, NY 10016

Himont
1313 N. Market St.
Wilmington, DE 19894

Hodag Chem. Corp.
7247 North Central Park Avenue
Stokie, IL 60076

Hoechst AG
D-6230 Frankfurt am Main 80
Frankfurt, West Germany

Hoechst Japan
10-33, 4-Chome-Akasaka
Minato-ku
Tokyo, Japan

Hoechst-Roussel
Route 202-206 North
Somerville, NJ 08876

Hoffman-La Roache Inc.
340 Kingsland Street
Nutley, NJ 07110

Holloway
230 Oxmoor Circle
S 1111
Birmingham, AL 35209

Hooker Chemicals & Plastics Corp.
345 Third Street
Niagara Falls, NY 14302

Hoover-Hanes Rubber Corporation
10 Peguanoc Dr.
Tallopoosa, GA 30176

Hopkins Agricultural Chem. Co.
P.O. Box 7532
Madison, WI 53707

Hormel-Geo. A.
P.O. Box 800
Austin, MN 55912

Hormmel
P.O. Box 475
Pittsburgh, PA 15230

E.F. Houghton & Co.
Madison & Van Buren Aves.
P.O. Box 930
Valley Forge, PA 19482

Howard Hall & Co.
223 E. Putnam Ave.
Cos Cob, CT 06807

Hoyt
Division of Colgate-Palmolive Co.
575 University Ave.
Norwood, MA 02062

J.M. Huber Corp.
Box 310
Havre de Grace, MD 21078

Humko Chem./Div. Witco Chem. Corp.
P.O. Box 125
Memphis, TN 38101

Humko Products
Chem. Div./Witco Chem.
755 Crossover Lane
Memphis, TN 38117

Humko Sheffield Chem.
Div. Kraftco
P.O. Box 398
Memphis, TN 38101

Humphrey Chemical
Devine Street
North Haven, CT 06473

Hung Ltd.
Hong Kong, UK

Huntington Alloy Products
Div. International Nickel Co. Inc.
1 First Canadian Pl.
Toronto, Ontario
M5X 1C4 Canada

Hyland Therapeutics
Travenol Laboratories, Inc.
444 W. Glenoaks Blvd.
Glendale, CA 91202

Hynson, Westcott, & Dunning, Inc.
Charles and Chase Sts.
Baltimore, MD 21201

Hyrex
3494 Democrat Rd.
Memphis, TN 38118

Hysan
919 West 38th Street
Chicago, IL 60609

ICI Americas Inc.
Concord Pike & New Murphy Rd.
Wilmington, DE 19897

ICI Australia Ltd.
1 Nicholson Street
Melbourne 3000, Australia

ICI England
Imperial Chemical House
Millbank, London
SW1P 3JF, UK

ICI England Petrochemicals & Plastics Div.
P.O. Box 90
Wilton Middlesbrough, Cleveland
TS6 8JE, UK

ICI Organics Div.
P.O. Box 42
Hexagone House
Blackley, Manchester
M9 3DA, UK

ICI Petrochemicals
Wilmington, DE 19897

ICI Plant Protection Division
Fernhurst, Haslemere
Surrey
GU27 3JE, UK

ICI United States
New Murphy Rd. & Concord Pike
Wilmington, DE 19897

ICN Pharmaceuticals
222 N. Vincent Ave.
Covina, CA 91722

Ikeda Bussan Kaisha
New Tokyo Building
3-1-Marunouchi, 3-Chome
Chiyoda-Ku
Tokyo 100 Japan

Illinois Minerals
2035 Washington Ave.
Cairo, IL 62914

IMC Chem. Group
666 Garland Pl.
Des Plaines, IL 60016

IMC Chem.-NP Div.
2315 Sanders Rd.
Northbrook, IL 60062

Independent Chemical
79-51 Cooper Ave.
Glendale, NY 11385

Industrial Minerals
Industrial Mineral Ventures, Inc.
1800 E. Sahara Ave. Suite 107
Las Vegas, NV 89104

Inmont Corp.
1255 Broad Street
Clifton, NJ 07015

Inolex Chem. Co.
Jackson & Swanson Sts.
Philadelphia, PA 19148

International Minerals & Chem. Corp. NV Div
2315 Sanders Rd.
Northbrook, IL 60062

International Dioxide, Inc.
136 Central Ave.
Clark, NJ 07066

International Minerals & Chemicals
P.O. Box 207
Terre Haute, IN 47808 or

666 Garland Pl.
Des Plaines, IL 60016

International Rustproof Co.
Div. of Lubrizol Corp.
Box 17100
Cleveland, OH 44117

International Wax
P.O. Box 221
181 E. Jamaica Ave.
Valley Stream, NY 11582

Interstab Chem. Inc.
500 Jersey Ave.
New Brunswick, NJ 08903

Intex Products
P.O. Box 6648
Greenville, SC 29606

Isochem Corp.
99 Cook St.
Lincoln, RI 02865

Isochem Resins
99 Cook Street
Lincoln, RI 02865

Ives
685 Third Ave.
New York, NY 10017

Jacobus
37 Cleveland Lane
Princeton, NJ 08540

Jahres Fabrikker
P.O. Box 235
3201 Sandefijord
Norway

James River Limestone
Box 617
Buchanan, VA 24066

Janssen
40 Kingsbridge Rd.
Piscataway, NJ 08854

Jayco
895 Poplar Church Rd.
Camp Hill, PA 17011

Jean D'Avez

Jefferson Chemical Co., Inc.
P.O. Box 4128
Austin, TX 78765

Jetco Chemicals
P.O. Box 1898
Corsicana, TX 75110

Johns-Manville
Ken Caryl Ranch
Denver, CO 80217

Jojoba Commodities Group
11304 Chandler Blvd.
Box 953
North Hollywood, CA 91603

Jojoba Growers
515 Madison Ave.
Suite 1700
New York, NY 10022

Jordan Chemical
1830 Columbia Ave.
Folcroft, PA 19032

Kaiser Chemicals
Div. Kaiser Aluminum
300 Lakeside Dr.
Oakland, CA 94643

Kaken Chemical Co.
28-8, 2-Chome
Honkomagome
Bunkyo-Ku
Tokyo 113, Japan

Kalama Chemical, Inc.
Suite 1110 Bank of California Center
Seattle, WA 98164

Kalo Agricultural Chem. Inc.
4550 West 109 St.
Suite 222
Overland Park, KS 66211

Kao Corp.
1-Chome Kayabacho
Nihonbashi Chuo-Ku
Tokyo 103, Japan

Kao Food
14-10, Nihonbashi
Kayabacho 1-Chome
Chuo-Ku
Tokyo 103, Japan

Kaopolite
P.O. Box 277
511 Westminster Ave.
Elizabeth, NJ 07207

Kao Soap
1, 1 Chome Kayabacho
Nihonbashi Chuo-Ku
Tokyo 103, Japan

Katsura
4-2 Nihonbashi-Honcho
Chuo-Ku
Tokyo, Japan

Kay-Fries Inc.
10 Link Dr.
Rockleigh, NJ 07647

Keil Chemical
3000 Sheffield Ave.
Hammond, IN 46320

Kelco Co.
Sub. Merck & Co., Inc.
126 East Lincoln Ave.
Rahway, NJ 07065

Kelco/Div. of Merck & Co. Inc.
8355 Aero Dr.
San Diego, CA 92123

Kemira Oy
P.O. Box 330
SF-00101
Helsinki 10, Finland

Keno Gard VT AB
P.O. Box 11033
S-10061 Stockholm, Sweden

Keno Gard S.A
rue Gachard 88, bte. 9
B-1050 Bruxelles, Belgium

Kenrich Petrochemicals
140 E. 22nd St.
Bayonne, NJ 07002

Kenwood
490-A Main St.
New Rochelle, NY 10801

Kerr-McGee
P.O. Box 25861
Oklahoma City, OK 73125

Key Pharmaceuticals
18425 N.W. 2nd Ave.
Miami, FL 33169

Kennametal Inc.
Latrobe, PA 15650

Klinger Engineering Plastics Div.
Sidcup Kent
DA14 5AG, UK

Knapp Products Inc.
Lodi, NJ 07644

Knoll Pharmaceutical Co.
30 North Jefferson Rd.
Whippany, NJ 07981

H. Kohnstamm & Co.
161 Ave. of the Americas
New York, NY 10013

Koppers, Co. Inc.
1900 Koppers Bldg.
Pittsburgh, PA 15219

Kramer
8778 S.W. 8th Street
Miami, FL 33174

Kremers-Urban
P.O. Box 2038
Milwaukee, WI 53201

Laboratoires Serobiologiques
P.O. Box 670
54010 Nancy, France

Laboratori Vevy, Inc.
One Dag Hammarskjold Plaza
New York, NY 10017

Lact Aid
600 Fire Rd.
P.O. Box 111
Pleasantville, NJ 08232-011

Lafayette
4200 South Hulen St.
Fort Worth, TX 76109

Lake States Chemical
Div. St. Regis Paper Co.
237 Park Ave.
New York, NY 10017

Lambda
P.O. Box 813
Gurabo, PR 00658

La Motte Chemical Products Inc.
P.O. Box 329
Chesterton, MD

Lanaetex
151-157 Third Ave.
Elizabeth, NJ 07206

Lancaster
Sub. AZ Products, Co.
2525 S. Cumbee Rd.
Lakeland, FL 33801

Lankro Chem Ltd.
Eceles
Manchester M30 0BH

Lannett
9000 State Rd.
Philadelphia, PA 19136

Laporte Industries Ltd.
Hanover House
14 Hanover Square
London W1R 0BE, UK

La Salle
Subsidiary of Mallard
3021 Wabash Ave.
Detroit, MI 48216

Laser
2000 N. Main Street
P.O. Box 905
Crown Point, IN 46307

La Tassilchimica
Bergamo, Italy

Laur Silicone Rubber Compounding, Inc.
P.O. Box 509A
Beaverton, MI 48612

Lederle
One Cyanamid Plaza
Wayne, NJ 07470

Leeming
Pfizer Inc.
100 Jefferson Rd.
Parsippany, NJ 07054

Legere
7326 E. Evans Rd.
Scottsdale, AZ 85260

Lemmon
Post Office Box 30
Sellersville, PA 18960

Lensfield
Maulden Road-Flitwick
Bedford MK 45 5, UK

Lever Industriel
56 Rue de Paris
Noisy-le-Sec 93130, France

Lilachim S.A.
Rue de Science
37 Wetenschapsstraat
1040 Brussels, Belgium

Eli Lilly & Co.
307 E. McCarty St.
Indianapolis, IN 46285

Lion Corp.
2-22.1 Chome, Yokoami
Sumida-Ku
Tokyo, Japan

Lipo Chemicals
207 19th Ave.
Paterson, NJ 07504

Liquid Nitrogen Processing Corp.
412 King Street
Malvern, PA 19355

LNP Corp.
412 King Street
Malvern, PA 19355

Lonza Inc.
22-10 Route 208 Fairlawn, NJ 07504

Los Angeles Chemical Co.
4545 Ardine St.
P.O. Box 1987
South Gate, CA 90280

Lowenstein Dyes & Cosmetics, Inc.
420 Morgan Ave.
Brooklyn, NY 11222

Lubrisol Corp.
29400 Lakeland Blvd.
Wickliffe, OH 44092

Lucas Meyer
D-2000 Hamburg 28
Postfach 280-246
West Germany

Lucas Meyer GmbH & Co.
D-2000 Hamburg 28
Postfach 280-246
West Germany

Lucidol Div./Penmualt Corp.
1740 Military Rd.
Buffalo, NY 14217

Lucidol Div., Ram Div.
1740 Military Rd.
Buffalo, NY 14217

Lyndal Chemical
P.O. Box 1740
Dalton, GA 30720

Maag Agrochemicals (Switzerland)
CH 8157
Dielsdorf, Switzerland

Macfarlan Smith
Wheatfield Rd.
Edinburgh EH11 2QA, Scotland

Magna Corp.
7505 Fannin St.
Houston, TX 77054

Makhteshim Agan
P.O. Box 60
Beer-sheva, Israel 84100

Mallard
3021 Wabash Ave.
Detroit, MI 48216

I. Mallinckrodt Inc.
P.O. Box 5439
St. Louis, MO 63147

Malvern Minerals
220 Runyon St.
Box 1246
Hot Springs Natl. Park, AR 71901

Manchem Ltd.
Aston New Rd.
Manchester M11 4AT, UK

Manro Products Ltd.
Bridge St.
Stalybridge
Cheshire SK15 1PH, UK

Marathon Marco Co.
1502 Pine Ave.
Dickinson, TX 77539

Marchon
Leconfield House
Curzon St.
London W1Y 8JR, UK

Marine Colloids
P.O. Box 70
2 Edison Place
Springfield, NJ

Marion Labs
10236 Bunker Ridge Rd.
Kansas City, MO 64137

Marlowe-Van Loan Corp.
P.O. Box 1851
Hight Point, NC 27261

Marlyn
350 Pauma Place
Escondido, CA 92025

Mars Chemical Corp.
762 Marietta Blvd., N.W.
Atlanta, GA 30318

Martin Marietta
Executive Plaza 11
Hunt Valley, MD 21030

Marubishi Oil Chemical Co.
3-7-12 Tomobuchi-Cho
Miyakojima-Ku
Osaka, Japan

Maruzen Fine Chemicals
P.O. Box 389
353 Crider Avenue
Moorestown, NJ 08057

Mason Chemical Co.
5352 W. Belmont Ave.
Chicago, IL 60641

Mason Pharmaceutical
P.O. Box 8330
Newport Beach, CA 92660

Masonite Corp.
29 N. Wacker Dr.
Chicago, IL 60606

Mastar Pharmaceutical
P.O. Box 3144
Bethlehem, PA 18017

Matsumoto Yushi-Seiyaku Co.
1-3, 2-Chome
Shibukawa-Cho
Yao City
Osaka, Japan

May & Baker
Dagenham
Essex RM10 7XS, UK

Maybrook
600 Broadway
Lawrence, MA 01842

Mayrand
P.O. Box 8869
4 Dundas Circle
Greensboro, NC 27419

Mazer Chemicals Inc.
3938 Porett Drive
Gurnee, IL 60031

McDanel Refractory Porcelain Co.
Box 560
Beaver Falls, PA 15010

McGregor Pharma.
32580 Grand River Ave.
Farmington, MI 48024

McIntyre Chem. Co. Ltd.
4851 S. St. Louis Ave.
Chicago, IL 60632

McLauglin Gormley King Co.
8810 10th Ave. N.
Minneapolis, MN 55427

McNeil Consumer Products
McNEILAB, Inc.
Fort Washington, PA 19034

McNeil Pharmaceutical
McNEILAB, Inc.
Spring House, PA 19477

Mead Johnson Nutritional
Mead Johnson and Company
2404 W. Pennsylvania St.
Evansville, IN 47721

Mead Johnson Pharmaceutical
Mead Johnson & Company
2404 Pennsylvania St.
Evansville, IN 47721

Mearl Corp.
217 N. Highland Ave.
Ossining, NY 10562

Med-Chem Labs
P.O. Box 1117
Monroe, MI 48161

Medicone Comp.
225 Varick St.
New York, NY 10014

Meer Corp.
9500 Railroad Ave.
North Bergen, NJ 07047

Merck & Co. Inc.
P.O. Box 2000
Rahway, NJ 07065

Merck Sharp & Dohme
Div. of Merch & Comp. Inc.
West Point, PA 19486

Merieux Inst., Inc.
1200 N.W. 78th Ave.
Suite 109
Miami, FL 33126

Merrell Dow Phar., Inc.
Subsid. of Dow Chemical Co.
Cincinnati, OH 45215

M. Michel & Co. Inc.
90 Broad St.
New York, NY 10004

Michigan Chemical Corp.
Chicago, IL

Midox Ltd.
Glaston Park
Glaston, Oakham
Leicestershire LE15 9BX, UK

Miles Laboratories Inc.
P.O. Box 340
Elkhart, IN 46515

Miles Pharmaceuticals
Div. of Miles Laboratories, Inc.
400 Morgan Lane
West Haven, CT 06516

Milex Products, Inc.
5915 Northwest Highway
Chicago, IL 60631

Miller Chemical & Fertilizer Corp.
P.O. Box 333
Hanover, PA 17331

Milliken Chemical
P.O. Box 1927
Spartenburg, SC 29304

MineraLab Inc.
3501 Breakwater Ave.
Hayward, CA 94545

Minerals & Chemicals
Div. Engelhard Minerals & Chemicals Corp.
Menlo Park
Edison, NJ 08837

Mineral Research & Development Corp.
4 Woodlawn Green, Suite 232
Charlotte, NC 28210

Miranol Chem. Co. Inc.
68 Culver Road
P.O. Box 411
Dayton, NJ 08810

Mission Phar. Co.
1325 E. Durango
San Antonio, TX 78210

Mitsubishi Chem. Ind. Ltd.
5-2 Maruniuchi
2-Chome
Chiyoda-Ku
Tokyo, Japan

Mitsui Toatsu Chemicals Inc.
2-5 Kasumigaseki 3-Chome
Chiyoda-Ku
Tokyo 100, Japan

M & J Chemicals
2200 One Riverway Dr.
Houston, TX 77056

Mobay Chem. Corp.
Mobay Rd.
Pittsburgh, Pa 15205

Mobay Chemical Corp.
Agricultural Chemicals Division
P.O. Box 4913
Hawthorn Rd.
Kansas City, MO 64120

Mobil Chem. Co.
Chem. Prod. Div.
P.O. Box 26683
Richmond, VA 23261

Mona Industries
76 E. 24th St.
Paterson, NJ 07544

Mono-Chem Corp.
P.O. Drawer 830-C
Atlanta, TX 75551

Monsanto Agricultural Products Co.
800 N. Lindbergh Blvd.
St. Louis, MO 63167

Monsanto Co.
800 N. Lindbergh Blvd.
St. Louis, MO 63166

Monsanto Industrial Chemicals Co.
800 N. Lindbergh Blvd.
St. Louis, MO 63166

Monsanto Plastics & Resins
800 N. Lindbergh Blvd.
St. Louis, MO 63166

Montedison USA, Inc.
1114 Ave. of Americas
New York, NY 10036

Mooney
2301 Scranton Rd.
Cleveland, OH 44113-9988

Moores Lime Co.
Box 878
Springfield, OH 45501

Morton Chemical
Div. Morton-Norwich
2 N. Riverside Plaza
Chicago, IL 60606

Morton Thiokol, Inc.
101 Carnegie Center
Princeton, NJ 08540

Motomco, Inc.
29 N. Fort Harrison
Clearwater, FL 33515

M & T Chemicals
Woodbridge Road & Randolph Ave.
Rahway, NJ 07065

H. Muehlstein
591 W. Putnam Ave.
Greenwich, CT 06836

Muro Phar. Inc.
890 East Street
Tewksburg, MA 01876

Murphy-Phoenix Co.
P.O. Box 22930
Beachwood, OH 44122

National Dermaceutical
8749 Surrey Place
Maineville, OH 45039

National Lead Co.
1230 Ave. of the Americas
New York, NY 10020

National Lead Co. Baroid Div.
1230 Ave. of the Americas
New York, NY 10020

National Lead Co. De Lore Division
1230 Ave. of the Americas
New York, NY 10020

National Lead Co. Evans Lead Div.
1230 Ave. of the Americas
New York, NY 10020

National Polychemicals Inc.
51 Eames St.
Wilmington, MA

National Starch & Chem. Corp.
10 Finderne Ave.
Bridgewater, NJ 08807

Natrochem
Exley Ave.
Savannah, GA

Nease Chemical Co. Inc.
P.O. Box 221
State College, PA 16801

Neutrogena Dermatologics
Div. of Neutrogena Corp.
5755 West 96th Street
P.O. Box 45036
Los Angeles, CA 90045

Neville Chemicals
Neville Island
Pittsburgh, PA 15225

Neville-Synthese Organics, Inc.
Neville Island
Pittsburgh, PA 15225

New Japan Chemical Co. Ltd.
51 Bingo-Machi 2-Chome
Higashi-Ku
Osaka, Japan

New Jersey Zinc
1 Commerce Place
Nashville, TN 37239

Niacet Corp.
47th Street & Pine Ave.
Niagara Falls, NY 14302

Nihon Emulsion Co. Ltd.
32-7, 5-Chome, Koenji-Minami
Suginami-Ku
Tokyo 116, Japan

Nihon Nohyaku Co.
2-5 Nihonbashi, 1-Chome
Chuoku
Tokyo 103, Japan

Nikka Chem. Ind. Co., Ltd.
23-1, 4-Chome, Bunkyo
Fukui City, Japan

Nikko Chemicals Co. Ltd.
1-4-8 Nihonbashi-Bakurocho
Chuo-Ku
Tokyo 103, Japan

Nipa Laboratories Ltd.
Nipa Industrial Estate
Llantwit Fardre, Near Pontyprid
Mid-Glamorgan CF38 2S, UK

Nippon Chemical
3-1 Iwamot-Cho, 2-Chome
Chiyoda-Ku
Tokyo, Japan

Nippon Kayaku Co.
1-2-1 Marunouchi, Chiyoda-Ku
Tokyo 100, Japan

Nippon Nyukazai Co.
19-9, 3-Chome, Ginza Chuo-Ku
Tokyo 104, Japan

Nippon Oil & Fats Co.
10-1, Yuraku-Cho, 1-Chome
Chiyoda-Ku
Tokyo 100, Japan

Nippon Senka Chemical Industries, Ltd.
17-34 Hanaten-Higashi, 1-Chome
Tsurumi-Ku
Osaka, Japan

Nippon Soda Co.
2-2-1, Ohtemachi, Chiyoda-Ku
Tokyo 100, Japan

Nissan Chem. Ind.
Kowa-Hitotsubashi Bldg.
7-1, 3-Chome, Kanda-Nishiki-Cho
Chiyoda-Ku
Tokyo, Japan

Nisshin Oil Mills
23-1, 1-Chome, Shinkawa
Chuo-Ku
Tokyo 104, Japan

NL Chemicals
P.O. Box 700
Wycoff Mills Road
Hightstown, NJ 08520

NL Ind./Baroid Div.
P.O. Box 1675
2404 Southwest Freeway
Houston, TX 77001

NL Industries Inc.
P.O. Box 700
Hightstown, NJ 08520

N.L. Ind./Ind. Chem. Div.
P.O. Box 700 Hightstown, NJ 08520

Nopco Division Shamrock Corp.
P.O. Box 321
London, Ontario
N6A 5G7 Canada

Norac Co. Inc.
405 S. Motor Ave.
Azusa, CA 91702

Nor-Am Agricultural Products Inc.
350 W. Shuman Blvd.
Naperville, IL 60566

Norcliff Thayer
303 South Broadway
Tarrytown, NY 10591

Norda
140 Route 10
East Hanover, NJ 07936

Nordisk-USA
6500 Rock Spring Drive
Suite 304
Bethesda, MD 20817

Norgine Laboratories, Inc.
2 Overhill Road
Scarsdale, NY 10583

Norman, Fox & Co.
5611 So. Boyle Ave.
Vernon, CA 90058

Norsolor/Co. of C.D.F. Chemie
950 3rd Ave.
New York, NY 10022

Norwich Eaton
13-27 Eaton Ave.
Norwich, NY 13815

Noury Chem. Corp.
2153 Lockport & Olcott Rd.
Burt, NY 14028

Nova Industri A/S
Novo Alle
DK-2880
Bagsvoerd, Denmark

Nyacol Products Inc.
Megunco Rd.
Ashland, MA 01721

Ohio Lime
Sub. Steetley Holdings Ltd.
605 James St. N
Box 2029 St'n. A
Hamilton, Ontario
L8N 3S9 Canada

Ocumed
194 Hanse Ave.
Freeport, NY 11520

OLC Laboratories
99 N.W. Miami Gardens Dr.
Miami, FL 33169

Oleofina S.A.
Rue de Science 37 Wetenschapsstraat
1040 Brussels, Belgium

Olin Corp.
First National Bldg.
P.O. Box 991
Little Rock, AR 72203

Omya
61 Main Street
Proctor, VT 05765

O'Neal, Jones & Feldman
2510 Metro Blvd.
Maryland Heights, MO 63043

Onyx Chemical Co.
Millmaster Onyx Group
190 Warren Street
Jersey City, NJ 07302

Organon
375 Mount Pleasant Ave.
West Orange, NJ 07052

Ortho Diagnostic Systems
Route 202
Raritan, NJ 08869

Ortho Pharmaceutical
(Dermatological Div.)
Route 202
Raritan, NJ 08869

Osaka Kasei Co., Ltd.
6-11 2-Chrome
Nakajima
Nishi-Yodogawa-Ku
Osaka, Japan

Otsuka Chem. Co.
200 Park Ave.
Suite 2617
New York, NY 10166

Ottawa Chem. Co.
700 North Wheeling Street
Toledo, Ohio 43605

Ozark Mahoning Co.
A Pennwalt Subsidiary
1870 S. Boulder Ave.
Tulsa, OK 74119

Pamol Ltd. Arad
Luxembarg Chemicals
P.O. Box 13
Tel Aviv 61000, Israel 61000

Paisley Products
Edison, NJ

Paniplus
100 Paniplus Roadway
Olathe, KS 66061

Pantasote, Inc.
26 Jefferson Street
Passaic, NJ 07055

Parke Davis & Co.
201 Taber Road
Morris Plains, NJ 07950

Patco Cosmetic Products
3947 Broadway
Kansas City, MO 64111

Patco Products Div., C.J.P.
3947 Broadway
Kansas City, MO 64111

C.J. Patterson Co.
3947 Broadway
Kansas City, MO 64111

PBI/Gordon Corp.
P.O. Box 4090
Kansas City, MO 64101

Pearsall Chem. Div./Witco Chem. Corp.
P.O. Box 437
2519 Fairway Park Drive
Houston, TX 77001

Pecten Chemical Inc.
One Shell Plaza
P.O. Box 4407
Houston, TX 77210

Pedinol Phar.
110 Bell Street
W. Babylon, NY 11704

S.B. Penick Div.
CPC Int.
158 Mt. Olivet Ave.
Newark, NJ
New York, NY

Pennsylvania Glass Sand Corp.
431 N. Craig St.
Pittsburgh, PA 15213

Pennwalt Corp.
3 Parkway
Philadelphia, PA 19102

Penreco Div. of Pennzoil
106 South Main Street
Bolter, PA 16001

Pennwalt Holland B.V.
P.O. Box 7120
3000 H C Rotterdam, Holland

Permutit Co.
E49 Midland Ave.
Paramus, NJ 07652

Person & Covey
616 Allen Ave.
Glendale, CA 91201

Pestcon Systems Inc.
2221 Poplar Blvd.
P.O. Box 469
Alhambra, CA 91802

Petrochemicals Co. Inc.
P.O. Box 2199
Ft. Worth, TX 76101

Pfanstiehl Laboratories
1219 Glenrock Ave.
Waukegan, IL 60085

Pfipharmecs Div.
Pfizer Inc.
235 E. 42nd St.
New York, NY 10017

Pfister
235 E. 42nd St.
New York, NY 10017

Pfizer Inc.
2110 High Point Road
Greensboro, NC 27403

PFW
33 Sprague Ave.
Middletown, NY 10940

Pharmacia Fine Chemicals
800 Centennial Ave.
Piscataway, NJ 08854

Philadelphia Quarts Co.
P.O. Box 5407
HWY 56 Loop
Augusta, GA

Phillips Chem. Co.
1501 Commerce Dr.
Stow, OH 44224

Pigment Dispersions Inc.
54 Kellog Court
Edison, NJ 08817

Pillar International Co.
P.O. Box 70211
Taipei, Taiwan, ROC

Pilot Chem. Co.
11756 Burke Street
Santa Fe Springs, CA 90670

Poco Graphite Inc.
Sub. Poco Oil
13103 Holston Hills Dr.
Houston, TX 77069

Polaroid Corp.
784 Memorial Dr.
Cambridge, MA

Polyfil DL
Sub. Reed International PLC
Reed House
83 Piccadilly
London W1A 1EJ, UK

Polymel
Baltimore, MD

Polymer Applications, Inc.
3445 River Road
Tonawanda, NY 14150

Polyvinyl Chem. Industries
730 Main Street
Wilmington, MA 01887

Wm. P. Poythness & Co., Inc.
16 N. 22nd St.
P.O. Box 26946
Richmond, VA 23261

PPF International
Lindtsedjik 8
3336 LE Zwijndrecht
The Netherlands

P.P.G. Industries, Inc.
One Gateway Ctr.
Pittsburgh, PA 15222

The PQ Corp.
P.O. Box 840
Valley Forge Executive Mall
Valley Forge, PA 19482

Premier Malt Products Inc.
1137 North 8th Street
Milwaukee, WI 53201

Prochinite Avenches SA
Elsasserstr.
229-231
CH-4013 Basel
Switzerland

Proctor & Gamble Co.
P.O. Box 171
Cincinnati, OH 45201

Proctor Chem. Co.
P.O. Box 399
Salisbury, NC 28144

Produits Chimiques de la Montagne Noire
145, Avenue de Lavaur
81105 Castres
Cedex, France

Produits Chimiques Ugine Kullmann
(France)
Tour Manhailan
Cedex 21
92087 Paris La Defense

Protameen Chem., Inc.
409 Minnisink Road
Totowa, NJ 07511

Purdue Frederick
100 Connecticut Ave.
Norwalk, CT 06854

PVO Int'l.
416 Division Street
Boonton, NJ 07005

Quaker Oats Co.
Merchandise Mart Plaza
Chicago, IL 60654

RAM Chem.
Div. Whittaker Corp.
210 E. Alondra Blvd.
Gardena, CA 90248

RAM Laboratories
P.O. Box 350952
Miami, FL 33135

Reed & Carnick
1 New England Ave.
Piscataway, NJ 08854

Reed Lignin, Inc.
81 Holly Hill Lane
Greenwich, CT 06830

Reed Ltd. Chemical Div.
P.O. Box 2025
Quebec, P.Q. Canada

Reheis Chem. Co.
235 Snyder Avenue
Berkeley Heights, NJ 07922

Reichhold Chem. Inc.
RCI Building
White Plains, NY 10602

Reid-Provident, Direct Div.
640 Tenth St., N.W.
Atlanta, GA 30318

Reid Provident Labs.
25 Fifth Street, N.W.
Atlanta, GA 30308

Reilly-Whiteman Inc.
Washington and Righter Sts.
Conshohocken, PA 19428

Rewo Chemical Group
9th Floor, Crown House
London Rd.
Morden Surrey SM4 5DU, UK

Rewo Chemische Werke GmbH.
D-6497 Steinau an der Strasse
P.O. Box 1160
West Germany

Rexar Phar. Corp.
396 Rockaway Ave.
Valley Stream, NY 11581

Rexolin Chem. AB
Box 622
S-251 06 Helsingborg
Sweden

Rhein Chemie
D-6900 Heidelberg 1
Bergstr. 116
Postfach 104620
West Germany

Rhode Island Laboratories Inc.
West Warwick, RI 02893

Rhone-Poulenc Inc.
22 Avenue Montaigne
75360 Paris Cedex 08
France

Rhone-Poulenc Inc., Agrochemical Div.
Black Horse Lane
P.O. Box 125
Monmouth Junction, NJ 08552

Riceland Foods
Sub. Witco Chem. Corp., Organics Div.
520 Madison Ave.
New York, NY 10022

The Richardson Co.
2400 Devon Ave.
Des Plains, IL 60018

Riken Vitamin Oil
3-8-10 Nishi-Kanda
Chiyoda-Ku
Tokyo, Japan

Riker Lab., Inc.
19901 Nordhoff Street
Box #1
Northridge, CA 91328

R.I.T.A. Corp.
P.O. Box 556
Crystal Lake, IL 60014

Robeco Chem., Inc.
99 Park Ave.
New York, NY 10016

A.H. Robins Co.
Pharmaceutical Division
1407 Cummings Ave.
Richmond, VA 23220

Roche Labs.
Div. of Hoffmann-La Roche Inc.
Nutley, NJ 07110

Roerig
Div. of Pfizer Phar.
235 E. 42nd St.
New York, NY 10017

Rogers Corp.
Rogers, CT 06263

Rohm & Haas (Australia) Pty. Ltd.
969 Burke Rd.
P.O. Box 115 Camberwell
Victoria, 3124 Australia

Rohm & Haas Co.
Independence Mall West
Philadelphia, PA 19105

Rohm & Haas France SA
Independence Mall West
Philadelphia, PA 19105

Rora Pearl
P.O. Box 81
Bayonne, NJ 07002

Ronsheim & Moore
Ings Lane, Castleford
West Yorkshire WF10 2J, UK

W.H. Rorer, Inc.
500 Virginia Drive
Fort Washington, PA 19034

Ross Labs.
Div. Abbott Labs.
Columbus, OH 43216

Rotex Phar., Inc.
P.O. Box 19283
Orlando, FL 32814

Roussel Uclaf.
163, Avenue Gambetta
75020 Paris, France

Rowell Labs. Inc.
210 Main Street, W.
Baudette, MN 56623

Roxane Labs., Inc.
330 Oak Street
Columbus, OH 43216

H.M. Royal, Inc.
689 Pennington Avenue
Trenton, NJ 08601

Royce Chem. Co.
17 Carlton Ave.
E. Rutherford, NJ 07073

Ruetgers Nease Chem. Co.
Box 221
State College, PA 16801

Rumford Chemical Works
Essex Chem. Corp.
Clifton, NJ

Ryco, Inc.
801 Washington St.
Conshohocken, PA 19428

Rydelle Labs., Inc.
1525 Howe Street
Racine, WI 53403

Rystan Co., Inc.
47 Center Ave,
P.O. Box 214
Little Falls, NJ 07424

Salomon & Bros.
P.O. Box 828
Port Washington, NY 11050

Sandoz, Agro Div.
480 Camino de Rio South
San Diego, CA 92108

Sandoz Colors & Chemicals
Route 10
East Hanover, NJ 07936

Sandoz Ltd.
Lichtstrasse 35
4002 Basel, Switzerland

Sandoz Ltd., Agro Division
Lichtstrasse 35
4002 Basle, Switzerland

Sankyo Co. Ltd.
Agrochemicals Planning Dept.
No. 7-12 Ginza, 2-Chome
Chuo-Ku
Tokyo 104, Japan

Sanyo Chem. Ind. Ltd.
11-1 Ikkyo Nomoto-Cho
Higashiyama-Ku
Kyoto 605, Japan

Saron Phar. Corp.
1640 Central Ave.
St. Petersburg, FL 33712

Sartomer Co.
P.O. Box 799
West Chester, PA 19380

Savage Labs.
60 Baylis Road
P.O. Box 2006
Melville, NY 11747

Saytech Inc.
25 Kimberley Rd.
East Brunswick, NJ 08816

Santech Inc.
5238 Belfield Ave.
Philadelphia, PA 19144

Henry Schein, Inc.
5 Harbor Park Drive
Port Washington, NY 11050

Schenectady Chem. Inc.
P.O. Box 1046
Schenectady, NY 12301

Scher Chemicals, Inc.
1 Styertowne Rd.
Clifton, NJ 07012

Schering AG
Agrochemical Div.
P.O. Box 650311
D-1000 Berling 65, F.R.G

Schmid Products Co.
Route 46 West
Little Falls, NJ 07424

J.C. Schumacher Co.
D-7120
Bietigheim-Bissingen
Postfach 207
West Germany

Schuylkill Chem. Co.
2364 West Sedgley Ave.
Philadelphia, PA 19132

Scientific Chemicals, Inc.
15564 Producer Lane
Huntington Beach, CA

Sclavo Inc.
5 Mansard Court
Wayne, NJ 07470

Scot-Tussin Phar. Co., Inc.
50 Clemence Street
P.O. Box 8217
Cranston, RI 02920-0217

SDS BioTech Corp.
7528 Auburn Rd.
P.O. Box 348
Painesville, OH 44077

Seamless Hospital Products Co.
P.O. Box 828
Barnes Industrial Park
Wallingford, CT 06492

Searle & Co.
San Juan, PR 00936

Searle Pharmaceuticals, Inc.
Box 5110
Chicago, IL 60680

Secol, Inc.
849 Broadway
Newark, NJ 07104

Sederma
24 bis, Bd Verd de St-Julien
92190 Meudon, France

SEPPIC
70 Champs-Elysees
75008 Paris, France

Seres Labs., Inc.
3331 Industrial Drive
Box 470
Santa Rosa, CA 95402

Serono Labs., Inc.
280 Pond Street
Randolph, MA 02368

Servo BV
P.O. Box 1
7490 AA Delden, Holland

Shamokin Filler
Bear Valley Rd.
Shamokin, PA

Shawinigan Products
Sub. Lavalin Inc.
1130 Sherbrooke St. West
Montreal P.Q. H3A 2R5

Shell Chemicals Co.
One Shell Plaza
Houston, TX 77001

Shell Chemicals UK Ltd.
Shell Chimie France
1 Northumberland Ave.
London WC2N 5LA, UK

Shell Chimie France
27 Rue de Berri
7539, Paris Cedex 08
France

Shell Int'l Chem. Co. Ltd.
Shell Centre
London SE1 7PG, UK

Shell Nederland Chemie N.V.
P.O. Box 187
2501 CD The Hague
The Netherlands

Sherex Chemical Co.
P.O. Box 646
Dublin, OH 43017

Sherwin Williams Chemicals
P.O. Box 6520
Cleveland, OH 44101

Shin-Ei Chem. Co., Ltd.
19 Doshomachi, 1-Chome
Higashi-Ku
Osaka 541, Japan

Shinung Corp.
Shinung Bldg.
45 Wuchuan Center St.
Taichung, Taiwan, R.O.C.

Shroffs Industrial Chemicals Pvt. Ltd.
167 Dr. Annie Besant Rd.
Worli, Bombay 400 018 India

Shulton, Inc.
One Cyanamid Plaza
Wayne, Nj 07470

Richardson Co.
2400 Devon Ave.
Des Plains, IL 60018

Siror Kao, S.A.
Calle Aragon
383 Atico
Barcelona (13) Spain

E.W. Smith Chemical Co.
15020-T E. Proctor Ave.
Industry, CA 91746

Smith Kline Corpl
1 Franklin Plaza
Philadelphia, PA 19101

Smith Kline & French
1500 Spring Garden St.
P.O. Box 7929
Philadelphia, PA 19101

Smith Labs Inc.
2211 Sanders Rd.
P.O. Box 3044
Northbrook, IL 60062

Soitem-Societa Ital. Emulsionanti
Via R. Cuzzi
34, 20125 Milano, Italy

Solem Industries Inc.
5824--D Peachtree Corners East
Norcross, GA 30092

Solvay
Rue de Prince Albert
33D-1050 Brussels, Belgium

Southeastern Clay
P.O. Box 1055
Aiken, SC 29801

Southland Corp. Chem. Div.
P.O. Box 134
Argo, IL 60501

Spaulding Fibre Co., Inc.
1 American Dr.
Buffalo, NY 14225

Spencer-Kellogg
120 Delaware Ave.
Buffalo, NY 14240

Springbok
12502 South Garden St.
Houston, TX 77071

Squibb/Connaught
330 Alexander Street
Princeton, NJ 08540

Squibb-Novo
120 Alexander Street
Princeton, NJ 08540

E.R. Squibb & Sons Inc.
P.O. Box 4000
Princeton, NJ 08540

Staley Chemical Co.
2200 Eldorado St.
Decatur, IL 62525

Standard Process Labs., Inc.
2023 West Wisconsin Ave.
Milwaukee, WI 53201

Star Phar., Inc.
1990 N.W. 44th St.
P.O. Box 50228
Pompano Beach, FL 33074-0228

Stauffer Chemical Co.
Weston, CT 06880

Staveley Chemical
Staveley Work, Chesterfield
Derbyshire S43 2PB, UK

F. Steinfels AG
Heinrichstrasse 255
P.O. Box CH-8023
Zurich, Switzerland

Stein Hall & Co. Inc.
Sub. Celanese Corp.
1211 Ave. of the Americas
New York, NY 10036

Stepan Chemical Co.
Edens & Winnetka
Northfield, IL 60093

Stepan Europe
BP 12
Voreppe France 38340

Stevenson Bros. & Co.
1039 West Venango St.
Philadelphia, PA 19140

Stiefel Labs., Inc.
2801 Ponce de Leon Blvd.
Coral Gables, FL 33134

St. Joe Resources
250 Park Ave.
New York, NY 10177

Stokely-Van Camp, Inc.
1530 S. Jackson St.
Janesville, WI 53545

Strahl & Pitsch
P.O. Box 1098
230 Great East Neck Road
West Babylon, NY 11704

Stuart Phar.
Div. of ICI Americas Inc.
Wilmington, DE 19897

Sumitomo Chemical Americal Inc.
345 Park Ave.
New York, NY 10154

Summit Chemical
15 5-Chome, Kitahama Higashi-Ku
Osaka, Japan

Sunkist Growers
720 East Sunkist St.
Ontario, CA 91761

Sun Petroleum
P.O. Box 4783
Philadelphia, PA

Sutton Labs
116 Summit Ave.
Chatham, NJ 07928

Swastik Household & Industrial Products
Div. of Ambalal Sara Bnai Ent. Ltd.
Shahibag House
PB 362, 13 Walchard Hirochand Morg
Ballard Estate
Bombay 400 038, India

Sween
Sween Building
P.O. Box 980
Lake Crystal, MN 56055

Swift Chemical
P.O. Box 1720
Guelph, Ontario
N1H 6Z9 Canada

SWS Silicones
Sutton Road
Adrian, MI 49221

Sybron Corp. Chem. Div.
P.O. Box 125
Wellford, SC 29385

Sylacauga Calcium Prod.
Box 330
Sylacauga, AL 35150

Syntex Labs., Inc.
3401 Hillview Ave.
P.O. Box 10850
Palo Alto, CA 94303

Synthetic Products Co.
1409 16th Street
Racine, WI 53403

TAC Industries
821 Tilton Bridge Rd SE
Dalton, GA 30720

Taiyo Kagaku
9-5 Akahori-Shinmachi
Yokkaichi, Mie-Pref. Japan

Takeda Chemicals Industries
12-10 Nihonbashi 2-Chome
Chuo-Ku, Tokyo, Japan

Takemoto Oil & Fat Co.
No. 5, Sec. 2, Minato-Machi
Gamagori Aichi, Japan

Tenneco Chemicals Inc.
Park 80 Plaza W-1
Saddle Brook, NJ 07662

Tenneco Chemicals Inc. Intermediates
Division
Turner Place
P.O. Box 365
Piscataway, NJ 08854

Tennessee Corp.
ICD, Cities Service Bldg.
3445 Peachtree Rd. N.E.
Atlanta, GA 30326

Terry Corp.
P.O. Box 2348
Satellite Beach, FL 32937

Texaco Chemical Co.
P.O. Box 15730
Austin, TX 78761

Texo Corp.
2801 Highland Ave.
Cincinnati, OH 45212

Theodor Leonhard Wax Co.
136 Church Street
Haledon, NJ 07508

Th. Goldschmidt AG
Goldschmidtstr. 100, 4300 Essen 1
Postfach 101461
West Germany

Thiokol Chemical Corp.
110 N. Wacker Dr.
Chicago, IL 60606

Thomas Triantaphyllou S.A.
405 Tatoiou Av.
TK 13671
Acharnes, Athens, Greece

Thompson-Hayward Chemical Co.
5200 Speaker Road
P.O. Box 2383
Kansas City, KS 66110

Thompson Medical
919 Third Avenue
New York, NY 10022

Thompson, Weinman & Co.
P.O. Box 130
Cartersville, GA 30120

3 (Three) M
Commercial Chem. Div.
3M Center
St. Paul, MN 55101

(3) Three V Chem.
3M Center
St. Paul, MN 55101

Titan Chem. Prod. Inc.
P.O. Box 20
Short Hills, NJ 07078

Toho Chem. Industry Co.
1-14-9 Nihonbashi
Kakigara-Cho, Chuo-Ku
Tokyo, Japan

Tokai Seiyu Ltd.
67, 2-Chome
Yamadahigashimachi
Higashi-Ku
Nagoya, Japan

Tomah Products Inc.
1012 Terra Drive
Milton, WI 53563

Tri-K Ind.
99 Kinderkamack Road
Westwood, NJ 07675

Trimen Labs., Inc.
80-26th Street
Pittsburgh, PA 15222

Troy Chemical Corp.
1 Ave. L
Newark, NJ 07105

Tuco/Div. of Upjohn Co.
9540-190-14
Kalamazoo, MI 49001

Tulco, Inc.
Faulkner Street
No. Billerica, MA

Tyson and Assoc., Inc.
1661 Lincoln Blvd.
Suite 300
Santa Monica, CA 90404

U.C.T. Inc.
226 Production Court
Louisville, KY 40299

Unger Fabrikker AS
N-1601
Fredrikstad, Norway

Unichema Internat.
Postfach 1280
D-4240 Emmerich
West Germany

Union Camp Corp.
1600 Valley Road
Wayne, NJ 07470

Union Carbide
Agricultural Products Co.
P.O. Box 12014
T.W. Alexander Drive
Research Triangle Park, NC 27709

Union Carbide Corp.
Old Ridgebury Road
Danbury, CT 06817

Union Carbide Corp.
Ethylene Oxide Derivatives Div.
Old Ridgebury Road
Danbury, CT 06770

Union Carbide Corp.
Linde Division
Old Ridgebury Rd.
Danbury, CT 06817

Uniroyal Chemical
Div. of Uniroyal Inc.
Elm Street Naugatuck, CT 06770

Universal Chemicals
4660 Spring Grove Ave.
Cincinnati, OH 45232

Universal Oil Products Co.
20 U.O.P. Plaza
Des Plaines, IL 60016

UOP
20 U.O.P. Plaza
Des Plaines, IL 60016

Upjohn Co.
7000 Portage Road
Kalamazoo, MI 49001

Upjohn Polymer Chem.
Box 685
La Porte, TX 77571

U.S. Borax & Chemical Corp.
3075 Wilshire Blvd.
Los Angeles, CA 90010

U.S. Chemical
Marketing Group, Inc.
203 Rio Circle
Decatur, GA 30030

U.S. Industrial Chemicals
99 Park Avenue
New York, NY 10016

U.S. Peroxygen Div./Witco
Dept. A2
850 Morton Ave.
Richmond, CA 94804

USV Pharmaceutical
303 South Broadway
Tarrytown, NY 10591

Upsher-Smith Labs., Inc.
14905 23rd Ave. North
Minneapolis, MN 55441

Valchem (Australia) Pty. Ltd.
P.O. Box 255
Wangaratla, Victoria 3677
Australia

Valchem Chem.
Div. of United Merchants &
Manufacturers Inc.
Box 38
Langley, SC 29834

R.T. Vanderbilt Co. Inc.
39 Winfield St.
Norwalk, CT 06855

Van Dyk & Co.
11 Williams St.
Belleville, NJ 07109

Van Schuppen Chemie
3900 AB Veenendaal
(P.O. Box 70)
The Netherlands

V-C. Chemical Co.
401 E. Main St.
Richmond, VA 23208

Velsicol Chemical Corp.
341 E. Ohio
Chicago, IL 60611

Ventron Corp.
150 Andover St.
Danvers, MA 01923

Verex Labs, Inc.
5241 South Quebec St.
Englewood, CO 80111

Vevy
One Dag Hammarskjold Plaza
New York, NY 10017

Vicks Pharmacy Products
Richardson-Vicks, Inc.
10 Westport Road
Wilton, CT 06897

Vikon Chemical Co. Inc.
P.O. Box 1520
Burlington, NC 27215

Vineland Chem. Co.
P.O. Box 745
Vineland, NJ 08360

Viobin Corp.
226 W. Livingston St.
Monticello, IL 61856

Virginia Chemical
3340 W. Norfolk Rd.
Portsmoth, VA 23703

Vitaline Formulas
P.O. Box 6757
Incline Village, NV 89450

Vitamins, Inc.
200 E. Randolph Dr.
Chicago, IL 60601

Vulcan Chemicals
P.O. Box 7689
Birmingham, AL 35253

Wacker-Chemie GmbH
Prinzregentenstrasse 22
D-8000 Munchen 22
West Germany

Vulnax
Sub. Imperial Chemical Ind. PLC
Imperial Chemical House
Millbank, London
SW1P 3JF, UK

Walker, Corp.
Easthampton Pl. & N. Collingwood Ave.
Syracuse, NY 13206

Wallace Labs.
P.O. Box 1
Cranbury, NJ 08512

Wallace & Tierman Inc.
Harchem Div.
25 Main St.
Belleville, NJ 07109

Wallerstein Co.
P.O. Box 510
Kingstree, SC

Washington Penn Plastic Co. Inc.
2080 N. Main St.
Washington, PA 15301

Webcon Phar.
P.O. Box 3000
Humacao, PR 00661

Werner G. Smith, Inc.
1730 Train Ave.
Cleveland, OH

Westinghouse Electric Corp.
Micarta Div.
P.O. Box 248
Hampton, SC 29924

Westvaco-Oleochemical Div.
P.O. Box 70848
N. Charleston Heights, SC 29405

Westvaco Polychemicals
P.O. Box 70848
Charleston Heights, SC 29405

West Virginia Pulp & Paper Co.
818 Magazine St.
Covington, VA

Westwood Phar.
100 Forest Ave.
Buffalo, NY 14213

Wharton Labs., Inc.
37-02 48th Ave.
Long Island City, NY 11101

Whitehall Labs., Inc.
685 Third Ave.
New York, NY 10017

Whitestone Chemical
P.O. Box 2108
Spartanburg, SC 29304

Whittaker, Clark & Daniels
1000 Coolidge Street
South Plainfield, NJ 07080

Wickhen Products, Inc.
Big Bond Rd.
Huguenot, NY 12746

Wilbur B. Driver Co.
1875 McCarter Hwy.
Newark, NJ 07104

Wilkinson Minerals
106 Paper Mill Rd.
Gordon, GA

Willen Drug Co.
18 North High St.
Baltimore, MD 21202

Williams Ltd.
Greville House Hibernia Rd.
Hounslow Middlesex
TW3 3RX, UK

Wilmington Chem. Corp.
P.O. Box 66
Pyles Lane
Wilmington, DE 19899

Winthrop Laboratories Sterling Drug Co.
90 Park Ave.
New York, NY 10016

Wisconsin Alumni Research Fdn.
614 N. Walnut St.
Madison, WI

Witco Chem.
633 Court St.
Brooklyn, NY 11231-2193

Witco Chem. Co. Organics Division
520 Madison Ave.
New York, NY 10022

Witco Chem. Corp.
Sonneborn Division
520 Madison Ave.
New York, NY 10022

Witco Chem. Ltd., Israel
P.O. Box 975
Haifa 31000, Israel

Witco Chemial S.A.
10 Rue Cambaceres
Paris 75008, France

Witco Chem. U.K.
Union Lane
Droitwich, Worcester
WR9 9BB, UK

Witco/Cyclo Div.
10, Rue Cambaceres
75008 Paris, France

Witco Golden Bear Div.
10100 Santa Monica Blvd.
Century City
Los Angeles, CA 90067

Woburn Chem. Corp.
190 19th Ave.
Paterson, NJ 07504

Wyeth Labs.
P.O. Box 8299
Philadelphia, PA 19101

XCEL Corp.
290 Ferry St.
Newark, NJ 07105

Yoshimura Oil Chem. Co. Ltd.
Minami 5-Chrome-1-1
Honan-Cho, Toyonaka-Shi
Osaka 561, Japan

Youngs Drug Products Corp.
P.O. Box 385
865 Centennial Ave.
Piscataway, NJ 08854

Yuen Fa Chemical Co.
225 Wan Ta Road
Taipei, Taiwan, R.O.C. 109

Zohar Detergent Factory
Kibbutz Dalia
Israel 18920

Zschimmer & Schwarz
Postfach 2179
4-5 Max-Schwarz-Strasse
D-5420 Lahnstein
West Germany

Books For Consultation

Blue Book
Lippincott & Peto
1867 W. Market St.
Akron, OH 44313

Concise Chemical & Technical Dictionary - Bennett
Chemical Publishing Co.
80 Eighth Ave.
New York, NY 10011

The Condensed Chemical Dictionary - Hawley
Tenth Edition
Van Nostrand Reinhold
115 Fifth Ave.
New York, NY 10003

CTFA Cosmetic Ingredient Dictionary
Third Edition
The Cosmetic, Toiletry and Fragrance Assoc.
1110 Vermont Ave., N.W.
Washington, D.C. 20005

Encyclopedia of Chemical Trademarks and Synonyms - Bennett
Vols. I-III
Chemical Publishing Co.
80 Eighth Ave.
New York, NY 10011

Encyclopedia of Industrial Chemical Additives - Ash
Vols. I-III
Chemical Publishing Co.
80 Eighth Ave.
New York, NY 10011

Encyclopedia of Plastics, Polymers, and Resins - Ash
Vols. I-III
Chemical Publishing Co.
80 Eighth Ave.
New York, NY 10011

Encyclopedia of Surfactants - Ash
Vols. I-IV
Chemical Publishing Co.
80 Eighth Ave.
New York, NY 10011

Farm Chemicals Handbook
Meister Publishing
37841 Euclid Ave.
Willoughby, OH 44094

McCutcheon's Detergents & Emulsifiers, N.A.
McCutcheon's Publications
175 Rock Rd.
Glen Rock, NJ 07452

McCutcheon's Detergents & Emulsifiers, Int'l.
McCutcheon's Publications
175 Rock Rd.
Glen Rock, NJ 07452

McCutcheon's Functional Materials, N.A.
McCutcheon's Publications
175 Rock Rd.
Glen Rock, NJ 07452

McCutcheon's Functional Materials, Int'l.
McCutcheon's Publications
175 Rock Rd.
Glen Rock, NJ 07452

Modern Plastics Encyclopedia
McGraw-Hill
1221 Ave. of the Americas
New York, NY 10020

Physicians' Desk Reference
Medical Economics Co.
Oradell, NJ 07649

Soap Cosmetics Chemical Specialties
Mac Nair - Dorland Co.
101 W. 31st St.
New York, NY 10001